The Mathematical Legacy of Eduard Čech

Edited by

Miroslav Katětov
Petr Simon

1993

Birkhäuser Verlag
Basel · Boston · Berlin

Editors

Miroslav Katětov
Matematický ústav UK
Sokolovská 83
186 00 Praha 8
Czech Republic

Petr Simon
Matematický ústav UK
Sokolovská 83
186 00 Praha 8
Czech Republic

Reviewers
Prof. RNDr. Věra Trnková, DrSc.
Prof. RNDr. Oldřich Kowalski, DrSc.

Co-edition by Birkhäuser Verlag AG, Basel, Switzerland, and Academia, Publishing House of the Academy of Sciences of the Czech Republic, Prague, Czech Republic

Exclusive distribution rights worldwide:
Birkhäuser Verlag AG, Basel, Switzerland

with the exception of Albania, Bulgaria, China, Cuba, Czech Republic, Hungary, Mongolia, North Korea, Poland, Rumania, Slovak Republic, Vietnam, and countries of the former USSR and Yugoslavia, for which rights are held by Academia, Publishing House of the Academy of Sciences of the Czech Republic, Prague, Czech Republic

Library of Congress Cataloging-in-Publication Data

The Mathematical legacy of Eduard Čech / edited by Miroslav Katětov and Petr Simon.
p. cm.
Includes bibliographical references and index.

1. Algebraic topology. 2. Geometry, Differential. 3. Dimension theory (Topology) 4. Stone–Čech compactifications. 5. Čech, Eduard, 1893–1960. I. Katětov, Miroslav. II. Simon, Petr, 1944–

QA612.M378 1993
514′.2—dc20

Deutsche Bibliothek Cataloging-in-Publication Data

The mathematical legacy of Eduard Čech/ed. by Miroslav Katětov and Petr Simon. - Basel; Boston; Berlin; Birkhäuser, 1993

NE: Katětov, Miroslav [Hrsg.]

Softcover reprint of the hardcover 1st edition 1993

Camera-ready copy prepared by the authors in $\mathcal{A}_{\mathcal{M}}\mathcal{S}$-$\TeX$

ISBN 978-3-0348-7526-4 ISBN 978-3-0348-7524-0 (eBook)
DOI 10.1007/978-3-0348-7524-0

9 8 7 6 5 4 3 2 1

EDUARD ČECH 1893–1960

Foreword

The work of Professor Eduard Čech had a significant influence on the development of algebraic and general topology and differential geometry. This book, which appears on the occasion of the centenary of Čech's birth, contains some of his most important papers and traces the subsequent trends emerging from his ideas. The body of the book consists of four chapters devoted to algebraic topology, Čech-Stone compactification, dimension theory and differential geometry. Each of these includes a selection of Čech's papers, a brief summary of some results which followed from his work or constituted solutions to the problems he posed, and several selected papers by various authors concerning the areas of study he initiated.

The book also contains a concise biography borrowed with minor changes from the book *Topological papers of E. Čech*, a list of Čech's publications and a very brief note on his activity in the didactics of mathematics.

The editors wish to express their sincere gratitude to all who contributed to the completion and publication of this book.

The volume, with the exception of reprinted papers, has been typeset in AMS-TEX.

Miroslav Katětov and Petr Simon

Prague, February 24, 1993

Contents

Life and Work of Eduard Čech

Eduard Čech, professor of Charles University, and member of the Czechoslovak Academy of Sciences, was the greatest Czechoslovak mathematician and one of the leading world specialists in the fields of differential geometry and topology. To these fields he contributed works of basic importance.

He was born on June 29, 1893, in Stračov in northeastern Bohemia. He attended the secondary school in Hradec Králové. In 1912 he began to study mathematics at Charles University in Prague. He learnt most of his mathematics in the library of the Union of Czech Mathematicians and Physicists. Within the period of five semesters he studied thoroughly a considerable amount of mathematical literature of his own choice and acquired knowledge in a number of mathematical disciplines without any guidance. While studying some treatises on elementary mathematics he discovered logical gaps in the proofs; he took a special liking for correcting and completing them. This was the origin of his interest in didactic questions of mathematics. As at that time two fields of study were required for the position of the secondary school teacher, he chose as the other subject descriptive geometry and devoted himself to the study of different branches of geometry.

Eduard Čech spent only five semesters at Charles University. In 1915 he had to interrupt his studies and leave for service in the army. After the war he completed his studies by passing State examinations and for a short period he taught mathematics at a secondary school in Prague.

In 1920 he received the degree of Doctor of Philosophy from Charles University for his thesis "On curve and plane elements of the third order". Thereafter Čech became deeply interested in research. He started studying in a systematic manner the differential projective properties of geometric objects. He became acquainted with papers by the outstanding Italian geometer G. Fubini and, having obtained a scholarship, he spent the school year 1921–22 in Turin. Professor Fubini saw the extraordinary capabilities of young Čech and offered him a coauthorship of a monograph. As a result of the cooperation two volumes of *Geometria proiettiva differenziale* appeared in 1926 and 1927. The authors afterwards wrote another book under the title *Introduction à la géométrie projective différentielle des surfaces*, published in Paris in 1931.

In 1922, Čech submitted a habilitation thesis on projective differential geometry and became Docent at Charles University. A year later, not yet 30, he was appointed

Extraordinary Professor at the Faculty of Sciences of the Masaryk University in Brno, where the chair previously held by Matyáš Lerch had become vacant. As the chair of geometry was occupied, he was asked to teach courses in mathematical analysis and algebra. Therefore he started an intensive study of these disciplines. In a short space of time he mastered the appropriate literature and for twelve years lectured on analysis and algebra at this university. This work seems to have had important implications for his interest in topology.

In 1928 he was appointed Full Professor. At that time he manifested a deep interest in topology. The principal sources he found on the subject were papers published in *Fundamenta mathematicae.* He was also influenced by the papers of outstanding American and Soviet topologists. After 1931 he published no more papers on differential geometry and devoted himself to research in the field of general and combinatorial topology. Let us mention the first two papers of pioneer character published in 1932. One of them is concerned with the general theory of homology in arbitrary spaces and the other with the general theory of manifolds and theorems of duality; these papers established Čech's reputation as one of the best specialists in the field of combinatorial topology. In September 1935 he was invited to a conference on combinatorial topology held in Moscow. This meeting was attended by a number of the foremost European and American topologists. Professor Čech reported there on the results of his research, which met with such attention that he was invited to lecture at the Institute for Advanced Study in Princeton.

After his return from the U.S.A. in 1936, Čech gave new impulses to the mathematical research in Brno. With a group of young people deeply interested in mathematics he founded a topological seminar where at the beginning the papers of P. S. Alexandrov and P. Urysohn were systematically discussed. The atmosphere of the seminar, as well as the personality of Professor Čech, who continued to encourage the participants in their work, had a favourable influence on all its members. Many problems raised by Čech were solved and, within a period of three years, 26 scientific papers originated in the seminar. Čech's paper on bicompact spaces was among them. In this paper he investigated the compactification of completely regular topological spaces now known as the Čech-Stone compactification. The topological seminar continued till 1939, when after the German occupation of Bohemia and Moravia all Czech universities were closed. Nevertheless, even after that Čech met regularly with his closest students, B. Pospíšil and J. Novák, in Pospíšil's flat until the arrest of B. Pospíšil by the Gestapo in 1941. Čech's topological seminar holds an important place in the history of Czechoslovak mathematics. He introduced there a team-work form of mathematical research.

After twenty-two years of teaching and scientific activity in Brno, Professor Čech moved to Charles University in Prague in 1945. He became the leading personality in the organization of Czechoslovak mathematical activities. In 1947 he was appointed the director of the Mathematical Research Institute of the Czech Academy of Sciences and Arts. In 1950 the Central Mathematical Institute was established to which Čech was also appointed director. When the Czechoslovak Academy of Sciences was founded in 1952, this institute was incorporated into the Academy as the Mathematical Institute of the Czechoslovak Academy of Sciences, again with Čech as its first director. He laid the foundations of the structure and research orientation of the Institute, aiming at a balanced development of Czechoslovak mathematics both in the theory and in the applications in technical as well as biological sciences. In 1954 he returned to Charles University as a director of the newly founded Mathematical Institute of Charles University. After 1949 he resumed his own research work and thereafter published 17 papers on differential geometry. Nevertheless, he continued to be interested in topology; he wrote the book *Topologické prostory* and, just before his death on March 15, 1960, he initiated the first of the Prague Topological Symposia.

The scientific, teaching and organizational activities of Professor Čech contributed substantially to the development of mathematics in Czechoslovakia. In addition to this rich involvement, he was deeply interested in the problems of teaching mathematics. He was one of those mathematicians who understood that there should exist a close cooperation between university professors and secondary school teachers. Led by this conviction, he wrote textbooks for secondary schools. In these textbooks he focused his attention on fixing mathematical concepts in the mind of the pupils and on the development of abstract logical reasoning.

In a series of pedagogical seminars held in Brno since 1938, Professor Čech devoted much of his time and energy to the problems of high school mathematics. After 1945 these seminars on elementary mathematics were held both in Prague and Brno.

Professor Čech took part in a number of international mathematical congresses. He lectured as visiting professor at several European and American universities such as the University of Warsaw, the University of Lvov, Moscow State University, Princeton University, the University of Michigan, the State University of New York, Harvard University, to name only some. He was member of the Czechoslovak Academy of Sciences, the Czech Academy of Sciences and Arts, the Royal Czech Society of Sciences, the Moravian Society of Sciences, an honorary member of the Union of Czechoslovak Mathematicians and Physicists, member of the Polish Academy of Sciences and member of the learned society "Towarzystwo Naukowe" of Wrocław. He received honorary degrees from the University of Warsaw and from the University of Bologna. His scientific publications encompassed ninety-six papers and nine books. Moreover, he published seven textbooks for secondary schools. Professor Čech had a considerable influence on a number of Czechoslovak mathematicians. Many Czech mathematicians considered him as their teacher, among them Bedřich Pospíšil, Miroslav Katětov and Josef Novák. He founded a school both in topology and in differential geometry. Numerous mathematicians throughout the world were influenced by his ideas and made use of his results. In Czechoslovakia, the government acknowledged the significance of his work by awarding him the State Prize twice, in 1951 and in 1954, for his fundamental research results. In his activities after 1945, Čech wished to assist the changes which he considered progressive and endeavoured to prepare Czechoslovak mathematics for accomplishing the important mission which the science should have in the life of this country. For his merits in science as well as in this endeavour he was awarded the Order of the Republic.

The scientific activity of Professor Čech was extremely rich. Starting in 1925 he concentrated on topology, both general and algebraic — a closer connection of the two branches constituted a substantial part of the program he set himself — and in 1930, his first paper on topology appeared. By 1938 he had published about 30 topological papers. Later on, he again took up his research in differential geometry. Nevertheless, even in that period he was interested in topology and in addition to a paper written in 1947 (jointly with J. Novák), he published the book *Topologické prostory* (Topological Spaces) in 1959.

Eduard Čech wrote 12 papers on general topology, or rather, topological papers not using algebraic methods; in fact, most of his papers on algebraic topology also refer to very general spaces; this is one of their characteristic features. Among his papers let us mention the one on compact spaces [29] (instead of the original term "bicompact", we use the current term "compact"). For the first time the so-called maximal compactification $\beta(S)$ (that is, a compact space containing S as a dense subset and such that each bounded continuous function on S can be extended to $\beta(S)$) of a completely regular space is systematically studied. Certain properties of the space $\beta(S)$ were investigated at the same time by both Čech and M. H. Stone, although the latter used a different approach; nevertheless Čech was the first to show the importance of this space and the potentialities of its use. The compactification $\beta(S)$, called the Čech-

Stone or Stone-Čech compactification in the literature, has become a very important tool in general topology and in certain fields of functional analysis. Numerous other concepts of general topology (realcompact spaces etc.) have their origin in the theory of the Čech-Stone compactification; one of them, namely the topologically complete space (now called Čech complete) was studied in the paper [29] quoted above. Three other papers [28, 30, 31] are related to this paper in general character. The paper "Topological spaces" [28] had origin in Čech's lectures at the topological seminar of Brno; it contains fundamental concepts of the theory of topological spaces presented according to an original and very general conception. One paper [30] (written jointly with B. Pospíšil) is concerned with various questions of general topology, especially with the character of points in spaces of continuous functions and with the number of incomparable L-topologies having certain other properties. Another [31] (written jointly with J. Novák) analyses in detail some concepts connected with the Wallman compactification (which for a normal space coincides with that of Čech-Stone).

Two of his works [6, 11] concern dimension theory as does the preliminary communication [3]. In the first the concept now called "large" inductive dimension was studied; for perfectly normal spaces the so-called sum theorem (the dimension of a countable union of closed sets is equal to the supremum of their dimensions) and the theorem on monotonicity were proved. In the second paper, the dimension defined by means of covering was studied; in particular, the sum theorem for normal spaces was proved.

Further papers by Čech on general topology concern connected spaces, one of which [4] studies the irreducible connectedness between several points and the generalized concept of "dendrite" for arbitrary topological spaces. A short paper [5] deals with continua that can be mapped onto a segment in such a way that the inverse images of points are finite sets; the paper based on the results obtained by Menger and Nöbeling [18] treats the problems connected with so-called "n-Bogensatz". Finally, "Une démonstration du théorème de Jordan" [1], the first topological paper to be published by Čech, contains a new demonstration of the Jordan theorem.

Of considerable importance for Czechoslovak mathematics was Čech's book *Bodové množiny* (Point sets) [IV] with a supplement by V. Jarník. Published in 1936, the book was a pioneering work in Czech mathematical literature and even now it is still not obsolete. Its first part is devoted to the topology of metric spaces, especially to complete spaces and to compactification; this subject, considered standard nowadays, is treated with admirable exactness and elegance and is presented according to an original methodological conception. Čech's last book, *Topologické prostory* (Topological spaces) with two supplements by J. Novák and M. Katětov, appeared in 1959 but had been ready in fact for several years. The theory of topological spaces is treated in a way considerably more general than is common; special attention is given to problems studied at the Brno seminar. Among the characteristic features of this book, which is written in the precise manner characteristic for Čech, two should be mentioned in particular: the spaces for which the axiom $\overline{\overline{A}} = \overline{A}$ is not necessarily valid are studied in a fairly detailed manner; various properties of the mappings such as "exact continuity" and "inverse continuity" are studied in a general situation. (It should be mentioned here that a revised edition of this book appeared in English in 1966.) Among unpublished papers by Čech a complete manuscript of *Bodové množiny II* was found. This book with some parts of *Bodové množiny I* (Chapters I, II, III) was published under the title *Bodové množiny* in 1966; the English translation *Point sets* was published in 1968.

Cech's first papers on algebraic (combinatorial) topology deal in the first place with the theory of homology and general manifolds. As indicated by Eduard Čech himself in the introductory part to the report [20], the aim was to unite the methods and the way of reasoning used in the set topology and in the classical combinatorial topology, or rather, to discover the general substance of the classical theory of homology, of the theory of manifolds etc. and to incorporate it organically into the general theory of topological spaces. Čech made an important contribution to this program.

In his fundamental paper "Théorie générale de l'homologie dans un espace quelconque" [7], Čech formulates in detail, for completely general spaces, the theory of homology based on finite open coverings. As a matter of fact he does not even suppose, at least at the beginning, that a topological space is considered; the concepts studied are, in fact, in modern terminology, projective limits of homological objects on finite complexes. Of the whole of Čech's work in algebraic topology, the results of this paper are probably known best. The theory formulated here constitutes a part of the foundations of contemporary algebraic topology; later it turned out that it is particularly appropriate for compact spaces and in the literature it is commonly designated by Čech's name. However, it should be noted that the idea of the so-called projective sequence of complexes and particularly that of the nerve of a finite open covering of a compact space was introduced by P. S. Alexandrov as early as 1925 and was treated in detail by him in his paper of 1929.

Related to this paper [7] is another one [14] in which some results of the former are improved, and where the study of local Betti numbers (introduced independently also by P. S. Alexandrov in his paper of 1934) and of other concepts also considered in two later studies [19, 21] is started. The second of these papers is devoted to a detailed study of local connectedness (or local acyclicity) of higher orders defined in terms of the theory of homology (local connectedness in this sense was also introduced by P. S. Alexandrov in 1929, but was not studied in detail until Čech's paper appeared). Local Betti numbers and local acyclicity are studied in different relations in "Sur les nombres Betti locaux" [19] resting equally on the fundamental treatise in "Théorie générale de l'homologie dans un espace quelconque" [7]; it contains a methodological novelty with a bearing on the papers on manifolds, namely the deduction of a number of theorems on the sphere without triangulation by means of a certain theorem concerning the relationship of homotopical and homological concepts. Finally, in "Sur les continus Péaniens unicohérents" [10], the relations between unicoherence (defined in set theoretical terms) and the first Betti number are studied by means used earlier [7].

Seven papers [8, 13, 15, 17, 19, 23, 25] are devoted to different aspects of manifolds. These papers constitute an important chapter of algebraic topology and must be numbered among the most important achievements of Czechoslovak mathematics. The principal aim was to introduce a general concept of manifold so as to include all connected spaces locally homeomorphic to E_n and defined uniquely by general topological properties as well as by assumptions expressed in terms of general homology theory; of course, it is desirable that the theorems on duality be true, with necessary modifications, for these general manifolds. This aim was in fact attained (S. Lefschetz obtained analogous results independently and at about the same time); moreover, many of the theorems were new even for the classical case of duality (for sets in E_n or in S_n). Later on, R. Wilder and other authors started developing the results obtained by Eduard Čech and succeeded in simplifying them considerably by new means; however, it seems that the theory of general manifolds in Čech's sense is far from being complete.

Three papers [2, 22, 26] are in a loose relationship to the main directions taken by Čech's work in algebraic topology discussed above. In one important paper [26] cohomological concepts (in the terminology of that time dual cycles etc.) were studied. A short time later they were formulated explicitly by J. W. Alexander and A. N. Kolmogorov in 1935. The multiplication of cocycles and of a cycle and cocycle were introduced here. In another paper [22] theorems concerning the unique determination of Betti groups with arbitrary coefficients by means of ordinary Betti groups are proved for infinite complexes. "Trois théorèmes sur l'homologie" [2] is the first of Čech's papers on algebraic topology. It contains theorems of considerable generality concerning, among other things, the cutting of a space between two points; some classical theorems on the topology of surfaces are just special cases of these theorems.

We should mention in addition two papers containing only results without proofs: One on Betti groups of compact spaces [25] (these are in general continuous groups) and the one on the accessibility of the points of a closed set in E_n [27]. Finally, at the International Congress of Mathematicians held in Zürich in 1932, Čech presented a communication on higher homotopy groups. Unfortunately, he never returned to this subject again and in the Proceedings of the Congress, his communication (see [9]) was formulated in a very brief and not completely clear manner; nevertheless, W. Hurewicz, who formulated in an outstanding way a systematic theory of higher groups of homotopy in his publications from 1935 onwards, says in one of his papers (Akad. Wetensch. Amsterdam, Proc. 38 (1935), p. 521) that Čech's definition of these groups is equivalent to his.

At the Topological Symposium held in Prague in 1961, P. S. Alexandrov said the following about Čech's definition:

"This definition did not meet with the attention it merited; in fact, the commutativity of these groups for dimensions exceeding one was criticised. (This was unfounded, as we now know.)

"Thus, Professor Čech's definition of the homotopy groups was, in 1932, simply not understood — a situation extremely rare in modern mathematics. We must express our admiration at the intuition and talent of Professor Čech, who defined the homotopy groups several years before W. Hurewicz."

Čech's papers on mathematical analysis are connected to a considerable extent with his teaching activity at the university and have the character of brief notes. In one paper [90], he derived by an original method properties of the functions x^s, e^x, $\log x$, $\sin x$, $\cos x$; in another [91], he generalized the elementary method of K. Petr for the examination of Fourier series for the functions of bounded variation; in two others [92] and [93], he gave a simple proof of Cauchy's theorem and Gauss' formula. Methodologically, the paper "Sur les fonctions continues qui prennent chaque leur valeur un nombre fini de fois" [94] is related to Čech's papers on general topology. It deals with continuous functions on an interval which are nonconstant on any infinite set. The second part of the book *Bodové množiny I* [IV] deals with measure and integral; the approach to these subjects is remarkably original and some specific results were probably new when it appeared.

The papers by Professor Čech on differential geometry were written during two periods; from 1921 to 1930 and in the years after World War II. Čech is one of the founders of projective differential geometry and his work not only brought many valuable results, but also influenced substantially the entire development of this discipline. His work was continued primarily in Italy, but also in Germany, and of course in this country; his papers received considerable attention in the U.S.S.R. Čech succeeded in developing three analytical approaches which appear distinctly in his work and which are of essential importance to research in differential geometry: a systematic attention given to the contact of manifolds, the study of correspondences (as opposed to the study of individual manifolds) and a systematic use of duality in projective spaces. To appraise the value of Čech's work would be to write the history of projective differential geometry; here only an account of concrete results achieved will be given.

Čech's very first papers [32, 33] deal with the association of certain geometric objects and correspondences with the elements of lowest order of curves and surfaces in a three-dimensional projective space; as a matter of fact, they present a geometric determination of these elements by a minimal number of objects. A similar problem is dealt with in one paper [36] studying the element of fourth order of a surface and in another [37], where the results previously obtained are applied to ruled surfaces and where the neighborhood of the straight line generating the surface in question is obtained. In "Sulle omografie e correlazioni che conservano l'elemento del terzo ordine di una superficie in S_3" [42], Čech studies collineations of a projective space onto itself preserving the element of third order of the surface. Starting from these considerations, he gave in the years after World War I (in an unpublished paper) an absolute definition of the canonical straight lines of the surface. A 1928 paper [59] deals with the geometric significance of the index of Darboux quadrics.

In an earlier paper [34], he proves, among other things, that the osculating planes of three curves of Segre have one canonical straight line in common. In "Sur les surfaces dont toutes les courbes de Segre sont planes" [39] with a preliminary communication [38], Čech discovers all surfaces for which all these straight lines go through a fixed point, in other words, for which the curves of Segre are plane; another paper [40] determines the surfaces with plane Darboux curves. It should be noted that these computations required a very difficult integration of a system of partial differential equations.

It is well known that the study of a surface in a Euclidean three-dimensional space is equivalent to the study of two fundamental differential forms on the surface.

The main idea of G. Fubini was to establish a similar procedure for a surface and hypersurface in a projective space using a quadratic and cubic form. Čech contributed to this theory in six separate papers [41, 43, 49, 50, 52, 62]. He found the geometric significance of different normalizations of homogeneous coordinates of the points of the surface, the geometric significance of the projective linear element (playing the same role as ds^2 in Euclidean geometry) and a complete system of its invariants; he further studied its extremals (projective geodesics).

Many works [34, 35, 55, 60, 63, 64, 67 and 68] are devoted to the theory of correspondences between surfaces. Eduard Čech contributed here substantially to the theory of the projective deformation of the surface in a three-dimensional space. He gave a new characterization of projective deformation by means of osculating planes corresponding to each other and he further studied different generalizations of projective deformation as well as the general asymptotic or semi-asymptotic correspondence between surfaces; he found the solution to the main existence questions for different types of these asymptotic correspondences. He finally used all these techniques to study and to find the congruence of the straight lines, the focal surfaces of which are in projective deformation or on which Darboux curves correspond to one another. Later on, the same problem was studied by different methods by S. P. Finikov. Of considerable importance for the theory of projective deformations is, furthermore, the discovery of surfaces admitting ∞^1 projective deformations in themselves or on which there exist ∞^1 R-nets, one of which has the same invariants.

A new method of study of ruled surfaces, applicable mainly to projective spaces of odd dimension, was introduced in three papers [48, 53, 54]. Other authors, primarily the Czechoslovak ones, exploited these results and proved the advantage of Čech's procedure.

Two papers of fundamental importance [58, 66] deal with the contact of two curves in projective spaces of an arbitrary dimension and with the possibility of increasing this contact by projection from a suitably chosen centre. Eduard Čech came back to this problem in his last paper [87], where similar problems for two manifolds are studied. These papers not only contained concrete results of basic importance, but also constituted a starting point for the formulation of the theory of correspondences which will be discussed later.

Some papers [46, 47, 56, 57] are devoted to the study of the strips of contact elements on a surface in a three-dimensional or affine space, i.e. of a system of plane elements in the points of a curve situated on the considered surface. Special attention is given to pairs of surfaces having contact of a certain order along the whole curve and the conditions are studied under which this curve is a curve of Darboux or Segre on both surfaces at the same time, as well as other problems of this character. Čech strongly emphasized the importance of his procedure which consists in considering a whole strip of elements instead of a curve (which in practice we do in Euclidean geometry without being aware of it); the potential of these papers has not yet been exploited.

Finally, two papers [61] and [65] are devoted to the projective differential geometry of plane nets.

This first period of Čech's active interest in differential geometry culminates with the publication of three books [I, II, III], two of which were written in cooperation with G. Fubini. It should be mentioned that the second and third are the first systematic books on projective differential geometry. Both books originated from long written discussions on the conception of the topics as a whole; and a specialist who recognizes Čech's geometric lucidity combined with extremely complicated computations may easily trace the contribution of each author in the whole work. Thanks to Čech's initiative, a chapter on the use of Cartan's methods was put into the French book; nowadays we clearly see that at that time it was a very sagacious act. *Projektivní diferenciální geometrie* (Projective differential geometry), his first book, is an isolated work in the world literature; it deals in an exact, formal manner with one-parameter objects and so shows that differential geometry can be explained in a perfectly precise way.

After World War II, Eduard Čech continued to work in classical differential geometry. He again achieved very important and highly appreciated results.

Two papers [69] and [71] give a systematic theory of correspondences between projective spaces studied from the point of view of the possibility of their best approximation by means of tangent homographies. This determines the natural classification of special types of correspondences which are either contructed directly geometrically or at least whose general character is given. In a very detailed way, projective deformations of the layer of hypersurfaces are studied. Čech found a great number of secondary results (from the point of view of the theory of correspondences) which play, nonetheless, an important role in other theories. In this manner all asymptotic transformations of the congruence of the straight lines L (i.e. all transformations $S_3' \to S_3$ for which every ruled surface in L passes asymptotically into the ruled surface of the corresponding congruence L') were found. This problem is in fact equivalent to the classical problem posed by Fubini concerning the discovery of projective deformations on the surface. Čech's theory met with attention abroad and influenced substantially the group of Italian geometers in Bologna, who had been working intensively in the geometry of correspondences under the leadership of Professor M. Villa.

It became apparrent that in the development of the theory of correspondences, the congruences of straight lines are of essential importance. It is therefore natural that later on, Čech started to study them systematically; the results were published in four papers [72, 75, 78, 84]. He paid attention to correspondences between congruences which transfer in themselves their developable surfaces and he analysed in detail the problem of their projective deformation; he achieved outstanding results especially for W-congruences. In this field, which has also been studied by P. S. Finikov and his Moscow school, existence questions and geometrical constructions by Čech rank among the best results. Our geometers achieved by these methods a number of deep and sometimes definitive results in the theory of Segre congruences and of congruences and surfaces with a conjugate set in higher-dimensional spaces.

Four papers [80, 81, 82, 86] are devoted to various subjects; they study relations between differential classes of spaces of the points of the curve and of associated objects, n-frame of Frenet and the osculating circle and sphere in Euclidean spaces of dimension three or four. These results are partly definitive and rather surprising. Nevertheless, vigorous efforts must be made to formulate a systematic theory in this new part of differential geometry of curves and to find, eventually, more effective methods of investigation.

In conclusion, let us mention "Sur la déformation projective des surfaces développables" [83] dealing with projective deformation of developable surfaces and two papers [39] and [74] which have the character of summary reports on the theory of correspondences and on some fundamental questions of differential geometry.

This enumeration of Čech's papers in differential geometry is of course incomplete not only by virtue of its brevity, but also due to the fact that a number of Čech's ideas and methods were dealt with in the papers by his direct or indirect students. In addition, a number of manuscripts (often very incomplete) of new papers were found after his death. Some of them have been published in "Quelques travaux de géométrie différentielle" [89].

M. Katětov, J. Novák, A. Švec

Bibliography of Eduard Čech

Topological papers

[1] *Une démonstration du théorème de Jordan*, Atti Accad. Naz. Lincei. Rend. Cl. Sci. Fis. Mat. Nat. **(6) 12** (1930), 386 - 388.

[2] *Trois théorèmes sur l'homologie*, Spisy Přírod. Fak. Univ. Brno **144** (1931), 21 pp.

[3] *Sur la théorie de la dimension*, C. R. Acad. Sci. Paris **193** (1931), 976 - 977.

[4] *Množství ireducibilně souvislá mezi n body (Sur les ensembles connexes irréductibles entre n points)*, Časopis Pěst. Mat. **61** (1932), 109 - 129.

[5] *Une nouvelle classe de continus*, Fund. Math. **18** (1931), 85 - 87.

[6] *Dimense dokonale normálních prostorů (Sur le dimension des espaces parfaitement normaux)*, Rozpr. Čes. Akad. Věd **(13) 42** (1932), 22 pp.

[7] *Théorie générale de l'homologie dans un espace quelconque*, Fund. Math. **19** (1932), 149 - 183.

[8] *La notion de variété et les théorèmes de dualité*, Verh. des int. Kongr. Zürich **2** (1932), 194.

[9] *Höherdimensionale Homotopiegruppen*, Verh. des int. Kongr. Zürich **2** (1932), 203.

[10] *Sur les continus Péaniens unicohérents*, Fund. Math. **20** (1933), 232 - 243.

[11] *Příspěvek k theorii dimense (Contribution à la théorie de la dimension)*, Časopis Pěst. Mat. **62** (1933), 277 - 291.

[12] *Über einen kurventheoretischen Satz von Ayres*, Erg. Koll. Wien **5** (1933), 24 - 25.

[13] *Eine Verallgemeinerung des Jordan-Brouwerschen Satzes*, Erg. Koll. Wien **5** (1933), 29 - 31.

[14] *Úvod do theorie homologie (Introduction à la théorie de l'homologie)*, Spisy Přírod. Fak. Univ. Brno **184** (1933), 36 pp.

[15] *Théorie générale des variétés et de leurs théorèmes de dualité*, Ann. of Math. **(2) 34** (1933), 29 - 31.

[16] *Užití theorie homologie na theorii souvislosti (Application de la théorie de l'homologie à la théorie de la connexité)*, Spisy Přírod. Fak. Univ. Brno **188** (1933), 40 pp.

[17] *Sur la décomposition d'une pseudovariété par un sous-ensemble fermé*, C. R. Acad. Sci. Paris **198** (1934), 1342 - 1345.

[18] *Sur les arcs indépendants dans un continu localement connexe*, Spisy Přírod. Fak. Univ. Brno **193** (1934), 10 pp.

[19] *Sur les nombres Betti locaux*, Ann. of Math. **(2) 35** (1934), 678 - 701.

[20] *Les théorèmes de dualité en topologie*, Časopis Pěst. Mat. Fys. (1935), 17 - 25.

[21] *Sur la connexité locale d'ordre supérieur*, Compositio Math. **2** (1935), 1 - 25.

[22] *Les groupes de Betti d'un complexe infini*, Fund. Math. **25** (1935), 33 - 44.

[23] *On general manifolds*, Proc. Nat. Acad. Sci. U.S.A. **22** (1936), 110 - 111.

[24] *On pseudomanifolds*, Lectures at the Inst. Adv. St., Princeton, (mimeographed) (1935), 17 pp.

[25] *Über die Bettischen Gruppen kompakter Räume*, Erg. Koll. Wien **7** (1936), 47 - 50.

[26] *Multiplication on a complex*, Ann. of Math. **37** (1936), 681 - 697.

[27] *Accessibility and homology*, Mat. Sb. **1 (43)** (1936), 661.

[28] *Topologické prostory (Topological spaces)*, Časopis Pěst. Mat. **66** (1937), D 225 - D 264.

[29] *On bicompact spaces*, Ann. of Math. **38** (1937), 823 - 844.

[30] *I. Sur les espaces compacts. – II. Sur les caractères des points dans les espaces $\mathfrak{L}$. (Avec B. Pospíšil)*, Spisy Přírod. Fak. Univ. Brno **258** (1938), 14 pp.

[31] *On regular and combinatorial imbedding (Jointly with J. Novák)*, Časopis Pěst. Mat. **72** (1947), 7 – 16.

Geometrical papers

[32] *O křivkovém a plošném elementu třetího řádu projektivního prostoru (Sur l'élément curviligne et superficiel du troisième ordre de l'espace projectif)*, Časopis Pěst. Mat. **50** (1921), 219 – 249, 305 – 306.

[33] *K diferenciální geometrii prostorových křivek (Sur la géométrie différentielle de courbes gauches)*, Rozpr. Čes. Akad. Věd **(15) 30** (1921), 16 pp.

[34] *O trilineárních systémech čar na ploše a o projektivní aplikaci ploch (Systèmes trilinéaires des lignes sur une surface et déformation projective des surfaces)*, Rozpr. Čes. Akad. Věd **(23) 30** (1921), 6 pp.

[35] *O obecné příbuznosti mezi dvěma plochami (Sur la correspondance générale de deux surfaces)*, Rozpr. Čes. Akad. Věd **(36) 30** (1921), 4 pp.

[36] *Moutardovy kvadriky (Les quadriques de Moutard)*, Spisy Přírod. Fak. Univ. Brno **3** (1921), 17 pp.

[37] *Projektivní geometrie pěti soumezných mimoběžek (Géometrie projective de cinq droites infinément voisines)*, Spisy Přírod. Fak. Univ. Brno **4** (1921), 37 pp.

[38] *Sur les surfaces dont toutes les courbes de Segre sont planes*, Atti Accad. Naz. Lincei. Rend. Cl. Sci. Fis. Mat. Nat. **(5)** 30_2 (1921), 491 – 492.

[39] *Sur les surfaces dont toutes les courbes de Segre sont planes*, Spisy Přírod. Fak. Univ. Brno **11** (1922), 35 pp.

[40] *Sur les surfaces dont toutes les courbes de Darboux sont planes*, Atti Accad. Naz. Lincei. Rend. Cl. Sci. Fis. Mat. Nat. **(5)** 31_1 (1922), 154 – 156.

[41] *Sur les formes différentielles de M. Fubini*, Atti Accad. Naz. Lincei. Rend. Cl. Sci. Fis. Mat. Nat. **(5)** 31_1 (1922), 350 – 352.

[42] *Sulle omografie e correlazioni che conservano l'elemento del terzo ordine di una superficie in S_3*, Atti Accad. Naz. Lincei. Rend. Cl. Sci. Fis. Mat. Nat. **(5)** 31_1 (1922), 496 – 498.

[43] *Sur la géométrie d'une surface et sur le facteur arbitraire des coordonnées homogènes*, Atti Accad. Naz. Lincei. Rend. Cl. Sci. Fis. Mat. Nat. **(5)** 31_1 (1922), 475 – 478.

[44] *L'intorno d'un punto d'una superficie considerato dal punto di vista proiettivo*, Ann Mat. Pura Appl. **(3) 31** (1922), 191 – 206.

[45] *I fondamenti della geometria proiettiva differenziale secondo il metodo di Fubini*, Ann. Mat. Pura Appl. **(3) 31** (1922), 251 – 278.

[46] *Nouvelles formules de la géométrie affine*, Atti Accad. Naz. Lincei. Rend. Cl. Sci. Fis. Mat. Nat. **(5)** 32_1 (1923), 311 – 315.

[47] *Courbes tracées sur une surface dans l'espace affine*, Spisy Přírod. Fak. Univ. Brno **28** (1923), 47 pp.

[48] *O jedné třídě ploch zborcených (Sur une classe des surfaces réglées)*, Čas. Pěst. Mat. **52** (1923), 18 – 24.

[49] *Sur les invariants de l'élément linéaire projectif d'une surface*, Atti Accad. Naz. Lincei. Rend. Cl. Sci. Fis. Mat. Nat. **(5)** 32_2 (1923), 335 – 338.

[50] *Sur les géodésiques projectives*, Atti Accad. Naz. Lincei. Rend. Cl. Sci. Fis. Mat. Nat. **(5)** 33_1 (1924), 15 – 16.

[51] *Algebraické formy o proměnných koeficientech (Formes algébriques à coefficients variables)*, Rozpr. České Akad. Věd **(9) 33** (1924), 2 pp.

[52] *Étude analytique de l'élément linéaire projectif d'une surface*, Spisy Přírod. Fak. Univ. Brno **36** (1924), 24 pp.

[53] *Projektivní geometrie přímkových ploch v prostorech o jakémkoli počtu dimenzí, I. (Géométrie projective des surfaces réglées dans les espaces à un nombre quelconque de dimensions, I.)*, Rozpr. České Akad. Věd **(13) 33** (1924), 9 pp.

[54] *Nová methoda projektivní geometrie zborcených ploch (Une méthode nouvelle dans la géométrie projective des surfaces réglées)*, Časopis Pěst. Mat. **53** (1924), 31 - 37.

[55] *Sur les surfaces qui admettent ∞^1 déformations projectives en elles mêmes*, Spisy Přírod. Fak. Univ. Brno **40** (1924), 47 pp.

[56] *Courbes tracées sur une surface dans l'espace projectif, I.*, Spisy Přírod. Fak. Univ. Brno **46** (1924), 35 pp.

[57] *Géométrie projective des bandes d'éléments de contact de troisième ordre*, Atti Accad. Naz. Lincei. Rend. Cl. Sci. Fis. Mat. Nat. **(6)** 1_1 (1925), 200 - 204.

[58] *Propriétés projectives du contact, I.*, Spisy Přírod. Fak. Univ. Brno **91** (1928), 26 pp.

[59] *Osservazioni sulle quadriche di Darboux*, Atti Accad. Naz. Lincei. Rend. Cl. Sci. Fis. Mat. Nat. **(6)** 8_2 (1928), 371 - 372.

[60] *Sur les correspondances asymptotiques entre deux surfaces*, Atti Accad. Naz. Lincei. Rend. Cl. Sci. Fis. Mat. Nat. **(6)** 8_2 (1928), 484 - 486, 552 - 554.

[61] *Déformation projective de réseaux plans*, C. R. Acad. Sci. Paris **188** (1929), 291 - 292.

[62] *Quelques remarques relatives à la géométrie différentielle projective des surfaces*, C. R. Acad. Sci. Paris **188** (1929), 1331 - 1333.

[63] *Sur les correspondances asymtotiques entre deux surfaces*, Rozpr. Čes. Akad. Věd **(3) 38** (1929), 38 pp.

[64] *Sur une propriété carastéristique des surfaces F de M. Fubini*, Atti Accad. Naz. Lincei. Rend. Cl. Sci. Fis. Mat. Nat. **(6)** 9_1 (1929), 975 - 977.

[65] *Projektive Differentialgeometrie der Kurvennetze in der Ebene*, Jber. Deutsch. Math. Verein. **(5-8) 39** (1930), 31 - 34.

[66] *Propriétés projectives du contact, II.*, Spisy Přírod. Fak. Univ. Brno **121** (1930), 21 pp.

[67] *Una generalizzazione della deformazione proiettiva*, Atti del Congr. int. dei Matem. Bologna, 1928 **4** (1931), 299 - 300.

[68] *Réseaux R à invariants égaux*, Spisy Přírod. Fak. Univ. Brno **143** (1931), 29 pp.

[69] *Géométrie projective différentielle des correspondances entre deux espaces: I.*, Časopis Pěst. Mat. **74** (1949), 32 - 46. *II.*, Časopis Pěst. Mat. **75** (1950), 123 - 136. *III.*, Časopis Pěst. Mat. **75** (1950), 137 - 158. The same title published in Russian: *Proektivnaja differencial'naja geometrija sootvetstvij meždu dvumja prostranstvami: I.*, Czechoslovak Math. J. **2 (77)** (1952), 91 - 107. *II.*, Czechoslovak Math. J. **2 (77)** (1952), 109 - 123. *III.*, Czechoslovak Math. J. **2 (77)** (1952), 125 - 148.

[69a] The same title, parts IV - VIII published in Russian: *Proektivnaja differencial'naja geometrija sootvetstvij meždu dvumja prostranstvami: IV.*, Czechoslovak Math. J. **2 (77)** (1952), 149 - 166. *V.*, Czechoslovak Math. J. **2 (77)** (1952), 167 - 188. *VI.*, Czechoslovak Math. J. **2 (77)** (1952), 297 - 331. *VII.*, Czechoslovak Math. J. **3 (78)** (1953), 123 - 137. *VIII.*, Czechoslovak Math. J. **4 (79)** (1954), 143 - 174.

[70] *Quadriques osculatrices à centre donné et leur signification projective*, C. R. de la Soc. des Sci. et des Lettr. Wrocław **7** (1952), 9 pp.

[71] *Deformazione proiettiva di strati d'ipersuperficie*, Convegno int. di geom. diff. 1953, Ediz. Cremonese, Roma (1954), 266 - 273.

[72] *O točečnych izgibanijach kongruencij prjamych (Déformation ponctuelle des congruences de droites)*, Czechoslovak Math. J. **5 (80)** (1955), 234 - 273.

[73] *Remarques au sujet de la géométrie différentielle projective*, Acta Math. Sci. Hungar. **5** (1954), 137 - 144.

[74] *Deformazioni proiettive nel senso di Fubini e generalizzazioni*, Conf. Sem. Mat. Univ. Bari (1955), 1 - 12.

[75] *Deformazioni di congruenze di rette*, Rend. Semin. Mat. Univ. e Politechn. Torino **14** (1954/55), 55 - 66.

[76] *Transformations développables des congruences de droites*, Czechoslovak Math. J. **6** **(81)** (1956), 260 - 286.

[77] *Deformazioni proiettive di congruenze e questioni connesse*, Ist. Mat. Univ. Roma (1956), 44 pp.

[78] *Déformation projective des congruences* W, Czechoslovak Math. J. **6** **(81)** (1956), 401 - 414.

[79] *Zur projektiven Differentialgeometrie*, Deutsch. Akad. Wiss. Berlin. Schr. Forschungsinst. Math. **1** (1957), 138 - 142.

[80] *Détermination du type différentiel d'une courbe de l'espace à deux, trois ou quatre dimensions*, Czechoslovak Math. J. **7** **(82)** (1957), 599 - 631.

[81] *Classe différentielle des courbes. Sections et projections*, Rev. Math. Pures Appl. **2** (1957), 151 - 159.

[82] *Sur le type différentiel anallagmatique d'une courbe plane ou gauche*, Colloq. Math. **6** (1958), 141 - 143.

[83] *Sur la déformation projective des surfaces développables*, Izv. na mat. inst. Sofija 3_2 (1959), 81 - 97.

[84] *Compléments au Mémoire Déformation projective des congruences* W, Czechoslovak Math. J. **9** **(84)** (1959), 289 - 296.

[85] *Sulla differenziabilità del triedro di Frenet*, Ann. Mat. Pura Appl. **49** (1960), 91 - 96.

[86] *Classe différentielle des courbes. Circles osculateurs et sphères osculatrices*, Bul. Inst. Polit. Iassy **5** **(9)** (1959), 1 - 4.

[87] *Propriétés projectives du contact III.*, Comment. Math. Univ. Carolinae **1** (1960), 1 - 19.

[88] *Déformation projective des congruences paraboliques*, Publ. Math. Debrecen **7** (1960), 108 - 121.

[89] *Quelques travaux de géométrie différentielle. (Avec A. Švec)*, Czechoslovak Math. J. **12** **(87)** (1962), 169 - 222.

Other papers

[90] *O funkcích* x^s, e^x, $\log x$, $\cos x$, $\sin x$. *(Sur les fonctions* x^s, e^x, $\log x$, $\cos x$, $\sin x$.*)*, Časopis Pěst. Mat. **57** (1928), 208 - 216.

[91] *Petrova elementární methoda vyšetřování Fourierových řad. (Sur la méthode élémentaire de M. Petr dans la théorie des séries de Fourier.)*, Časopis Pěst. Mat. **59** (1930), 145 - 150.

[92] *Une démonstration du théorème de Cauchy et de la formule de Gauss*, Atti. Accad. Naz. Lincei. Rend. Cl. Sci. Fis. Mat. Nat. **(6)** **11** (1930), 884 - 887.

[93] *Encore sur le théorème de Cauchy*, Atti. Accad. Naz. Lincei. Rend. Cl. Sci. Fis. Mat. Nat. **(6)** **12** (1930), 286 - 289.

[94] *Sur les fonctions continues qui prennent chaque leur valeur un nombre fini de fois*, Fund. Math. **17** (1931), 32 - 39.

Books

[I] *Projektivní diferenciální geometrie (Projective differential geometry)*, JČMF, Praha, 1926, 406 pp.

[II] *Geometria proiettiva differenziale. Con G. Fubini*, Zanichelli, Bologna, I, 1926; II, 1927, 794 pp.

[III] *Introduction à la géométrie projective différentielle des surfaces. Avec G. Fubini*, Gauthier-Villars, Paris, 1931, 290 pp.

[IV] *Bodové množiny I. S dodatkem V. Jarníka: O derivovaných číslech funkcí jedné proměnné. (Point sets I. With a supplement by V. Jarník: On the derived numbers of functions of one variable)*, JČMF, Praha, 1936, 275 pp.

[V] *Co je a nač je vyšší matematika (What is and what is the use of higher mathematics)*, JČMF, Praha, 1942, 124 pp.

[VI] *Elementární funkce (Elementary functions)*, JČMF, Praha, 1944, 86 pp.

[VII] *Základy analytické geometrie (Foundations of analytical geometry)*, Přírodovědecké vydavatelství, Praha, I, 1951, 218 pp.; II, 1952, 220 pp.

[VIII] *Čísla a početní výkony (Numbers and operations with them)*, SNTL, Praha, 1954, 248 pp.

[IX] *Topologické prostory. S dodatky: J. Novák: Konstrukce některých význačných topologických prostorů; M. Katětov: Plně normální prostory. (Topological spaces. With supplements: J. Novák: Construction of certain important topological spaces; M. Katětov: Fully normal spaces)*, NČSAV, Praha, 1959, 524 pp.

[X] *Topological spaces. Revised edition by Z. Frolík and M. Katětov*, Academia, Praha, 1966, 893 pp.

[XI] *Bodové množiny (Point sets)*, Academia, Praha, 1966, 284 pp.

Čech-Stone Compactification

Petr Simon

Introduction. In 1937, two papers opened a new field in general topology. These were E. Čech's paper "On bicompact spaces" [ČECH 1937] and M. H. Stone's "Application of Boolean rings to general topology" [STONE 1937]; the field is nowadays recognized as the theory of Čech-Stone compactification.

Recall that a topological space is a set X equipped with a family $\mathcal{O}$ of its subsets, which are called *open sets*; the empty set is open, X itself is open, any union of open sets is open and every intersection of finitely many open sets is open. A topological space is called Hausdorff, if for any pair of distinct points there are two disjoint open sets, each containing one point from the pair. A topological space is compact, if every family of open sets, whose union is the whole of X (such a family *covers* X), contains a finite subfamily which covers X as well.

The notion of compactness had been recognized years before the general topology was already established. Namely, the Heine-Borel-Lebesgue theorem states that every closed bounded subset of an Euclidean space is compact. The first book treating compact spaces systematically, was P. S. Alexandrov's and P. S. Uryson's "Mémoire sur les espaces topologiques compacts" [1929]. Written in 1922, it became widely popular long before its publication in 1929. This slim book influenced the research in topology for more than a decade.

In 1930, another major contribution to general topology appeared. It was Tychonoff's paper "Über die topologische Erweiterung von Räumen" [TYCHONOFF 1930]. He introduces here the product of topological spaces, defines a completely regular space and proves two important theorems. The first of them states that completely regular spaces are just subspaces of compact Hausdorff spaces; the second one asserts that the product of non-empty compact spaces is compact. The difficult implication in the characterization theorem for completely regular spaces is proved using an embedding into the product of sufficiently many intervals.

This was the state of art when Čech's and Stone's papers appeared. We included the paper by E. Čech in the present book; nevertheless, let us emphasize the main result.

Theorem. *For every completely regular space X there is a space βX such that:*

(i) *βX is a compact Hausdorff space;*

(ii) *$X \subset \beta X$;*

(iii) *X is dense in βX;*

(iv) *every bounded real-valued function defined on X extends to βX.*

The space βX is unique up to a homeomorphism, i.e. any space satisfying the conditions (i) – (iv) *is homeomorphic to βX.*

Let us call any space bX satisfying (i), (ii) and (iii) above by a *compactification* of the space X. In Stone's paper, (iv) is replaced by

(iv′) *every mapping defined on X and ranging in a compact Hausdorff space extends to βX.*

The Čech-Stone compactification was not the only notion studied by E. Čech in his article. Topologically complete spaces, nowadays called Čech complete spaces, were the second one. However, we shall omit it here.

E. Čech himself made a remark in his paper, stating that the remainder of a countable discrete space, $\beta\omega \setminus \omega$, is an example of a compact Hausdorff space, which contains no $\varkappa$-points. (A *$\varkappa$-point* is a point, which is a limit point of some convergent sequence in the space.) Alexandrov and Uryson asked for the existence of such a space and they noted: "Does there exist a compact Hausdorff space containing no $\varkappa$-point (and, consequently, containing no convergent sequence consisting from pairwise distinct points)? If such a compact space were constructed, it would have to be of an essentially different nature than all examples of compact spaces known up to now." The prophecy was fulfilled. We shall see later how fascinating was the study of topological properties of Čech-Stone compactifications.

The very next paper in the same volume of *Annals of Mathematics* was Pospíšil's solution of a problem posed by Čech on the cardinality of $\beta\omega$. We include this paper in our book. It is difficult to understand why both Čech and Pospíšil attributed the authorship of β-compactification to Tychonoff; in fact, Tychonoff's article does not support such a view. On the other hand, the reader can undoubtedly recognize that in Pospíšil's paper [1937], the essential step in showing the lower estimation for the size of Čech-Stone compactification of a discrete space of size κ is to prove that the product of 2^κ many copies of a discrete two-point space is of density κ. Nevertheless, the papers of Hewitt, Marczewski and Pondiczery appeared as late as in 1944 – 47.

When constructing βX, E. Čech used the canonical embedding of X into the product of unit intervals, $I^{C(X,I)}$, where $C(X,I)$ stands for the family of all continuous functions from X to I. M. H. Stone's construction is based on an approach formerly developed for Boolean algebras. There are other possibilities; let us mention, for example, the characterization of βX as a maximal compactification in the partially ordered set of all compactifications of X; the order is given by the existence of a continuous onto mapping. If there is a continuous mapping from a compactification aX onto a compactification bX, then aX is larger than bX. Gelfand and Kolmogoroff's paper [1939], the third paper reprinted in this chapter, gives yet another characterization of βX. It shows that βX is homeomorphic to the space of all maximal ideals in the ring of all continuous real-valued bounded functions on X. The reader willing to know more on this subject is recommended to find the book *Rings of Continuous Functions* by Gillman and Jerrison.

Generalizations. Product theorem. The essential features of Čech's proof were soon pointed out. As early as 1948, E. Hewitt [1948] defined realcompact spaces — a Tychonoff space is *realcompact* if it can be embedded into a suitable product of real lines as a closed subset — and found an analogy of the Čech-Stone theorem for them. The resulting realcompact space, containing the given completely regular space X as a dense subset, and such that every real valued continuous function on X has a continuous extension onto it, is called Hewitt's realcompactification and denoted by υX. It is possible to describe Hewitt's realcompactification υX intrinsically. According to the description of βX via maximal ideals in the ring of all bounded continuous real-valued functions defined on X, the points of βX can be identified with all prime filters on X, consisting of zero-sets. (A subset Z of a topological space X is called a *zero-set* if Z is a preimage of the point 0 under some continuous real-valued function defined on X.) Then points of υX are just those points from βX which are prime filters of zero-sets with a countable intersection property. Ten years later, R. Engelking and S. Mrówka defined E-compact spaces as closed subspaces of some product of copies of a topological space E [1958]. It turned out that the good behaviour of classes of spaces defined in such a manner has a general categorical reason: all such classes are epireflective subcategories of the category of all Hausdorff spaces, similarly as compact Hausdorff (or realcompact) spaces. Any class of Hausdorff spaces, which is closed under Tychonoff products and hereditary with respect to closed subspaces, is an epireflective subcategory of the category of all Hausdorff spaces. Hence there is an epireflection. However, not all properties of an epireflective functor like β follow from the general statements of category theory.

The question, whether β commutes at all or under what conditions with the product operation is an example of such a property. Let us be reminded that a topological space is *pseudocompact*, if every continuous real-valued function defined on it is bounded. In 1959, I. Glicksberg proved that for infinite Tychonoff spaces X and Y, $\beta(X \times Y) = \beta X \times \beta Y$ if and only if the product $X \times Y$ is pseudocompact. (Paper [GLICKSBERG 1959] is reprinted in the present book.) An analogous question for Hewitt's realcompactification υ is still open. The main difficulty lies in the role of a measurable cardinal, which enters almost all considerations on realcompact spaces. The most striking dissimilarity to the case of compact spaces is demonstrated perhaps in the following statement. Whenever a space X of cardinality smaller than the first measurable cardinal is not realcompact, then there is a realcompact space Y with $\upsilon X \times Y \neq \upsilon(X \times Y)$ ([HUŠEK 1970]).

Nonhomogeneity of remainders. A fascinating chapter in the study of Čech-Stone compactification was written by mathematicians who studied the intrinsic topological properties of remainders. This is the field in which general topology, set theory, Boolean algebras and mathematical logic meet. It was soon clear that $\beta\omega \setminus \omega$, the Čech-Stone remainder of a countable discrete space, would play a prominent role. The reason is rather simple. The copy of $\beta\omega \setminus \omega$ can be found in any $\beta X \setminus X$, provided X is not pseudocompact; thus, it is, in a sense, a "typical" remainder. On the other hand, when also considering pseudocompact spaces, one finds the class of remainders too wide: For every completely regular space S there is a completely regular space X such that S is homeomorphic to $\beta X \setminus X$ ([GILLMAN and JERISON 1960, ex. 9.K]).

We included to the present book W. Rudin's paper [RUDIN 1956], where a proof that $\beta\omega \setminus \omega$ is not homogeneous appeared for the first time. (A topological space is *homogeneous*, if for any two points there is a homeomorphism of the space onto itself, mapping one of the points to the other.) W. Rudin needed to assume the continuum hypothesis, and he showed that $\beta\omega \setminus \omega$ contains a P-point, i.e. the point, which is contained in the interior of any G_δ set, which contains it. He also proved that $\beta\omega \setminus \omega$ has $2^{\mathfrak{c}}$ autohomeomorphisms, again under CH. The assumption of CH was removed by Z. Frolík 11 years later in [FROLÍK 1967a]. It turned out that the proof admits a generalization to the widest class of spaces possible: We present another Frolík's paper in our book, where the main result states that the remainder $\beta X \setminus X$ is not homogeneous, provided X is not pseudocompact.

There is a sharp distinction between the proofs given by Rudin and Frolík. W. Rudin exhibits two points of obviously distinct topological properties (clearly, every infinite compact space contains a point, which is not a P-point); Z. Frolík defines (in today's terminology) Rudin-Frolík preorder $\leqslant_{RF}$ of ultrafilters. For $x, y \in \beta\omega$, $x \leqslant_{RF} y$ if there is an embedding $h : \omega \to \beta\omega$ such that for its Čech-Stone extension $\beta f(x) = y$. He shows then that there must be two points with distinct sets of predecessors. The points in $\beta\omega \setminus \omega$, which are minimal in Rudin-Frolík's preorder, are easy to describe: such a point is never an accumulation point of a countable discrete subset of $\beta\omega \setminus \omega$. It turned out that these points really exist, and even more is true. There is a point in $\beta\omega \setminus \omega$, which is not an accumulation point of any countable subset of $\beta\omega \setminus \omega$ (such points are called *weak P-points*), and this, again simply by compactness of the space, cannot be a property shared by all points. The last result was published in 1978 by K. Kunen; his paper [KUNEN 1978] is reprinted here. Kunen's method has a wider use, as showed by J. van Mill in [1982]. J. van Mill proves that if X is a non-pseudocompact space which is either nowhere ccc (i.e. every non-void open set contains uncountably many disjoint non-void open subsets) or nowhere of weight $\leq 2^{\omega}$, then there is a weak P-point in $\beta X \setminus X$.

It should be remarked here that there is no hope of proving Rudin's results only in ZFC. S. Shelah showed the consistency of "ZFC + there is no P-point in $\beta\omega \setminus \omega$" as well as the consistency of "ZFC + all autohomeomorphisms of $\beta\omega \setminus \omega$ are trivial" [SHELAH 1982].

The remainder of reals $\beta\mathbb{R} \setminus \mathbb{R}$ allowed the consideration of another topological property of points. Call a point $x \in \beta X \setminus X$ a *remote point*, if x does not belong to a closure of any nowhere dense subset of X. The definition is due to N. J. Fine and L. Gillman [1962], who also proved the existence of remote points in $\beta X \setminus X$ under continuum hypothesis for any non-pseudocompact separable space X — in particular, in $\beta\mathbb{R} \setminus \mathbb{R}$.

Here it is useful to recall a notion of a π-base. A family $\mathcal{G}$ of non-void open subsets of a topological space is called a *π-base*, if every non-void open set contains a member of $\mathcal{G}$; *π-weight* of a topological space is the minimal cardinality of a π-base. It turned out that some restrictions imposed on a π-weight rather than on the density allow us to remove the set theoretical assumption from Fine and Gillman's result. E. K. van Douwen [1978, 1981] and independently S. B. Chae and J. H. Smith [1980] showed the existence of remote points in $\beta X \setminus X$ for X non-pseudocompact with countable π-weight. The result has been strengthened to ccc spaces with π-weight ω_1 and, with an additional set theoretical assumption, even to all spaces with π-weight ω_1 in [DOW 1984a]. However, there is also a consistent example of a space with π-weight ω_1 without remote points ([VAN MILL 1979]); therefore the assumption of $\pi(X) \leq \omega_1$ alone is not strong enough for a ZFC result. Nevertheless, also non-pseudocompact spaces without remote points really exist. The first examples appeared almost simultaneously in [VAN DOUWEN and VAN MILL 1983] and [DOW 1983]. A. Dow's example is particularly easy to describe: If X is an arbitrary compact space with cellularity $> \omega_1$, then there is no remote point in $\beta Y \setminus Y$, where $Y = \omega \times X^\omega$. Another result in this direction is contained in [KUNEN, VAN MILL and MILLS 1980]: CH is equivalent with the statement that every non-pseudocompact space X which has at most $2^{\aleph_0}$ bounded continuous functions, has a remote point. A. Dow proved recently the consistency of a separable space without remote points [DOW 1989].

It should however be noted that spaces without remote points are always somehow exotic. This opinion may be supported by the following theorem due to A. Dow and T. Peters [1987]: Consider a family of spaces which have a σ-locally finite π-base – any discrete or any metrizable space is such. If the Tychonoff product of the family is not compact, then it has a remote point.

Parovičenko spaces. A paper by I. I. Parovičenko [1963] (also reprinted here) added another point to the study of topological properties of Čech-Stone compactification. It gave a characterization of $\beta\omega \setminus \omega$ and of its continuous images. Both main theorems are valid under the assumption of the continuum hypothesis. The reader may recognize that the result follows by the fact that Boolean algebras form a Jónsson class (see [JÓNSSON 1956, 1960]); this fact, however, was unknown at the time of origin of Parovičenko's paper and has been verified by various authors later. The role of the continuum hypothesis (and of other set-theoretical assumptions) in Parovičenko's theorem and in related problems has been extensively discussed since then. A *Parovičenko space* is a space described as $\beta\omega \setminus \omega$ in Parovičenko's theorem, i.e., it is a compact zero-dimensional space without isolated points, of weight 2^ω, satisfying: a) every non-void G_δ set has a non-empty interior, and b) every pair of disjoint F_σ sets can be separated by a clopen set. E. K. van Douwen a J. van Mill in [1978] completed Parovičenko's theorem by proving the implication in the opposite direction: the statement "All Parovičenko spaces are homeomorphic to $\beta\omega \setminus \omega$" implies CH.

Two compact spaces X and Y are called *coabsolute*, if the Boolean algebras of all regular open subsets of X and Y are isomorphic. One can easily observe that homeomorphic compact spaces must be coabsolute as well as that there are coabsolute compact spaces which are not homeomorphic. Since Parovičenko spaces need not be homeomorphic, the question, whether they are coabsolute, was natural. They need not be: S. Broverman and W. Weiss [1981] were able to exhibit two Parovičenko spaces, which were not coabsolute. They assumed the negation of CH together with the following consequence of Martin's axiom: for all infinite $\kappa < 2^{\aleph_0}$, $2^\kappa \leqslant 2^{\aleph_0}$. This result has been strengthened by J. van Mill and S. Williams [1983]. They proved the same result, i.e. the existence of two not coabsolute Parovičenko spaces, assuming only $2^{\aleph_0} = 2^{\aleph_1}$.

On the other hand, all Parovičenko spaces are coabsolute, provided the cofinality of $2^{\aleph_0}$ is ω_1 ([DOW 1984b]). Thus, the coabsoluteness of Parovičenko spaces depends on set theory and, up to now, no necessary and sufficient condition (like van Douwen-van Mill's theorem above) in terms of cardinal arithmetic is known.

κ^+-points in $U(\kappa)$. A large number of papers was devoted to the Čech-Stone compactification of a discrete space. In this case, βX is nothing other than the space of all ultrafilters on X equipped with the Stone topology. Since the Boolean algebra of all subsets of X is complete, βX is extremally disconnected. (A space is *extremally disconnected*, if the closure of any open set is open.) But the remainder $\beta X \setminus X$ is not extremally disconnected, so there are points in $\beta X \setminus X$ belonging to the closures of two disjoint open subsets of $\beta X \setminus X$. In [1967], R. S. Pierce showed that under CH, more is true: there are points in $\beta\omega \setminus \omega$ belonging to the closures of three disjoint open sets. If τ is a cardinal number, X a topological space, a point $x \in X$ is called a *τ-point*, if there is a pairwise disjoint family $\mathcal{U}$ of size $\geqslant \tau$ of open subsets of X such that x is in the closure of every $U \in \mathcal{U}$. Pierce's result has been soon improved by N. B. Hindman in [1969], who showed that there are 2^ω-points in $\beta\omega\setminus\omega$, and, under CH, that all points are such.

After partial results of E. K. van Douwen, J. Roitman, A. Szymański and others, the definitive answer was given by B. Balcar and P. Vojtáš in [1980]. Without any set-theoretical assumptions they showed that every point in $\beta\omega \setminus \omega$ is a 2^ω-point.

If κ is an infinite cardinal number, consider κ as a discrete topological space. Call an ultrafilter $\mathcal{U}$ on κ *uniform*, if every $U \in \mathcal{U}$ is of full size κ, and denote by $U(\kappa)$ the set of all uniform ultrafilters on κ. Thus, $U(\kappa)$ is nothing other than the set of all complete accumulation points of the set κ in $\beta\kappa$. Clearly, $U(\omega) = \beta\omega \setminus \omega$; for $\kappa > \omega$, $U(\kappa)$ is a proper subset of $\beta\kappa \setminus \kappa$. Pierce's (or Hindman's) question naturally extends from $U(\omega)$ to $U(\kappa)$; it was explicitly posed in [COMFORT and HINDMAN 1976] as two questions: Is every point in $U(\kappa)$ a κ^+-point? a 2^κ-point? Comfort and Hindman gave an equivalent formulation of the problem by means of a refinement property. The same year J. Baumgartner showed that there is no hope of answering the second question in the affirmative, because there need not be room for such a large family of disjoint open sets. He proved that it is consistent with ZFC that the cellularity of $U(\omega_1)$ equals ω_2, but 2^{ω_1} is arbitrarily large [BAUMGARTNER 1976]. K. Prikry, assuming GCH, proved that every point in $U(\kappa)$ is a κ^+-point in two consecutive papers, the case of κ regular is treated in [1975], of singular κ in [1976]. Later, the same result, but not depending on additional axioms of set theory, was obtained in [BALCAR and SIMON 1981] for regular κ, and in [SIMON 1984] for a singular κ with countable cofinality. The case of a singular cardinal with uncountable cofinality is, except for Prikry's GCH result, still open.

The theorems on the existence of κ^+-points in $U(\kappa)$ as well as the theorems concerning Parovičenko spaces need some special knowledge of the structure of $U(\kappa)$ or, equivalently, of Boolean algebra $\mathcal{P}(\kappa)/[\kappa]^{<\kappa}$. Here $\mathcal{P}(\kappa)$ stands for the power set of κ and $[\kappa]^{<\kappa}$ for the ideal of all subsets of size $< \kappa$. Let us briefly mention a few results relevant in this context. If κ is an uncountable regular cardinal and $2^\kappa = \kappa^+$, then $U(\kappa)$ has a Noetherian π-base, i.e. a π-base such that any two elements of it are either disjoint or comparable by inclusion [BALCAR and VOPĚNKA 1972]. ZFC results are the following: $\beta\omega \setminus \omega$ $(= U(\omega))$ has a Noetherian π-base [BALCAR, PELANT and SIMON 1981]. If κ is an uncountable cardinal, then the Boolean algebra $\mathcal{P}(\kappa)/[\kappa]^{<\kappa}$ is nowhere ω-distributive, provided cofinality of κ is uncountable [BALCAR and VOPĚNKA 1972] and is nowhere ω_1-distributive, if κ is of countable cofinality [BALCAR and SIMON 1989].

The continuum $\beta\mathbb{H} \setminus \mathbb{H}$. A (non-metric) *continuum* is a compact, connected set. A continuum is called *indecomposable*, if it cannot be covered by two proper subcontinua. Denote by $\mathbb{H}$ the half-line $[0, +\infty) \subset \mathbb{R}$. It is easy to show that $\beta\mathbb{R} \setminus \mathbb{R}$ is not connected — the Čech-Stone extension of a function $\operatorname{arctg} x$ maps $\beta\mathbb{R} \setminus \mathbb{R}$ onto a two-point set. The situation is different with $\mathbb{H}$. In 1968, Bellamy and Woods proved in their PhD theses that $\beta\mathbb{H} \setminus \mathbb{H}$ is a non-metrizable indecomposable continuum (see [BELLAMY 1971]). Bellamy also showed that all non-degenerate subcontinua of $\beta\mathbb{H} \setminus \mathbb{H}$ can be continuously mapped onto $\beta\mathbb{H} \setminus \mathbb{H}$ and, consequently, are of the full cardinality $2^{\mathfrak{c}}$.

A *composant* of a topological space in a point x is the union of all proper subcontinua, containing x. In an indecomposable continuum, distinct composants are disjoint. Now, the problem was, how many composants in $\beta\mathbb{H} \setminus \mathbb{H}$ are there? In the case of an indecomposable metrizable continuum, the number of composants must be uncountable. Bellamy conjectured, that $\beta\mathbb{H} \setminus \mathbb{H}$ could have only one composant.

We have already mentioned the Rudin-Frolík's preorder of ultrafilters. A Rudin-Keisler preorder of ultrafilters on ω is defined as follows. For $x, y \in \beta\omega \setminus \omega$, $x \leqslant_{RK} y$ provided there is a mapping $f : \omega \to \omega$ such that $\beta f(y) = x$. Two ultrafilers x, y are *incompatible*, if there is no $z \in \beta\omega\setminus\omega$ with $z \leqslant_{RK} x$, $z \leqslant_{RK} y$. Now, M. E. Rudin [1971] and J. Mioduszewski [1974, 1980] proved, that the number of composants in $\beta\mathbb{H} \setminus \mathbb{H}$ equals to the maximal size of a set of ultrafilters in $\beta\omega \setminus \omega$ pairwise incompatible under $\leqslant_{RK}$. Consequently, if CH is assumed, then $\beta\mathbb{H} \setminus \mathbb{H}$ has $2^{\mathfrak{c}}$ composants.

However, the positive answer to Bellamy's conjecture is consistent as well. In [Blass and Shelah 1987], the authors proved that the following principle is consistent with ZFC: For any $x, y \in \beta\omega \setminus \omega$ there exists a finite-to-one mapping $f : \omega \to \omega$ that $\beta f(x) = \beta f(y)$. Therefore $\beta\mathbb{H} \setminus \mathbb{H}$ may have only one composant. In [Blass 1987], the topics is discussed in a great detail.

References

P. S. Alexandrov and P. S. Uryson

[1929] *Mémoire sur les espaces topologiques compacts*, Verh. Akad. Wetensch. Amsterdam **14** (1929), 1 - 96.

B. Balcar and R. Frankiewicz

[1979] *Ultrafilters and ω_1-points in $\beta N \setminus N$*, Bull. Acad. Polon. Sci. **27** (1979), 593 - 598.

B. Balcar, J. Pelant and P. Simon

[1980] *The space of ultrafilters on N covered by nowhere dense sets*, Fund. Math. **110** (1980), 11 - 24.

B. Balcar and P. Simon

[1982] *Strong decomposability of ultrafilters I*, Logic Colloquium 1980, North Holland Publ. Co. 1982, 1 - 10.

[1989] *Disjoint refinement*, in: Handbook of Boolean algebras, Elsevier Science Publishers B.V. (1989), 335 - 386.

B. Balcar and P. Vojtáš

[1980] *Almost disjoint refinement of families of subsets of N*, Proc. Amer. Math. Soc. **79** (1980), 465 - 470.

B. Balcar and P. Vopěnka

[1972] *On systems of almost disjoint sets*, Bull. Acad. Polon. Sci., Sér. Sci. Math. **20** (1972), 421 - 424.

J. E. Baumgartner

[1976] *Almost-disjoint sets, the dense set problem and the partition calculus*, J. Symb. Logic **10** (1976), 401 - 439.

A. Blass

[1987] *Near coherence of filters II. Applications to operator ideals, the Stone-Čech remainder of a half-line, order ideals of sequences, and slenderness of groups*, Trans. Amer.

Math. Soc. **300** (1987), 557 - 581.

A. BLASS and S. SHELAH

[1987] *There may be simple* $P_{\aleph_1}$ *and* $P_{\aleph_2}$ *points and the Rudin-Keisler order may be downward directed*, Ann. Pure Appl. Logic **33** (1987), 213 - 243.

D. P. BELLAMY

[1971] *A non-metric indecomposable continuum*, Duke Math. J. **38** (1971), 15 - 20.

S. BROVERMAN and W. WEISS

[1981] *Spaces coabsolute with* $\beta\mathbb{N} \setminus \mathbb{N}$, Topology Appl. **12** (1981), 127 - 133.

E.ČECH

[1937] *On bicompact spaces*, Ann. of Math. **38** (1937), 823 - 844.

W. W. COMFORT and N. B. HINDMAN

[1976] *Refining families for ultrafilters*, Math. Z. **149** (1976), 189 - 199.

S. B. CHAE and J. H. SMITH

[1980] *Remote points and G-spaces*, Topology Appl. **11** (1980), 243 - 246.

E. K. VAN DOUWEN

[1978] *Existence and applications of remote points*, Bull. Amer. Math. Soc. **841** (1978), 161 - 163.

[1981] *Remote points*, Dissertationes Math. CLXXXVIII (1981).

E. K. VAN DOUWEN and J. VAN MILL

[1978] *Parovičenko's characterization of* $\beta\omega \setminus \omega$ *implies CH*, Proc. Amer. Math. Soc. **72** (1978), 539 - 541.

[1983] *Spaces without remote points*, Pacific J. Math. **105** (1983), 69 - 75.

A. DOW

[1983] *Products without remote points*, Topology Appl. 15 (1983), 239 - 246.

[1984a] *Remote Points in Spaces with* π*-weight* ω_1, Fund. Math. **124** (1984), 197 - 205.

[1984b] *Co-absolutes of* $\beta\mathbb{N} \setminus \mathbb{N}$, Topology Appl. **18** (1984), 1 - 15.

[1989] *A Separable Space with no Remote Points*, Trans. Amer. Math. Soc. **312** (1989), 335 - 353.

A. DOW and T. PETERS

[1987] *Game strategies yield remote points*, Topology Appl. 27 (1987), 245 - 256.

R. ENGELKING and S. MRÓWKA

[1958] *On E-compact spaces*, Bull. Acad. Polon. Sci. Sér. Sci. Math. Astronom. Phys. **6** (1958), 429 - 436.

N. J. FINE and L. GILLMAN

[1962] *Remote points in* $\beta\mathbb{R}$, Proc. Amer. Math. Soc. **13** (1962), 29 - 36.

Z. FROLÍK

[1967a] *Sums of ultrafilters*, Bull. Amer. Math. Soc. **73** (1967), 87 - 91.

[1967b] *Non-homogeneity of* $\beta P \setminus P$, Comment. Math. Univ. Carolinae **8** (1967), 705 - 709.

I. GELFAND and A. KOLMOGOROFF

[1939] *On the rings of continuous functions on topological spaces*, Dokl. Akad. Nauk SSSR **22** (1939), 11 - 15.

L. GILLMAN and M. JERISON

[1960] *Rings of continuous functions*, Van Nostrand, Princeton, 1960.

I. GLICKSBERG

[1959] *Stone-Čech compactifications of products*, Trans. Amer. Math. Soc. **90** (1959), 369 - 382.

E. HEWITT

[1946] *A remark on density characters*, Bull. Amer. Math. Soc. **52** (1946), 641 - 642.

[1948] *Rings of real-valued continuous functions I.*, Trans. Amer. Math. Soc. **64** (1948), 45 - 99.

N. B. HINDMAN

[1969] *On the existence of* $\mathfrak{c}$*-points in* $\beta\mathbb{N}\setminus\mathbb{N}$, Proc. Amer. Math. Soc. **21** (1969), 277 - 280.

M. HUŠEK

[1970] *The Hewitt realcompactification of a product*, Comment. Math. Univ. Carolinae **11** (1970), 393 - 395.

B. JÓNSSON

[1956] *Universal relational systems*, Math. Scand. **4** (1956), 193 - 208.

[1960] *Homogeneous universal relational systems*, Math. Scand. **8** (1960), 137 - 142.

K. KUNEN

[1978] *Weak P-points in* $\mathbb{N}^*$, Colloquium on Topology, Budapest (1978), 741 - 749.

K. KUNEN, J. VAN MILL and C. MILLS

[1980] *On nowhere dense closed P-sets*, Proc. Amer. Math. Soc. **78** (1980), 119 - 123.

E. MARCZEWSKI

[1947] *Séparabilité et multiplication cartésienne des espaces topologiques*, Fund. Math. **34** (1947), 127 - 143.

J. VAN MILL

[1979] *More on remote points*, Rapport nr. 91, Wiskundig Seminarium, Vrije Universiteit, Amsterdam (1979), 15 pp.

[1982] *Weak P-point in Čech-Stone compactifications*, Trans. Amer. Math. Soc. **273** (1982), 657 - 678.

J. VAN MILL and S. WILLIAMS

[1983] *A compact F-space not coabsolute with* $\beta\mathbb{N}-\mathbb{N}$, Topology Appl. **15** (1983), 59 - 64.

J. MIODUSZEWSKI

[1974] *On composants of* $\beta R\setminus R$, Proc. of the Conference Topology and Measure, I (Zinnowitz, 1974), Part 2, 257 - 283, Ernst-Moritz-Arndt-Univ., Greifswald, 1978.

[1980] *An approach to* $\beta R\setminus R$, Topology, Colloq. Math. Soc. János Bolyai 23, North Holland 1980, 257 - 283.

I. I. PAROVIČENKO

[1963] *On a universal bicompactum of weight* $\aleph$, Dokl. Akad. Nauk SSSR **150** (1963), 36 - 39.

R. S. PIERCE

[1967] *Modules over commutative regular rings*, Memoirs of Amer. Math. Soc., No 70, Providence (1967).

E. S. PONDICZERY

[1944] *Power problems in abstract spaces*, Duke Math. J. **11** (1944), 835 - 837.

B. POSPÍŠIL

[1937] *Remark on bicompact spaces*, Ann. of Math. **38** (1937), 845 - 846.

K. PRIKRY

[1974] *Ultrafilters and almost disjoint sets*, Gen. Top. and Appl. **4** (1974), 269 - 282.

[1975] *Ultrafilters and almost disjoint sets II*, Bull. Amer. Math. Soc. **81** (1975), 209 - 212.

J. ROITMAN

[1975] *Almost disjoint strong refinements*, Notices Amer. Math. Soc. **22** (1975), A 328.

M. E. RUDIN

[1970] *Composants and* $\beta\mathbb{N}$, Proc. Wash. State Univ. Conf. on Gen. Topology (Pullman, Wash., 1970), 117 -119.

[1971] *Partial orders on the types of* $\beta\mathbb{N}$, Trans. Amer. Math. Soc. **155** (1971), 353 - 362.

W. RUDIN

[1956] *Homogeneity problems in the theory of Čech compactifications*, Duke Math. J. **23** (1956), 409 - 419, 633.

S. SHELAH
[1982] *Proper Forcing*, Lecture Notes in Mathematics 940, Springer-Verlag, Berlin Heidelberg New York, 1982.

P. SIMON
[1984] *Strong decomposability of ultrafilters on cardinals with countable cofinality*, Acta Univ. Carol. Math. et Phys. **25** (1984), 11 - 26.

M. H. STONE
[1937] *Application of Boolean rings to general topology*, Trans. Amer. Math. Soc. **41** (1937), 375 - 481.

A. SZYMAŃSKI
[1977] *On the existence of $\aleph_0$-points*, Proc. Amer. Math. Soc. **66** (1977), 128 - 130.

R. C. WALKER
[1974] *The Stone-Čech compactification*, Springer-Verlag, Berlin Heidelberg New York, 1974.

A. N. TYCHONOFF
[1930] *Über die topologische Erweiterung von Räumen*, Math. Ann. **102** (1929), 544 - 561.

Annals of Mathematics
Vol. 38, No. 4, October, 1937

ON BICOMPACT SPACES

By Eduard Čech

(Received February 3, 1937)

The theory of bicompact spaces was extensively studied by P. Alexandroff and P. Urysohn in their paper *Mémoire sur les espaces topologiques compacts*, Verhandlingen der Kon. Akademia Amsterdam, Deel XIV, No. 1, 1929; I shall refer to this paper with the letters AU. An important result was added by A. Tychonoff in his paper *Über die topologische Erweiterung von Räumen*, Math. Annalen 102, 1930, who proved that complete regularity is the necessary and sufficient condition for a topological space to be a subset of some bicompact Hausdorff space. As a matter of fact, Tychonoff proves more, viz. that, given a completely regular space S, there exists a bicompact Hausdorff space $\beta(S)$ such that (i) S is dense in $\beta(S)$, (ii) any bounded continuous real function defined in the domain S admits of a continuous extension to the domain $\beta(S)$. It is easily seen that $\beta(S)$ is uniquely defined by the two properties (i) and (ii). The aim of the present paper is chiefly the study of $\beta(S)$.

The paper is divided into four chapters. In chapter I, I briefly resume some well known definitions adding a few simple remarks. In particular I show that an *arbitrary* topological space S determines a completely regular space $\rho(S)$ such that a good deal of topology of S reduces to the topology of $\rho(S)$, this being true in particular for the theory of real valued continuous and Baire functions. Chapter II contains the theory of the bicompact space $\beta(S)$ mentioned above. Here I shall recall only a few results of chapter II. First, if the space S is *normal*, then $\beta(S)$ may be defined without any reference to continuous real function since property (ii) may be replaced by the following: if two closed subsets of S have no common point, then their closures in $\beta(S)$ have no common point either. Second, if the space S satisfies the first countability axiom, then S is completely determined by $\beta(S)$, S being simply the set of all points of $\beta(S)$ where the first countability axiom holds true. This implies that in this case (embracing the case of metrizable spaces) the whole topology of S may be reduced to the topology of the bicompact space $\beta(S)$. Hence it is evident that it is highly desirable to carry further the study of bicompact spaces and in particular of $\beta(S)$. Of course it must be emphasized that $\beta(S)$ may be defined only formally (not constructively) since it exists only in virtue of Zermelo's theorem. If I denotes the space of integer numbers, then I think it is impossible to determine effectively (in the sense of Sierpiński) a point of $\beta(I) - I$. I was even unable to determine the cardinal number of $\beta(I)$. (The paper contains several other unsolved problems.) The space $\beta(I) - I$ furnishes incidentally a positive solution of a problem proposed by Alexandroff and Urysohn (AU, p. 54:

Existe-t-il un espace bicompact ne contenant aucun point (κ)? The authors write in this connection: La résolution affirmative de ce problème nous donnerait un exemple des espace bicompacts d'une nature toute différente de celle des espaces connus jusqu'à présent). In chapter III, I call a completely regular space S *topologically complete* if S is a G_δ in $\beta(S)$. The reason for this designation lies in the fact that, if S is metrizable, it has this property if and only if it is homeomorphic with a metric complete space. The proof is an easy adaptation of Hausdorff's well known proof of the theorem that a G_δ in a metric complete space is a homeomorph of a metric complete space. In chapter IV, I consider *locally normal* spaces and I prove that a locally normal space S is always an open subset of some normal space. This was of course to be expected but I think it would be difficult to prove without the theory of $\beta(S)$.

I

A set S is called a *topological space* (and its elements are called *points*) if there is given a class $\mathfrak{F}$ of subsets of S (called *closed* subsets of S) such that (1) the whole space S and the vacuous set 0 are closed, (2) the intersection of any family of closed sets is closed, (3) the sum of two closed sets is closed. A set $G \subset S$ is called *open*, if the complementary set $S - G$ is closed. A *neighborhood* of a set $A \subset S$ (A may consist of a single point) is an open set containing A.

The intersection of all closed sets containing a given set A is called the *closure* of A and is denoted by $\bar{A}$. The closure operation has the following properties: (1) $\bar{0} = 0$, (2) $A \subset \bar{A}$, (3) $\overline{A + B} = \bar{A} + \bar{B}$, (4) $\bar{\bar{A}} = \bar{A}$. Conversely, it is possible to define the general notion of a topological space starting with an operation $\bar{A}$ subject only to conditions (1)-(4) and defining closed sets by the condition $\bar{A} = A$.

An *open base* of a topological space S is a class $\mathfrak{B}$ of open sets such that any open set is the sum of some of the elements of $\mathfrak{B}$. The class $\mathfrak{H}$ of *all* open sets is a particular open base. Any open base $\mathfrak{B}$ has the following properties: (1) given a point $x \in S$, there exists a $U \in \mathfrak{B}$ such that $x \in U$, (2) given a point $x \in S$ and two sets U and V such that $U \in \mathfrak{B}$, $V \in \mathfrak{B}$, $x \in UV$, there exists a set W such that $W \in \mathfrak{B}$, $x \in W$, $W \subset UV$. Conversely it is possible (and the possibility is utilized very frequently in practice) to define a topological space starting with a class $\mathfrak{B}$ subject only to condition (1) and (2); the closure $\bar{A}$ of a set $A \subset S$ consists then of all the points x such that

$$U \in \mathfrak{B},\ x \in U \text{ implies } UA \neq 0.$$

A fixed subset T of a topological space S is always considered as a topological space, defining a set $A \subset T$ to be *relatively closed* (i.e. closed in the space T) whenever A is the intersection of T with some closed subset of S. A set $A \subset T$ is *relatively open* whenever A is the intersection of T with some open subset of S. The *relative closure* of a set $A \subset T$ is the intersection $T\bar{A}$ of T with the closure of A in the space S. Any open base $\mathfrak{B}$ of S determines an open base $\mathfrak{B}_0$ of T; the elements of $\mathfrak{B}_0$ are the intersections of T with the elements of $\mathfrak{B}$.

A *mapping* f of a topological space S_1 into a topological space S_2 is an operation attaching to each point $x \in S_1$ a definite point $f(x) \in S_2$; we always suppose that, given any point $y \in S_2$, there exists at least one point $x \in S_1$ such that $f(x) = y$. The space S_1 is the *domain* of f, S_2 is its *range*. The *image* $f(A)$ of a set $A \subset S_1$ is the set of all points $f(x)$, x running over A. The *inverse image* $f^{-1}(B)$ of a set $B \subset S_2$ is the set of all points $x \in S_1$ such that $f(x) \in B$. The mapping f is *one-to-one* if

$$x_1 \in S_1,\ x_2 \in S_1,\ x_1 \neq x_2 \text{ implies } f(x_1) \neq f(x_2).$$

If f is one-to-one, then the inverse operation f^{-1} is a one-to-one mapping of S_2 into S_1. The mapping f will be called a *function* if its range consists of real numbers. The function f is *bounded* if its range is a bounded set.

The mapping f is called *continuous at a point* $x \in S_1$ if, given any neighborhood V of $f(x)$, there exists a neighborhood U of x such that $f(U) \subset V$. f is called *continuous* (simply) if it is continuous at any point $x \in S_1$. f is called *homeomorphic* if it is one-to-one and if both f and f^{-1} are continuous. f is continuous, if and only if the inverse image of any closed subset of S_2 is a closed subset of S_1.

A set $A \subset S$ is called a G_δ-set if there exists a countable sequence $\{G_n\}$ of open sets such that $A = \prod_1^\infty G_n$; A is called an F_σ-set if there exists a countable sequence $\{F_n\}$ of closed sets such that $A = \sum_1^\infty F_n$. The complement of a G_δ-set is an F_σ-set and vice-versa.

S is called a *Kolmogoroff space*[1] if the closures of any two distinct points are distinct. S is called a *Riesz space*[2] if any single point is closed. S is a Riesz space if and only if the intersection of all the neighborhoods of any point x consists of x only. S is called a *Hausdorff space* if the intersection of the closures of all the neighbhorhoods of any point x consists of x only. Any Riesz space is a Kolmogoroff space. Any Hausdorff space is a Riesz space. Any subset of a Kolmogoroff space is a Kolmogoroff space. Any subset of a Riesz space is a Riesz space. Any subset of a Hausdorff space is a Hausdorff space. Let $\mathfrak{B}$ be any open base of S. S is a Kolmogoroff space if and only if, given two distinct points x and y, there exists a set $U \in \mathfrak{B}$ containing precisely one of the points x and y. S is a Riesz space if and only if, given two distinct points x and y, there exists a set $U \in \mathfrak{B}$ containing x and not containing y. S is a Hausdorff space if and only if, given two distinct points x and y, there exist sets U and V such that $U \in \mathfrak{B}$, $V \in \mathfrak{B}$, $x \in U$, $y \in V$, $UV = 0$.

Now we proceed to prove that *the theory of general topological spaces* (in the sense precised above) *can be completely reduced to the theory of Kolmogoroff spaces.* Let S be a topological space. Two points $x \in S$ and $y \in S$ will be called equivalent (for the time being) if $\bar{x} = \bar{y}$. Let F be any closed subset of S and let x and y be two equivalent points; if $x \in F$, then $\bar{x} \subset F$, since F is closed, but $y \in \bar{y}$ and $\bar{y} = \bar{x}$, so that $y \in F$. It follows that any closed subset of S consists of complete

[1] See P. Alexandroff and H. Hopf, *Topologie* I, p. 58.

[2] See G. Birkhoff, *On the combination of topologies*, Fund. Math. 26, p. 162.

classes of mutually equivalent points. Now let us attach to each point $x \in S$ a new symbol $\tau(x)$ chosen in such manner that $\tau(x) = \tau(y)$ if and only if x and y are equivalent; let us call S_0 the set of the symbols $\tau(x)$, so that τ is a mapping of S into S_0. A set $A_0 \subset S_0$ will be considered as closed if and only if its inverse image $\tau^{-1}(A_0)$ is a closed subset of S. It is evident that S_0 is a topological space and that τ is a continuous mapping. Further it is evident that for any set $A \subset S$ we have $\tau(\bar{A}) = \overline{\tau(A)}$; in particular $\tau(\bar{x}) = \overline{\tau(x)}$ for any $x \in S$. If $\tau(x) \neq \tau(y)$, we have $\bar{x} \neq \bar{y}$; since the sets $\bar{x}$ and $\bar{y}$ are closed, it easily follows that $\tau(\bar{x}) \neq \tau(\bar{y})$, or $\overline{\tau(x)} \neq \overline{\tau(y)}$, so that S_0 is a Kolmogoroff space. Conversely, let S_0 be a Kolmogoroff space. Let τ be a mapping of a set S into S_0. Let us call closed in S the inverse image of any closed subset of S_0. Then S is the most general topological space and τ has the previous meaning. Evidently the topology of S is quite completely described by that of S_0.

S is called a *regular space* if it is a Kolmogoroff space having the following property: given a neighborhood U of a point x, there exists a neighborhood V of x such that $\bar{V} \subset U$.[3] We shall prove that any regular space S is a Hausdorff space.[4] Let x and y be two distinct points of S. If we had both $x \in \bar{y}$ and $y \in \bar{x}$, it would follow, since $\bar{x}$ and $\bar{y}$ are closed, that $\bar{x} \subset \bar{y}$ and $\bar{y} \subset \bar{x}$, i.e. $\bar{x} = \bar{y}$, which is impossible. The argument being symmetrical, we may suppose that x does not belong to $\bar{y}$, so that $S - \bar{y}$ is a neighborhood of x. Hence there exists a neighborhood U of x such that $\bar{U} \subset S - \bar{y}$. Putting $V = S - \bar{U}$, we have two open sets U and V such that $x \in U$, $y \in V$, $UV = 0$, so that S is a Hausdorff space.

Any subset of a regular space is a regular space.

S is called a *completely regular space* if it is a Kolmogoroff space having the following property: given a closed set F and a point $a \in S - F$, there exists a continuous function f (in the domain S) such that $f(a) = 0$ and $f(x) = 1$ for any $x \in F$.[5] It is easy to see that a completely regular space is regular and that any subset of a completely regular space is a completely regular space.

Now we shall start with an arbitrary topological space S and we shall attach to it a uniquely defined completely regular space $\rho(S)$ in such manner that a great deal of topology of S may be reduced to that of $\rho(S)$. Two points x and y of S will be called equivalent (for the time being) if $f(x) = f(y)$ for every continuous function f (in the domain S). To each point $x \in S$ let us attach a new symbol $\rho(x)$ chosen in such a manner that $\rho(x) = \rho(y)$ if and only if x and y are equivalent;[6] let us call S_1 the set of all the symbols $\rho(x)$, so that ρ is a mapping of S into $S_1 = \rho(S)$. We shall introduce a topology in S_1 by defining an open

[3] The neighborhoods may here be restricted to a given open base of S.

[4] This is usually done assuming *a priori* that S is a Riesz space; for this point I am indebted to Dr. K. Koutský.

[5] We may assume that $0 \leqq f(x) \leqq 1$ for every $x \in S$, since we could replace f with φ by defining $\varphi(x) = f(x)$ if $0 \leqq f(x) \leqq 1$, $\varphi(x) = 0$ if $f(x) < 0$, and $\varphi(x) = 1$ if $f(x) > 1$.

[6] It is evident that $\tau(x) = \tau(y)$ implies $\rho(x) = \rho(y)$, but of course we may restrict ourselves to Kolmogoroff spaces.

base $\mathfrak{B}$ for S_1. An element $[f, I]$ of $\mathfrak{B}$ will be defined by a continuous function f in the domain S and an open interval I, $[f, I]$ consisting of the points $\rho(x)$ of S_1 such that $f(x) \in I$. To prove that S_1 is a topological space we have to verify two things. First, for any $a \in S$, there evidently exists an $[f, I]$ containing $\rho(a)$. Second, let $\rho(a)$ belong both to $[f_1, I_1]$ and to $[f_2, I_2]$; we have to prove that there exists an $[f, I]$ such that $\rho(a) \in [f, I]$ and $[f, I] \subset [f_1, I_1] \cdot [f_2, I_2]$. There exists a number $\varepsilon > 0$ such that, for $i = 1$ and for $i = 2$, the interval $f_i(a) - \varepsilon < t < f_i(a) + \varepsilon$ is a subset of I_i. It is easy to see that we may put $f(x) = |f_1(x) - f_1(a)| + |f_2(x) - f_2(a)|$, choosing I to be the interval $-\varepsilon < t < \varepsilon$. Hence S_1 is a topological space.

Since the topology of S_1 was defined by means of *continuous* functions in the domain S, it is easy to see that ρ is a continuous mapping of S into S_1 so that, if φ is any continuous function in the domain S_1, $f(x) = \varphi[\rho(x)]$ is a continuous function in the domain S. Moreover, in our case the converse is also true: *any continuous function in the domain S has the form* $f(x) = \varphi[\rho(x)]$, φ *being a continuous function in the domain* S_1.

If $\rho(a)$ and $\rho(b)$ are two distinct points of S_1, then there exists a continuous function f in the domain S such that $f(a) \neq f(b)$. There exist two disjoined open intervals I_1 and I_2 such that $f(a) \in I_1$ and $f(b) \in I_2$. Then $[f, I_1]$ and $[f, I_2]$ are two disjoined open subsets of S_1 and $\rho(a) \in [f, I_1]$, $\rho(b) \in [f, I_2]$. It follows that S_1 is a Hausdorff space. As a matter of fact, S_1 is a completely regular space. Let Φ be a closed subset of S_1 not containing the point $\rho(a)$. There exists an $[f, I]$ such that $\rho(a) \in [f, I] \subset S_1 - \Phi$; we may suppose that I consists of all numbers t such that $|t - f(a)| < \varepsilon (\varepsilon > 0)$. If $|f(x) - f(a)| \geqq \varepsilon$, put $g(x) = 1$; if $|f(x) - f(a)| < \varepsilon$, put $g(x) = \varepsilon^{-1}. |f(x) - f(a)|$. Then g is a continuous function in the domain S, so that there exists a continuous function φ in the domain S_1 such that $g(x) = \varphi[\rho(x)]$. It is easy to see that $\varphi[\rho(a)] = 0$ and $\varphi(x) = 1$ for each $x \in \Phi$.

Let F be a closed subset of S. We shall prove that *a necessary and sufficient condition for the set* $\rho(F)$ *to be closed in* S_1 *is that for any point*

$$a \in S - \rho^{-1}[\rho(F)]$$

there exists a continuous function f in the domain S such that $f(a) = 0$ *and* $f(x) = 1$ *for each* $x \in F$. First suppose the condition satisfied. If $\rho(F)$ were not closed in S_1, we could choose a point a such that

$$\rho(a) \in \overline{\rho(F)} - \rho(F).$$

Since $\rho(a) \in S_1 - \rho(F)$, there would exist a continuous function f in the domain S such that $f(a) = 0$ and $f(x) = 1$ for each $x \in F$. There would exist a continuous function φ in the domain S_1 such that $f(x) = \varphi[\rho(x)]$. For $x \in \rho(F)$ we would have $\varphi(x) = 1$; since φ is continuous, it easily follows that $\varphi(x) = 1$ for $x \in \overline{\rho(F)}$, in particular $\varphi[\rho(a)] = 1$, i.e. $f(a) = 1$, which is a contradiction. Secondly, suppose $\rho(F)$ closed in S_1. Let $a \in S - \rho^{-1}[\rho(F)]$. Then $\rho(a) \in S_1 - \rho(F)$. Since S_1 is completely regular, there exists a continuous function φ in the domain

S_1 such that $\varphi[\rho(a)] = 0$ and $\varphi(x) = 1$ for each $x \in \rho(F)$. Putting $f(x) = \varphi[\rho(x)]$, we have a continuous function f in the domain S such that $f(a) = 0$ and $f(x) = 1$ for each $x \in F$.

As a corollary, we obtain that, if the space S itself is completely regular, the mapping ρ is homeomorphic.

The following property is characteristic for completely regular spaces S: *Let σ be a continuous mapping of S into a topological space R such that each continuous function f in the domain S has the form $f(x) = \varphi[\sigma(x)]$, φ being a continuous function in the domain R. Then the mapping σ is homeomorphic.* The property cannot be true if S is not completely regular, as is seen by putting $\sigma = \rho$. Hence suppose that S is completely regular. If $a \in S$, $b \in S$, $a \neq b$, there exists a continuous function f in the domain S such that $f(a) \neq f(b)$; since $f(x) = \varphi[\sigma(x)]$, we have $\sigma(a) \neq \sigma(b)$, i.e. the mapping σ is one-to-one. It remains to show that if F is a closed subset of S the set $\sigma(F)$ is closed in R. If $\sigma(F)$ is not closed, there exists a point $a \in S$ such that

$$\sigma(a) \in \overline{\sigma(F)} - \sigma(F).$$

There exists a continuous function f in the domain S such that $f(a) = 0$ and $f(x) = 1$ for each $x \in F$. We may put $f(x) = \varphi[\sigma(x)]$ and we have $\varphi[\sigma(a)] = 0$ and $\varphi(x) = 1$ for each $x \in \sigma(F)$. Since φ is continuous, we must have $\varphi(x) = 1$ for each $x \in \overline{\sigma(F)}$, hence for $x = a$, which is a contradiction.

Consider the following three properties of a topological space S: (1) If F_1 and F_2 are two closed sets such that $F_1F_2 = 0$, there exist two open sets G_1 and G_1 such that $F_1 \subset G_1$, $F_2 \subset G_2$, $G_1G_2 = 0$. (2) If F_1 and F_2 are two closed sets such that $F_1F_2 = 0$, there exists a continuous function f in the domain S such that $f(x) = 0$ for each $x \in F_1$ and $f(x) = 1$ for each $x \in F_2$.[5] (3) If F is a closed set and if φ is a bounded[7] continuous function in the domain F, there exists a continuous function f in the domain S such that $f(x) = \varphi(x)$ for each $x \in F$. It is easily seen that (2) is formally stronger than (1) and that (3) is formally stronger than (2). But Urysohn proved[8] that all three properties are equivalent to one another. A space having these properties is called *normal*. Property (2) shows that a normal Riesz space is a completely regular space (hence a regular space, therefore a Hausdorff space).

If the space S is normal, then $\rho(S)$ is normal as well. Let Φ_1 and Φ_2 be two closed subsets of $\rho(S)$ such that $\Phi_1\Phi_2 = 0$. Then $F_1 = \rho^{-1}(\Phi_1)$ and $F_2 = \rho^{-1}(\Phi_2)$ are two closed subsets of S such that $F_1F_2 = 0$. Since S is normal, there exists a continuous function f in the domain S such that $f(x) = 0$ for each $x \in F_1$ and $f(x) = 1$ for each $x \in F_2$. There exists a continuous function φ in the domain $\rho(S)$ such that $f(x) = \varphi[\rho(x)]$. Evidently $\varphi(x) = 0$ for each $x \in \Phi_1$ and $\varphi(x) = 1$ for each $x \in \Phi_2$.

If the space S is normal, then for $a \in S$, $b \in S$ we have $\rho(a) = \rho(b)$ if and only if

---

[7] It is easy to prove that the word *bounded* may be omitted.

[8] P. Urysohn, *Über die Mächtigkeit zusammenhängender Mengen*, Math. Annalen 94, 1925.

$\bar{a}\cdot\bar{b} \neq 0$. Suppose first that $c \in \bar{a}\cdot\bar{b}$. If f is a continuous function in the domain S, it is easy to see that $f(a) = f(c) = f(b)$, whence $\rho(a) = \rho(b)$. Secondly, suppose that $\bar{a}\cdot\bar{b} = 0$. Since S is normal, there exists a continuous function f in the domain S such that $f(x) = 0$ for each $x \in \bar{a}$ and $f(x) = 1$ for each $x \in \bar{b}$, whence $f(a) = 0, f(b) = 1$.

If the space S is normal and if F is a closed subset of S, then $\rho(F)$ is a closed subset of $\rho(S)$. Let $a \in S - \rho^{-1}[\rho(F)]$. For $x \in F$ we have $\rho(a) \neq \rho(x)$, whence $\bar{a}\cdot\bar{x} = 0$; therefore $\bar{a}\cdot F = 0$. Hence there exists a continuous function f in the domain S such that $f(x) = 1$ for each $x \in F$ and $f(x) = 0$ for each $x \in \bar{a}$, in particular $f(a) = 0$. We know that this implies that $\rho(F)$ is closed in $\rho(S)$.

The last two theorems show that, if S is normal, the space $\rho(S)$ and its topology may be completely described without any explicit reference to continuous functions: The space $\rho(S)$ consists of symbols $\rho(x)$ attached to single points $x \in S$, $\rho(x)$ and $\rho(y)$ being identical if and only if $\bar{x}\cdot\bar{y} \neq 0$; and a set $\Phi \subset \rho(S)$ is closed in $\rho(S)$ if and only if the set $\rho^{-1}(\Phi)$ is closed in S. It is an interesting problem to give a similar description of $\rho(S)$ in the general case.

If the space S is normal, then a necessary and sufficient condition for a set $A \subset S$ to be both closed and a G_δ is the existence of a continuous function f in the domain S such that $f(x) = 0$ if and only if $x \in A$. Suppose first that such a function f exists. Then $A = \{f(x) = 0\}$ is a closed set and $G_n = \{\,|f(x)| < 1/n\}$ are open sets and $A = \prod G_n$. Conversely let $A = \bar{A} = \prod G_n$, G_n being open. Since S is normal, there exist continuous functions f_n in the domain S such that $f_n(x) = 0$ for $x \in A$, $f_n(x) = 1$ for $x \in S - G_n$, $0 \leqq f_n(x) \leqq 1$ for $x \in S$. It is sufficient to put $f(x) = \sum 2^{-n}\cdot f_n(x)$.

A point x of a topological space S is called a *complete limit point* of a set $A \subset S$ if, for any neighborhood U of x, the cardinal number of the set AU is equal to the cardinal number of the set A. A family $\mathfrak{C}$ of subsets of S is called *monotonic* if for any two sets $A \in \mathfrak{C}$, $B \in \mathfrak{C}$ we have either $A \subset B$ or $B \subset A$. A family $\mathfrak{C}$ of subsets of S is called a *covering* of S if each point of S belongs to some set of $\mathfrak{C}$.

Consider the following three properties of a topological space S: (1) Every infinite subset possesses at least one complete limit point. (2) A monotonic family of non-vacuous closed subsets has a non-vacuous intersection. (3) Any covering of S consisting of open sets contains a finite covering of S. It is known that all three properties are equivalent to one another.[9] A space having these properties is called *bicompact*. It is known that *a bicompact Hausdorff space is normal*[10] (hence completely regular). *A closed subset of a bicompact space is a bicompact space.* Conversely, *a bicompact subset of a Hausdorff space is closed.*[11] It easily follows that *a one-to-one continuous mapping of a bicompact Hausdorff space is homeomorphic.*

Let $\{S_\iota\}$ be a family of sets; the subscript ι runs over an arbitrarily given set I. The cartesian product $\mathfrak{P}_\iota S_\iota$ of the family $\{S_\iota\}$ is the set of all families $x = \{x_\iota\}$,

---

[9] AU, p. 8.
[10] AU, p. 26.
[11] AU, p. 47.

each x_ι belonging to S_ι. The x_ι's are called the coordinates of x. If every S_ι is a topological space, we introduce a topology into $S = \mathfrak{P}_\iota S_\iota$ by means of the following open base $\mathfrak{B}$: The elements of $\mathfrak{B}$ are sets of the form $\mathfrak{P}_\iota G_\iota$ where (1) each G_ι is an open subset of S_ι, (2) $G_\iota = S_\iota$ except for a finite number of subscripts ι. It is easy to see that S is a Kolmogoroff space, a Riesz space, a Hausdorff space, a regular space, a completely regular space, if and only if *every* factor space S_ι belongs to the corresponding category of spaces. If S is normal, every S_ι is normal as well; but the converse is false.

The cartesian product $S = \mathfrak{P}_\iota S_\iota$ of any family of bicompact spaces ia a bicompact space. Using Zermelo's theorem, we may suppose that the set I consists of all ordinal numbers less than a given ordinal number. Let there be given an infinite subset A of S. We have to construct a complete limit point $z = \{z_\iota\}$ of S. According to the way the topology of S was introduced, it is sufficient to construct the coordinates z_ι by transfinite induction, choosing each $z_\iota \epsilon S_\iota$ in such a way that it have the following property π_ι: If there is given a finite number of subscripts $\iota_n \leqq \iota$ and, for each ι_n, a neighborhood G_n of z_{ι_n} (in the space S_{ι_n}), then the cardinal number of the intersection of A with the set of those points $x = \{x_\iota\}$ for which $x_{\iota_n} \epsilon G_n$ (for each of the given subscripts ι_n) is equal to the cardinal number of A. We need only prove that the definition of the z_ι's by transfinite induction may be carried through. Hence suppose that, for a definite value $\lambda \epsilon I$, the points z_ι (with property π_ι) having already been constructed for $\iota < \lambda$, it is impossible to choose $z_\lambda \epsilon S_\lambda$ with property π_λ. Then, for every point $y_\lambda \epsilon S_\lambda$, there exist: a neighborhood $T(y_\lambda)$ of the point y_λ (in the space S_λ), a finite (perhaps vacuous) set $M(y_\lambda)$ of subscripts $\iota < \lambda$ and, for each $\iota \epsilon M(y_\lambda)$, a neighborhood $G(z_\iota, y_\lambda)$ of the point z_ι (in the space S_ι) such that the cardinal number of the set $A \cdot H(y_\lambda) \cdot K(y_\lambda)$ is less than the cardinal number of A, where $H(y_\lambda)$ is the set of all points $x = \{x_\iota\}$ for which $x_\lambda \epsilon T(y_\lambda)$ and $K(y_\lambda)$ is the set of all points $x = \{x_\iota\}$ for which $x_\iota \epsilon G(z_\iota, y_\lambda)$ for every $\iota \epsilon M(y_\lambda)$. Since the space S_λ is bicompact, there exists a finite set of points $y_\lambda^{(i)} \epsilon S_\lambda (1 \leqq i \leqq m < \infty)$ such that

(1) $$\sum_{i=1}^{m} T(y_\lambda^{(i)}) = S_\lambda.$$

The cardinal number of the set

(2) $$\sum_{i=1}^{m} A \cdot H(y_\lambda^{(i)}) \cdot K(y_\lambda^{(i)})$$

is less than the cardinal number of A. On the other hand, it follows from (1) that

$$\sum_{i=1}^{m} H(y_\lambda^{(i)}) = S$$

so that the set (2) contains the set

(3) $$A \cdot \prod_{i=1}^{m} K(y_\lambda^{(i)}).$$

It follows that the cardinal number of the set (3) is less than the cardinal number of A. But it is easy to see that this is in contradiction with property π_μ, choosing $\mu < \lambda$ and $\mu \geqq \iota$ for every $\iota \in \sum_i M(y_\lambda^{(i)})$.

II

Since a bicompact Hausdorff space is completely regular, every subset of a bicompact Hausdorff space is also completely regular. Following Tychonoff, we shall prove conversely that *every completely regular space is a subset of some bicompact Hausdorff space.*

Let S be given completely regular space. Let T denote the interval $0 \leqq t \leqq 1$. Let Φ denote the set of all continuous functions f in the domain S such that $f(S) \subset T$. Choose a set I having the same potency as the set Φ, so that there exists a one-to-one mapping of I into Φ; let f_ι be the function corresponding to $\iota \in I$. For $\iota \in I$, put $T_\iota = T$ and let R be the cartesian product $\mathfrak{P}_\iota T_\iota$. Since every T_ι is a bicompact Hausdorff space, R is also a bicompact Hausdorff space. For any $x \in S$, put $g(x) = \xi = \{\xi_\iota\} \in R$, where $\xi_\iota = f_\iota(x)$. Then g is a mapping of the space S into the space $S^* = g(S) \subset R$. It is easy to see that the mapping g is homeomorphic. For $\iota \in I$ and $\xi \in R$, put $\varphi_\iota(\xi) = \xi_\iota$. Then φ_ι is a continuous function in the domain R such that $\varphi_\iota(R) = T$. Moreover, we see that $\varphi_\iota[g(x)] = f_\iota(x)$ for $x \in S$.

If S is a completely regular space, let $\beta(S)$ designate any topological space having the following four properties: (1) $\beta(S)$ *is a bicompact Hausdorff space,* (2) $S \subset \beta(S)$, (3) S *is dense in* $\beta(S)$ (i.e. the closure of S in the space $\beta(S)$ is the whole space $\beta(S)$), (4) *every bounded continuous function f in the domain S may be extended*[12] *to the domain* $\beta(S)$ (i.e. there exists a continuous function φ in the domain $\beta(S)$ such that $\varphi(x) = f(x)$ for every $x \in S$).

The space $\beta(S)$ exists for every completely regular S. Using the above notation, we easily see that the closure of S^* in the space R has the properties (1)-(4) relatively to S^*, so that $\beta(S^*)$ exists. Since S and S^* are homeomorphic, $\beta(S)$ exists as well.

Given a completely regular space S, the space $\beta(S)$ is essentially unique. More precisely: If B_1 and B_2 both have properties (1)-(4) of $\beta(S)$, then there exists a homeomorphic mapping h of B_1 into B_2 such that $h(x) = x$ for each $x \in S$. This is but a particular case of the following theorem: *Let S be a completely regular space. Let B be a space having properties* (1)-(3) of $\beta(S)$ (but not necessarily property (4)). *Then there exists a continuous mapping h of $\beta(S)$ into B such that*: (i) $h(x) = x$ *for each* $x \in S$, (ii) $h[\beta(S) - S] = B - S$. *The mapping h is one-to-one (and consequently homeomorphic) if and only if B also possesses property* (4). Let I, T, R, g and S^* have the above meaning. Divide the set I into two disjoined subsets I_1 and I_2, putting $\iota \in I_1$ if and only if the continuous function f_ι may be extended to the domain B. Let R_1 denote the cartesian product $\mathfrak{P}_\iota T_\iota$ where ι runs over I_1 and $T_\iota = T$ for each ι. For any $x \in B$, put $g_1(x) = \xi = \{\xi_\iota\}_{\iota \in I_1} \in R_1$, where $\xi_\iota = \varphi_\iota(x)$, φ_ι being the extension of f_ι to the domain B.

[12] It follows easily from property (3) that the extended function is uniquely defined by f.

Then g_1 is a homeomorphic mapping of the space B into the space $B^* = g_1(B) \subset R_1$, just as g was a homeomorphic mapping of S into the space S^*. For any point $\xi = \{\xi_\iota\}_{\iota \epsilon I} \epsilon R$, put $k(\xi) = \{\xi_\iota\}_{\iota \epsilon I_1} \epsilon R_1$. Evidently k is a continuous mapping of R into R_1. For $x \epsilon S$, it is easy to see that $k[g(x)] = g_1(x)$ so that $k(S^*) \subset B^*$. Since k is continuous, it follows that $k(\overline{S^*}) \subset \overline{B^*}$, where $\overline{S^*}$ is the closure of S^* in the space R and $\overline{B^*}$ is the closure of B^* in the space R_1. Since B^* is a homeomorph of B, B^* is a bicompact Hausdorff space, whence $\overline{B^*} = B^*$. Therefore $k(\overline{S^*}) \subset B^*$, i.e. k defines a continuous mapping k_0 of $\overline{S^*}$ into a subset of B^*. Since $\overline{S^*}$ was homeomorphic with $\beta(S)$, and B^* was homeomorphic with B, k_0 defines a continuous mapping h of $\beta(S)$ into a subspace $h[\beta(S)]$ of B; evidently $h(x) = x$ for every $x \epsilon S$. The space $h[\beta(S)]$, as a continuous image of the bicompact space $\beta(S)$, must be bicompact. It follows that $h[\beta(S)]$ is closed in B. On the other hand, $h[\beta(S)] \supset S$ must be dense in B. Therefore, $h[\beta(S)] = B$, i.e., h is a continuous mapping of $\beta(S)$ into B. If B possesses property (4) of $\beta(S)$, we have $I_1 = I$, whence $R_1 = R$ and k is the identity. This readily implies that the mapping h is homeomorphic.

Returning to the general case, we still have to prove that $h[\beta(S) - S] = B - S$. Of course $h[\beta(S) - S] \supset B - S$. It remains to arrive at a contradiction in supposing the existence of a point $b \epsilon \beta(S) - S$ such that $a = h(b) \epsilon S$. Since $\beta(S)$ is a bicompact Hausdorff space, it is completely regular. Hence there exists a continuous function φ in the domain $\beta(S)$ such that $\varphi(a) = 0$, $\varphi(b) = 1$. Let Q be the set of all points $x \epsilon S$ such that $\varphi(x) \geqq \frac{1}{2}$. Then Q is a closed subset of S, so that there exists a closed subset P of the space B such that $Q = SP$. Since B is a bicompact Hausdorff space, it is completely regular. Hence there exists a continuous function ψ in the domain B such that $\psi(a) = 0$, $\psi(x) = 1$ for each $x \epsilon P$ and $0 \leqq \psi(x) \leqq 1$ for each $x \epsilon B$. From property (4) of $\beta(S)$ it follows that there exists a continuous function χ in the domain $\beta(S)$ such that $\chi(x) = \psi(x)$ for each $x \epsilon S$, whence $\chi(a) = 0$. Since h is a continuous mapping of $\beta(S)$ into B, $\psi[h(x)]$ is a continuous function in the domain $\beta(S)$. The set C of all points $x \epsilon \beta(S)$ such that $\psi[h(x)] = \chi(x)$, is closed in $\beta(S)$ and contains the set S which is dense in $\beta(S)$; therefore $C = \beta(S)$, whence $\chi(b) = \psi[h(b)] = \psi(a) = 0$. The set D of all points $x \epsilon \beta(S)$ such that both $\varphi(x) > \frac{1}{2}$ and $\chi(x) < \frac{1}{2}$ is open in $\beta(S)$ and is not vacuous, since $b \epsilon D$. Since S is dense in $\beta(S)$, there exists a point $c \epsilon S \cdot D$. Since $c \epsilon D$, we have $\chi(c) < \frac{1}{2}$; since $c \epsilon S$, we have $\chi(c) = \psi(c)$. Therefore $\psi(c) < \frac{1}{2}$ so that $c \epsilon S \cdot (B - P) = S - Q$. From the definition of Q it follows that $\varphi(c) < \frac{1}{2}$; since $c \epsilon D$, this is a contradiction.

Two subsets A_1 and A_2 of a topological space S will be called *completely separated* if there exists a continuous function f in the domain S such that $f(x) = 0$ for each $x \epsilon A_1$ and $f(x) = 1$ for each $x \epsilon A_2$.[5] It is easy to see that A_1 and A_2 are completely separated if and only if the closed sets $\bar{A}_1$ and $\bar{A}_2$ are completely separated. We know that S is completely regular if and only if any single point x and any closed set not containing x are always completely separated. We know that S is normal if and only if two closed sets without common points are always completely separated.

Let S be a completely regular space. We characterized the space $\beta(S)$ by the properties (1)-(4) given above. We will now show that $\beta(S)$ may be also characterized by the properties (1), (2), (3) and (4′), where (4′) means the following: *If A_1 and A_2 are two completely separated subsets of S, then the closures of A_1 and A_2 in the space $\beta(S)$ are disjoint.* Suppose first that A_1 and A_2 are two completely separated subsets of S. Then there exists a continuous function f in the domain S such that $f(x) = 0$ for each $x \in A_1$ and $f(x) = 1$ for each $x \in A_2$. We may suppose that $0 \leqq f(x) \leqq 1$ for each $x \in S$, so that there exists a continuous extension φ of f to the domain $\beta(S)$. Letting the bar denote closures in the space $\beta(S)$, we have $\varphi(x) = 0$ for each $x \in \bar{A}_1$ and $\varphi(x) = 1$ for each $x \in \bar{A}_2$, so that indeed $\bar{A}_1\bar{A}_2 = 0$. Conversely, let the space B have properties (1), (2), (3), (4′). There exists a continuous mapping h of the space $\beta(S)$ into the space B such that $h(x) = x$ for each $x \in S$. It is sufficient to prove that the mapping h is one-to-one. Suppose the contrary. Then there exist two points $a \in \beta(S)$, $b \in \beta(S)$ such that $a \neq b$, $h(a) = h(b)$. There exists a continuous function f in the domain $\beta(S)$ such that $f(a) = 0$, $f(b) = 1$. Let A_1 denote the set of all points $x \in S$ such that $f(x) \leqq \frac{1}{3}$; let A_2 denote the set of all points $x \in S$ such that $f(x) \geqq \frac{2}{3}$. It is easy to see that A_1 and A_2 are two completely separated subsets of S so that $\bar{A}_1\bar{A}_2 = 0$ where the bar designates closures in the space B. Since $h(a) = h(b)$, we shall have a contradiction if we shall prove that $h(a) \in \bar{A}_1$, $h(b) \in \bar{A}_2$. Let U be any neighborhood of $h(a)$ in the space B. Then $h^{-1}(U)$ is a neighborhood of a in the space $\beta(S)$. Since $f(a) = 0$ and since S is dense in $\beta(S)$, it is easy to see that $h^{-1}(U) \cdot A_1 \neq 0$, whence $U \cdot A_1 \neq 0$. Since U was an arbitrary neighborhood of $h(a)$ in the space B, we have indeed $h(a) \in \bar{A}_1$ and similarly we prove that $h(b) \in \bar{A}_2$.

In the particular case when *S is a normal Riesz space*, it follows from the result just proved that $\beta(S)$ may characterized by the properties (1), (2), (3) and (5) where (5) means the following: *If F_1 and F_2 are two closed subsets of S without common points, then the closures of F_1 and F_2 in the space $\beta(S)$ have no common points.* Conversely, *if there exists a space B having properties* (1), (2), (3) *and* (5), *then S is normal and $B = \beta(S)$.* Indeed, it is easy to see that property (5) is stronger than property (4′) so that $B = \beta(S)$. If F_1 and F_2 are two closed subsets of S and $F_1F_2 = 0$, then $\bar{F}_1\bar{F}_2 = 0$, the bar indicating closures in B. Since B is a bicompact Hausdorff space, it is normal, so that there exists a continuous function φ in the domain $\beta(S)$ such that $\varphi(x) = 0$ for each $x \in \bar{F}_1$ and $\varphi(x) = 1$ for each $x \in \bar{F}_2$. Hence it follows that S is normal.

Let S be a completely regular space. Let T be a closed subset of S; let $\bar{T}$ denote the closure of T in the space $\beta(S)$. Then *we have* $\bar{T} = \beta(T)$ (i.e. $\bar{T}$ possesses the properties (1)-(4) of $\beta(T)$) *if and only if every bounded*[7] *continuous function in the domain T admits of a continuous extension to the domain S.* Suppose first that $\bar{T} = \beta(T)$ and let f be a continuous function in the domain T such that e.g. $0 \leqq f(x) \leqq 1$ for each $x \in T$. Since $\bar{T} = \beta(T)$, there exists a continuous extension g of f to the domain $\bar{T}$; of course $0 \leqq g(x) \leqq 1$ for each $x \in \bar{T}$. Since $\beta(S)$ is a bicompact Hausdorff space, it is normal; since $\bar{T}$ is closed in

$\beta(S)$, there exists a continuous extension φ of g to the domain $\beta(S)$. Hence f may be continuously extended to the domain $\beta(S)$ and therefore also to the domain $S \subset \beta(S)$. Conversely suppose that every bounded continuous function in the domain T may be continuously extended to the domain S. Of course $\bar{T}$ has always properties (1)-(3) (relatively to T); therefore to prove that $\bar{T} = \beta(T)$ it is sufficient to prove that $\bar{T}$ has property (4′) (again relatively to T). Hence suppose that $A_1 \subset T$ and $A_2 \subset T$ are completely separated in the space T. Then there exists a continuous function f in the domain T such that $f(x) = 0$ for each $x \in A_1$, $f(x) = 1$ for each $x \in A_2$ and $0 \leqq f(x) \leqq 1$ for each $x \in T$. There exists a continuous extension φ of f to the domain S, whence it readily follows that A_1 and A_2 are completely separated in the space S. Since $\beta(S)$ has property (4′) (relatively to S), we have $\bar{A}_1\bar{A}_2 = 0$, the bar indicating closures in the space $\beta(S)$. But of course $\bar{A}_1$ and $\bar{A}_2$ are closures of A_1 and A_2 in the space $\bar{T}$, so that $\bar{T}$ has indeed property (4′) relatively to T.

The theorem just proved has the following consequence: *If S is a normal Riesz space, then $\bar{T} = \beta(T)$* (the bar indicating closure in $\beta(S)$) *for every closed subset T of S. If the completely regular space S is not normal, then there exists a closed subset T of S such that $\bar{T} \neq \beta(T)$.*

If Φ is a family of neighborhoods of a point x of a topological space S, then we say that Φ is *complete* if, given an arbitrary neighborhood G of x, there exists a neighborhood U of x such that both $U \in \Phi$ and $U \subset G$. The least cardinal number of a complete family of neighborhoods of x is called the *character*[13] of x (in the space S) and is denoted by $\chi(x) = \chi_S(x)$. If $T \subset S$ and $x \in T$, it is easy to see that

$$\chi_T(x) \leqq \chi_S(x).$$

Let S be a completely regular space. Then for every point $a \in S$ we have

$$\chi_S(a) = \chi_{\beta(S)}(a).$$

Let Φ be a complete family of neighborhoods of a in the space S whose cardinal number is equal to $\chi_S(a)$. It is sufficient to construct a complete family Ψ of neighborhoods of a in the space $\beta(S)$ such that the cardinal number of Ψ does not exceed $\chi_S(a)$. The family Ψ will be constructed as a transform of the family Φ, each $U \in \Phi$ determining a $\tau(U) \in \Psi$, in the following way,

$$\tau(U) = \beta(S) - \overline{S - U}$$

(the bar indicating closures in the space $\beta(S)$). Of course Ψ is a family of neighborhoods of a in the space $\beta(S)$ and the cardinal number of Ψ does not exceed $\chi_S(a)$. Hence we have only to prove that, given a neighborhood G of a in the space $\beta(S)$, there exists a $U \in \Phi$ such that $\tau(U) \subset G$. There exists a continuous function f in the domain $\beta(S)$ such that $f(a) = 0$ and $f(x) = 1$ for each $x \in \beta(S) - G$. Let H denote the set of all points $x \in S$ such that $f(x) < \frac{1}{2}$. Then H is a neighborhood of a in the space S, so that there exists a $U \in \Phi$ such that $U \subset H$.

[13] AU, p. 2.

It remains to prove that $\tau(U) \subset G$. Supposing the contrary, there exists a point $b \in \tau(U) - G$. Since $b \in \beta(S) - G$, we have $f(b) = 1$. Let V be an arbitrary neighborhood of b in the space $\beta(S)$. Since $f(b) = 1$ and since S is dense in $\beta(S)$, there exists a point $c \in SV$ such that $f(c) > \frac{1}{2}$. Since $U \subset H$, we cannot have $c \in U$. Therefore $c \in S - U$ so that $(S - U)\,V \neq 0$. Since V was an arbitrary neighborhood of b in the space $\beta(S)$, we have $b \in \overline{S - U} = \beta(S) - \tau(U)$, which is a contradiction.

Let S be a completely regular space. Let $A \subset \beta(S) - S (A \neq 0)$ be both closed and a G_δ in $\beta(S)$. Then the cardinal number of A is $\geqq 2^{\aleph_0}$. Since A is both closed and a G_δ in the normal space $\beta(S)$, there exists a continuous function f in the domain $\beta(S)$ such that $f(x) = 0$ for each $x \in A$ and $f(x) > 0$ for each $x \in \beta(S) - A$. The set of all points $x \in \beta(S)$ such that $f(x) < n^{-1} (n = 1, 2, 3, \cdots)$ is open and not vacuous. Since S is dense in $\beta(S)$, there exists a point $a_n \in S$ such that $f(a_n) < n^{-1}$. Since $AS = 0$, we have $f(a_n) > 0$. It is evident that the points a_n may be chosen is such a manner that $f(a_{n+1}) < f(a_n)$. Let us arrange the rational numbers of the interval $0 < t < 1$ in a simple sequence $\{r_n\}$. There exists a continuous function φ in the domain $0 < t < \infty$ such that $0 < \varphi(t) < 1$ and $\varphi[f(a_n)] = r_n (n = 1, 2, 3, \cdots)$. Since $f(x) > 0$ for each $x \in S$, we obtain a bounded continuous function g in the domain S such that $g(x) = \varphi[f(x)]$ for each $x \in S$. There exists a continuous extension h of g to the domain $\beta(S)$. Choose a real number $\alpha, 0 \leqq \alpha \leqq 1$. There exists a sequence $i_1 < i_2 < i_3 < \cdots$ such that $r_{i_n} \to \alpha$ for $n \to \infty$. Let M_n designate the set of points $a_{i_n}, a_{i_{n+1}}, a_{i_{n+2}}, \cdots$ so that $M_n \subset S$, $M_n \supset M_{n+1}$, $M_n \neq 0$. Since the space $\beta(S)$ is bicompact, there exists a point $b \in \prod \bar{M}_n$. Since the functions f and h are continuous, we have $f(\bar{M}_n) \subset \overline{f(M_n)}$, $h(\bar{M}_n) \subset \overline{h(M_n)} = \overline{g(M_n)}$, whence $f(b) \in \prod \overline{f(M_n)}$, $h(b) \in \prod \overline{g(M_n)}$. Since $f(a_{i_n}) \to 0$, $g(a_{i_n}) \to \alpha$ for $n \to \infty$, we easily see that $f(b) = 0$, $h(b) = \alpha$. Since $f(b) = 0$, we have $b \in A$. Therefore, for each α such that $0 \leqq \alpha \leqq 1$, the set A contains a point b such that $h(b) = \alpha$. Hence the cardinal number of A is at least $2^{\aleph_0}$.

Let S_1 and S_2 be two completely regular spaces satisfying the first countability axiom. Let the spaces $\beta(S_1)$ and $\beta(S_2)$ be homeomorphic. Then the spaces S_1 and S_2 are homeomorphic. We may assume that $\beta(S_1) = \beta(S_2)$. According to the preceding theorem no point $x \in \beta(S_1) - S_1$ is a G_δ in $\beta(S_1)$. But every point $x \in S_2$ satisfies the first countability axiom relatively to S_2 and, therefore, after the theorem last but one, relatively to $\beta(S_2)$ as well and hence x is a G_δ in $\beta(S_2) = \beta(S_1)$. Therefore $S_2 \subset S_1$ and similarly $S_1 \subset S_2$, so that $S_1 = S_2$.

Let I denote an infinite countable isolated space (e.g. the space of all natural numbers). It is an important problem to determine the cardinal number $\mathfrak{m}$ of $\beta(I)$. All I know about it is that

$$2^{\aleph_0} \leqq \mathfrak{m} \leqq 2^{2^{\aleph_0}}.$$

It is easily seen that each point of I is an isolated point of $\beta(I)$ so that the set I is open in $\beta(I)$. Since I is countable, it is an F_σ in $\beta(I)$. Hence $\beta(I) - I$ is both closed and a G_δ in $\beta(I)$ so that the cardinal number of $\beta(I) - I$ is $\geqq 2^{\aleph_0}$.

On the other hand, since the set I is dense in the Hausdorff space $\beta(I)$, it is easy to see that a point $x \in \beta(I)$ is uniquely determined knowing the family of all sets $A \subset I$ such that $x \in \bar{A}$, so that the cardinal number of $\beta(I)$ is at most equal to the cardinal number $2^{2^{\aleph_0}}$ of all families of subsets of I.

A topological space S is called *compact* if, given any infinite subset A of S, there exists a point $x \in S$ such that $x \in \overline{A - x}$.

Let the normal Riesz space S be not compact. Then the cardinal number of $\beta(S) - S$ is at least equal to the cardinal number of $\beta(I)$ (hence at least equal to $2^{\aleph_0}$). Since S is not compact, it is well known that S contains a closed subset F homeomorphic with I. Since S is normal, we have $\beta(I) = \bar{I} \subset \beta(S)$, so that $\beta(I) - I \subset \beta(S) - S$. But the sets $\beta(I) - I$ and $\beta(I)$ have the same cardinal number.

I do not know whether this theorem remains true if we replace normality by complete regularity. It may be shown that the assumption of normality may be replaced by the following weaker assumption[14]: If F_1 and F_2 are two closed subsets of S such that F_1 *is countable* and $F_1F_2 = 0$, there exist two open sets G_1 and G_2 such that $G_1 \supset F_1$, $G_2 \supset F_2$, $G_1G_2 = 0$.

If the space S is compact, then the set $\beta(S) - S$ may consist of a single point. Let S be the set of all ordinal numbers $<\omega_1$, ω_1 being the first uncountable ordinal number. Let S_0 be the set of all ordinal numbers $\leqq \omega_1$. The topology of S and S_0 is the usual topology of an ordered set, an open base being given by the family of all open intervals. It is well known that S is a compact normal Riesz space and that S_0 is a bicompact Hausdorff space. We shall prove that $S_0 = \beta(S)$. Since it is evident that S_0 possesses properties (1)–(3) of $\beta(S)$, it is sufficient to prove that a continuous function f in the domain S admits of a continuous extension to the domain S_0. This is an easy consequence of the following theorem. *If f is a continuous function in the domain S, then there exists a point $\xi \in S$ such that f is constant for $x \geqq \xi$.* It is sufficient to prove that, given a number $\varepsilon > 0$, there exists a point $\xi(\varepsilon) \in S$ such that $| f(x) - f(y) | < \varepsilon$ for $x \in S, y \in S, x > \xi(\varepsilon), y > \xi(\varepsilon)$. Supposing the contrary, there would exist in S two sequences $\{a_n\}$ and $\{b_n\}$ such that $a_n < b_n < a_{n+1}$ and $| f(a_n) - f(b_n) | \geqq \varepsilon$. But this is impossible, because f would then be discontinuous at α, α being the first ordinal number greater than each a_n.

We say that $x \in S$ is a *κ-point*[15], if there exists a sequence $\{x_n\} \subset S - (x)$ such that $\lim x_n = x$, i.e. that, given any neighborhood U of x, we have $x_n \in U$ except for a finite number of subscripts n. Alexandroff and Urysohn raised the question[16] whether there exists a bicompact Hausdorff space which is dense in itself and which contains no κ-point. We shall prove that the space $\beta(I) - I$ has this property. Supposing the contrary, there exists a point $c \in \beta(I) - I$ and a sequence $\{a_n\} \subset \beta(I) - I - (c)$ such that $\lim a_n = c$. We may suppose that the points a_n are all distinct from one another. Let A_n be the set of the points

[14] AU, p. 58.
[15] AU, p. 53.
[16] AU, p. 54.

$a_n, a_{n+1}, a_{n+2} \cdots$ together with the point c. It is easy to see that A_n is a closed subset of $\beta(I)$. We shall construct successively open subsets U_n of the space $\beta(I)$ as follows. U_1 contains the point a_1, but $\overline{U}_1 A_2 = 0$. If, for a certain value of n, we have already constructed the set U_n so that $\overline{U}_n \cdot A_{n+1} = 0$, let U_{n+1} be an open subset containing a_{n+1}, but such that $\overline{U}_{n+1} \cdot \overline{U}_i = 0$ for $1 \leqq i \leqq n$ and $\overline{U}_{n+1} \cdot A_{n+2} = 0$. It is easy to see that the successive construction of the sequence $\{U_n\}$ may be carried through. Now put $\Phi = I \cdot \sum U_{2n-1}$, $\Psi = I \cdot \sum U_{2n}$. Then $\Phi\Psi = 0$ and the sets Φ and Ψ are of course closed in I, since I is an isolated space. Since I is normal, we must have $\bar{\Phi}\bar{\Psi} = 0$, the bars indicating closures in $\beta(I)$. On the other hand, since I is dense in $\beta(I)$ and U_n is open in $\beta(I)$, it is easy to see that $\overline{IU_n} = \overline{U}_n$, so that $a_n \in \overline{IU_n}$, whence we easily get the contradiction $c \in \bar{\Phi}\bar{\Psi}$.

III

We shall say that the space S is *topologically complete* if there exists a bicompact Hausdorff space $B \supset S$ such that S is a G_δ in B. Of course S is then completely regular. *A G_δ in a topologically complete space is a topologically complete space. A closed subset of a topolologically complete space is a topologically complete space.*

A topological space S is topologically complete if and only if it is completely regular and a G_δ in $\beta(S)$. If S is a G_δ in $\beta(S)$, then it is topologically complete, since $\beta(S)$ is a bicompact Hausdorff space. Conversely suppose that S is topologically complete. Then there exists a bicompact Hausdorff space $B \supset S$ such that S is a G_δ in B. Let B_0 be the closure of S in the space B. Then B_0 is a bicompact Hausdorff space and S is dense in B_0 and a G_δ in B_0. We know that there exists a continuous mapping h of $\beta(S)$ into B_0 such that $h^{-1}(S) = S$. Since S is a G_δ in B_0, it is easy to see that $h^{-1}(S) = S$ is a G_δ in $\beta(S)$.

Let T be a completely regular[17] *space. Let $S \subset T$ be a topologically complete space. Then S is a G_δ in the closure of S in the space T.* Let S_0 be the closure of S in the space $\beta(T)$. It is sufficient to prove that S is a G_δ in S_0. Since S_0 is a bicompact Hausdorff space and since S is dense in S_0, there exists a continuous mapping h of $\beta(S)$ into S_0 such that $h[\beta(S) - S] = S_0 - S$. Since S is topologically complete, it is a G_δ in $\beta(S)$, so that $\beta(S) - S$ is an F_σ in $\beta(S)$. Hence there exist closed subsets F_n of $\beta(S)$ such that $\sum F_n = \beta(S) - S$, whence $S_0 - S = \sum h(F_n)$. Every F_n is a bicompact space, so that every $h(F_n)$ is a bicompact space. Since $h(F_n)$ is a bicompact subset of the Hausdorff space S_0, it is closed in S_0, so that $S_0 - S$ is an F_σ in S_0 and finally S is a G_δ in S_0.

Let T be a topologically complete space. Let $S \subset T$. Then S is a topologically complete space if and only if it is the intersection of a closed subset of T and a G_δ in T. If $S = FH$, where F is closed in T and H is a G_δ in T, then F is a topologically complete space and S is a G_δ in F, so that S is a topologically complete space. Conversely let S be topologically complete. Then S is a G_δ in the closure $\bar{S}$ of S in T, so that $S = \bar{S}H$, H being a G_δ in T.

---

[17] I do not know whether this assumption is necessary.

Let $S \neq 0$ be a topologically complete space[18]. *Let $\{G_n\}$ be a sequence of open and dense subsets of S. Let $H = \prod G_n$. Then $H \neq 0$ and, moreover, H is dense in S.* There exists a regular compact (as a matter of fact, bicompact) space $K \supset S$ such that S is a G_δ in K. We may suppose that $\bar{S} = K$, the bar denoting closure in K. The sets G_n being open in S, there exist sets Γ_n open in K and such that $G_n = S \cdot \Gamma_n$. Since S is a G_δ in K, there exist sets Δ_n open in K and such that $S = \prod \Delta_n$. Since S is dense in K and G_n are dense in S, the sets G_n are dense in K. Choose an arbitrary point $a_0 \epsilon S$ and an arbitrary neighborhood V of a_0 in the space S. All we have to prove is that $HV \neq 0$. There exists a neighborhood U_0 of a_0 in the space K such that $V = SU_0$. Since the set G_1 is dense in K, there exists a point $a_1 \epsilon G_1 U_0 = S \cdot \Gamma_1 U_0 \subset \Delta_1 \Gamma_1 U_0$. Hence $\Delta_1 \Gamma_1 U_0$ is a neighborhood of a_1 in the space K. Since K is regular, there exists a neighborhood U_1 of a_1 (in the space K) such that $\bar{U}_1 \subset \Delta_1 \Gamma_1 U_0$. Generally, let there be given for a certain value of n a point $a_n \epsilon G_n$ and its neighborhood U_n (in the space K) such that $\bar{U}_n \subset \Delta_n \Gamma_n U_{n-1}$. Then $a_n \epsilon G_n \subset S$ and SU_n is a neighborhood of a_n in the space S; since G_{n+1} is dense in S, there exists a point $a_{n+1} \epsilon G_{n+1} U_n = S \cdot \Gamma_{n+1} U_n \subset \Delta_{n+1} \Gamma_{n+1} U_n$. Hence $\Delta_{n+1} \Gamma_{n+1} U_n$ is a neighborhood of a_{n+1} in the regular space K, so that there exists a neighborhood U_{n+1} of a_{n+1} (in the space K) such that $\bar{U}_{n+1} \subset \Delta_{n+1} \Gamma_{n+1} U_n$. Thus we construct a sequence $\{a_n\}$ of points and a sequence $\{U_n\}$ of open sets so that $a_n \epsilon G_n U_n$, $\bar{U}_{n+1} \subset \Delta_{n+1} \Gamma_{n+1} U_n$. Since $a_n \epsilon U_n$, we have $U_n \neq 0$. Since K is compact and $\bar{U}_{n+1} \subset U_n$, there exists a point $b \epsilon \prod U_n = \prod \bar{U}_n$. Since $\bar{U}_{n+1} \subset \Delta_{n+1} \Gamma_{n+1} U_n$, we have $b \epsilon \prod \Delta_n$. $\prod \Gamma_n = S \cdot \prod \Gamma_n = \prod G_n = H$. Moreover $b \epsilon U_0$, so that $b \epsilon HU_0 = HV$.

Let S be a metric space. A *Cauchy sequence* in S is a sequence $\{x_n\} \subset S$ such that, given a number $\varepsilon > 0$, there exists a number p such that the distance of x_m and x_n is less than ε, whenever both m and n are greater than p. A metric space S is called *metrically complete* if, given any Cauchy sequence $\{x_n\}$ in S, there exists a point $x \epsilon S$ such that $\lim x_n = x$. A topological space is called *completely metrizable*, if it is homeomorphic with a metrically complete space.

We next prove our principal theorem: *A metrizable space S is topologically complete if and only if it is completely metrizable.*

Let S be a metrically complete space and let ρ be its distance function. We may suppose that $\rho(x, y) \leqq 1$ for every pair of points, since otherwise we may replace ρ by ρ_1, putting $\rho_1(x, y) = \rho(x, y)$ if $\rho(x, y) \leqq 1$, $\rho_1(x, y) = 1$ if $\rho(x, y) > 1$ Since S is metric, it is completely regular, so that $\beta(S)$ exists. For any given $a \epsilon S$, $\rho(a, x)$ is a bounded continuous function in the domain S so that there exists a continuous function $\varphi_a(x)$ in the domain $\beta(S)$ such that $\varphi_a(x) = \rho(a, x)$ for each $x \epsilon S$. If $a \epsilon S$, $b \epsilon S$, then the set $T(a, b)$ of all points $x \epsilon \beta(S)$ such that $\varphi_a(x) + \varphi_b(x) \geqq \rho(a, b)$ is closed in $\beta(S)$ and contains S. Since S is dense in $\beta(S)$, we must have $T(a, b) = \beta(S)$, i.e. $\varphi_a(x) + \varphi_b(x) \geqq \rho(a, b)$ for each $x \epsilon \beta(S)$.

---

[18] It is evident from the proof that it is possible to replace this by the weaker assumption that S is a G_δ in some regular compact space.

For $a \in S$ and $n = 1, 2, 3, \cdots$ let $\Gamma(a, n)$ be the set of all points $x \in \beta(S)$ such that $\varphi_a(x) < n^{-1}$. Since the function $\varphi_a(x)$ is continuous, $\Gamma(a, n)$ is an open subset of $\beta(S)$. Therefore

$$G_n = \sum_{a \in S} \Gamma(a, n)$$

is an open set. We shall prove that $S = \prod G_n$, so that the set S is a G_δ in $\beta(S)$ and thus topologically complete. Evidently $\prod G_n \supset S$. Conversely let $b \in \prod G_n$. We have to prove that $b \in S$. According to the definition of G_n, there exist points $a_n \in S$ such that $\varphi_{a_n}(b) < n^{-1}$. Therefore

$$\rho(a_n, a_m) \leqq \varphi_{a_n}(b) + \varphi_{a_m}(b) < \frac{1}{n} + \frac{1}{m},$$

so that $\{a_n\}$ is a Cauchy sequence in S. Since S is metrically complete, there exists a point $a \in S$ such that $a = \lim a_n$. It is sufficient to prove that $a = b$. Suppose that $a \neq b$. Since $\beta(S)$ is a Hausdorff space, there exist two open subsets U and V of $\beta(S)$ such that $a \in U$, $b \in V$, $UV = 0$. Since US is a neighborhood of a in the metric space S, there exists an integer $n > 0$ such that U contains every point $x \in S$ such that $\rho(a, x) < 2 \cdot n^{-1}$. This can be written in the form $SW \subset U$, W being the set of all points $x \in \beta(S)$ such that $\varphi_a(x) < 2 \cdot n^{-1}$. Since φ_a is continuous, W is an open subset of $\beta(S)$. Since S is dense in $\beta(S)$ and U, V and W are open in $\beta(S)$, we have $W \subset \overline{W} = \overline{SW} \subset \overline{U} \subset \beta(S) - V$, or $WV = 0$. Hence for each $x \in V$ we have $\varphi_a(x) \geqq 2 \cdot n^{-1}$; in particular $\varphi_a(b) \geqq 2 \cdot n^{-1}$. Since $\rho(a_n, a_m) < n^{-1} + m^{-1}$ and $\lim a_n = a$, we have $\rho(a, a_n) \leqq n^{-1}$. Hence for each $x \in S$ we have $\rho(a, x) \leqq \rho(a, a_n) + \rho(a_n, x) \leqq n^{-1} + \rho(a_n, x)$, whence it easily follows that for each $x \in \beta(S)$ we have $\varphi_a(x) \leqq \varphi_{a_n}(x) + n^{-1}$, in particular $\varphi_a(b) \leqq \varphi_{a_n}(b) + n^{-1} < n^{-1} + n^{-1} = 2 \cdot n^{-1}$, which is a contradiction.

Now suppose that the metric space S is topologically complete. Let ρ denote the distance function of S; again, we shall suppose that $\rho(x, y) \leqq 1$ for every couple of points. Since S is topologically complete, there exists a sequence $\{F_n\}$ of closed subsets of $\beta(S)$ such that $\beta(S) - S = \sum F_n$. If $S = \beta(S)$, then S is a bicompact metric space, and then it is well known that S is metrically complete. Hence let us suppose that $S \neq \beta(S)$; we may then assume that $F_n \neq 0$ for every n. Given any point $a \in S$, $\rho(a, x)$ is a bounded continuous function in the domain S, which admits of a continuous extension φ_a to the domain $\beta(S)$. If the point $b \in \beta(S)$ is different from a, then there exist open subsets U and V of $\beta(S)$ such that $a \in U$, $b \in V$, $UV = 0$. Since SU is a neighborhood of a in the metric space S, there exists a number $\varepsilon > 0$ such that U contains every point $x \in S$ such that $\rho(a, x) < \varepsilon$. Since S is dense in $\beta(S)$, it easily follows that $\overline{U}$ contains every point $x \in \beta(S)$ such that $\varphi_a(x) < \varepsilon$. Since $U \subset \beta(S) - V = \overline{\beta(S) - V}$, we have $\overline{U} \subset \beta(S) - V$ so that $b \in \beta(S) - \overline{U}$, whence $\varphi_a(b) \geqq \varepsilon$. Thus we proved that $\varphi_a(b) > 0$ for every $b \in \beta(S)$ except for $b = a$. Since the set $F_n \neq 0$ is closed in the bicompact space $\beta(S)$, it is easy to see that the function $\varphi_a(x)$, x running over F_n, admits of a minimum value $\sigma(a, F_n)$. Since $a \in S$, $F_n S = 0$, we have $\sigma(a, F_n) > 0$.

If $a \in S$, $b \in S$, then we have $\rho(a, x) \leqq \rho(a, b) + \rho(b, x)$ for every $x \in S$, whence $\varphi_a(x) \leqq \rho(a, b) + \varphi_a(x)$ for every $x \in \beta(S)$. Therefore $\sigma(a, F_n) \leqq \rho(a, b) + \sigma(b, F_n)$, and similarly $\sigma(b, F_n) \leqq \rho(a, b) + \sigma(a, F_n)$. Hence

$$| \sigma(a, F_n) - \sigma(b, F_n) | \leqq \rho(a, b).$$

Now let us put for $x \in S$, $y \in S$

$$f_n(x, y) = \rho(x, y) + \sigma(x, F_n) + \sigma(y, F_n),$$

$$g_n(x, y) = \frac{\rho(x, y)}{f_n(x. y)},$$

$$\rho_0(x, y) = \rho(x, y) + \sum_1^\infty 2^{-n} \cdot g_n(x, y).$$

Since $\rho(x, y) \geqq 0$, $\sigma(x, F_n) > 0$, $\sigma(y, F_n) > 0$, we have $f_n(x, y) > 0$. Hence $g_n(x, y)$ exists and $0 \leqq g_n(x, y) \leqq 1$, so that the series $\sum 2^{-n} \cdot g_n(x, y)$ is convergent. It is evident that $\rho_0(x, y) = \rho_0(y, x)$ and that $\rho_0(x, x) = 0$, whereas $\rho_0(x, y) > 0$ if $x \neq y$. Next we shall prove that $\rho_0(x, z) \leqq \rho_0(x, y) + \rho_0(y, z)$ for $x \in S$, $y \in S$, $z \in S$. Since

$$\frac{t_1}{c + t_1} \leqq \frac{t_2}{c + t_2} \quad \text{for } c > 0,\ 0 \leqq t_1 \leqq t_2$$

and since $0 \leqq \rho(x, z) \leqq \rho(x, y) + \rho(y, z)$, we have

$$g_n(x, z) \leqq \frac{\rho(x, y) + \rho(y, z)}{\rho(x, y) + \rho(y, z) + \sigma(x, F_n) + \sigma(z, F_n)}.$$

Since

$$\sigma(y, F_n) \leqq \rho(x, y) + \sigma(x, F_n),$$

$$\sigma(y, F_n) \geqq \rho(y, z) + \sigma(z, F_n),$$

we have

$$\rho(x, y) + \rho(y, z) + \sigma(x, F_n) + \sigma(z, F_n) \geqq \begin{cases} \rho(x, y) + \sigma(x, F_n) + \sigma(y, F_n), \\ \rho(y, z) + \sigma(y, F_n) + \sigma(z, F_n), \end{cases}$$

whence

$$g_n(x, z) \leqq g_n(x, y) + g_n(y, z),$$

so that indeed

$$\rho_0(x, z) \leqq \rho_0(x, y) + \rho_0(y, z).$$

Hence ρ_0 has all the properties of a distance function. Next we prove that ρ and ρ_0 are equivalent metrics in S, i.e. that for $x \in S$ and $\{x_n\} \subset S$ we have

$$\lim \rho(x_n, x) = 0 \text{ if and only if } \lim \rho_0(x_n, x) = 0.$$

If $\lim \rho_0(x_n, x) = 0$, then $\lim \rho(x_n, x) = 0$, since $0 \leqq \rho(x_n, x) \leqq \rho_0(x_n, x)$. Conversely suppose that $\lim \rho(x_n, x) = 0$. Choose a number $\varepsilon > 0$ and an integer $k > 0$ such that $2^{-k+1} < \varepsilon$. Then we have for all values of n

$$\sum_{i=k+1}^{\infty} 2^{-i} g_i(x_n, x) \leqq \sum_{i=k+1}^{\infty} 2^{-i} = 2^{-k} < \tfrac{1}{2}\varepsilon,$$

whence

$$\rho_0(x_n, x) < \rho(x_n, x) + \sum_{i=1}^{k} 2^{-i} g_i(x_n, x) + \tfrac{1}{2}\varepsilon$$

$$\leqq \rho(x_n, x) + \sum_{i=1}^{k} 2^{-i} \frac{\rho(x_n, x)}{\rho(x_n, x) + \sigma(x, F_i)} + \tfrac{1}{2}\varepsilon.$$

Since $\lim \rho(x_n, x) = 0$, we must have

$$\lim_{n \to \infty} \sum_{i=1}^{k} 2^{-i} \frac{\rho(x_n, x)}{\rho(x_n, x) + \sigma(x, F_i)} = 0,$$

so that there exists an integer p such that for $n > p$ we have

$$0 \leqq \sum_{i=1}^{k} 2^{-i} \frac{\rho(x_n, x)}{\rho(x_n, x) + \sigma(x, F_i)} < \tfrac{1}{2}\varepsilon.$$

Therefore

$$\rho_0(x_n, x) < \rho(x_n, x) + \varepsilon$$

for every $n > p$. Since $\lim \rho(x_n, x) = 0$ and the number $\varepsilon > 0$ was arbitrary, we have indeed $\lim \rho_0(x_n, x) = 0$. Thus we proved that ρ and ρ_0 are equivalent metrics in S, i.e. that the metric spaces $S = (S, \rho)$ and (S, ρ_0) are homeomorphic.

It remains to be shown that the metric space (S, ρ_0) is metrically complete. Hence suppose that $\{x_n\}$ is a Cauchy sequence in (S, ρ_0). We have to prove that there exists a point $x \in S$ such that $\lim \rho_0(x_n, x) = 0$, or, what we already know to be equivalent, that $\lim \rho(x_n, x) = 0$. Since the space $\beta(S)$ is bicompact, it is easy to see that there exists a point $x \in \beta(S)$ such that, given any neighborhood U of x (in the space $\beta(S)$), we have $x_n \in U$ for an infinite number of values of n. It is sufficient to prove that $x \in S$, for then, since $\{x_n\}$ is a Cauchy sequence, it is easy to show that $\lim \rho(x_n, x) = 0$. Suppose, on the contrary, that the point x belongs to the set $\beta(S) - S = \sum F_n$. Hence there exists an integer $k > 0$ such that $x \in F_k$.

We shall prove that $\sigma(x_n, F_k) \to 0$ for $n \to \infty$. Choose a number $\varepsilon > 0$. There exists an integer $p > 0$ such that for $n > p$, $m > p$ we have $\rho(x_n, x_m) \leqq \rho_0(x_n, x_m) < \varepsilon$. Let n be greater than p. The number $\sigma(x_n, F_k)$ is the minimum value of $\varphi_{x_n}(y)$ for $y \in F_k$. Since $x \in F_k$, we must have $0 < \sigma(x_n, F_k) \leqq \varphi_{x_n}(x)$. There exists a neighborhood Ω_n of x in $\beta(S)$ such that $|\varphi_{x_n}(z) - \varphi_{x_n}(x)| < \varepsilon$ for every $z \in \Omega_n$. There exists an integer $m_n > p$ such that $x_{m_n} \in \Omega_n$, whence $|\varphi_{x_n}(x_{m_n}) - \varphi_{x_n}(x)| < \varepsilon$, i.e. $|\rho(x_n, x_{m_n}) - \varphi_{x_n}(x)| < \varepsilon$. Since $n > p$, $m_n > p$, we must have $\rho(x_n, x_{m_n}) < \varepsilon$, whence $\varphi_{x_n}(x) < 2\varepsilon$. Therefore $0 < \sigma(x_n, F_k) < 2\varepsilon$ for $n > p$, so that indeed $\sigma(x_n, F_k) \to 0$ for $n \to \infty$.

Since $\{x_n\}$ is a Cauchy sequence in (S, ρ_0), there exists an integer p such that $\rho_0(x_n, x_p) < 2^{-k-2}$ for each $n > p$. But

$$\rho_0(x_n, x_p) \geqq 2^{-k} g_k(x_n, x_p) = 2^{-k} \frac{\rho(x_n, x_p)}{\rho(x_n, x_p) + \sigma(x_n, F_k) + \sigma(x_p, F_k)}.$$

Since

$$\sigma(x_p, F_k) \leqq \rho(x_n, x_p) + \sigma(x_n, F_k),$$

it follows that

$$\rho_0(x_n, x_p) \geqq 2^{-k-1} \frac{\rho(x_n, x_p)}{\rho(x_n, x_p) + \sigma(x_n, F_k)} \geqq 0,$$

so that for every $n > p$ we have

$$0 \leqq \frac{\rho(x_n, x_p)}{\rho(x_n, x_p) + \sigma(x_n, F_k)} < \tfrac{1}{2},$$

whence $\rho(x_n, x_p) < \sigma(x_n, F_k)$. But $\sigma(x_n, F_k) \to 0$ f [illegible] ∞. Therefore $\rho(x_n, x_p) \to 0$ for $n \to \infty$. Hence there exists an integer $q >$ [illegible] h that for every $n > q$ we have $\rho(x_n, x_p) < \frac{1}{2} \varphi_{x_p}(x)$. [Since $x_p \in S$, $x \in \beta(S) - S$, we know that $\varphi_{x_p}(x) > 0$.] There exists a neighborhood U of x in the space $\beta(S)$ such that $\varphi_{x_p}(z) > \frac{1}{2}\varphi_{x_p}(x)$ for any $z \in U$. There exists an integer $n > q$ such that $x_n \in U$, whence $\rho(x_n, x_p) = \varphi_{x_p}(x_n) > \frac{1}{2}\varphi_{x_p}(x)$, which is a contradiction.

IV

Let S be a completely regular space. Let $\lambda(S)$ be the set of all points $x \in \beta(S)$ such that x possesses a neighborhood U(in the space $\beta(S)$) such that $S \cdot \bar{U}$ is a normal space. [$\bar{U}$ is the closure of U in $\beta(S)$]. It is easy to see that $\lambda(S)$ is an open subset of $\beta(S)$.

Let F_1 and F_2 be two closed subsets of a completely regular space S such that $F_1F_2 = 0$. Then

$$\bar{F}_1 \cdot \bar{F}_2 \cdot \lambda(S) = 0,$$

the bars indicating closures in $\beta(S)$. Supposing the contrary, there exists a point $a \in \bar{F}_1 \cdot \bar{F}_2 \cdot \lambda(S)$. Since $a \in \lambda(S)$, there exists a neighborhood U of a (in the space $\beta(S)$) such that $S \cdot \bar{U}$ is a normal space. There exists a neighborhood V of a such that $\bar{V} \subset U$. Put

$$\Phi_1 = \bar{V} \cdot F_1, \Phi_2 = \bar{U} \cdot F_2 + S(\bar{U} - U).$$

Then Φ_1 and Φ_2 are two closed subsets of $S\bar{U}$ such that $\Phi_1 \Phi_2 = 0$. Moreover, it is easy to see that $a \in \bar{\Phi}_1 \cdot \bar{\Phi}_2$. Since $S\bar{U}$ is a normal space, there exists a bounded continuous function f in the domain $S\bar{U}$ such that $f(x) = 0$ for each $x \in \Phi_1$ and $f(x) = 1$ for each $x \in \Phi_2$. For $x \in S$ put (i) $g(x) = f(x)$ if $x \in SU$, (ii) $g(x) = 1$ if $x \in S - U$. Then it is easy to see that g is a bounded continuous extension of f to the domain S. According to the definition of $\beta(S)$, there exists a continuous extension φ of g (hence of f) to the domain $\beta(S)$. We have

$\varphi(x) = f(x) = 0$ for each $x \in \Phi_1$ and $\varphi(x) = f(x) = 1$ for each $x \in \Phi_2$. Since φ is continuous, we must have $\varphi(x) = 0$ for each $x \in \bar{\Phi}_1$ and $\varphi(x) = 1$ for each $x \in \bar{\Phi}_2$, so that $\bar{\Phi}_1\bar{\Phi}_2 = 0$, which is a contradiction.

The topological space S will be called *locally normal* if each point $x \in S$ possesses a neighborhood U such that $\bar{U}$ is a normal space. Any normal space is locally normal; more generally, any open subset of a locally normal space is locally normal.

A locally normal Riesz space S is completely regular. Let a be a given point of a locally normal space S and let V be a given neighborhood of a. There exists a neighborhood U of a such that $\bar{U}$ is a normal space. Also $\overline{UV}$ is a normal space, since it is a closed subset of $\bar{U}$. Since (a) and $\overline{UV} - UV$ are two closed subsets of the normal space $\overline{UV}$ without a common point, there exists a continuous function f in the domain $\overline{UV}$ such that $f(a) = 0$ and $f(x) = 1$ for each $x \in \overline{UV} - UV$. For $x \in S$ put (i) $g(x) = f(x)$ if $x \in UV$, (ii) $g(x) = 1$ if $x \in S - UV$. Then it is easy to see that g is a continuous function in the domain S such that $g(a) = 0$ and $g(x) = 1$ for each $x \in S - V$. Therefore S is completely regular.

A completely regular space S need not be locally normal. Let ω be the first infinite ordinal number. Let ω_1 be the first uncountable ordinal number. Let S_1 be the space of all ordinal numbers $\leqq \omega$. Let S_2 be the space of all ordinal numbers $\leqq \omega_1 \cdot \omega$. The topology in S_1 and in S_2 is defined in the usual way by means of intervals. Let S_{12} be the cartesian product of the two spaces S_1 and S_2. Let T be the set of all points $(x, y) \in S_{12}$, for which $x = \omega$ and $y = \omega_1 \cdot n (n = 1, 2, 3, \cdots)$. Let $S = S_{12} - T$. Then S is a completely regular space, but it is not locally normal.

It is easy to see that a completely regular space S is locally normal if and only if $S \subset \lambda(S)$. I do not know whether there exists a completely regular space $S \neq 0$ such that $S \cdot \lambda(S) = 0$.

A Riesz space S is locally normal if and only if it is homeomorphic with an open subset of a normal Riesz space.[19] We know that an open subset of a normal Riesz space is a locally normal Riesz space. Conversely let S be a locally normal Riesz space. Let S_0 be a new space consisting of all points of S and of a single new point ω. The topology of S_0 is defined as follows. If $\omega \in A \subset S_0$, then A is closed in S_0 if and only if $A - (\omega)$ is closed in S. If $A \subset S_0 - (\omega) = S$, then A is closed in S_0 if and only if (i) A is closed in S, (ii) $\bar{A} \subset \lambda(S)$, the bar indicating closure in $\beta(S)$. It is easy to see that S_0 is a Riesz space and that S is an open subset of S_0. It remains to be shown that the space S_0 is normal. Let F_1 and F_2 be two closed subsets of S_0 such that $F_1F_2 = 0$. Since the point ω belongs at most to one of the two sets F_1 and F_2, we may suppose that $F_1 \subset S$. Since F_1 is closed in S_0, the closure $\bar{F}_1$ of F_1 in the space $\beta(S)$ is a subset of $\lambda(S)$. Put $F_3 = F_2 - (\omega)$. Then F_1 and F_3 are two closed subsets of S and $F_1F_3 = 0$. We know that $\bar{F}_1 \cdot \bar{F}_3 \cdot \lambda(S) = 0$ (the closures being formed again in $\beta(S)$). But

[19] I do not know whether the restriction to Riesz spaces is really necessary in this theorem.

$\bar{F}_1 \subset \lambda(S)$ so that $\bar{F}_1$ and $\bar{F}_2 + \beta(S) - \lambda(S)$ are two closed subsets of $\beta(S)$ without a common point. Since $\beta(S)$ is a bicompact Hausdorff space, it is normal, so that there exists a continuous function φ in the domain $\beta(S)$ such that $\varphi(x) = 0$ for each $x \in \bar{F}_1$ and $\varphi(x) = 1$ for each $x \in \bar{F}_2$ and for each $x \in \beta(S) - \lambda(S)$. Let us define a function f in the domain S_0 in the following way. If $x \in S$, then $f(x) = \varphi(x)$; moreover $f(\omega) = 1$. Then it is easy to see that f is a continuous function in the domain S_0 such that $f(x) = 0$ for each $x \in F_1$ and $f(x) = 1$ for each $x \in F_2$.

I conclude with two more unsolved questions. A topological space S is called *completely normal* if every subset of S is a normal space. S may be called *locally completely normal* if every point $x \in S$ possesses a neighborhood U such that $\bar{U}$ is a completely normal space. S may be called *completely locally normal* if every subset of S is a locally normal space. It is easy to see that a locally completely normal space is completely locally normal. I do not know whether the converse holds true. Any open subset of a completely normal space is a locally completely normal space. I do not know whether a locally completely normal space must be homeomorphic with an open subset of a completely normal space.

BRNO, CZECHOSLOVAKIA.

ANNALS OF MATHEMATICS
Vol. 38, No. 4, October, 1937

REMARK ON BICOMPACT SPACES

BY BEDŘICH POSPÍŠIL

(Received April 26, 1937)

We write $\exp \mathfrak{x} = 2^{\mathfrak{x}}$ for each cardinal number $\mathfrak{x}$.

THEOREM I. *For each infinite cardinal number* $\mathfrak{h}$, *there exists a bicompact Hausdorff space* S *and a subspace* $T \subset S$ *such that* (i) T *is an isolated space,* (ii) T *is dense in* S, (iii) *the cardinal number of* T *is* $\mathfrak{h}$, (iv) *the cardinal number of* S *is* exp exp $\mathfrak{h}$.

PROOF. Let H denote a set of $\mathfrak{h}$ elements. Let X denote the set of all functions φ defined over H and assuming only values 0 and 1. Let Ω denote the family of all subsets of X such that any $A \in \Omega$ is the set of all functions $\varphi \in X$ assuming a given value at each of a given finite number of given elements of H. Then the cardinal number of X is exp $\mathfrak{h}$ and the cardinal number of Ω is $\mathfrak{h}$. We consider X as a topological space, a neighborhood of $\varphi \in X$ being any $A \in \Omega$ such that $\varphi \in A$. It is easy to see that we may choose a set E dense in X of cardinal number $\mathfrak{h}$. Let F denote the set of all mappings of X into a subset of X. Let T denote the set of all finite families t of pairs (C_k, e_k) where $C_k \in \Omega$, $e_k \in E$ and any two sets C_k have a vacuous intersection; the pairs (C_k, e_k) will be termed the *coordinates* of $t \in T$.

We put $S = T + F$, the topology of S being defined by neighborhoods as follows. A neighborhood of $t \in T$ consists of t only. Neighborhoods of $f \in F$ will be defined in a somewhat complicated manner. Choose a finite number of different points $x_k \in X (1 \leqq k \leqq n)$. For $1 \leqq k \leqq n$, choose sets $A_k \in \Omega$ and $B_k \in \Omega$ such that $x_k \in A_k$ and $f(x_k) \in B_k$. Let Q be the set of all $g \in F$ such that $g(x_k) \in B_k$ for $1 \leqq k \leqq n$. Define a set $P \subset T$ in the following manner. An element t of T belongs to P if and only if, for $1 \leqq k \leqq n$, there exists a coordinate (C_k, e_k) of t such that $x_k \in C_k$, $C_k \subset A_k$, $e_k \in B_k$. [Of course, t may possess *more* than n coordinates.] Let K be any finite subset of P. Put $O = P - K + Q$. Any such set O is a neighborhood of f in our space S.

The properties (i)-(iv) being evident, to prove theorem I it suffices to show that S is a bicompact Hausdorff space. This is indeed true but a quicker way of completing our proof is as follows. It is easy to see that all our neighborhoods are both closed and open in our space S, whence it readily follows that S is a completely regular space. Therefore, using a result of Tychonoff,[1] there exists a bicompact Hausdorff space B containing S. If S_0 denotes the closure of S in B, then the spaces S_0 and $T \subset S_0$ satisfy our theorem. Indeed, it is easy to prove

[1] A. Tychonoff, *Über die topologische Erweiterung von Räumen*, Math. Annalen, 102, 1930, 544–561.

that the cardinal number of S_0 cannot exceed exp exp $\mathfrak{h}$, since S_0 is a Hausdorff space containing a dense subset of cardinal number $\mathfrak{h}$.

In a recent paper,[2] Čech attached to any completely regular space T a unique bicompact space $\beta(T) \supset T$ and posed the problem of determining the cardinal number of $\beta(T)$ in case T is an isolated countable space. The answer to Čech's question is a particular case of the following

THEOREM II. *Let T denote an infinite isolated space of cardinal number $\mathfrak{h}$; then the cardinal number of $\beta(T)$ is* exp exp $\mathfrak{h}$.

PROOF. Again applying the final sentence in the proof of theorem I, the cardinal number of $\beta(T)$ cannot exceed exp exp $\mathfrak{h}$. By theorem I, there exists a bicompact space $S \supset T$ of cardinal number exp exp $\mathfrak{h}$ such that T is dense in S. In his paper, already quoted, Čech proved that there exists a continuous mapping of $\beta(T)$ into S. Hence the cardinal number of $\beta(T)$ is not less than exp exp $\mathfrak{h}$.

BRNO, CZECHOSLOVAKIA.

---

[2] E. Čech, *On bicompact spaces*, Annals of Math., 38, 1937, 823–844.

Comptes Rendus (Doklady) de l'Académie des Sciences de l'URSS
1939. Volume XXII, № 1

MATHEMATICS

ON RINGS OF CONTINUOUS FUNCTIONS ON TOPOLOGICAL SPACES

By I. GELFAND and A. KOLMOGOROFF

(Communicated by I. M. Vinogradow, Member of the Academy, 17. XI. 1938)

The present note deals with the same subject as the investigations by M. Stone (²) and the note of G. Šilov published above. In difference from this last note, we consider a ring of continuous functions defined on a certain topological space as a purely algebraical formation, without introducing in it any topological relations. It turns out that in the case of bicompact spaces considered by M. Stone, as well as in considerably more general cases, the algebraical structure of the ring of continuous functions already defines the topological space up to a homeomorphism.

For any topological space S we shall consider two rings: the ring $C(S)$ of all real continuous functions defined on S and the ring $C'(S)$ of all bounded functions from $C(S)$.

It is natural that in studying these rings we should confine ourselves to the case of totally regular spaces S. This may be justified by the fact that the rings $C(S)$ and $C'(S)$ of an arbitrary topological space S are respectively isomorphic to the rings $C(\rho S)$ and $C'(\rho S)$ of a definite totally regular space ρS introduced by E. Čech (¹), which is connected with S in a unique manner. Together with the space ρS E. Čech introduces a certain continuous mapping $y = \rho(x)$ of the space S on the space ρS, which possesses the following property: the real continuous functions on S coincide with functions of the type $f(x) = \varphi[\rho(x)]$, where $\varphi(y)$ is a continuous real function on the space ρS. Hence it follows that the ring $C(S)$ is isomorphic to the ring $C(\rho S)$ and the ring $C'(S)$— to the ring $C'(\rho S)$.

Thus, we may suppose that in what follows the original space S is totally regular (in this case ρS coincides with S).

We shall say that an ideal of the ring C is maximum, if it does not coincide with the whole ring and is not contained in any greater ideal except the ring C itself. Let us form from the set γ of maximum ideals of the ring C a topological space by means of the following definition: the maximum ideal α is called a point of contact of the set $\mathfrak{M}$ of maximum ideals, if it contains the intersection of all maximum ideals contained in $\mathfrak{M}$. It is easily verified that for any ring C this definition of points of contact makes the set γ to a space of the type T_1 (according to the terminology of Alexandroff-Hopf, Topologie, 1). Observe that this method of introducing topology in the set of maximum ideals was already used by M. H. Stone ([2]).

We shall denote the space γ corresponding to the ring $C(S)$ by $\gamma(S)$ and the space γ corresponding to the ring $C'(S)$ by $\gamma'(S)$.

1. The Case of a Bicompact Space S

If the space S is bicompact, the rings $C(S)$ and $C'(S)$ coincide (all continuous functions on S are bounded) and it becomes unnecessary to consider them separately.

Theorem I. *If the space S is bicompact, it is homeomorphic to the space $\gamma(S)$.*

Before proving this theorem, let us establish the following

Lemma. For any ideal A of the ring $C(S)$ not coinciding with the whole ring there exists a point a of the space S, at which all functions belonging to A vanish.

Proof of the Lemma. Assume that the assertion is false. Then for every point ξ of the space S we can find a function $f_\xi(x)$ from A, which does not vanish at the point ξ. The function $f_\xi(x)$ is different from zero in a certain neighbourhood $U(\xi)$ of the point ξ. By Borel-Lebesgue's theorem we can choose a finite set of points,

$$\xi_1, \xi_2, \ldots, \xi_n,$$

so that the neighbourhoods

$$U(\xi_1), U(\xi_2), \ldots, U(\xi_n)$$

cover the whole space S. The function

$$\varphi(x) = f_{\xi_1}^2(x) + f_{\xi_2}^2(x) + \ldots + f_{\xi_n}^2(x)$$

belongs to the ideal A and is everywhere different from zero. Hence every function $f(x)$ of the ring $C(S)$ may be represented in the form $f(x) = \psi(x)\varphi(x)$, and, consequently, also belongs to A, contrary to assumption. The obtained contradiction proves the lemma.

Proof of Theorem I. From the lemma just proved it follows that any ideal A not coinciding with the whole ring $C(S)$ is contained in a certain ideal $I(a)$, consisting of all fnnctions vanishing at the point a. Consequently, only ideals of this last type can be maximal. It is also plain that any ideal of the type $I(a)$ is in fact maximal. In virtue of the total regularity of the space S the ideals $I(a)$ and $I(a')$, corresponding to two different points a and a', are different: there exists a function from $C(S)$ vanishing at the point a and different from zero at the point a'. Thus, correlating to every point a of the space S the maximum ideal $I(a)$, we obtain a one-to-one correspondence between the space S and the set $\gamma(S)$ of maximum ideals of the ring $C(S)$.

We have now to prove that the correspondence obtained between S and $\gamma(S)$ is a homeomorphism. Suppose that to a set M of points of the space S corresponds the set $\mathfrak{M}$ of maximum ideals. If a point a is a point of contact of the set M, all functions contained in the intersection of all ideals from $\mathfrak{M}$, i. e. vanishing at all points of the set M, vanish also at the point a, i. e. belong to the ideal $I(a)$. According to definition, this means, however, that $I(a)$ is a point of contact of the set $\mathfrak{M}$. Conversely, if a point a is not a point of contact for M, then, in virtue of the total regularity of S, we may construct a function $f(x)$ from $C(S)$, vanishing on M and different from zero at the point a; this function

12

belongs to the intersection of all ideals from $\mathfrak{M}$, but does not belong to $I(a)$; consequently, in this case $I(a)$ will not be a point of contact for $\mathfrak{M}$. The homeomorphism between the spaces S and $\gamma(S)$ is thus proved.

Theorem II. *In order that two bicompact spaces S and S_1 should be homeomorphic, it is necessary and sufficient that the rings $C(S)$ and $C(S_1)$ should be algebraically isomorphic.*

Proof. The necessity of the condition is obvious: any homeomorphism between S and S_1 automatically generates an isomorphic mapping of the ring $C(S)$ on the ring $C(S_1)$. That the condition is also sufficient, follows from Theorem I: from the isomorphism of $C(S)$ and $C(S_1)$ plainly follows the homeomorphism of the spaces $\gamma(S)$ and $\gamma(S_1)$; but these last are, by Theorem I, homeomorphic to the spaces S and S_1.

2. The Ring $C'(S)$ in the General Case

The question of the structure of the ring $C'(S)$ of an arbitrary totally regular space S may be reduced to the case of a bicompact space S. To do this we use the space βS introduced by E. Čech. According to E. Čech, to every totally regular space S corresponds a space βS uniquely determined up to a homeomorphism and characterized by the following conditions: 1) βS is bicompact; 2) βS contains S; 3) S is dense on βS; 4) every continuous bounded real function defined on S may be continued into βS, the continuity being preserved.

Since S is dense on βS, the continuation of continuous functions from S into βS is defined uniquely. Thus, there arises a one-to-one correspondence between bounded continued functions defined on S and on βS. It is easily seen that in virtue of this correspondence the ring $C'(S)$ is isomorphically mapped on the ring $C'(\beta S)$. We have thus proved the following

Theorem III'. *The ring $C'(S)$ is isomorphic to the ring $C'(\beta S)$* [or, which is the same, to the ring $C(\beta S)$].

From Theorems I and III' follows

Theorem IV'. *The space $\gamma'(S)$ is homeomorphic to the space βS.*

In virtue of Theorem III' it is clear that two non-homeomorphic spaces S and S_1 may have isomorphic rings $C'(S)$ and $C'(S_1)$. To see this it is sufficient to take for S any non-bicompact space and put $S_1=\beta S$; then, by Theorem III', the ring $C'(S)$ will be isomorphic to the ring $C'(S_1)$. We have, however, the following

Theorem V'. *In order that the spaces S and S_1 satisfying the first axiom of enumerability should be homeomorphic, it is necessary and sufficient that the rings $C'(S)$ and $C'(S_1)$ should be algebraically isomorphic.*

Proof. That the condition is sufficient, may be proved in the same way, as in the case of Theorem I. The necessity follows from the fact, established by E. Čech, that for two spaces S and S_1 satisfying the first axiom of enumerability the homeomorphism of βS and βS_1 implies the homeomorphism of the original spaces S and S_1. Indeed, from the isomorphism of $C'(S)$ and $C'(S_1)$ follows the homeomorphism of $\gamma'(S)$ and $\gamma'(S_1)$, i. e., by Theorem IV', the homeomorphism of βS and βS_1 and, consequently, in virtue of the just mentioned result of E. Čech, also the homeomorphism between S and S_1.

3. The Ring $C(S)$ in the General Case

For the ring $C(S)$ the theorem analogous to Theorem III' concerning the ring $C'(S)$ is not true. For instance, if S is the space of integers

(i. e. enumerable space of isolated points) and $S_1 = \beta S$, then $C(S)$ and $C(S_1)$ are not isomorphic (we omit the proof of this fact). Theorems analogous to Theorems IV′ and V′ remain, however, valid.

Theorem IV. *The space $\gamma(S)$ is homeomorphic to the space βS.*

Theorem V. *In order that two spaces S and S_1 satisfying the first axiom of enumerability should be homeomorphic, it is necessary and sufficient that the rings $C(S)$ and $C(S_1)$ should be algebraically isomorphic.*

Theorem V may be deduced from Theorem IV in the same way as Theorem V′ was deduced from Theorem IV′. Hence we have only to prove Theorem IV.

Proof of Theorem IV. To every function $f(x)$ from $C(S)$ we let correspond the function $f_1(x) = \frac{2}{\pi} \operatorname{arctg} [f(x)]$. The function $f_1(x)$ is continuous and bounded on S. Consequently, it may be continued with preservation of continuity into the space βS. This enables us to expand, in a certain sense, the definition of the function $f(x)$ to the whole space βS. Namely, for any point of the set $\beta S - S$ we put

$$
\begin{aligned}
f(x) &= \operatorname{tg} \left[\frac{\pi}{2} f_1(x) \right], && \text{if } |f_1(x)| \neq 1; \\
f(x) &= +\infty, && \text{if } f_1(x) = +1; \\
f(x) &= -\infty, && \text{if } f_1(x) = -1.
\end{aligned}
$$

When the functions $f(x)$ are thus extended to the whole space βS, the proof of Theorem V proceeds, with certain complications, on the lines of the proof of Theorem I. We state here only the fundamental points of this proof. In the first place we establish the following

Lemma 1. For any ideal A of the ring $C(S)$, different from the whole ring, there exists a point a of the space βS, at which all functions belonging to A vanish.

Any point a of the space βS defines the ideal $I(a)$ of the ring $C(S)$ consisting of all functions $f(x)$ of this ring, for which all products

$$\varphi(x) = \psi(x) f(x)$$

with factors $\psi(x)$ from $C(S)$ vanish at the point a (if the point a belongs to $\beta S - S$, under $\varphi(a)$ we understand the value in the sense defined above).

On ground of Lemma 1 we next prove that all ideals of the type $I(a)$ are maximal and that there are no other maximum ideals in the ring $C(S)$.

Lemma 2. If the point a belongs to S, the ideal $I(a)$ consists of all functions $f(x)$ of the ring $C(S)$, vanishing at the point a. If the point a belongs to $\beta S - S$, the ideal $I(a)$ consists of all functions $f(x)$ of the ring $C(S)$, vanishing on the set N_f of points of the original space S having a for its point of contact.

We shall omit the proof of Lemma 2. From this lemma it follows that the ideals $I(a)$ and $I(a')$, corresponding to two different points a and a', are different. Thus, we can establish a one-to-one correspondence between the points a of the space βS and the maximum ideals $I(a)$. Similarly as in the case of Theorem I we prove that this correspondence between βS and $\gamma(S)$ is a homeomorphism.

Remark. From the isomorphism of the rings $C(S)$ and $C(S_1)$ follows (by Theorem IV) the homeomorphism of the spaces βS and βS_1, and consequently (by Theorem III′), also the isomorphism of the rings $C'(S)$ and $C'(S_1)$. But, conversely, from the isomorphism of the rings $C'(S)$ and $C'(S_1)$ the isomorphism of the rings $C(S)$ and $C(S_1)$, generally speaking, does not follow. For instance, if S is the space of integers and $S_1 = \beta S$, the rings $C'(S)$ and $C'(S_1)$ are isomorphic, but the rings $C(S)$ and $C(S_1)$ are not. Thus, the ring $C(S)$ provides us with a more elastic means of investigation of the topological properties of the space S than the ring $C'(S)$. However, there still exist spaces S and S_1 with isomorphic rings $C(S)$ and $C(S_1)$, which are not homeomorphic to each other. Such are, for instance, the space S of transfinite numbers $< \Omega$ and the space S_1 of transfinite numbers $\leqslant \Omega$, taken with limiting relations defined in them in the usual manner.

Received
20. XI. 1938.

REFERENCES

[1] E. Čech, Annals of Mathematics, 38, 823—844 (1937). [2] M. H. Stone, Trans. American Math. Soc., 41, 375—481 (1937).

STONE-ČECH COMPACTIFICATIONS OF PRODUCTS[1]

BY

IRVING GLICKSBERG

1. **Introduction.** As is well known from the work of Tychonoff [10], Stone [8], and Čech [1], every completely regular space X can be imbedded as a dense subspace of a compact Hausdorff space $\beta(X)$ in such a way that continuous (real valued and bounded) functions on X extend continuously to $\beta(X)$; indeed the resulting compactification of X, the Stone-Čech compactification, is uniquely determined by just these properties. For a set of completely regular spaces, the naturally induced imbedding of their product PX_α as a dense subspace of $P\beta(X_\alpha)$ yields a compactification of their product, and the question arises as to when one can identify(2) this with the Stone-Čech compactification. The main purpose of this paper is to show that aside from a trivial case, this identification is possible if and only if PX_α is pseudo-compact(3) [5], i.e., if and only if every real valued continuous function on it is bounded, or, equivalently, every bounded continuous function assumes its bounds(4). A side result of the investigation is the fact that every product of uncountably many compact spaces, each having at least two points, is the Stone-Čech compactification of certain proper subspaces, yielding a fairly accessible body of nontrivial Stone-Čech compactifications. Finally we shall give several conditions sufficient to insure that a product of pseudo-compact spaces be pseudo-compact, and briefly discuss a related question.

2. **Pseudo-compact spaces.** Several facts concerning pseudo-compact spaces will be used repeatedly in the following pages, and we shall collect them here. One of our main tools is a characteristic property, that *Ascoli's theorem holds* [3, Theorem 2], that is, every bounded equicontinuous set of functions on a pseudo-compact space X has compact closure in the Banach

Received by the editors November 2, 1956 and, in revised form, May 6, 1957.

(1) Some of the results of this paper (essentially the necessity in Theorem 1, and Theorem 3) were originally included in a separate note submitted to the Proc. Amer. Math. Soc. in 1955, and were obtained while the writer was at The RAND Corporation. The same part of Theorem 1 has been obtained by M. Henriksen and J. R. Isbell [4] who also obtained some results in the converse direction. The writer would like to express his thanks to Professors Henriksen and Isbell for allowing him to read their manuscript.

(2) We shall simply write $\beta(PX_\alpha)=P\beta(X_\alpha)$ to express this identity (rather than the identity of the compact spaces involved (cf. §6)); when the meaning is clear we shall speak of $\beta(X)$ as the compactification. It will be convenient to always consider X as a subspace of $\beta(X)$, and thus PX_α as a subspace of $P\beta(X_\alpha)$.

(3) It should perhaps be noted that this corrects an erroneous assertion made by Hewitt [5, Theorem 14].

(4) In addition pseudo-compactness requires complete regularity.

space $C(X)$ of all bounded real valued continuous functions on X. A direct corollary is the fact that, for a bounded set of functions on such a space, *equicontinuity is equivalent to equicontinuity of each countable subset*, since conditional compactness and conditional countable compactness coincide in $C(X)$.

Another characteristic property of these spaces which will prove useful is purely topological in form [3, Theorem 2]: *every sequence of nonvoid, pairwise disjoint open sets has a cluster point* (which by definition has the property that each of its neighborhoods meets infinitely many elements of the sequence). Of course such a cluster point prevents the sequence from forming a locally finite collection of open sets. On the other hand, given an infinite locally finite collection C of open sets, on any space, we can produce a sequence of the type described with no cluster point: we simply choose an open neighborhood N_1 contained in one element of C and meeting only finitely many; deleting these from C to form C_1 we obtain a similar neighborhood N_2 for C_1, and continue. Clearly the sequence $\{N_i\}$ we obtain, being locally finite, has no cluster point, and is easily seen to satisfy our other requirements. Consequently, if a space is pseudo-compact then every locally finite collection of open sets is finite. The converse of this assertion is obvious, so we have another characteristic property, which in turn yields the following strengthened form of our original topological condition: *every sequence of nonvoid open sets has a cluster point if and only if the space is pseudo-compact.* We need only verify that a pseudo-compact space has the required property. But a sequence of nonvoid open sets yields a collection of sets which is locally finite or not; in the first case the collection is finite so that some point lies in infinitely many sets of the sequence, and is thus a cluster point. In the second case some point prohibits local finiteness, and clearly this point is the required cluster point.

Pseudo-compactness is also reflected in the relationship between $\beta(X)$ and its subspace X: *X is pseudo-compact if and only if there is no nonvoid closed G_δ in $\beta(X)\backslash X$.* For, given a nonvoid closed G_δ in $\beta(X)\backslash X$, we can produce by Urysohn's lemma a continuous function on $\beta(X)$ that does not assume its maximum value on the dense subspace X; conversely, given a continuous bounded function on X which does not assume its least upper bound, the set where the continuous extension of this function to $\beta(X)$ assumes its maximum is a nonvoid closed G_δ in $\beta(X)\backslash X$.

If a product PX_α is pseudo-compact, clearly each factor space and each partial product is also pseudo-compact, since continuous functions on these spaces may be considered as continuous functions on the full product. Further, *pseudo-compactness of the product is equivalent to pseudo-compactness of every countable partial product $P_{i=1}^{\infty} X_{\alpha_i}$.* For, since it suffices to show that every sequence of (nonvoid) canonical neighborhoods in PX_α has a cluster point, and each neighborhood places restrictions on only finitely many coordinates, only countably many coordinates are involved, say those corresponding to $\alpha_1, \alpha_2, \cdots$. Choosing any cluster point of the projection of our sequence into PX_{α_i}, we can (by any arbitrary choice of all other coordinates) clearly extend this point to a cluster point of the original sequence.

NOTATION. We shall allow x to represent a generic element of $\beta(X)$ rather than of X. For $f \in C(X)$, f^* will denote its unique continuous extension to $\beta(X)$, $f|F$ its restriction to a subset F of X. For such a subset F^- and F' will denote closure and complement, respectively, in X. Where no confusion can arise we shall refer to $\beta(X)\backslash X$ as the set of ideal points of $\beta(X)$. All spaces will be assumed completely regular, and all functions real valued.

3. **The main result.** If, for some α_0, $P_{\alpha\neq\alpha_0}X_\alpha$ forms a finite set, then every $f \in C(PX_\alpha)$ clearly has a continuous extension to $P\beta(X_\alpha)$, and we may write $\beta(PX_\alpha) = P\beta(X_\alpha)$. This is precisely the trivial case mentioned in the introduction.

THEOREM 1. *Let* $\{X_\alpha\}$ *be a set of completely regular spaces, and suppose the set* $P_{\alpha\neq\alpha_0} X_\alpha$ *is infinite for every* α_0. *Then a necessary and sufficient condition that* $\beta(PX_\alpha) = P\beta(X_\alpha)$ *is that* PX_α *be pseudo-compact.*

Proof of the necessity. A moment's reflection shows that the property $\beta(PX_\alpha) = P\beta(X_\alpha)$ is inherited by all partial products[6]; consequently, for any subset B of our index set, we may evidently write

$$\beta\left(\underset{\alpha\in B}{P} X_\alpha \times \underset{\alpha\notin B}{P} X_\alpha\right) = P\beta(X_\alpha) = \underset{\alpha\in B}{P} \beta(X_\alpha) \times \underset{\alpha\notin B}{P} \beta(X_\alpha)$$

$$= \beta\left(\underset{\alpha\in B}{P} X_\alpha\right) \times \beta\left(\underset{\alpha\notin B}{P} X_\alpha\right).$$

Thus it will suffice to prove that if X and Y are infinite spaces such that $\beta(X\times Y) = \beta(X)\times\beta(Y)$, then $X\times Y$ is pseudo-compact. We shall first show that each of the spaces X and Y is pseudo-compact. In showing Y is pseudo-compact, we can assume X is compact. For if $f \in C(\beta(X)\times Y)$, then $(f|X\times Y)^*$ clearly must coincide with f on $\beta(X)\times Y$, and thus continuously extends f to $\beta(X)\times\beta(Y)$; thus $\beta(\beta(X)\times Y) = \beta(X)\times\beta(Y)$, and we may replace X by $\beta(X)$.

Suppose then that Y is not pseudo-compact and that X is compact. Then there is a g in $C(Y)$ which never vanishes on Y but has zero as its greatest lower bound, and hence has a sequence of values $g(y_n)\to 0$. Since X is infinite,

[6] For we may identify the partial product $P_{\alpha\in B}X_\alpha$ with the subspace $P_{\alpha\in B}X_\alpha \times \{x\}$ of $P\beta(X_\alpha)$ (where $x \in P_{\alpha\notin B}X_\alpha$), and similarly $P_{\alpha\in B}\beta(X_\alpha)$ with $P_{\alpha\in B}\beta(X_\alpha) \times \{x\}$; since $f \in C(P_{\alpha\in B}X_\alpha \times \{x\})$ has an obvious continuous extension to PX_α, it has one to $P\beta(X_\alpha)$, hence to $P_{\alpha\in B}\beta(X_\alpha) \times \{x\}$.

there is an $f \in C(X)$ having infinitely many values(6), and we may assume that a sequence of these, $\{f(x_n)\}$, strictly decreases to zero. By linear interpolation, we can construct a non-negative, continuous, bounded function h on the reals for which $h(f(x_n)) = g(y_n)$, and setting

$$\phi(x, y) = 2h(f(x))g(y)/(h(f(x))^2 + g(y)^2),$$

we obtain an element ϕ of $C(X \times Y)$: for g never vanishes and $0 \leqq \phi \leqq 1$. But consider ϕ^*. If (x, y) is a cluster point of the sequence $\{(x_n, y_n)\}$ in the compact space $X \times \beta(Y)$, then $\phi^*(x, y) = 1$ since $\phi^*(x_n, y_n) = \phi(x_n, y_n) = 1$. On the other hand, x is a cluster point of $\{x_n\}$ and, since $\phi^*(x_n, y') = \phi(x_n, y') \to 0$ for each y' in Y, we have $\phi^*(x, y') = 0$ for each y' in Y. Since Y is dense in $\beta(Y)$, it follows that $\phi^*(x, y) = 0$, which is the desired contradiction.

Returning to the general case, we see that both X and Y must be pseudo-compact. In order to see that this is also the case for $X \times Y$, we need only prove that any closed G_δ in the set of ideal points of $\beta(X \times Y)$ must be void (cf. §2). But if F is such a subset of the set of ideal points of $\beta(X \times Y)$, then, for each x in X, $F \cap (\{x\} \times \beta(Y))$ is a closed G_δ in the set of ideal points of $\beta(\{x\} \times Y) = \{x\} \times \beta(Y)$, and thus is void since $\{x\} \times Y$ is pseudo-compact. Hence $F \cap (X \times \beta(Y))$ is void and $F \subset (\beta(X) \setminus X) \times \beta(Y)$. But now for every y in $\beta(Y)$, $F \cap (\beta(X) \times \{y\})$ is a closed G_δ in the set of ideal points of $\beta(X \times \{y\}) = \beta(X) \times \{y\}$, hence void, so that $F = F \cap (\beta(X) \times \beta(Y))$ must be void, and $X \times Y$ must be pseudo-compact.

Our proof of the sufficiency is based on the possibility of coordinate-wise extension of an element of $C(PX_\alpha)$, which results, in the case of finitely many factors, from the following lemmas.

LEMMA 1. *If $X \times Y$ is pseudo-compact and $f \in C(X \times Y)$, then the family $\{f(x, \cdot) : x \in X\}$ is equicontinuous on Y. Consequently the mapping $y \to f(\cdot, y)$ of Y into $C(X)$ is continuous.*

Proof. Since the second statement of the lemma is an obvious consequence of the first, and since Y is pseudo-compact, it is sufficient to prove that any sequence $\{f(x_n', \cdot)\}$ is equicontinuous on Y (cf. §2). Suppose not. Then for some y_0 in Y and $\epsilon > 0$, no neighborhood W of y_0 satisfies the condition

$$(*) \qquad |f(x_n', y) - f(x_n', y_0)| < \epsilon \qquad \text{for } y \in W,$$

for all n. We now select a subsequence $\{x_n\}$ of $\{x_n'\}$ and open neighborhoods V_n of x_n, W_n of y_0 as follows. Let $W_1 = Y$ and x_1 be the first x_n' for which $(*)$ fails for $W = W_1$. We choose a neighborhood $V_1 \times W_2$ of (x_1, y_0) on which f varies by <1; having chosen $x_1, \cdots, x_k, V_1, \cdots, V_k, W_1, \cdots, W_{k+1}$ we

(6) The following argument is due to the referee. If continuous functions with only finitely many values suffice to separate every pair of points, then it is easy to see that X contains a strictly decreasing sequence of open and closed subsets. Taking f_n as the characteristic function of the nth subset, then $f = \sum 2^{-n} f_n$ is continuous and assumes infinitely many values.

select x_{k+1} as the first x_n' for which (*) fails for $W=W_{k+1}$, and select an open neighborhood $V_{k+1}\times W_{k+2}$ of (x_{k+1}, y_0) on which f varies by $<1/(k+1)$, with $W_{k+2}^{-}\subset W_{k+1}$.

From our choice of x_n we have a y_n in W_n for which $|f(x_n, y_n)-f(x_n, y_0)| \geqq \epsilon$; consequently we may choose an open neighborhood $\tilde{V}_n\times\tilde{W}_n$ of (x_n, y_n) lying within $V_n\times W_n$ which yields $|f(x, y)-f(x, y_0)|>\epsilon/2$ for $(x, y)\in\tilde{V}_n\times\tilde{W}_n$. But by one of our topological conditions for pseudo-compactness, $\{\tilde{V}_n\times\tilde{W}_n\}$ has a cluster point $(\bar{x}, \bar{y})$ in $X\times Y$, so that $|f(\bar{x}, \bar{y})-f(\bar{x}, y_0)|\geqq\epsilon/2$ by continuity. On the other hand, as a cluster point of $\{\tilde{W}_n\}$, $\bar{y}\in\bigcap_{j=1}^{\infty} W_j$; for $\bar{y}\notin W_k$ implies $\bar{y}\notin W_{k+1}^{-}$ so that $W_{k+1}^{-\prime}$ is a neighborhood of $\bar{y}$ meeting only finitely many $\tilde{W}_n$. Thus for $x\in\tilde{V}_n\subset V_n$, $|f(x, \bar{y})-f(x, y_0)|<1/n$ since both $(x, \bar{y})$ and (x, y_0) lie in $V_n\times W_{n+1}$. Since $\bar{x}$ is a cluster point of $\{\tilde{V}_n\}$, $0=|f(\bar{x}, \bar{y})-f(\bar{x}, y_0)|\geqq\epsilon/2$, a contradiction establishing equicontinuity, and completing the proof.

LEMMA 2. *Let X and Y be completely regular and $f\in C(X\times Y)$. If the mapping $y\rightarrow f(\cdot, y)$ of Y into $C(X)$ is continuous, then f has a continuous extension to $\beta(X)\times Y$.*

Proof. In view of the natural isomorphism of $C(X)$ and $C(\beta(X))$, continuity of the mapping $y\rightarrow f(\cdot, y)$ implies continuity of the mapping $y\rightarrow\bar{f}(\cdot, y)$ of Y into $C(\beta(X))$, where $\bar{f}(\cdot, y)$ is the unique continuous extension of $f(\cdot, y)$ to $\beta(X)$. But as a function on $\beta(X)\times Y$, $\bar{f}$ is continuous. For given $(x_0, y_0)\in\beta(X)\times Y$ and $\epsilon>0$, we have a neighborhood V of x_0 satisfying $|\bar{f}(x, y_0)-\bar{f}(x_0, y_0)|<\epsilon$ for $x\in V$, since $\bar{f}(\cdot, y_0)\in C(\beta(X))$; since $y\rightarrow\bar{f}(\cdot, y)$ is continuous, y_0 has a neighborhood W satisfying $|\bar{f}(x, y)-\bar{f}(x, y_0)|<\epsilon$ for $y\in W$ and all x in $\beta(X)$. Thus for $(x, y)\in V\times W$, $|\bar{f}(x, y)-\bar{f}(x_0, y_0)|\leqq|\bar{f}(x, y)-\bar{f}(x, y_0)|+|\bar{f}(x, y_0)-\bar{f}(x_0, y_0)|<2\epsilon$, which establishes continuity and completes the proof of Lemma 2.

Proof of the sufficiency in Theorem 1. If $X\times Y$ is pseudo-compact, then by Lemmas 1 and 2 we may extend $f\in C(X\times Y)$ continuously to $\beta(X)\times Y$. Now any space containing a dense pseudo-compact subspace is pseudo-compact (since any continuous function on the space, being bounded on the dense subspace, is bounded); thus $\beta(X)\times Y$ is pseudo-compact and applying the lemmas once more we obtain a continuous extension to $\beta(X)\times\beta(Y)$, and the proof is complete for the case of two factor spaces. Since the result now follows immediately for a finite set of factors, we may turn to the infinite case.

Suppose that PX_α is pseudo-compact and $f\in C(PX_\alpha)$. Then for every $\epsilon>0$, there is a finite set $\alpha_1, \cdots, \alpha_n$ of indices for which $x, y\in PX_\alpha$, $x_{\alpha_i}=y_{\alpha_i}$, $i=1, \cdots, n$, imply $|f(x)-f(y)|<\epsilon$. Suppose not, and let F_0 be any finite set of coordinate indices. Then we have points x^1 and y^1, agreeing in those coordinates with $\alpha\in F_0$, with $|f(x^1)-f(y^1)|\geqq\epsilon$. By continuity and the form of neighborhoods in the product we may clearly assume, if we replace ϵ by

$\epsilon/2$, that x^1 and y^1 differ in only finitely many coordinates, say those corresponding to $\alpha \in F_1$. Considering the finite set $F_0 \cup F_1$, we can arrive in the same fashion at two points x^2 and y^2 with $|f(x^2)-f(y^2)| \geqq \epsilon/2$, agreeing in those coordinates with $\alpha \in F_0 \cup F_1$ and differing in just those coordinates with $\alpha \in F_2$, F_2 finite. Continuing, we obtain sequences $\{x^i\}$, $\{y^i\}$, $\{F_i\}$ for which $|f(x^i)-f(y^i)| \geqq \epsilon/2$, x^i and y^i differ in only those coordinates with $\alpha \in F_i$, and the F_i are finite and pairwise disjoint.

Now, by continuity, we can enlarge each x^i to a canonical open neighborhood U^i and each y^i to a canonical open neighborhood V^i so that $|f(x)-f(y)| > \epsilon/4$ for $x \in U^i$, $y \in V^i$. Indeed we can clearly insist that U^i and V^i agree in all their components except those in which x^i and y^i disagree, i.e., for $\alpha \in F_i$. But $\{U^i\}$ has a cluster point $\bar{x}$, and if W is a canonical open neighborhood of $\bar{x}$, placing restrictions on only those coordinates corresponding to $\alpha_1, \cdots, \alpha_n$, then we have W meeting U^i if and only if W meets V^i for i large enough to yield $\{\alpha_1, \cdots, \alpha_n\} \cap \bigcup_{j=1}^{\infty} F_j \subset \bigcup_{j=1}^{i-1} F_j$ (since U^i and V^i then have the same α_1st, $\cdots$, α_nth components). Consequently each neighborhood of $\bar{x}$ contains a pair of points satisfying $|f(x)-f(y)| > \epsilon/4$, contradicting continuity, and establishing our assertion.

Now let us consider f as defined on the subspace PX_α of $P\beta(X_\alpha)$. If f has no continuous extension to $P\beta(X_\alpha)$, we have an element x^0 of this space for which $\lim\sup_{x\to x^0} f(x) - \lim\inf_{x\to x^0} f(x) = a > 0$, where x is taken from the dense subspace PX_α. Let $0 < 3\epsilon < a$ and $\alpha_1, \cdots, \alpha_n$ be the finite set of indices obtained for this ϵ in the previous paragraph. For brevity let us now write $X = P_{i=1}^n X_{\alpha_i}$, $Y = P_{\alpha \neq \alpha_1, \cdots, \alpha_n} X_\alpha$ and the values of f as for a function on $X \times Y$.

By Lemmas 1 and 2, f has a continuous extension $\bar{f}$ to $\beta(X) \times Y = P_{i=1}^n \beta(X_{\alpha_i}) \times Y$, a subspace of $P\beta(X_\alpha)$. Moreover since $\beta(X) \times Y$ is pseudocompact, by Lemma 1 $\{\bar{f}(\cdot, y)\}_{y \in Y}$ is equicontinuous on $\beta(X)$. In particular, this is the case at $(x^0_{\alpha_1}, \cdots, x^0_{\alpha_n}) \in P_{i=1}^n \beta(X_{\alpha_i}) = \beta(X)$, so that this point has a neighborhood V on which $\bar{f}(\cdot, y)$ varies by $<\epsilon$, for each y in Y. But consider the corresponding neighborhood $V \times P_{\alpha \neq \alpha_1, \cdots, \alpha_n} \beta(X_\alpha)$ of x^0 in $P\beta(X_\alpha)$. It contains elements (x^1, y^1) and (x^2, y^2) of $X \times Y$ for which $a - \epsilon < f(x^1, y^1) - f(x^2, y^2)$ so that

$$
\begin{aligned}
a - \epsilon &< |f(x^1, y^1) - f(x^1, y^2)| + |f(x^1, y^2) - f(x^2, y^2)| \\
&= |f(x^1, y^1) - f(x^1, y^2)| + |\bar{f}(x^1, y^2) - \bar{f}(x^2, y^2)| < 2\epsilon,
\end{aligned}
$$

and $3\epsilon < a < 3\epsilon$, contradicting our assumption that f has no continuous extension to $P\beta(X_\alpha)$, and completing the proof.

4. Some particular nontrivial Stone-Čech compactifications. The final portion of our proof of Theorem 1 reveals a class of Stone-Čech compactifications of a nontrivial, although certainly a very special, sort.

THEOREM 2. *Let $\{X_\alpha\}$ be a set of uncountably many compact Hausdorff spaces, each having at least two points. For $b \in PX_\alpha$ let X^b be the subspace of the* *product consisting of all points $x \in PX_\alpha$ with $x_\alpha \neq b_\alpha$ for at most countably many α. Then PX_α is the Stone-Čech compactification of the proper subspace X^b.*

The cardinality restrictions are only required to yield X^b a proper subspace. Trivially X^b is dense, and countably compact, since any sequence of points of X^b has a cluster point (in the full product) which must, of course, lie in X^b. Just as in the proof of Theorem 1 we may assert that for $f \in C(X^b)$ and $\epsilon > 0$ there are indices $\alpha_1, \cdots, \alpha_n$, for which $x, y \in X^b$, $x_{\alpha_i} = y_{\alpha_i}$, $i = 1, \cdots, n$ imply $|f(x) - f(y)| < \epsilon$. For if not, we may proceed as in the previous proof (since we may alter countably many coordinates of a point in X^b and remain in X^b) to sequences $\{x^i\}$, $\{y^i\}$ in X^b and a sequence $\{F_i\}$ with just the properties which held there. But here $\{x^i\}$ has a cluster point in X^b, and any of its neighborhoods must contain x^i and y^i simultaneously for i sufficiently large, so that we obtain the same contradiction.

Again, if f does not have a continuous extension to PX_α, we have an x^0 for which $\lim \sup_{x \to x^0} f(x) - \lim \inf_{x \to x_0} f(x) = a > 0$. Taking $0 < 3\epsilon < a$ and $\alpha_1, \cdots, \alpha_n$ the corresponding set of indices then, since we may write $X^b = X \times Y$ with $X = P_{i=1}^n X_{\alpha_i}$, and X^b is pseudo-compact, we arrive at the same final contradiction.

Some particular examples are intriguing. Take, for example, each X_α a compact topological group, b the identity of the product group, so that X^b and $\beta(X^b)$ are topological groups. Indeed take the simplest case with each X_α the two element group $\{0, 1\}$, and let $\mathbf{0}$, $\mathbf{1}$ be respectively the identity and $(1, 1, \cdots)$. Extending the subgroup X^0 of the full product to a maximal subgroup Z not containing $\mathbf{1}$, we have Z and $\mathbf{1} + Z$ complementary, since, for $x \notin Z$, $\mathbf{1}$ lies in the subgroup generated by x and Z, or $\mathbf{1} = x + z$, $z \in Z$, and $x = \mathbf{1} + z \in \mathbf{1} + Z$. Thus in this case Z and $\beta Z \backslash Z$ are homeomorphic, and all of the spaces Z, $\beta(Z) \backslash Z$ and $\beta(Z)$ are homogeneous (cf. [7])(7).

5. **Products of pseudo-compact spaces.** It is not true that a product of pseudo-compact spaces must be pseudo-compact. Indeed Novák [6] has recently given an example of a pair of countably compact completely regular spaces whose product is not even pseudo-compact(8). In this section we shall

(7) Numerous companions to Rudin's example of a homogeneous X with $\beta(X) \backslash X$ not homogeneous can also be constructed via Theorem 2. Let K be a compact space, not totally disconnected, with an isolated point k_0 and cardinality $\aleph > \aleph_0$. Let Y be the product of $\aleph$ replicas of K, which, since $\aleph^2 = \aleph$, contains the nonvoid subspace $X = \{y : y_\alpha = k$ for $\aleph$ α's, for each $k \in K\}$. Then X is homogeneous since to take x^1 into x^2 only requires a permutation of the α's. But $x \in X$ implies $Y^x \subset X$ and $x \in Y \backslash X$ implies $Y^x \subset Y \backslash X$ so that $\beta(X) = Y$ and $\beta(Y \backslash X) = Y$. Thus any homeomorphism of $\beta(X) \backslash X = Y \backslash X$ with itself extends to one of $\beta(Y \backslash X) = Y$ with itself. Consequently $\beta(X) \backslash X$ cannot be homogeneous since $(k_0, k_0, \cdots)$ is obviously its own component in Y, and Y is not totally disconnected.

(8) Novák's example was designed as a noncountably compact product of a pair of countably compact spaces. However, it contains an open and closed discrete infinite subspace, hence is clearly not pseudo-compact. A similar example has been given by Terasaka [9]. The writer is indebted to Dr. D. O. Ellis for reference to Novak's result.

give some conditions which are sufficient to insure that a product be pseudo-compact, and yield some analogous facts for other forms of compactness.

The generality of our results is increased somewhat by utilizing the notion of a P-point introduced by Gillman and Henriksen [2]. Such a point has the property that every countable intersection of its neighborhoods is a neighborhood (not necessarily open).

THEOREM 3. *Let X and Y be pseudo-compact. If X is locally $\aleph_\gamma$ compact*(⁹), *and each non-P-point of Y has a base of neighborhoods of cardinality $\leqq\aleph_\gamma$, then $X\times Y$ is pseudo-compact.*

In particular this yields the fact that a product of two pseudo-compact spaces is pseudo-compact if one of the spaces is locally compact.

Proof. We shall show directly that $\beta(X\times Y)=\beta(X)\times\beta(Y)$, so that the conclusion will follow from Theorem 1 (unless either space is finite, in which case the conclusion is obvious). Let $f\in C(X\times Y)$. For $x_0\in X$, we have an $\aleph_\gamma$ compact neighborhood V, and, as a consequence, we can assert that $\{f(x,\cdot)\}_{x\in V}$ is equicontinuous at any non-P-point y_0 in Y. Suppose not, so that for some $\epsilon>0$, $F_W=\{x: x\in V, \sup_{y\in W}|f(x,y)-f(x,y_0)|\geqq\epsilon\}$ is nonvoid for each neighborhood W of y_0. Allowing W to range over a base of neighborhoods B, of cardinality $\leqq\aleph_\gamma$, $\{F_W\}$ forms a filter base on V which has to have an adherent point x_1 in V. On the other hand, (x_1, y_0) has a neighborhood $V_1\times W_1$, $W_1\in B$, on which f varies by at most $\epsilon/2$ so that V_1 cannot meet F_{W_1}, and x_1 cannot be adherent to $\{F_W\}$.

Since equicontinuity of any countable subset of $C(Y)$ is automatic at P-points, every countable subset of $\{f(x,\cdot)\}_{x\in V}$ is equicontinuous on Y. As we have noted, this implies that the full set is equicontinuous on Y, and by Ascoli's theorem it thus has compact closure in $C(Y)$. But this implies that the mapping $x\to f(x,\cdot)$ of V into $C(Y)$ is continuous at x_0; for if the filter $\mathfrak{F}$ on V converges to x_0, then $\lim_{\mathfrak{F}} f(x,y)=f(x_0,y)$ for $y\in Y$ and the image of the filter has at most the single adherent point $f(x_0,\cdot)$, hence must converge to $f(x_0,\cdot)$ by compactness. Since V is a neighborhood of x_0, the mapping $x\to f(x,\cdot)$ of X into $C(Y)$ is continuous at x_0, hence everywhere since x_0 was arbitrary, and, by Lemma 2, f has a continuous extension $\bar{f}$ to $X\times\beta(Y)$. Since $\beta(Y)$ is locally compact, the same argument (with $\beta(Y)$ taking the rôle of X) yields the desired extension of $\bar{f}$ to $\beta(X)\times\beta(Y)$, and the proof is complete.

Because of the basic asymmetry of the hypotheses on X and Y in Theorem 3, its applicability is somewhat enhanced, as is easily seen by examples, by the trivial fact that a space which is a finite union of pseudo-compact subspaces is pseudo-compact (or the equally trivial fact that having a dense pseudo-compact subspace makes a space pseudo-compact, e.g. if X is not locally $\aleph_\gamma$ compact but contains a dense pseudo-compact subspace which is). One particular instance of the use of a finite decomposition in showing a

(⁹) That is, every point of X has a neighborhood V such that every open covering of V, of cardinality $\leqq\aleph_\gamma$, has a finite subcovering.

product $X\times Y$ is pseudo-compact should be mentioned, since it is a consequence of Theorem 3; we may disregard any open subset (of either space) which has compact closure. Specifically, *if X and Y are pseudo-compact, and $W\subset Y$ is open and has compact closure, then $X\times W'$ is pseudo-compact if and only if $X\times Y$ is pseudo-compact.* For, as one easily sees from the topological conditions for pseudo-compactness, the closure of an open subset inherits this property. Thus if $X\times Y$ is pseudo-compact, $X\times W^{-\prime-}$ inherits the property; since $X\times$(boundary W) is pseudo-compact by Theorem 3, the same is true of $X\times W'$ as the union of these two sets. On the other hand, if $X\times W'$ is pseudo-compact the same is true of $X\times Y$ as the union of this set and $X\times W^-$, which is pseudo-compact by Theorem 3[10].

In showing an infinite product is pseudo-compact we need only consider countable partial products (cf. §2). This suggests use of the diagonal process which leads to the following, somewhat restricted, extensions of Theorem 3.

THEOREM 4. *Let $\{X_\alpha\}$ be a set of pseudo-compact spaces. Then PX_α is pseudo-compact in each of the following cases:*

(a) *all but one X_α is locally compact;*

(b) *for all but one α, each non-P-point of X_α is a G_δ;*

(c) *for every α, X_α is $\aleph_\gamma$ compact and each non-P-point of X_α has a base of neighborhoods of cardinality $\leqq\aleph_\gamma$.*

Proof. (a) Since we need only consider, at worst, the countable case, let $X_0, X_1, \cdots$ be a sequence of spaces of the type described with X_0 the exceptional space. We shall show that any sequence $\{U^n\}$ of nonvoid canonical open neighborhoods in PX_j has a cluster point.

Let $U^n=PU_j^n$, with $U_j^n=X_j$ for $j\geqq m_n$. Set $X_j^*=X_j$ for $j=0$ and all compact factors, and, for X_j noncompact let X_j^* be the one-point Alexandroff compactification of X_j, with x_j^* the (nonisolated) point at infinity. Then U_j^n remains open in X_j^*, and if we set $V_j^n=U_j^n$ for $j<m_n$, $V_j^n=X_j^*$ for $j\geqq m_n$, and $V^n=PV_j^n$, then $\{V^n\}$ forms a sequence of nonvoid open subsets of $PX_j^*=X_0\times P_{j\geq 1}X_j^*$, which is pseudo-compact by Tychonoff's theorem and Theorem 3. Moreover U^n is dense in V^n so that any cluster point of $\{V^n\}$ is also a cluster point of $\{U^n\}$. Thus if we can produce a cluster point of $\{V^n\}$ lying in the subspace PX_j of PX_j^* our proof will be complete.

Let X_{j_1} be the first noncompact factor with $j\geqq 1$. $\{V_{j_1}^n\}$ has a cluster point in X_{j_1}, so that some closed neighborhood W_{j_1} of $x_{j_1}^*$ does not meet some neighborhood of this cluster point, and, for infinitely many n, $V_{j_1}^n\cap W_{j_1}'\neq\emptyset$. Let $\{n_i^1\}$ be the infinite sequence of integers n for which this holds. Considering the next noncompact factor X_{j_2} and $\{V_{j_2}^{n_i^1}\}$ we can obtain in the same fashion a closed neighborhood W_{j_2} of $x_{j_2}^*$ and a subsequence $\{n_i^2\}$ of $\{n_i^1\}$ for which $V_{j_2}^{n_i^2}\cap W_{j_2}'\neq\emptyset$. We continue; replacing $\{V^n\}$ by the diagonal subsequence (or the final subsequence, if the process terminates) and setting

[10] Indeed we only need to have W^- locally compact to conclude $X\times Y$ is pseudo-compact.

$W_j=\emptyset$ for all other j we have $V_j^n\cap W_j'\neq\emptyset$ for all n and j with $n\geqq j$. Let $\tilde{V}_j^n=V_j^n\cap W_j'$ for $j\leqq n$ and $=X_j^*$ otherwise; setting $\tilde{V}^n=P\tilde{V}_j^n$, $\{\tilde{V}^n\}$ again forms a sequence of nonvoid open sets in PX_j^*, and as such has a cluster point x in PX_j^*. But x must also be a cluster point of $\{V^n\}$; for if $V=PV_j$ is a neighborhood of x with $V_j=X_j^*$ for $j\geqq k$, then $\tilde{V}^n\cap V\neq\emptyset$ implies $V^n\cap V\neq\emptyset$ if n exceeds k, since $\tilde{V}_j^n\subset V_j^n$ for $j\leqq n$. Moreover x must lie in the subspace PX_j of PX_j^*. For if $x_k=x_k^*$ for some k then the neighborhood $\{x': x_k'\in W_k\}$ of x in PX_j^* meets infinitely many $\tilde{V}^n$ despite the fact that $W_k\cap\tilde{V}_k^n$ is void as soon as n exceeds k. Thus x is the required cluster point in PX_j of $\{U^n\}$.

(b) We note first that if a point y is a G_δ in a pseudo-compact space, then it has a countable base of neighborhoods. For by regularity: (i) $\{y\}=\bigcap W_j$ with W_j open and $\overline{W}_{j+1}\subset W_j$; (ii) in order to show $\{W_j\}$ is a base of neighborhoods it suffices to show each closed neighborhood W contains some W_j. But the collection $\{W_j\cap W'\}$ is locally finite so that only finitely many elements of the collection are nonvoid; since $\bigcap(W_j\cap W')=\emptyset$ and the sets decrease some element of our sequence $\{W_j\cap W'\}$ must be void.

Again we need only consider $X_0, X_1, \cdots$ with X_0 the exceptional space. Let $\{V^n\}$ be a sequence of canonical open neighborhoods in PX_j with $U^n=PV_j^n$. Consider $\{V_1^n\}$. If possible, we select an x_1 in X_1 which lies in the closure of infinitely many of these sets, say those corresponding to $n_1^1, n_2^1, \cdots$; if not, the corresponding collection of open sets in X_1 cannot be locally finite so that for some x_1 each neighborhood of x_1 meets infinitely many of the sets while x_1 lies in the closure of only finitely many. Clearly x_1 cannot be a P-point, and thus has a countable base of neighborhoods; by the diagonal process we may then select a subsequence $\{n_i^1\}$ of the integers for which each neighborhood of x_1 meets almost all $V_1^{n_i^1}$. Hence in either case we have an x_1 in X_1 and a subsequence $\{n_i^1\}$ with this property.

Considering $\{V_2^{n_i^1}\}$, we then select in the same fashion an x_2 in X_2 and a subsequence $\{n_i^2\}$ of $\{n_i^1\}$ for which each neighborhood of x_2 meets almost all $V_2^{n_i^2}$. Continuing, we obtain a point $(x_1, x_2, \cdots)$ of $P_{j\geq 1}X_j$ and a subsequence $\{V^{n_i^i}\}$ of our original sequence for which each neighborhood of $(x_1, x_2, \cdots)$ in $P_{j\geq 1}X_j$ meets almost all of the sets $\tilde{V}^{n_i^i}=P_{j\geq 1}V_j^{n_i^i}$. Hence taking x_0 to be any cluster point in X_0 of $\{V_0^{n_i^i}\}$ yields the desired cluster point $(x_0, x_1, x_2, \cdots)$ of $\{V^n\}$.

(c) Here any countable product is actually countably compact. Let $\{x^n\}$ be any sequence of points in PX_j; for notational convenience let us write the sequence of integers as $\{n_i^0\}$. Let j_1 be the first j, if there are any, for which sequence of jth coordinates $\{x_j^n\}$ has a cluster point x_j in X_j which is also a P-point. Then evidently x_{j_1} occurs infinitely often in $\{x_{j_1}^n\}$, and we have a subsequence $\{n_i^1\}$ of the integers for which $x_{j_1}^{n_i^1}=x_{j_1}$. If possible we select a least j_2 for which $\{x_{j_2}^{n_i^1}\}$ has a P-cluster point x_{j_2}, and thus obtain a subsequence $\{n_i^2\}$ of $\{n_i^1\}$ for which $x_{j_2}^{n_i^2}=x_{j_2}$, and continue. If the process terminates after k $(=0, 1, \cdots)$ stages then $\{x^{n_i^k}\}$ has the corresponding j_1st, $\cdots$, j_kth coordinates constant while, for each other j, $\{x_j^{n_i^k}\}$ can have only non-P-points as cluster points. If the process does not terminate, then for each $k \geqq 1$, $x_{j_k}^{n_i^i} = x_{j_k}$ for almost all i while, for every other j, $\{x_j^{n_i^i}\}$ has only non-P-points as cluster points. Thus, in either case we may replace our sequence $\{x^n\}$ by a subsequence $\{y^n\}$ with the property that for each i, $y_{j_i}^n = x_{j_i}$ for almost all n while, for every other j, $\{y_j^n\}$ has only non-P-points as cluster points.

Now the points $x_{j_1}, x_{j_2}, \cdots$ are natural candidates for certain of the coordinates of a cluster point of $\{y^n\}$; we may select the remaining coordinates as follows. Let $F_n = \{y^m\}_{m \geq n}$, $\mathfrak{F}_0$ be the filter base $\{F_n\}$, and p_j the projection of our product into X_j. Let j_1^* be the first j, if there are any, not included in $\{j_i\}$, and select an adherent point $x_{j_1^*}$ in $X_{j_1^*}$ of $p_{j_1^*}\mathfrak{F}_0$. Since $x_{j_1^*}$ is then a cluster point of $\{y_{j_1^*}^n\}$, we know it is not a P-point, hence has a base of neighborhoods, $B_{j_1^*}$, of cardinality $\leqq \aleph_\gamma$. Thus the collection $\mathfrak{F}_1 = \{F \cap V: F \in \mathfrak{F}_0,\ V = V_{j_1^*} \times P_{j \neq j_1^*} X_j,\ V_{j_1^*} \in B_{j_1^*}\}$ forms a filter base of cardinality $\leqq \aleph_\gamma \cdot \aleph_\gamma = \aleph_\gamma$, so that $p_{j_2^*}\mathfrak{F}_1$ has an adherent point $x_{j_2^*}$ in $X_{j_2^*}$. Since $x_{j_2^*}$ is also adherent to the less fine filter base $p_{j_2^*}\mathfrak{F}_0$, hence is a cluster point of $\{y_{j_2^*}^n\}$, $x_{j_2^*}$ cannot be a P-point. Again we obtain $B_{j_2^*}$ and form the filter base $\mathfrak{F}_2 = \{F \cap V: F \in \mathfrak{F}_1,\ V = V_{j_2^*} \times P_{j \neq j_2^*} X_j,\ V_{j_2^*} \in B_{j_2^*}\}$ of cardinality $\leqq \aleph_\gamma$; clearly we may continue in this fashion to obtain filter bases $\mathfrak{F}_i$ and non-P-points $x_{j_i^*}$ as long as any indices j are not included among the j_i and j_i^*. Whether this process terminates or not (or even starts) we obtain a point $(x_1, x_2, \cdots)$ in PX_j which must be a cluster point of $\{y^n\}$. For any canonical neighborhood of this point contains one given by constraints of the form $y_{j_i} \in V_{j_i}$, $i = 1, \cdots, k$, $y_{j_i^*} \in V_{j_i^*}$, $i = 1, \cdots, m$, with $V_{j_i^*} \in B_{j_i^*}$. But almost all y^n satisfy each of the constraints of the first set, hence almost all y^n satisfy all the constraints of this set. On the other hand, infinitely many y^n satisfy all constraints of the second set since, for each $F \in \mathfrak{F}_0$, $F \cap \{y: y_{j_i^*} \in V_{j_i^*},\ i \leqq m\}$ is an element of the filter base $\mathfrak{F}_{j_m^*}$, hence nonvoid. Thus we have obtained our cluster point and the proof is complete.

REMARKS. Simple modifications of the proof of (a) yield the following facts: (1) *if PY_α is pseudo-compact and X_α is an open pseudo-compact subspace of Y_α then PX_α is pseudo-compact*; (2) *a product of no more than $\aleph_\gamma$ spaces, each $\aleph_\gamma$ compact and all but one locally compact, is $\aleph_\gamma$ compact.* (1) contains (a) and follows by allowing X_α' to assume the role of x_α^*; for (2) one requires the simple result that the product of an $\aleph_\gamma$ compact space and a compact space is $\aleph_\gamma$ compact[11]. Indeed take the indices to be the ordinals

(11) Let X be compact, Y $\aleph_\gamma$ compact, and $\mathfrak{F}$ a filter base of closed subsets of $X \times Y$, of cardinality $\leqq \aleph_\gamma$. Assuming no point is adherent to $\mathfrak{F}$ we have, for each $y \in Y$ an $F \in \mathfrak{F}$ for which $F \cap (X \times \{y\}) = \emptyset$, since $X \times \{y\}$ is compact. By a standard compactness argument this continues to hold in a neighborhood of y. Thus $\{U_F\}_{F \in \mathfrak{F}}$, with $U_F = \{y: F \cap (X \times \{y\}) = \emptyset\}$ forms an open covering of Y of cardinality $\leqq \aleph_\gamma$; for a finite subcovering $U_{F_1}, \cdots, U_{F_n}$, we obviously have $F_1 \cap F_2 \cap \cdots \cap F_n = \emptyset$, the desired contradiction.

$\leqq\omega_\gamma$, $X_1=X_1^*$ to be the exceptional space, and X_α^* to be again the Alexandroff compactification of X_α for $\alpha>1$, so that PX_α^* is $\aleph_\gamma$ compact. Then given a filter base $\mathfrak{F}_1$ on PX_α of cardinality $\leqq\aleph_\gamma$ one chooses similarly restricted filter bases $\mathfrak{F}_\alpha$ and neighborhoods W_α of x_α^* (taking $W_1=\emptyset$) inductively by requiring that W_β avoid some neighborhood of some point in X_β adherent to $p_\beta(\bigcup_{\alpha<\beta}\mathfrak{F}_\alpha)$ (which is possible since $\bigcup_{\alpha<\beta}\mathfrak{F}_\alpha$ has cardinality $\leqq\aleph_\gamma\cdot\aleph_\gamma=\aleph_\gamma$) and selecting $\mathfrak{F}_\beta$ as the filter base generated by $\bigcup_{\alpha<\beta}\mathfrak{F}_\alpha$ and the set $\{x: x_\beta\notin W_\beta\}$. Then $\mathfrak{F}_{\omega_\gamma}$ has an adherent point in PX_α^* which obviously must lie in PX_α, and is adherent to $\mathfrak{F}_1$.

Similarly the proof of (b) shows that (3) *any countable product $P_{j=0}^\infty X_j$ of countably compact spaces for which, for $j>0$, we have each non-P-point of X_j a G_δ, is countably compact.* The argument here is simpler; starting with a sequence $\{x^n\}$ of elements of PX_j we choose x_1 as any cluster point of $\{x_1^n\}$; if x_1 is a P-point it must occur infinitely often in $\{x_1^n\}$. Otherwise x_1 has a countable base of neighborhoods, but in either case we may clearly begin an application of the diagonal process as before. Without reference to P-points (3) has an analogue for $\gamma>0$, whose proofs follows the final portion of that of (c): (4) *a product of no more than $\aleph_\gamma$ spaces, each $\aleph_\gamma$ compact and all but one having at each point a base of neighborhoods of cardinality $\leqq\aleph_\gamma$, is $\aleph_\gamma$ compact.* Again we take the indices to be the ordinals $\leqq\omega_\gamma$ and let X_1 be the exceptional space; given a filter base $\mathfrak{F}_1$ of cardinality $\leqq\aleph_\gamma$ one chooses, for $\alpha>1$, similarly restricted filter bases $\mathfrak{F}_\alpha$ and elements x_α of X_α inductively by selecting x_β adherent to $p_\beta(\bigcup_{\alpha<\beta}\mathfrak{F}_\alpha)$ and taking $\mathfrak{F}_\beta$ as the filter base formed by the sets $F\cap\{x': x'_\beta\in V_\beta\}$ where $F\in\bigcup_{\alpha<\beta}\mathfrak{F}_\alpha$ and V_β is taken from our base of neighborhoods at x_β. Choosing x_1 adherent to $p_1\mathfrak{F}_{\omega_\gamma}$, clearly $x=(x_1, x_2, \cdots)$ is adherent to $\mathfrak{F}_{\omega_\gamma}$, hence $\mathfrak{F}_1$.

Finally it should be noted that the argument used in (a) can be applied in other situations where a substitute for Tychonoff's theorem is available. For example (c) can be strengthened by replacing "X_α is $\aleph_\gamma$ compact" by "X_α is the union of $\aleph_\gamma$ open sets each having $\aleph_\gamma$ compact closure." For then each X_α is locally $\aleph_\gamma$ compact so that we can introduce, for each non $\aleph_\gamma$-compact factor X_α, its one-point $\aleph_\gamma$-compactification X_α^* in just the way one constructs the Alexandroff compactification (with "$\aleph_\gamma$-compact" replacing "compact"). Since each point at infinity has a base of neighborhoods of cardinality $\leqq\aleph_\gamma$, (c) implies PX_α^* is pseudo-compact. But now the remainder of the argument given for (a) applies.

6. A related question. Even if it fails to be true that every continuous function on the subspace PX_α of $P\beta(X_\alpha)$ has a continuous extension to $P\beta(X_\alpha)$, it still might be the case that this holds for some other dense subspace homeomorphic to PX_α, so that the Stone-Čech compactification of PX_α might be obtained from the compact space $P\beta(X_\alpha)$ via some other imbedding of PX_α therein. Some light is shed on this question by Theorem 4(a).

Call a point z inessential to the space Z if every $f\in C(Z\backslash\{z\})$ has a continuous extension to Z. Evidently any ideal point z of $\beta(Z)$ is inessential to $\beta(Z)$ and $\beta(\beta(Z)\setminus\{z\})=\beta(Z)$. By a theorem of Čech [1] which asserts that a nonvoid closed G_δ in the set of ideal points of $\beta(X)$ has high cardinality, we may conclude that $\beta(Z)\setminus\{z\}$ is then pseudo-compact; consequently any point x of $P\beta(X_\alpha)\setminus PX_\alpha$ is inessential to $P\beta(X_\alpha)$(12). For letting $Z_\alpha=\beta X_\alpha$ if $x_\alpha\in X_\alpha$, $Z_\alpha=\beta X_\alpha\setminus\{x_\alpha\}$ if $x_\alpha\notin X_\alpha$, we have PZ_α pseudo-compact by Theorem 4(a); thus $f\in C(P\beta(X_\alpha)\setminus\{x\})$, being continuous on PZ_α, has a continuous extension to $P\beta(Z_\alpha)=P\beta(X_\alpha)$ by Theorem 1, yielding the assertion.

Moreover any z_0 inessential to $\beta(Z)$ is inessential to Z. For suppose $f\in C(Z\setminus\{z_0\})$ has no continuous extension to Z. We can of course take $f\geqq 1$. Let $\bar{f}$ be the extension of f to Z obtained by setting $\bar{f}(z_0)=1$, and let $G=\{g: g\in C(\beta(Z)),\ 0\leqq g\leqq 1,\ g \text{ vanishes near } z_0\}$ so that $\bar{f}g\in C(Z)$ for $g\in G$. Finally, let $F(z)=\sup_G(\bar{f}g)^*(z)$ for $z\in\beta(Z)\setminus\{z_0\}$. For $z_1\in\beta(Z)\setminus\{z_0\}$ we have disjoint neighborhoods U of z_0 and V of z_1 in $\beta(Z)$, and for any $g_1\in G$ which is 1 outside U (such an element of G exists by Urysohn's lemma), we have $F(z)=(\bar{f}g_1)^*(z)$ for $z\in V$. Indeed if this is not the case, for some g_2 in G and z in V we have $(\bar{f}g_2)^*(z)>(\bar{f}g_1)^*(z)$; but some g in G satisfies $gg_i=g_i$, $i=1, 2$, again by Urysohn's lemma, so that $(\bar{f}g)^*(z)g_2(z)=(\bar{f}gg_2)^*(z)=(\bar{f}g_2)^*(z)>(\bar{f}g_1)^*(z)=(\bar{f}g)^*(z)g_1(z)$, and $g_2(z)>g_1(z)=1$, a contradiction. Thus we see that $F\in C(\beta(Z)\setminus\{z_0\})$; but F has no continuous extension to $\beta(Z)$ since clearly $F|(Z\setminus\{z_0\})=f$.

Suppose then that Z is a dense subspace of $P\beta(X_\alpha)$ with the property that all continuous functions on Z extend continuously to $P\beta(X_\alpha)$, i.e., with $\beta(Z)=P\beta(X_\alpha)$. Each point x of $Z\setminus PX_\alpha$ is inessential to $\beta(Z)=P\beta(X_\alpha)$ as an element of $P\beta(X_\alpha)\setminus PX_\alpha$, and thus is inessential to Z. If we now assume that the inessential points of Z form a discrete subspace, so does $Z\setminus PX_\alpha$. Since it is now simple to extend $f\in C(PX_\alpha)$ continuously to all of Z(13), we may further extend it to $\beta(Z)=P\beta(X_\alpha)$, so that $\beta(PX_\alpha)=P\beta(X_\alpha)$. Thus if the inessential points of PX_α form a discrete subspace, the existence of any imbedding of this space as a dense subspace of $P\beta(X_\alpha)$ which makes all elements of $C(PX_\alpha)$ continuously extendable yields the same property for the natural imbedding.

Added in proof (January 26, 1959). The reader may have noted that the example X^0 of §4 is a noncompact topological group on which all bounded continuous functions are almost periodic (being, essentially, continuous functions on the compact group $\beta(X^0)$). It is easily seen that any topological

---

(12) Of course every point of $P\beta(X_\alpha)$ is inessential to this space if $\{\beta(X_\alpha)\}$ satisfies the cardinality requirements of Theorem 2.

(13) To assign the appropriate value for this extension at $z\in Z\setminus PX_\alpha$ choose a $g\in C(P\beta(X_\alpha))$ which is 1 near z and vanishes on the remainder R of $Z\setminus PX_\alpha$, and extend fg to $Z\setminus\{z\}$ by assigning the value 0 to elements of R. This extension then lies in $C(Z\setminus\{z\})$ since $\{z': |g(z')|<\epsilon\}$ is a neighborhood of R, and thus has a further continuous extension to Z since z is inessential; the value at z of this last function obviously meets our requirements.

group G for which $C(G) = \mathrm{A.P.}(G)$ (=real almost periodic functions) must be pseudo-compact; while the converse question must be left open we can note that if $G \times G$ is pseudo-compact then $C(G) = \mathrm{A.P.}(G)$, and $C(G \times G) = \mathrm{A.P.}(G \times G)$. For $f \in C(G)$ implies $F: (x, y) \to f(xy)$ is an element of $C(G \times G)$ so that, by Lemma 1, the left translates of f are equicontinuous, and by Ascoli's theorem f is almost periodic. Since we may view $\beta(G)$ and the Bohr compactification of G as completions of G under the natural uniform structures provided by $C(G)$ and $\mathrm{A.P.}(G)$ respectively, these coincide as spaces and we may regard $\beta(G)$ as a compact group of which G is a (albeit not closed, topological) subgroup. Since we may thus view $\beta(G \times G) = \beta(G) \times \beta(G)$ as a compact group of which $G \times G$ is a subgroup, every $f \in C(G \times G)$ is also almost periodic. Consequently $C(G \times G) = \mathrm{A.P.}(G \times G)$ iff $G \times G$ is pseudo-compact.

References

1. E. Čech, *On bicompact spaces*, Ann. of Math. (2) vol. 38 (1937) pp. 823–844.

2. L. Gillman and M. Henriksen, *Concerning rings of continuous functions*, Trans. Amer. Math. Soc. vol. 77 (1954) pp. 340–362.

3. I. Glicksberg, *The representation of functionals by integrals*, Duke Math. J. vol. 19 (1952) pp. 253–261.

4. M. Henriksen and J. R. Isbell, *On the Stone-Čech compactification of a product of two spaces*, Bull. Amer. Math. Soc. Abstract 63-2-332.

5. E. Hewitt, *Rings of real valued continuous functions*, I, Trans. Amer. Math. Soc. vol. 64 (1948) pp. 45–99.

6. J. Novák, *On the cartesian product of two compact spaces*, Fund. Math. vol. 40 (1953) pp. 106–112.

7. W. Rudin, *Homogeneity problems in the theory of Čech compactifications*, Duke Math. J. vol. 23 (1956) pp. 409–419.

8. M. H. Stone, *Applications of the theory of Boolean rings to general topology*, Trans. Amer. Math. Soc. vol. 41 (1937) pp. 375–481.

9. H. Terasaka, *On the cartesian product of compact spaces*, Osaka Math. J. vol. 4 (1) (1952) pp. 11–15.

10. A. Tychonoff, *Ueber die topologische Erweiterungen von Räumen*, Math. Ann. vol. 102 (1930) pp. 544–561.

University of Notre Dame,
Notre Dame, Ind.

HOMOGENEITY PROBLEMS IN THE THEORY OF ČECH COMPACTIFICATIONS

By Walter Rudin

Introduction. If X is a completely regular topological space, there exists a space βX, the so-called Čech compactification of X, which is characterized by the following three properties: βX is a compact (bicompact, in the older terminology) Hausdorff space, X is a dense subset of βX, and every bounded continuous real-valued function on X can be extended to a continuous function on βX [1; 831].

A topological space X is homogeneous if to every pair of points p and q of X there exists at least one homeomorphism of X which carries p to q. In the seminar on function spaces which was part of the Institute on Set Theoretic Topology held at Madison during the summer of 1955, the question was raised whether the homogeneity of X implies the homogeneity of the space $X^* = \beta X - X$. In the present paper it is shown that the answer to this question is negative (under the assumption of the continuum hypothesis) even if X is a countable discrete space (Theorem 4.4). This in turn implies that X^* fails to be homogeneous whenever X is locally compact but not sequentially compact (Theorem 4.5).

We choose the set N of all positive integers as our countable discrete space, and construct βN as the set of all ultrafilters on N. This purely set-theoretic construction of βN leads to a fairly detailed description of that space, and in particular of the space $N^* = \beta N - N$. The latter is the main object of study and is found to be a compact Hausdorff space with 2^c points in which there are exactly c open-closed subsets (the letter c stands for the cardinal number of the continuum); these sets form a basis for the topology of N^*, and any two non-empty open-closed subsets are homeomorphic; the intersection of any countable family of open sets is either empty or contains a non-empty open set; and N^* is almost homogeneous, in the sense that if any point and any open set of N^* are given, there is a homeomorphism of N^* which carries the given point into the given open set. These properties do not depend on the continuum hypothesis.

If the continuum hypothesis is true, then N^* contains P-points, using the terminology of Gillman and Henriksen [2; 343] (see Definition 4.1), and it is this fact which leads almost immediately to the conclusion that N^* is not homogeneous.

The space βN has precisely c homeomorphisms; these are in natural one-to-one correspondence with the permutations of N. On the other hand, the con-

Received January 16, 1956

tinuum hypothesis implies that any of the 2^c P-points of N^* can be mapped into any other by a homeomorphism of N^*. It follows that N^* has precisely 2^c homeomorphisms (Theorem 4.7).

I. The ultrafilters on N. Some of the results of this section are not needed for the main purpose of the present paper but are included for their own sake. One such item is paragraph 1.7.

1.1. DEFINITION. A family Ω of non-empty subsets of N is an ultrafilter on N if Ω has the following properties:
(A) If $E_1 \in \Omega$ and $E_2 \in \Omega$, then $E_1 \cap E_2 \in \Omega$.
(B) Ω is maximal with respect to property (A).

If $n \in N$ and if Ω consists of all subsets of N which contain n, then Ω is evidently an ultrafilter, and we denote it by Ω^n; these ultrafilters are *fixed*. All other ultrafilters on N are *free*.

A collection $\mathfrak{F}$ of sets is said to have the finite intersection property if every finite subcollection of $\mathfrak{F}$ has non-empty intersection.

1.2. We now list some immediate properties of ultrafilters on N which will be used repeatedly, usually without explicit reference; the letter Ω will always stand for an ultrafilter on N.

(i) By Zorn's lemma, every family of subsets of N with the finite intersection property can be enlarged to an ultrafilter.

(ii) If $E_1 \in \Omega$ and $E_1 \subset E_2$, then $E_2 \in \Omega$.

(iii) For any Ω and any $E \subset N$, either $E \in \Omega$ or $N - E \in \Omega$. To prove this, note that if E intersects every member of Ω, then $E \in \Omega$. Thus if $E \notin \Omega$, there is a set $F \in \Omega$ such that $E \cap F = 0$, hence $F \subset N - E$, and $N - E \in \Omega$, by (ii).

(iv) If $\Omega_1 \neq \Omega_2$, there is a set $E \subset N$ such that $E \in \Omega_1$ and $N - E \in \Omega_2$.

(v) If Ω is free, then Ω contains no finite subset of N.

1.3. THEOREM. *There are precisely* 2^c *ultrafilters on* N.

Since the points of βN will be the ultrafilters on N, and since a more general result of Pospíšil [3] implies that βN has 2^c points, this theorem is known. The following elementary proof may nevertheless be of interest. We first need a lemma:

LEMMA. *There exists a family* $\mathfrak{F}$ *of subsets of* N, *which has* c *members, such that*

$$F_1 \cap \cdots \cap F_n \cap (N - F_{n+1}) \cap \cdots \cap (N - F_m) \neq 0$$

for every finite collection of distinct sets $F_i \in \mathfrak{F}$ $(i = 1, \cdots, m)$.

Proof. Let N be the union of disjoint sets A_p $(p = 1, 2, 3, \cdots)$, consisting of 2^{2^p} elements. For fixed p, let $q = 2^p$, and label the 2^q elements of A_p by ordered q-tuples $(x_1, \cdots, x_q)$, where $x_i = 0$ or 1; let E_i be the subset of A_p which consists of those q-tuples which have $x_i = 0$ $(i = 1, \cdots, q)$. Denote these 2^p sets E_i by $E(t_1, \cdots, t_p)$, where $t_k = 0$ or 1.

For each sequence $t_1, t_2, t_3, \cdots$ with $t_k = 0$ or 1, form the set

$$E(t_1, t_2, t_3, \cdots) = E(t_1) \cup E(t_1, t_2) \cup E(t_1, t_2, t_3) \cup \cdots,$$

and let $\mathfrak{F}$ be the family of all sets so obtained; $\mathfrak{F}$ evidently has c members.

If now $F_1, \cdots, F_n, F_{n+1}, \cdots, F_m$ are distinct members of $\mathfrak{F}$, there is an integer p such that no two of the corresponding sequences have the same initial p terms $t_1, \cdots, t_p$. This means that the sets $G_i = F_i \cap A_p$ $(i = 1, \cdots, m)$ are distinct, and it is easily verified that

$$G_1 \cap \cdots \cap G_n \cap (A_p - G_{n+1}) \cap \cdots \cap (A_p - G_m) \neq 0.$$

This proves the lemma.

Proof of the theorem. Since N has c subsets, there are at most 2^c ultrafilters on N. On the other hand, we can use the family $\mathfrak{F}$ of the lemma to construct 2^c collections $\mathfrak{G}$ of subsets of N, by considering each set $F \in \mathfrak{F}$, and admitting either F or $N - F$ into $\mathfrak{G}$. By the lemma, each $\mathfrak{G}$ so obtained has the finite intersection property; enlarging each $\mathfrak{G}$ to an ultrafilter, we obtain 2^c distinct ultrafilters.

1.4. Every Ω is a collection of subsets of N, and as such it can be regarded as a partially ordered set, with the partial order given by set inclusion. Let us say that two ultrafilters are of the same type if they are isomorphic as partially ordered sets. One may ask whether any two free ultrafilters on N are of the same type. The following theorem (pointed out to me by Hewitt Kenyon) shows that the answer is negative.

If π is a permutation of N (i.e., a one-to-one mapping of N onto N), let $\pi(\Omega)$ be the ultrafilter which contains the sets $\pi(E)$ for all $E \in \Omega$.

1.5. THEOREM. *If two free ultrafilters Ω_1 and Ω_2 are of the same type, then there is a permutation π of N such that $\Omega_2 = \pi(\Omega_1)$. There are 2^c types of free ultrafilters on N; each type contains c ultrafilters.*

Proof. If Ω_1 and Ω_2 are of the same type, there is a one-to-one mapping f of Ω_1 onto Ω_2 such that $A \subset B$ implies $f(A) \subset f(B)$. In particular, $f(N) = N$. Let H_n consist of all members of N, except n. The sets H_n are maximal proper subsets of N and are therefore permuted by f. Define π so that $\pi(n) = m$ if $f(H_n) = H_m$.

Choose $E \in \Omega_1$, let $F = N - E$. Then $E = \cap H_n$ $(n \in F)$. Let $E' = \cap f(H_n)$ $(n \in F)$. Since $f(E) \subset f(H_n)$, $f(E) \subset E'$. Similarly, $f^{-1}(E') \subset E$. Hence $f(E) = E'$. But $N - E'$ consists of the integers $\pi(n)$ $(n \in F)$; i.e., $N - f(E) = \pi(N - E)$. Taking complements, we obtain $f(E) = \pi(E)$, and the first part of the theorem follows.

Since N has precisely c permutations, the above shows that no type can contain more than c free ultrafilters, so that there are 2^c types.

To complete the theorem, choose a free Ω and a set $A \in \Omega$ whose complement is infinite, and let $\mathfrak{F}$ be a family of c infinite subsets of N such that any two members of $\mathfrak{F}$ have finite (possibly empty) intersection; such a family will be constructed below. For each $E \varepsilon \mathfrak{F}$ there is a permutation of N which carries Ω to an ultrafilter which contains E. Distinct members of $\mathfrak{F}$ give rise to distinct ultrafilters in this way, so that Ω belongs to the same type as c other ultrafilters.

To construct a family $\mathfrak{F}$ with the above properties, simply associate with each sequence $x_1, x_2, x_3, \cdots$ ($x_i = 0$ or 1) the set $E(x_1, x_2, x_3, \cdots)$ which consists of the integers

$$2^{n+1} + 2^n x_n + 2^{n-1} x_{n-1} + \cdots + 2x_1 \qquad (n = 1, 2, 3, \cdots)$$

and let $\mathfrak{F}$ be the family of all sets so obtained.

1.6. THEOREM. *If* $\Omega_1 \neq \Omega_2$, *there is a permutation* π *of* N *such that* $\pi(\Omega_1) = \Omega_1$, $\pi(\Omega_2) \neq \Omega_2$.

Proof. If $\Omega_1 \neq \Omega_2$, there exists a set $E_1 \subset N$ with infinite complement such that $E_1 \varepsilon \Omega_1$, $N - E_1 \varepsilon \Omega_2$. Choose $\Omega_3 \neq \Omega_2$ such that $N - E_1 \varepsilon \Omega_3$. There exist infinite sets E_2 and E_3 such that $E_2 \varepsilon \Omega_2$, $E_3 \varepsilon \Omega_3$, $N - E_1 = E_2 \cup E_3$, $E_2 \cap E_3 = 0$. Choose for π a permutation of N such that $\pi(n) = n$ for $n \varepsilon E_1$, $\pi(n) \varepsilon E_3$ if $n \varepsilon E_2$, $\pi(n) \varepsilon E_2$ if $n \varepsilon E_3$. This π has the desired properties.

1.7. The preceding theorem has an interesting corollary. Let Γ be the group of all permutations of N, and consider the subgroup of Γ which consists of those permutations which leave a given Ω fixed. By Theorem 1.6 different Ω's give rise to different subgroups, and Theorem 1.3 shows that Γ *has* 2^c *subgroups.*

It is known [4] that only two of these subgroups are proper normal subgroups of Γ, namely, the group of all finite permutations, and that of all even finite permutations.

II. The space βN.

2.1. Let S be the set of all ultrafilters on N. For every set $E \subset N$, let $V(E)$ be the set of all $\Omega \varepsilon S$ such that $E \varepsilon \Omega$. Define a topology on S by declaring each of the sets $V(E)$ to be open. The following relations will be useful:

(i) $$V(E_1 \cap E_2) = V(E_1) \cap V(E_2)$$

(ii) $$V(N - E) = S - V(E)$$

(iii) $$V(E_1 \cup E_2) = V(E_1) \cup V(E_2).$$

The first two of these are obvious, and (iii) follows from them.

2.2. THEOREM. *S is (homeomorphic to) βN.*

Proof. If $\Omega_1 \neq \Omega_2$, there exists $E \subset N$ such that $E \varepsilon \Omega_1$ and $N - E \varepsilon \Omega_2$; thus Ω_1 and Ω_2 have the disjoint neighborhoods $V(E)$ and $V(N - E)$, and S is a Hausdorff space.

Suppose $\mathcal{G}$ is an open covering of S. For every $\Omega \varepsilon S$ there exists $E_\Omega \subset N$

and $G_\Omega \varepsilon \mathcal{G}$ such that

$$\Omega \varepsilon V(E_\Omega) \subset G_\Omega .$$

If no finite collection of the sets $V(E_\Omega)$ covers S, then the complements $S - V(E_\Omega) = V(N - E_\Omega)$ have the finite intersection property. Hence the sets $N - E_\Omega$ have the finite intersection property, and there is an $\Omega_0 \varepsilon S$ such that $N - E_\Omega \varepsilon \Omega_0$ for all Ω, in particular for $\Omega = \Omega_0$; thus $\Omega_0 \varepsilon S - V(E_{\Omega_0})$. This contradicts the choice of $V(E_{\Omega_0})$. It follows that S is compact.

The mapping $n \to \Omega^n$ is a homeomorphism of N onto a dense subset of S. This is clear if we observe first that every fixed ultrafilter is an isolated point of S, and secondly that every $V(E)$ contains fixed ultrafilters.

From now on we identify n and Ω^n and regard N as a subset of S.

To complete the proof, let f be a bounded real-valued function on N. For any $\Omega \varepsilon S$ the sets $f(E)$ $(E \varepsilon \Omega)$ have the finite intersection property. Since their closures $\overline{f(E)}$ are compact, the set

$$K_\Omega = \bigcap_{E \varepsilon \Omega} \overline{f(E)}$$

is not empty. Define $f(\Omega)$ such that $f(\Omega) \varepsilon K_\Omega$.

This extended function f is continuous on S: Choose $\Omega_0 \varepsilon S$ and $\epsilon > 0$. Let E_0 be the set of all $n \varepsilon N$ such that $| f(n) - f(\Omega_0) | < \epsilon$. Since $f(E)$ has points in the segment $(f(\Omega_0) - \epsilon, f(\Omega_0) + \epsilon)$ for every $E \varepsilon \Omega_0$, we see that $E_0 \cap E \neq 0$ for every $E \varepsilon \Omega_0$, so that $E_0 \varepsilon \Omega_0$. In other words, $\Omega_0 \varepsilon V(E_0)$. It follows that $| f(\Omega) - f(\Omega_0) | \leq \epsilon$ for every $\Omega \varepsilon V(E_0)$.

Thus S has the three properties which characterize βN, and the theorem follows. The space S will from now on be called βN.

We note, incidentally, that the function f used above has only one continuous extension from N to βN, since N is dense in βN. Hence each of the sets K_Ω actually consists of only one point.

2.3. Let f be a bounded real function on N. The above construction shows that the extension of f to βN has the following properties:

(i) If Ω is free, then $f(\Omega)$ is a subsequential limit of $\{f(n)\}$.
(ii) Conversely, if x is a subsequential limit of $\{f(n)\}$, then there is a free Ω such that $f(\Omega) = x$.

2.4. We now turn our attention to the homeomorphisms of βN. Since the points of N are the only isolated points of βN, every homeomorphism of βN induces a permutation of N. On the other hand, if π is a permutation of N, and if $\pi(\Omega)$ is defined as in 1.4, then it is easily seen that this extension of π to βN is a homeomorphism of βN, and that it is the only homeomorphism of βN which coincides with π on N, since N is dense in βN.

It follows that there is a natural isomorphism between the group of all homeomorphisms of βN and the group of all permutations of N. Also, Theorem 1.5 implies that there is a homeomorphism of βN which carries Ω_1 to Ω_2 if and only if Ω_1 and Ω_2 are isomorphic as partially ordered sets.

III. The space N^*.

3.1. We recall that $N^* = \beta N - N$. If $E \subset N$, define

(i) $$W(E) = N^* \cap V(E),$$

with $V(E)$ as defined in 2.1. It is easily verified that $W(E_1) \subset W(E_2)$ if and only if $E_1 - E_2$ is a finite set (possibly empty); the inclusion is proper if and only if $E_2 - E_1$ is infinite. Consequently $W(E_1) = W(E_2)$ if and only if the symmetric difference $(E_1 - E_2) \cup (E_2 - E_1)$ is finite; it follows that there are c distinct sets $W(E)$.

Since the sets $V(E)$ form a basis for the topology of βN, the sets $W(E)$ form a basis for the topology of N^*. Note that $W(N - E) = N^* - W(E)$; hence the basis sets are open-closed, and N^* is totally disconnected.

The points of N are isolated points of βN, so that N is an open subset of βN. It follows that N^* is compact. Clearly N^* has no isolated points.

3.2. THEOREM. (a) *Every open-closed subset of N^* is of the form $W(E)$ for some $E \subset N$.*
(b) *If $W(E_1)$ and $W(E_2)$ are non-empty proper subsets of N^*, then there is a homeomorphism of βN which maps $W(E_1)$ onto $W(E_2)$.*

Proof. Suppose $N^* = A \cup B$, where $A \cap B = 0$, and A and B are open-closed. Since compact Hausdorff spaces are normal, there is a continuous function f on βN which has the value 1 on A and 0 on B. By 2.3, the only subsequential limits of the sequence $\{f(n)\}$ $(n \in N)$ are 1 and 0. Let E_1 be the set of all $n \in N$ such that $f(n) > 1/2$, E_2 the set of all $n \in N$ such that $f(n) < 1/2$. Then $N - E_1 - E_2$ is finite, and $A = W(E_1)$, $B = W(E_2)$, again by 2.3. This proves (a).

To prove (b), simply note that E_1, E_2, $N - E_1$, and $N - E_2$ are infinite, let π be a permutation of N which maps E_1 onto E_2, and extend π to a homeomorphism of βN, as in 2.4.

3.3. THEOREM. *The intersection of any countable family of open subsets of N^* is either empty or contains a non-empty open set.*

Proof. Let $\{G_i\}$ be a countable family of open subsets of N^* whose intersection contains a point Ω_0. There are sets $E_i \subset N$ such that

$$\Omega_0 \in W(E_i) \subset G_i \qquad (i = 1, 2, 3, \cdots).$$

The intersection of any finite collection of the sets $W(E_i)$ is non-empty and open, so that the intersection of any finite collection of the sets E_i is infinite. Hence there exists an increasing sequence of integers n_i such that $n_i \in E_1 \cap \cdots \cap E_i$. If E is the set of all n_i so chosen, then $E - E_i$ is finite for each i, so that $W(E) \subset W(E_i)$, and $W(E) \subset \bigcap_i G_i$. Since E is infinite, $W(E) \neq 0$.

IV. Consequences of the continuum hypothesis.

4.1. DEFINITION. We call a point p in a topological space X a P-point of X if every countable intersection of neighborhoods of p contains a neighborhood of p. If p is a P-point and f is a continuous real function on X, it is clear that f is constant in a neighborhood of p; conversely, if every continuous real function on X is constant in a neighborhood of p, and if X is completely regular, then p is a P-point of X [2; 344].

4.2. THEOREM. *If the continuum hypothesis is true, then N^* has 2^c P-points, and the set of all P-points of N^* is dense in N^*.*

Proof. The continuum hypothesis implies that there is a well-ordering $\{W_\alpha\}$ of the open-closed subsets of N^*, where α runs through the countable ordinals, and $W_1 = N^*$.

Let $A_1 = N^*$. Suppose α is a countable ordinal, and an open-closed set A_β has been selected for each $\beta < \alpha$, such that the set $B_\alpha = \bigcap_{\beta<\alpha} A_\beta$ is not empty. By Theorem 3.3 there exists an open-closed set $C_\alpha \neq 0$ such that $C_\alpha \subset B_\alpha$. Put $A_\alpha = C_\alpha$ if $C_\alpha \cap W_\alpha = 0$, and put $A_\alpha = C_\alpha \cap W_\alpha$ if $C_\alpha \cap W_\alpha \neq 0$.

Let $A = \bigcap A_\alpha$, the intersection being taken over all countable ordinals α. Being the intersection of a family of compact sets with the finite intersection property, $A \neq 0$. On the other hand, if $A \cap W_\alpha \neq 0$ for some α, then the choice of A_α shows that $A \subset W_\alpha$; since the sets W_α form a basis, we conclude that A consists of precisely one point Ω. If G is an open set containing Ω, then for some α we have $\Omega \in W_\alpha \subset G$, so that $A_\alpha \subset G$. Thus the family $\{A_\alpha\}$, which is linearly ordered by set-inclusion, forms a basis at the point Ω.

Now if $\{G_i\}$ $(i = 1, 2, 3, \cdots)$ is a countable collection of open sets containing Ω, there exist ordinals α_i such that $A_{\alpha_i} \subset G_i$. If α is the smallest ordinal which exceeds every α_i $(i = 1, 2, 3, \cdots)$, then $A_\alpha \subset \bigcap_i G_i$. Hence Ω is a P-point of N^*.

To prove that there are actually 2^c P-points in N^*, we merely note that at each stage of the preceding construction we had at least 2 (actually c) disjoint candidates for C_α. There were c stages, hence 2^c distinct possibilities. Finally, part (b) of Theorem 3.2 implies that the set of all P-points is dense in N^*.

4.3. THEOREM. *If every point of a compact Hausdorff space X is a P-point of X, then X is finite.*

Proof. If every point of X is a P-point, then the intersection of any countable family of open sets is easily seen to be open. Hence countable unions of closed sets are closed. It follows that every countable subset of X is closed and discrete. This is impossible in an infinite compact space. (See also [2; 346].)

4.4. THEOREM. *If the continuum hypothesis is true, then N^* is not homogeneous.*

Proof. By Theorem 4.2, N^* has a P-point Ω_1. Since N^* is infinite and compact, Theorem 4.3 shows that N^* also contains a point Ω_2 which is not a P-point of N^*. It is evident that no homeomorphism of N^* can carry Ω_1 to Ω_2.

4.5. THEOREM. *Suppose X is a locally compact Hausdorff space one of whose closed subsets is countably infinite and discrete. Then X^* is compact. If the continuum hypothesis is true, then X^* has P-points, and X^* is not homogeneous.*

Proof. It is convenient to break the proof up into several steps. To simplify the notation, we may as well assume that N is a closed subset of X.

(a) Locally compact Hausdorff spaces are completely regular, so that βX exists. The local compactness of X also implies that X is an open subset of βX, so that X^* is compact.

(b) *Every bounded real function on N can be extended to a bounded continuous real function on X.* To see this, suppose f is a bounded real function on N, and cover the points $n \varepsilon N$ with disjoint open sets $G_n \subset X$. Then find open sets H_n, containing n, whose closures are compact and lie in G_n. Since compact Hausdorff spaces are normal, there exist functions f_n, continuous on X, such that $f_n(n) = f(n)$, $f_n(x) = 0$ if $x \notin H_n$, and $| f_n(x) | \leq | f(n) |$ for all $x \varepsilon X$. The sum $\sum_1^\infty f_n$ is a continuous bounded extension of f to X.

(c) *The closure $\bar{N}$ of N in βX is homeomorphic to βN.* It is evident that N is a dense subset of the compact space $\bar{N}$. By step (b), every bounded real function on N can be extended continuously to X, hence to βX, hence to $\bar{N}$. Thus $\bar{N}$ has the characteristic properties of the Čech compactification of N.

If we identify the points of βN with their homeomorphic images in $\bar{N}$, we can put $\bar{N} = \beta N$, and thus embed βN in βX. Clearly $N^* = X^* \cap \beta N$. (See also [1; 833-834].)

(d) *Every P-point of N^* is a P-point of X^*.* Let Ω be a P-point of N^*, and let f be a continuous real function on X^* such that $f(\Omega) = 0$. Extend f to a continuous function on βX. Since Ω is a P-point of N^*, f vanishes on an N^*-neighborhood of Ω; this means that there is a set $E \subset N$ such that $E \varepsilon \Omega$ and such that $f(n) \to 0$ as $n \to \infty$ along E. To say that $E \varepsilon \Omega$ means that Ω is a limit point of E.

For every $n \varepsilon E$ let G_n be an open subset of X containing n, such that $| f(x) - f(n) | < 1/n$ for all $x \varepsilon G_n$. Find open sets H_n, for every $n \varepsilon E$, which contain n and whose compact closures lie in G_n. As in step (b), we can now construct a continuous bounded function g on X such that $g(n) = 1$ if $n \varepsilon E$ and $g(x) = 0$ in the complement of $\cup H_n$. Extend g to a continuous function on βX, and let H be the open subset of βX in which g is positive. Then $\Omega \varepsilon H$ (since $g(\Omega) = 1$).

Since g vanishes in $X - \cup H_n$, no point of $X^* \cap H$ is a limit point of $X - \cup H_n$. But every point of X^* is a limit point of X, so that every point of $X^* \cap H$ is a limit point of $\cup H_n$. Since each H_n has compact closure in X, every point of $X^* \cap H$ is actually a limit point of the set $K_p = H_p \cup H_{p+1} \cup H_{p+2} \cup \cdots$, for any positive integer p. But for every $\epsilon > 0$ our choice of G_n, together with the fact that $f(n) \to 0$ along E, shows that there is an integer p such that $| f(x) | < \epsilon$ for all $x \varepsilon K_p$. Hence $| f(x) | \leq \epsilon$ for all $x \varepsilon X^* \cap H$, so that f vanishes on $X^* \cap H$, i.e., on an X^*-neighborhood of Ω. Thus Ω is a P-point of X^*.

(e) Since X^* is an infinite compact space, not all of its points can be P-points (Theorem 4.3); hence (d) implies that X^* is not homogeneous.

4.6. *Remark.* If we consider spaces in which no neighborhood has compact closure, the situation is quite different. For instance, if R is the space of all rational numbers, with its natural topology, then R^* is not compact. In fact, R^* is a dense subset of βR, and it is seen quite easily that R^* has no P-points (although R^* does contain subsets homeomorphic to N^*). It is an open question whether R^* is homogeneous.

4.7. THEOREM. *If the continuum hypothesis is true and if* Ω_1 *and* Ω_2 *are* P-*points of* N^*, *then there is a homeomorphism of* N^* *which carries* Ω_1 *to* Ω_2. *Consequently* N^* *has precisely* 2^c *homeomorphisms.*

Proof. Since every homeomorphism of N^* induces a permutation of the c open-closed subsets of N^*, and since distinct homeomorphisms induce distinct permutations, N^* has at most 2^c homeomorphisms. Since N^* has 2^c P-points, the first part of the theorem implies the second.

To prove the first part, we use the continuum hypothesis to arrange the c open-closed subsets of N^* which contain Ω_1 as a well-ordered family $\{S_\alpha\}$, where α runs through the countable ordinals (this well-ordering has nothing to do with set inclusion). Let $T_\alpha = N^* - S_\alpha$. Similarly, denote the open-closed sets containing Ω_2 by X_α and let $Y_\alpha = N^* - X_\alpha$. Suppose also that $S_1 = X_1 = N^*$, $T_1 = Y_1 = 0$.

Call a family of subsets of N^* a *ring* if it is closed with respect to finite unions, finite intersections, and complementation. Clearly, each countable family of sets is contained in a countable ring.

We shall construct a permutation ψ of the family of all open-closed subsets of N^* such that

(i) $\psi(A) \subset \psi(B)$ if and only if $A \subset B$;
(ii) ψ maps $\{S_\alpha\}$ onto $\{X_\alpha\}$.

Once ψ is constructed, (i) implies that for any $\Omega \in N^*$ there is one and only one $h(\Omega) \in N^*$ which is contained in $\psi(A)$ for every open-closed set A which contains Ω, and it is easy to see that the mapping h so defined is a homeomorphism of N^*; (ii) implies that $h(\Omega_1) = \Omega_2$.

To construct ψ, put $\psi(S_1) = X_1$, $\psi(T_1) = Y_1$, and proceed by transfinite induction. Suppose α is the smallest ordinal for which $\psi(S_\alpha)$ has not yet been defined, and that the following induction hypothesis holds:

The sets for which ψ *has been defined form an at most countable ring, and* ψ *preserve finite unions, finite intersections, complementation, and inclusion.*

There is some redundancy in requiring all four of these properties, but this does no harm.

Since Ω_1 and Ω_2 are P-points, there are sets S_γ and X_γ (with $\alpha < \gamma$) such that $S_\gamma \subset S_\alpha$, S_γ lies in the intersection of all sets S_β for which ψ has been defined, and X_γ lies in the intersection of the corresponding sets $\psi(S_\beta)$. Define $\psi(S_\gamma) = X_\gamma$, $\psi(T_\gamma) = Y_\gamma$. Now ψ is again defined on a countable ring $\mathfrak{R}$. Divide the members of $\mathfrak{R}$ into three classes $\{F_i\}$, $\{G_j\}$, $\{H_k\}$, such that $S_\alpha \subset F_i$, $G_j \subset S_\alpha$, and no inclusion holds between S_α and H_k. Put $A_i = F_1 \cap \cdots \cap F_i$, $B_j = G_1 \cup \cdots \cup G_j$. Suppose for the moment (this will be proved below) that there is an open-closed set Z, different from any set so far in the range of ψ, such that $Z \subset \psi(A_i)$ for all i, $\psi(B_j) \subset Z$ for all j, and neither of the sets $\psi(H_k)$ and Z contains the other. Since $S_\gamma \subset S_\alpha$, we see that $X_\gamma \subset Z$, and Z is a member of the family $\{X_\alpha\}$.

We put $\psi(S_\alpha) = Z$, $\psi(T_\alpha) = N^* - Z$, and let $\mathfrak{R}'$ be the ring generated by $\mathfrak{R}$ and S_α. Note that $\mathfrak{R}'$ consists of all sets of the form $R \cap S_\alpha$, $R \cap T_\alpha$ ($R \in \mathfrak{R}$) and finite unions of these sets ($\mathfrak{R}$ contains both N^* and 0). Define $\psi(R \cap S_\alpha) = \psi(R) \cap \psi(S_\alpha)$, $\psi(R \cap T_\alpha) = \psi(R) \cap \psi(T_\alpha)$, and $\psi(P \cup Q) = \psi(P) \cup \psi(Q)$ if $\psi(P)$ and $\psi(Q)$ are already defined. In this way ψ is extended to the ring $\mathfrak{R}'$ so as to preserve finite unions, finite intersections, complementation, and inclusion. Also, for any set $R \in \mathfrak{R}'$, $\Omega_1 \in R$ if and only if $\Omega_2 \in \psi(R)$; this follows from our choice of S_γ and X_γ.

Having done this, the inverse mapping ψ^{-1} satisfies the same induction hypothesis, we can define ψ^{-1} for the first member of $\{X_\alpha\}$ which is not yet in the range of ψ, and extend to a ring, as above.

This process continues and yields the desired permutation ψ.

4.8. It remains to be shown that there always exists a set Z with the properties needed in the above proof. If $E \subset N$ and $F \subset N$, let us write $E < F$ if $E - F$ is finite and $F - E$ is infinite. If $E < F$ and $F < E$ are both false, we write $E \parallel F$. If $(E - F) \cup (F - E)$ is finite, we write $E \sim F$. Recalling the definition of the basis sets $W(E)$ and the fact that every open-closed subset of N^* is one of these basis sets, we see that the following lemma completes the proof.

LEMMA. *Suppose we are given three at most countable families* $\{A_i\}$, $\{B_j\}$, $\{C_k\}$ *of subsets of* N, *a function* f *defined for those sets which preserves the relations* $<$ *and* $\parallel$, *and a set* D *such that*

(i) $B_1 < B_2 < B_3 < \cdots < D < \cdots < A_3 < A_2 < A_1$;
(ii) $C_k \parallel D$ *for all* k.
Then $f(D)$ *can be defined so that*
(iii) $f(B_1) < f(B_2) < f(B_3) < \cdots < f(D) < \cdots < f(A_3) < f(A_2) < f(A_1)$;
(iv) $f(C_k) \parallel f(D)$ *for all* k.

Proof. To simplify the notation we assume that each of the three given families is infinite; the modifications needed in the contrary case are quite trivial.

Since $D < C_i$ is false, so is $A_n < C_i$, for any n and i, and $f(A_n) < f(C_i)$ is also false. Hence

(a) $f(A_n) - f(C_i)$ is an infinite set, for any n and i. Similarly we see that
(b) $f(C_i) - f(B_n)$ is an infinite set, for any n and i.

Let H_1 consist of $f(A_1) \cap f(B_1)$, plus one number $p_{11} \in f(A_1) - f(C_1)$.

Let K_1 consist of a number $q_{11} \in f(C_1)$, $q_{11} \notin H_1$; this is possible by (b), since $H_1 - f(B_1)$ is finite.

Suppose $H_1, \cdots, H_{n-1}$; $K_1, \cdots, K_{n-1}$ are constructed ($n \geq 2$), with every

K_i finite. Let t_n be an integer greater than any member of $K_1 \cup \cdots \cup K_{n-1}$, and put $L_n = f(A_1) \cap \cdots \cap f(A_n)$.

Let H_n consist of all $m \varepsilon L_n \cap f(B_n)$ such that $m > t_n$, plus numbers $p_{in} \varepsilon L_n - f(C_i)$ $(i = 1, \cdots, n)$ such that $p_{in} > t_n$; this is possible by (a), since $f(A_n) - L_n$ is finite.

Let K_n consist of numbers $q_{in} \varepsilon f(C_i)$ $(i = 1, \cdots, n)$ such that $q_{in} > t_n$ and $q_{in} \notin H_1 \cup \cdots \cup H_n$; this is possible by (b), since $(H_1 \cup \cdots \cup H_n) - f(B_n)$ is finite.

Put $f(D) = \bigcup_{n=1}^{\infty} H_n$. For fixed k we have $f(D) = P_k \cup Q_k$, where $P_k = H_1 \cup \cdots \cup H_k$ and $Q_k = H_{k+1} \cup H_{k+2} \cup \cdots$. Note that $H_n \sim f(B_n)$, so that $P_k \sim f(B_k)$, and since $P_k < P_{k+1}$, we have $f(B_k) < f(D)$. Also $Q_k \subset f(A_{k+1})$, so that $f(D) - f(A_{k+1})$ is finite; since $f(A_{k+1}) < f(A_k)$, we have $f(D) < f(A_k)$, and (iii) is proved.

Now consider some fixed C_i. There are infinitely many numbers $p_{in} \varepsilon f(D)$, $p_{in} \notin f(C_i)$, so that $f(D) - f(C_i)$ is infinite. The choice of the numbers q_{in} shows similarly that $f(C_i) - f(D)$ is infinite. This implies (iv) and completes the proof of the lemma as well as that of Theorem 4.7.

References

1. Eduard Čech, *On bicompact spaces*, Annals of Mathematics vol. 38(1937), pp. 823–844.
2. Leonard Gillman and Melvin Henriksen, *Concerning rings of continuous functions*, Transactions of the American Mathematical Society, vol. 77(1954), pp. 340–362.
3. Bedřich Pospíšil, *Remark on bicompact spaces*, Annals of Mathematics, vol. 38(1937), pp. 845–846.
4. J. Schreier and S. Ulam, *Über die Permutationsgruppe der natürlichen Zahlenfolge*, Studia Mathematica, vol. 4(1933), pp. 134–141.

University of Rochester

NOTE OF CORRECTION

BY WALTER RUDIN

The paper "Homogeneity problems in the theory of Čech compactifications", pp. 409–420 of this volume, contains an error. Normality of X should be added to the hypotheses of Theorem 4.5. That the theorem is false as it stands is shown by the following example (pointed out by Leonard Gillman).

Let ω and Ω be the first infinite ordinal and the first uncountable ordinal, respectively. Let A and B be the sets of all ordinals not exceeding ω and Ω, respectively, with the usual interval topology, and let S be the Cartesian product $A \times B$. Remove the point (ω, Ω) from S and let X be the resulting space. Then X is a locally compact Hausdorff space, and the set of points (n, Ω) $(n \in N)$ is closed, infinite, and discrete. But $\beta X = S$, so that X^* consists of just one point, and is therefore homogeneous.

Normality of X is needed to establish the result of step (b) in the proof of Theorem 4.5. The error in the proof occurs in the last sentence of (b).

Доклады Академии наук СССР
1963, Том 150, № 1, 36 – 39

MATHEMATICS

I. I. PAROVIČENKO

ON A UNIVERSAL BICOMPACTUM OF WEIGHT $\aleph$

(Presented by academician P. S. Alexandrov 20. I. 1962)

1. According to the well-known theorem by P. Alexandrov, the Cantor discontinuum $\Delta_{\aleph_0}$ has the universality property in the class of all compact spaces of weight $\leqslant \aleph_0$. The universality property means that every compact space of weight $\leqslant \aleph_0$ is its continuous image. P. Alexandrov also gave a simple topological definition of $\Delta_{\aleph_0}$ as a zero-dimensional perfect compact space of weight $\aleph_0$. In [1], A. Esenin-Vol'pin, assuming the generalized continuum hypothesis, proved the existence of a compact space of an arbitrary weight $\mathfrak{m}$, which is universal in the same sense.

In the present paper we shall prove (with the aid of continuum hypothesis, which will be in the sequel always assumed, if not stated otherwise) that the Čech remainder of the integers $\Delta_\aleph = \beta N \setminus N$ is a universal compact space for a class of compact spaces of weight $\leqslant \aleph$. We shall give also a topological definition of $\Delta_\aleph$ and obtain a theorem for $\Delta_\aleph$ which is completely analogous to that one for $\Delta_{\aleph_0}$.

2. Below we shall use the following notation in a boolean algebra L: $a+b$, ab, b', $ab' = a \setminus b = a - b$ (the last for $b \leqslant a$), the least element will be denoted by 0, the principal ideal $\{x | x \leqslant a\}$ will be denoted by L_a, the dual principal ideal $\{x | x \geqslant a\} = L^a$. As proved by Novák [2], the boolean algebra of all clopen subsets of $\Delta_\aleph$ is isomorphic to the boolean algebra $L_\aleph$ defined by the following.

The members of $L_\aleph$ are the equivalence classes in the system of all subsets of integers, where E is equivalent to G if $(E \setminus G) \cup (G \setminus E)$ is finite, and for $e, g \in L_\aleph$ $e < g$, if $G \setminus E$ is infinite and $E \setminus G$ finite, $E \in e$, $G \in g$.

We shall say that a partially ordered set T has a *simple separation property*, if for every $e < h$ in T there is a g such that $e < g < h$; *Cantor separation property*, if for an arbitrary subset $e_1 < \cdots < e_n < \cdots < h$ of type $\omega + 1$ there is a g such that $e_n < g < h$, and *du Bois-Reymond separation property*, if for an arbitrary set $e_1 < \ldots e_n < \cdots < h_n < \ldots h_1$ of type $\omega + \omega^*$ there is a g such that $e_n < g < h_n$. We shall say that a zero-dimensional compact space has the corresponding separation property, if it has the boolean algebra of its clopen subsets. In particular, the simple separation property coincides with the condition that the space is perfect. The remainder $\Delta_\aleph$, in addition to the simple separation property, has Cantor and du Bois-Reymond separation property, too, which is witnessed (using Novák's isomorphism) by the theorems of the same name on subsets of natural numbers ([3], p. 715). However in general, none of three separation properties implies another and no two of them imply the third one in the case of zero-dimensional compact spaces of weight $\aleph$. In fact, it follows from the Cantor separation property that the intersection of a strictly descending sequence of clopen subsets contains an interior point, but the compact space $\overline{\bigcup_n e_n}^{(\Delta_\aleph)}$ ($e_1 \subset e_2 \subset \ldots$ and clopen in $\Delta_\aleph$) does not have this property, though the other two separation properties hold in it.

Du Bois-Reymond separation property implies that there are no $\varkappa$-points in a compact space, but an ordered compact space of type $2^{\omega_1^*}$ (the notation from [5]) does not contain them, though it posses both the other separation properties.

Theorem 1. *Any perfect zero-dimensional compact space of weight $\aleph$ with separation properties of Cantor and du Bois-Reymond is homeomorphic to $\Delta_\aleph$, and independently of the continuum hypothesis it maps continuously onto an arbitrary compact space of weight $\leqslant \aleph_1$.*

Lemma 1. *In a boolean algebra L with Cantor separation property, every ideal, which is cofinal with at most countable set, is closed* (*see* [6], *p.* 95).

Let A be an ideal in L, which is cofinal with a sequence $a_1 < \cdots < a_n < \ldots$. In what follows we shall consider L as a set of all clopen subsets of some zero-dimensional compact space Δ, preserving at the same time the structural notations. Let $a \leqslant b$ for an arbitrary $b \supset \bigcup_n a_n$ and a fixed $a > 0$, and still for all a_n, a is not $\leqslant a_n$. Then the intersection of all b's as indicated is the set $\overline{\bigcup_n a_n}^{\Delta}$.

Consequently, $a \subset \overline{\bigcup_n a_n}$, and, since a is open, we have $a \cap (\bigcup_n a_n) \neq \emptyset$ and $aa_n > 0$ for $n \geqslant n_0$. But, by the condition, $a \setminus a_n > 0$ and a $a \setminus a_n$ descend. If the sequence $\{a \setminus a_n\}$ does not stabilize, then by the Cantor separation property there is a c such that $0 < c < a \setminus a_n$; hence $0 < c \leqslant a_n \subset \overline{\bigcup_n a_n)}$ and $c \cap (\bigcup_n a_n) = \emptyset$, which is impossible. The case of stabilization does not need the use of a separation property.

Lemma 2. *If an ideal in a boolean algebra with the Cantor and du Bois-Reymond separation properties is cofinal with at most countable set and if $C = \{c_i\}$ is at most countable set in $L \setminus A$, then there is a principal ideal L_g such that $A \subset L_g$ and $C \subset L \setminus L_g$.*

Let A be cofinal with $a_1 < \cdots < a_n < \ldots$. According to Lemma 1 for every c_i we choose e_i such that $A \leqslant e_i$ and c_i is not $\leqslant e_i$. Let $g_j = \bigwedge_{i=1}^{j} e_i$; then $g_1 \geqslant \cdots \geqslant g_j \geqslant \ldots$ together with $g_j \leqslant e_i$ $(i = 1, \ldots, j)$. If the sequence $\{g_j\}$ stabilizes at g_{j_0}, then $L_{g_{j_0}}$ is as required. If $\{g_j\}$ does not stabilize, then, according to du Bois-Reymond separation property there is a g such that $a_n < g < g_j$ and c_i is not $\leqslant g$, L_g is as required.

Lemma 3. *Suppose taht three sets are given in a boolean algebra L with simple, Cantor and du Bois-Reymond separation properties: $A = \{a_l\}$, $B = \{b_m\}$, $C = \{c_n\}$, where $a_1 < \cdots < a_l < \cdots < b_m < \cdots < b_1$, and for arbitrary l, m, n one has c_n is not $\leqslant a_l$ and b_m is not $\leqslant c_n$. Then there exists a d such that $a_l < d < b_m$ and d is not comparable with any c_n.*

Let $\tilde{A} = \bigcup_l \{x | x \leqslant a_l\}$, $\tilde{B} = \bigcup_m \{x | x \geqslant b_m\}$; then $\tilde{A}$ is an ideal ($\tilde{B}$ is a dual ideal), which is cofinal (coinitial) with at most countable set and $\tilde{A} < \tilde{B}$.

Using the Cantor and du Bois-Reymond separation properties, we find g_0 and g_1 such that $\tilde{A} \leqslant g_0 < g_1 \leqslant \tilde{B}$. By Lemma 2 and the statement dual to it we can choose h_0 and h_1 such that $\tilde{A} \subset L_{h_0}$, $B \subset L^{h_1}$, $C \subset L \setminus (L_{h_0} \cup L^{h_1})$. Let $t_0 = g_0 h_0$, $t_1 = g_1 + h_1$; then $\tilde{A} \leqslant t_0 < t_1 \leqslant \tilde{B}$, $C \subset L \setminus (L_{t_0} \cup L^{t_1})$. Let us set $c_n 0 = c_n t_1$, since $c_n \notin L^{t_1}$, we have $c_n^0 < t_1$. Let $e_p = \bigvee_{n=1}^{p} c_n^0$, and put $q_1 = e_1$, $q_2 = q_1 + (e_2 - q_1), \ldots, q_p = q_{p-1} + (e_p - q_{p-1}), \ldots, r_0 = t_0, \ldots, r_p = t_0 + q_p, \ldots$, then $t_0 = r_0 \leqslant r_1 \leqslant r_2 \leqslant \cdots < t_1$. Using either the simple or Cantor separation property, we shall find r such that $r_p < r < t_1$ $(p = 0, 1, \dots)$, then $d = t_0 + (t_1 - r)$ is the required one.

Now, Theorem 1 can be proved analogously to Rudin's theorem from [7], 4.7, on homeomorphisms of $\Delta_\aleph$ onto intself, but two boolean algebras are to be considered and lemma 4.8 from [7] has to be replaced by our Lemma 3 (compare with [1], too).

Corollary. *The class of compact spaces of weight $\leqslant \aleph$ coincides with the class of all compact remainders on natural numbers.*

2. Let $C_\aleph = \{y\}$ be the lexicographically ordered set of all sequences of type ω_1 consisting of real numbers y_ξ, $0 \leqslant y_\xi \leqslant 1$, and let $I_\aleph$ be the ordered set, obtained by deleting all $\varkappa$-points from $C_\aleph$.

Theorem 2 is analogous to the theorem on universality of Baire 0-space.

Theorem 2. *The ordered set $I_\aleph$ admits a one-to-one continuous mapping onto $\Delta_\aleph$, consequently it can be mapped continuously onto an arbitrary compact space of weight $\leqslant \aleph$.*

Since $\Delta_\aleph$ is a one-to-one continuous image of $T^1\Delta_\aleph$ [8], it is enough to prove that $I_\aleph$ and $T^1\Delta_\aleph$ are homeomorphic. Let $\mathfrak{G} = \{\Gamma_\nu\}_{\nu<\omega_1}$ be an enumeration of the collection of all nonvoid clopen subsets of $\Delta_\aleph$; assume moreover that if ξ is either zero or limit, then $\Gamma_{\xi+2k}$ and $\Gamma_{\xi+2k+1}$ are complementary in $\Delta_\aleph$ $(0 \leqslant k < \omega)$. Let $x = \{i_\xi\}_{\xi<\omega_1}$ be a sequence of 0's and 1's.

We shall define $\mathfrak{D}(x) = \{D_\xi(x)\}$ by induction: $D_0(x) = \Gamma_0$ for $i_0 = 0$ and $D_0(x) = 1$ for $i_0 = 1$. If $D_\xi(x)$ have been defined for $\xi < \eta$, then for η limit, $D_\eta(x) = \bigcap_{\xi<\eta} D_\xi(x)$, and for $\eta = \eta_0 + 1$, $D_\eta(x) = D_{\eta_0}(x) \cap \Gamma_{\nu_0}$ if $i_\eta = 0$, and $D_\eta(x) = D_{\eta_0}(x) \cap \Gamma\nu_0 + 1$ if $i_\eta = 1$, where $\nu_0 = \nu_0(\eta)$ is the smallest of all ν such that simultaneously $D_{\eta_0} \cap \Gamma_\nu \neq \emptyset$, $D_{\eta_0} \cap \Gamma_{\nu+1} \neq \emptyset$. It is clear that $\nu_0(\eta)$ strictly increases. The sequence $\mathfrak{D}(x)$ descends and it has a non-empty intersection due to the compactness of $\Delta_\aleph$. From the construction easily follows that the intersection contains a unique point, in the sequel we shall assume $\Delta_\aleph = \{x\}$. We shall prove that $\mathfrak{D} = \bigcup_x \mathfrak{D}(x)$ is an open base in $T^1\Delta_\aleph$. Indeed, let $Ox = \bigcap_{n<\omega} G_n(x)$, where $G_n(x)$ are neighborhoods of x in $\Delta_\aleph$. As $\bigcap_\xi D_\xi(x) = x \in G_n(x)$, then by compactness of $\Delta_\aleph$ there exists a ξ_n such that $D_{\xi_n}(x) \subset \bigcap G_n(x)$; taking $\xi' > \xi_n$, we have: $D_{\xi'}(x) \subset \bigcap_n D_{\xi_n}(x) \subset \bigcap_n G_n(x) = Ox$. Enumerate now all countable limit ordinals: $\tau_0 = \omega$, $\tau_1 = \omega \cdot 2, \dots, \tau_n, \dots$, and let $\mathfrak{D}_\pi = \{D_{\tau_p i}(x) | x \in \Delta_\aleph\}$; then $\mathfrak{D} = \bigcup_{\pi<\omega_1} \mathfrak{D}_p \dot{i}$, where $\{\mathfrak{D}_\pi\}$ is a sequence of clopen covers of $T^1\Delta_\omega$, while every element of $\mathfrak{D}_\pi$ contains $\aleph$ elements of $\mathfrak{D}_{\pi+1}$. Let $C(y_0, \dots, y_\xi, \dots)$ be a set of all points in $C_\aleph$ beginning with a complex $(y_1, \dots, y_\xi, \dots)_{\xi<\eta<\omega_1}$, and $\mathfrak{C}_\eta = \{C(y_0, \dots, y_\xi, \dots)\}$. Then the family $\{\mathfrak{C}_\eta\}_{\eta<\omega_1}$ is a sequence of covers of $C_\aleph$ by disjoint collections of intervals, and the order type of $\mathfrak{C}_\eta$ is θ^{η^*}. Thus $\varkappa$-points and only them are the ends of intervals in $\bigcup_{\eta<\omega_1} \mathfrak{C}_\eta$. If we remove them, we obtain a sequence of interval covers of $I_\aleph$. The union of those is a base of $I_\aleph$, which is isomorphic to $\mathfrak{D}$. The existence of the required homeomorphism follows.

We shall call a set M to be of the first (second) $\aleph$-category in S, if M can (cannot) be represented as a union of $\leqslant \aleph$ nowhere dense subsets in S. It is easy to see that $\Delta_\aleph$ is of the second $\aleph$-category, and if Π is the set of all P-points in $\Delta_\aleph$ [7], then $\Delta_\aleph \setminus \Pi$ is of the first $\aleph$-category, since $\Delta_\aleph \setminus \Pi$ is a union of boundaries of all countable intersections of clopen subsets of $\Delta_\aleph$. The last system is of size $\aleph^{\aleph_0} = \aleph$.

By a slight change of the proof of Theorem 2 one obtains

Theorem 3. *The set of all P-points of a space $\Delta_{\aleph}$ is homeomorphic to $I_{\aleph}$ and consequently the remainder is an ordered space up to a set of first $\aleph$-category.*

3. Theorem 4. *All maximal ordered subsets of $L_{\aleph}$ are similar and of type $1 + \eta_1 + 1$, where η_1 is a normal type in the sense of Hausdorff* ([9], *p.* 181).

This theorem immediately follows from the separability properties of $L_{\aleph}$ and from Theorem II in [9], p. 181.

Theorem 4 gives a real embodiment to somewhat indefinite ideas of Lusin concerning the "phenomenon of Pythagoras" on "transfinite reals" ([3], p. 721). In particular, since Dedekind completion of a set of type $1 + \eta_1 + 1$ is similar to $C_{\aleph}$ (this also easily follows from Hausdorff theorem quoted before), and since the cardinality of $C_{\aleph}$ is $2^{\aleph}$, there is more "transfinite irrationalities" in the sense of Lusin, similarly to irrationalities of the usual real line, than "real" (rational) points: they correspond to Dedekind cuts in the set of type $1 + \eta_1 + 1$.

In accordance we got an immediate solution of all Lusin's problems from [6], p. 721, which, however, is not new [2].

Kišinev State
University

Recieved
16 XI 1962

References

[1] A. Esenin-Vol'pin, *On the existence of a universal bicompactum of an arbitrary weight*, (Russian), Dokl. Akad. Nauk SSSR **68, 4** (1949), 649 – 652.

[2] J. Novák, *On some problems of Luzin concerning the subsets of natural numbers*, Czechoslovak Math. J. **3 (78)** (1953), 385 – 395.

[3] N. Lusin, *Collected works II.: Descriptive theory of sets*, (Russian), Izdat. Akad. Nauk SSSR, Moscow, 1958.

[4] P. S. Uryson, *Tr. po topologii i drugim oblastjam matematiki*, **2**, Izdat. Akad. Nauk SSSR, Moscow, 1951.

[5] F. Hausdorff, Teorija množestv (1937), ONTI, Moscow.

[6] G. Birkhoff, *Lattice theory*, New York, 1940.

[7] W. Rudin, *Homogeneity problems in the theory of Čech compactifications*, Duke Math. J. **23, 3** (1956), 409 – 419; 633.

[8] I. I. Parovičenko, *Certain special classes of topological spaces and δ_S-operation*, (Russian), Dokl. Akad. Nauk SSSR **115** (1957), 866 – 868.

[9] F. Hausdorff, *Grundzüge der Mengenlehre*, New York, 1949.

Commentationes Mathematicae Universitatis Carolinae
8,4 (1967)

NON–HOMOGENEITY OF $\beta P - P$

BY
Zdeněk Frolík, Praha

This is to answer a question I have been asked repeatedly by my colleagues since my proofs (independent of the continuum hypothesis) of non-homogeneity of $\beta\mathbb{N} - \mathbb{N}$ were preprinted. It turns out that both methods, that one depending on the author's estimate of the cardinal of the set of all relative types of a given point of $\beta\mathbb{N} - \mathbb{N}$ (see [1]), and also that one depending on the author's non-fixed point theorem (see [2] and [3] or [4]), work in general situation. What is needed in addition is not too much, however, perhaps a little bit surprising, see Lemma and its Corollary 2 below. Non-homogeneity problems for extremally disconnected spaces will be treated in a forthcoming paper.

The main results are two proofs of the following theorem, without any assumption concerning the continuum hypothesis.

Theorem 1. *If P is not pseudocompact then $\beta P - P$ is not homogeneous.*

Under an additional assumption that P is locally compact Theorem 1 was stated and proved by W. Rudin [5]; his proof heavily depends on the Continuum Hypothesis (the existence of P-points in $\beta\mathbb{N} - \mathbb{N}$). Using the same method (in particular, using the continuum hypothesis) T. Isiwata [6] formulated and proved Theorem 1 above. Our proof depends on Corollary 2 of the following lemma, and the theory of types of ultrafilters.

Lemma. *Let X be a completely normally embedded countable subset of a completely regular space P. Let $Y \subset \beta P$ be a countable set which is semi-separated to X (i.e. $(X \cap clY) \cup (Y \cap clX) = \emptyset$). Then X and Y are functionally separated in βP,* i.e.

$$clY \cap clX = \emptyset$$

Proof. The sets X and Y are countable (hence Lindelöf) and semi-separated, and hence separated, that means there exist two disjoint open sets U and V in βP such that $U \supset X$, $V \supset Y$. Put $F = P - U$. Since X is countable, there exists a zero set Z in P which is disjoint to X and contains F. Since X is completely normally embedded, X and Z are functionally normally separated in P. Hence X and F are functionally separated in P, which implies that $clX \cap clF = \emptyset$ (because of the characteristic property of Čech-Stone compactifications). Now observe that $clY \subset clF$.

ZDENĚK FROLÍK

Corollary 1. *Let X be a countable infinite completely normally embedded set in a space P. If $x \in clX - X$ (in βP) is in the closure of a countable set Y with $clY \subset \beta P - P$, then*

$$x \in cl(Y \cap clX)$$

Corollary 2. *Let X be a countably infinite completely normally embedded set in a space P, and let $x \in clX - X$. If Y is a countable discrete set which is with its closure contained in $\beta P - P$, and if $x \in clY - Y$, then $x \in clZ - Z$ for some $Z \subset Y \cap clX$.*

The proof of Corollary 1 is easy, and Corollary 2 is a particular case of Corollary 1.

Now we are going to derive immediate consequences of Corollary 2 for types of ultrafilters. It is not necessary to introduce any notation and definitions concerning types, however, it seems to be convenient to do that, and also, the theory of types might be of some interest in itself. For the definition of types see [1; 1.1]. Roughly speaking, two ultrafilters on countable sets, say $\mathcal{X}$ on X and $\mathcal{Y}$ on Y, are defined to be equivalent, if there exists a bijective mapping f between X and Y such that $f[\mathcal{X}] = \mathcal{Y}$; now, the types would be the equivalence classes if there were no set-theoretical troubles.

Definition. Let Q be a space, and let $x \in Q$. Consider the collection $\mathcal{M}_x$ of all countable normally embedded discrete sets M such that $x \in clM - M$. Thus the intersections of the neighborhoods of x with any M in $\mathcal{M}_x$ form an ultrafilter $\alpha_x M$ on M, and the type of this ultrafilter is called the type of x with respect to M. If $S \subset Q$, then the types with respect to subsets M lying with their closures in S are called the types in S; the types in $\beta P - P$ are called the ideal types.

Theorem 2. *Let X be a countable infinite discrete completely normally embedded set in a space P and let $x \in clX - X$ (in $Q = \beta P$). Then*

A. The cardinal of the set of all ideal types of x is at most $\exp \aleph_0$.

B. The type of x with respect to X is distinct from any ideal type of x.

Proof. Let t be any ideal type of x, say with respect to Y. By Corollary 2 there exists a $Z \subset Y \cap clX$ such that $x \in clZ$. Clearly the types with respect to Y and Z coincide. Thus every ideal type of x is equal to the ideal type of x in $clX - X$. Since clX is a Čech-Stone compactification of X, both statements follow from the corresponding statements for the particular case $P = \mathbb{N}$; for A see Theorem C in [1], for B see Theorem in [2] or Theorem B in [3], or Proposition 2 in [4].

Proof of Theorem 1. Assume that P is not pseudocompact. It follows that there exists a countably infinite discrete completely normally embedded X in P. For each x in $\beta P - P$ let T_x denote the set of all types of x in $\beta P - P$. If h is any homeomorphism of $\beta P - P$ onto itself, and if $hx = y$ then $T_x = T_y$. Now let $x \in clX - X$. We want to find a y in $clX - X$ such that $T_x \neq T_y$. Now if we want to apply the assertion A in Theorem 2, we pick a type t which is not in T_x (the cardinal of the set of all types is $exp\ exp \aleph_0$), and then we select a y in $clX - X$ such that $t \in T_y$. If we want to apply the assertion B, we pick an y in $clX - X$ such that the type of x with respect to X belongs to T_y. Since $t \notin T_x$ by B, necessarily $T_x \neq T_y$.

NON–HOMOGENEITY OF $\beta P - P$

References

[1] Z. Frolík, *Sums of ultrafilters*, Bull. Amer. Math. Soc. **73** (1967), 87 – 91.

[2] ———, *Types of ultrafilters*, Proc. 2nd Prague Symposium on General Topology in 1966 (Academia, Praha 1967).

[3] ———, *Fixed points of maps of* $\beta\mathbb{N}$, Bull. Amer. Math. Soc. **74** (1968), 187 – 191.

[4] M. Katětov, *A theorem on mappings*, Comment. Math. Univ. Carolinae **8** (1967), 431 – 433.

[5] W. Rudin, *Homogeneity problems in the theory of Čech compactifications*, Duke Math. J. **23** (1956), 409 – 419, 633.

[6] T. Isiwata, *A generalization of Rudin's theorem for the homogeneity problem*, Sci. Rep. Tokyo Kyoiku Daigaku Sect. A5 (1957), 300 – 303.

(Received September 13, 1967)

COLLOQUIA MATHEMATICA SOCIETATIS JÁNOS BOLYAI
23. TOPOLOGY, BUDAPEST (HUNGARY), 1978.

WEAK P-POINTS IN N^*

K. KUNEN*

§0. INTRODUCTION

N denotes the space of natural numbers (i.e., ω) with the discrete topology. βN is the Cech compactification of N, and $N^* = \beta N \setminus N$. We often identify the points of βN with the ultrafilters on ω, in which case the points of N^* are the non-principal ultrafilters.

If X is any topological space and $p \in X$, p is a *P-point* iff whenever $\{U_n : n \in \omega\}$ is a countable family of neighborhoods of p, $p \in \operatorname{int}\left(\bigcap_n U_n\right)$. p is a *weak* P-point iff p is not a limit point of any countable subset of $X \setminus \{p\}$. Clearly, every P-point is a weak P-point.

W. Rudin [5] showed that the Continuum Hypothesis (CH) implies that there are 2^{2^ω} P-points in N^*, and the same result holds under Martin's Axiom (MA), with the same proof. However, S. Shelah (see [4] or [6]) showed that it is consistent with the usual axioms of set theory that there are no P-points in N^*. The purpose of this note is to show

*Research partly supported by NSF Grant MCS 76-06541.

that there must be weak P-points. More precisely, we prove in ZFC that

Theorem 0.1. *There are* 2^{2^ω} *points in* N^* *which are weak* P-*points but not* P-*points.* ∎

The proof is a modification of the proof in [2] that there are points in N^* which are Rudin – Keisler incomparable. These two proofs may be amalgamated to prove

Theorem 0.2. *There is a set* $A \subset N^*$ *such that* $|A| = 2^\omega$, *every element of* A *is a weak* P-*point, and the elements of* A *are pairwise incomparable in the Rudin – Keisler order.* ∎

The proofs of Theorems 0.1 and 0.2 are given in §3.

§1. OK POINTS

We describe here the nature of the weak P-points we shall construct. X denotes any topological space. κ and λ range over cardinals, and other Greek letters range over ordinals.

Definition 1.1. If $p \in X$ and U_n $(n \in \omega)$ are neighborhoods of p, a *κ-refinement system* for $\langle U_n : n \in \omega \rangle$ is a κ-sequence of neighborhoods of p, $\langle V_\alpha : \alpha < \kappa \rangle$ such that for all $n \geqslant 1$,

$$\forall \alpha_1 < \alpha_2 < \ldots < \alpha_n < \kappa \qquad (V_{\alpha_1} \cap \ldots \cap V_{\alpha_n} \subset U_n). \blacksquare$$

Definition 1.2. A point $p \in X$ is κ-OK iff whenever U_n $(n \in \omega)$ are neighborhoods of p, $\langle U_n : n \in \omega \rangle$ has a κ-refinement system. ∎

Observe that the property of being κ-OK gets stronger as κ gets bigger (i.e., if p is λ-OK and $\kappa \leqslant \lambda$, then p is κ-OK). Every point is ω-OK (take $V_\alpha = \bigcap \{U_n : n \leqslant (\alpha + 1)\}$). We do not require the V_α of a refinement system to be distinct; in particular, if p is a P-point, then p is κ-OK for all κ, and the V_α may all be chosen to be the same. Conversely, if p is κ-OK and $\kappa > \chi(p, X)$, then p is a P-point.

Lemma 1.3. *If* p *is* ω_1-*OK and* X *is* T_1, *then* p *is a weak* P-*point.*

Proof. Let $\{q_n : n \in \omega\} \subset X \setminus \{p\}$. For each n, let U_n be a neighborhood of p with $q_n \notin U_n$. Let $\langle V_\alpha : \alpha < \omega_1 \rangle$ be a refinement system for $\langle U_n : n \in \omega \rangle$. Then for each n, there can be at most $n - 1$ values of α such that $q_n \in V_\alpha$. Thus, for all but countably many α, $\forall n (q_n \notin V_\alpha)$, so $p \notin \mathrm{cl}(\{q_n : n \in \omega\})$. ∎

The weak P-points constructed in §3 will all be 2^ω-OK.

We conclude this section with some further remarks on OK points. These remarks are not needed in the rest of this paper.

Lemma 1.4. *If X is T_3 and has the countable chain condition (c.c.c.), and $p \in X$ is not isolated, then p is not ω_1-OK.*

Proof. First, observe that p is not a P-point, since if it were, we could inductively find a decreasing sequence of open neighborhoods of p, $\langle U_\alpha : \alpha < \omega_1 \rangle$, such that $\mathrm{cl}(U_\beta)$ is a proper subset of U_α whenever $\alpha < \beta$. But then $\{U_\alpha \setminus \mathrm{cl}(U_{\alpha+1}) : \alpha < \omega_1\}$ would contradict the c.c.c.

Now, let U_n $(n \in \omega)$ be open neighborhoods of p such that $p \notin \mathrm{int}\left(\bigcap_n \mathrm{cl}(U_n)\right)$. Let $\langle V_\alpha : \alpha < \omega_1 \rangle$ be a refinement system for $\langle U_n : n \in \omega \rangle$. For each α, there is an n such that $V_\alpha \setminus \mathrm{cl}(U_n) \neq \phi$, so fix $n \in \omega$ so that $B = \{\alpha < \omega_1 : V_\alpha \setminus \mathrm{cl}(U_n) \neq \phi\}$ is uncountable. For $\alpha \in B$, let $W_\alpha = V_\alpha \setminus \mathrm{cl}(U_n)$. Then the W_α are non-empty, but any n of them have empty intersection, which is easily seen to contradict the c.c.c. ∎

Theorem 0.1 implies in particular that there is a weak P-point in N^* which is not a P-point. Under CH or MA, this was known by [3], but the points produced there were in a copy of the Stone space of a measure algebra embedded in N^*. Thus, these points were not ω_1-OK, since by 1.4, if $p \in N^*$ is ω_1-OK then p is isolated in every c.c.c. subspace of N^*.

Our notion of an OK point was motivated by Keisler's notion of a good ultrafilter (see [1], p. 307). Let p be an ultrafilter on a set D, and think of p as a point in βD (where D is discrete). If p is κ^+-good, then p is κ-OK in βD (but not conversely). In [2] (see also [1],

Theorem 6.1.4) we showed that there is a point in D^* which is $|D|^+$-good (a result due to Keisler under GCH). Thus, if $|D| > \omega$, then there is a point in D^* which is ω_2-good, and hence ω_1-OK and in particular a weak P-point in βD (and thus also in D^*). No non-principal ultrafilter on N can be ω_2-good, and of course no point of N^* can be a weak P-point in βN. So, we weaken ω_2-good to ω_1-OK and require that property of p in N^* rather than in βN. The proof in §3 is similar to, but more complicated than, the proof of the existence of good ultrafilters.

Keisler's definition of good generalizes in an obvious way to define the notion of a point p is any space X being κ-good; so the κ-good ultrafilters on D are the κ-good points in βD. It is then true in general that κ^+-good implies κ-OK. One cannot prove in ZFC that there is a $(2^\omega)^+$-good point in N^* (add ω_2 Cohen reals to a model of CH). We do not know whether there must be an ω_2-good point in N^*. Observe that every P-point is ω_2-good.

§2. INDEPENDENT LINKED FAMILIES

This generalization of the notion of a family of independent sets will be used in §3 to construct 2^ω-OK points.

In the following, *fin* refers to the ideal of finite subsets of ω. The letter A with various subscripts and superscripts ranges over subsets of ω. $[I]^n = \{\sigma \subset I: |\sigma| = n\}$, and $[I]^{<\omega} = \bigcup_{n\in\omega} [I]^n$. $\mathscr{P}(I)$ is the set of all subsets of I.

Definition 2.1. Let $\mathfrak{F}$ be a filter on ω, $\mathfrak{J}$ the dual ideal, and assume $\mathfrak{J} \supset fin$.

(a) If $1 \leqslant n < \omega$, an indexed family $\{A_i: i \in I\}$ is *precisely n-linked with respect to (w.r.t.)* $\mathfrak{F}$ iff for all $\sigma \in [I]^n$, $\bigcap_{i\in\sigma} A_i \notin \mathfrak{J}$, but for all $\sigma \in [I]^{n+1}$, $\bigcap_{i\in\sigma} A_i \in fin$.

(b) An indexed family $\{A_{in}: i \in I,\ 1 \leqslant n < \omega\}$ is a *linked system w.r.t.* $\mathfrak{F}$ iff for each n, $\{A_{in}: i \in I\}$ is precisely n-linked w.r.t. $\mathfrak{F}$, and for each n and i, $A_{in} \subset A_{i,n+1}$.

(c) An indexed family $\{A^j_{in}: i \in I,\ 1 \leq n < \omega,\ j \in J\}$ is an *I by J independent linked family w.r.t.* $\mathfrak{F}$ iff for each $j \in J$, $\{A^j_{in}: i \in I,\ 1 \leq n < \omega\}$ is a linked system w.r.t. $\mathfrak{F}$, and:

$$\bigcap_{j \in \tau} \Big(\bigcap_{i \in \sigma_j} A^j_{in_j} \Big) \notin \mathfrak{I}$$

whenever $\tau \in [J]^{<\omega}$, and for each $j \in \tau$, $1 \leq n_j < \omega$ and $\sigma_j \in [I]^{n_j}$. ∎

Lemma 2.2. *There is a* 2^ω *by* 2^ω *independent linked family w.r.t. fin.*

Proof. Let

$$S = \{\langle k, f\rangle: k \in \omega \ \&\ f \in \mathcal{P}(\mathcal{P}(k))^{\mathcal{P}(k)}\}.$$

Clearly $|S| = \omega$. Our family will be of the form $\{A^Y_{Xn}: X \in \mathcal{P}(\omega),\ 1 \leq n < \omega,\ Y \in \mathcal{P}(\omega)\}$, where each A^Y_{Xn} is a subset of S (rather than ω) defined by

$$A^Y_{Xn} = \{\langle k, f\rangle \in S: |f(Y \cap k)| \leq n \ \&\ X \cap k \in f(Y \cap k)\}. ∎$$

Our original proof of 2.2 produced a linked system as the set of paths through a tree and an independent linked family was obtained via a tree of trees. We are grateful to P. Simon for pointing out the simple proof here.

§3. PROOFS OF THE MAIN THEOREMS

Since Theorems 0.1 and 0.2 are clearly variations on the same theme, we shall first describe in detail the proof of a simpler result (3.1), and then point out what modifications are necessary to obtain 0.1 and 0.2.

Theorem 3.1. *There is a* $p \in N^*$ *which is* 2^ω*-OK.*

Proof. We think of p as an ultrafilter on ω, and we construct p by transfinite induction in 2^ω stages. At each stage $\mu < 2^\omega$ we will have constructed a filter $\mathfrak{F}_\mu$, and p will be $\bigcup_\mu \mathfrak{F}_\mu$. At the stages wher

μ is even (i.e., $\exists\nu(\mu = 2 \cdot \nu)$, we ensure that p becomes an ultrafilter, and at the stages when μ is odd (i.e., $\exists\nu(\mu = 2 \cdot \nu + 1)$), we ensure that p becomes 2^ω-OK.

Let $\{B_\mu : \mu < 2^\omega \ \& \ \mu \text{ is even}\}$ enumerate all subsets of ω. Also, for $n \in \omega$ and μ an odd ordinal less than 2^ω, let $C_{\mu n} \subset \omega$ be such that for each μ, $\langle C_{\mu n} : n \in \omega\rangle$ is decreasing (i.e., $\forall n(C_{\mu, n+1} \subset C_{\mu n})$), and such that every decreasing sequence of subsets of ω is listed cofinally often. Finally, let $\{A^\beta_{\alpha n} : \alpha < 2^\omega, 1 \leqslant n < \omega, \beta < 2^\omega\}$ be an independent linked family w.r.t. *fin*.

By induction on $\mu < 2^\omega$, we construct $\mathfrak{F}_\mu$ and K_μ so that:

(1) $\mathfrak{F}_\mu$ is a filter on ω, $K_\mu \subset 2^\omega$, and $\{A^\beta_{\alpha n} : \alpha < 2^\omega, 1 \leqslant n < \omega, \beta \in K_\mu\}$ is an independent linked family w.r.t. $\mathfrak{F}_\mu$.

(2) $K_0 = 2^\omega$ and $\mathfrak{F}_0$ is the cofinite filter.

(3) $\nu < \mu \rightarrow \mathfrak{F}_\nu \subset \mathfrak{F}_\mu \ \& \ K_\nu \supset K_\mu$.

(4) If μ is a limit ordinal, $\mathfrak{F}_\mu = \bigcup_{\nu<\mu} \mathfrak{F}_\nu$ and $K_\mu = \bigcap_{\nu<\mu} K_\nu$.

(5) For each μ, $K_\mu \setminus K_{\mu+1}$ is finite.

(6) If μ is even, then either B_μ or $\omega \setminus B_\mu$ is in $\mathfrak{F}_{\mu+1}$.

(7) If μ is odd and each $C_{\mu n} \in \mathfrak{F}_\mu$, then there are $D_{\mu\alpha} \in \mathfrak{F}_{\mu+1}$ for $\alpha < 2^\omega$ such that for all $n \geqslant 1$ and all $\alpha_1 < \alpha_2 < \ldots < \alpha_n < 2^\omega$, $D_{\mu\alpha_1} \cap \ldots \cap D_{\mu\alpha_n} \setminus C_{\mu n}$ is finite.

Assuming that the $\mathfrak{F}_\mu$ and K_μ have been so constructed, we set $p = \bigcup_\mu \mathfrak{F}_\mu$. Condition (6) ensures that p is an ultrafilter, and condition (7) ensures that p is 2^ω-OK in N^*. To prove that this construction can indeed be carried out, we fix $\mu < 2^\omega$, assume that the $\mathfrak{F}_\nu, K_\nu$ have been constructed for $\nu \leqslant \mu$, and describe how to obtain $\mathfrak{F}_{\mu+1}$, and $K_{\mu+1}$.

If μ is even, let $\mathfrak{R}$ be the filter generated by $\mathfrak{F}_\mu$ and B_μ. If $\mathfrak{R}$ is a proper filter and $\{A^\beta_{\alpha n} : \alpha < 2^\omega, 1 \leqslant n < \omega, \beta \in K_\mu\}$ is independent w.r.t. $\mathfrak{K}$, we set $\mathfrak{F}_{\mu+1} = \mathfrak{K}$ and $K_{\mu+1} = K_\mu$. If not, then there is an $E \in \mathfrak{F}_\mu$ such that

$$B_\mu \cap E \cap \bigcap_{\beta \in \tau} \Big(\bigcap_{\alpha \in \sigma_\beta} A^\beta_{\alpha n_\beta} \Big) = \phi$$

for some $\tau \in [K_\mu]^{<\omega}$, $n_\beta \in \omega$, and $\sigma_\beta \in [2^\omega]^{n_\beta}$. Then let $K_{\mu+1} = K_\mu \setminus \tau$, and let $\mathfrak{F}_{\mu+1}$ be the filter generated by $\mathfrak{F}_\mu$ and

$$\bigcap_{\beta \in \tau} \Big(\bigcap_{\alpha \in \sigma_\beta} A^\beta_{\alpha n_\beta} \Big);$$

so $\omega \setminus B_\mu \in \mathfrak{F}_{\mu+1}$.

If μ is odd and some $C_{\mu n}$ is not in $\mathfrak{F}_\mu$, let $\mathfrak{F}_{\mu+1} = \mathfrak{F}_\mu$ and $K_{\mu+1} = K_\mu$. If each $C_{\mu n} \in \mathfrak{F}_\mu$, then fix $\beta \in K_\mu$; observe that by (5), $K_\mu \neq \phi$. Let $K_{\mu+1} = K_\mu \setminus \{\beta\}$ and let $\mathfrak{F}_{\mu+1}$ be the filter generated by $\mathfrak{F}_\mu$ and the $D_{\mu\alpha}$, for $\alpha < 2^\omega$, where

$$D_{\mu\alpha} = \Big(\bigcap_n C_{\mu n}\Big) \cup \Big(\bigcup_{1 < n < \omega} (A^\beta_{\alpha n} \cap C_{\mu n} \setminus C_{\mu, n+1})\Big).$$

To verify condition (7), let $\alpha_1 < \alpha_2 < \ldots < \alpha_n < 2^\omega$, and let $Y = D_{\mu\alpha_1} \cap \ldots \cap D_{\mu\alpha_n} \setminus C_{\mu n}$; if $n = 1$ then $Y = \phi$, while if $n > 1$ then $Y \subset A^\beta_{\alpha_1, n-1} \cap \ldots \cap A^\beta_{\alpha_n, n-1}$, which is finite because these $A^\beta_{\alpha, n-1}$ are precisely $(n-1)$-linked. To verify condition (1), observe that $D_{\mu\alpha} \supset C_{\mu n} \cap A^\beta_{\alpha n}$ for each n. ∎

Proof of Theorem 0.1. We construct 2^{2^ω} distinct 2^ω-OK points, $p^h \in N^*$, for h a function from 2^ω into 2. Each p^h is obtained by constructing $\mathfrak{F}^h_\mu$ and K^h_μ precisely as in the proof of 3.1 to satisfy (1)-(7), except that $\mathfrak{F}^h_0$ is no longer the cofinite filter. Instead, the $\mathfrak{F}^h_0$ are chosen so that whenever $h \neq g$, $\mathfrak{F}^h_0$ contains a set which is disjoint from some set in $\mathfrak{F}^g_0$; this ensures that the various p^h will be distinct. We furthermore choose each $\mathfrak{F}^h_0$ so that no ultrafilter extending it is a P-point. Once the $\mathfrak{F}^h_0$ are so determined, the construction of the $\mathfrak{F}^h_\mu$ proceeds separately for each fixed h as in 3.1.

We now describe explicitly how to choose the $\mathfrak{F}^h_0$. If $E \subset \omega$, let $E^0 = E$ and $E^1 = \omega \setminus E$. We begin with a collection of the form

$$\{A^\beta_{\alpha n}: \alpha < 2^\omega,\ 1 \leq n < \omega,\ \beta < 2^\omega\} \cup \{E_\gamma: \gamma < 2^\omega\}$$

such that the whole system is independent in the sense that any intersection of the form

$$\bigcap_{\beta \in \tau} \Big(\bigcap_{\alpha \in \sigma_\beta} A^\beta_{\alpha n_\beta}\Big) \cap \bigcap_{\gamma \in \rho} E^{f(\gamma)}_\gamma$$

is infinite, where $\rho \in [2^\omega]^{<\omega}$, $f: \rho \to 2$, $\tau \in [2^\omega]^{<\omega}$, and for each $\beta \in \tau$, $1 \leq n_\beta < \omega$ and $\sigma_\beta \in [2^\omega]^{n_\beta}$. To obtain such a family, let

$$\{A^\beta_{\alpha n}: \alpha < 2^\omega,\ 1 \leq n < \omega,\ \beta < 2^\omega + 2^\omega\}$$

be an independent linked family w.r.t. *fin*, and set $E_\gamma = A^{2^\omega + \gamma}_{\alpha 1}$ for any $\alpha < 2^\omega$. Now, let $\mathfrak{F}^h_0$ be the filter generated by *fin*, all the $E^{h(\gamma)}_\gamma$ for $\gamma < 2^\omega$, and all sets $\omega \setminus Y$ such that

$$(*) \qquad \forall \gamma < \omega(|Y \setminus E^{h(\gamma)}_\gamma| < \omega).$$

The inclusion of these Y guarantees that no extension of $\mathfrak{F}^h_0$ is a P-point. ∎

We could have replaced $(*)$ by

$$|\{\gamma < 2^\omega: |Y \setminus E^{h(\gamma)}_\gamma| < \omega\}| \geq \omega.$$

This would ensure in addition that each p^h has character 2^ω in N^* (see [2], Theorem 2.8).

Proof of Theorem 0.2′. We construct ultrafilters p^δ for $\delta < 2^\omega$. We again begin with a family

$$\{A^\beta_{\alpha n}: \alpha < 2^\omega,\ 1 \leq n < \omega,\ \beta < 2^\omega\} \cup \{E_\gamma: \gamma < 2^\omega\}$$

which is independent as above. Now each $\mathfrak{F}^\delta_0$ is the cofinite filter, and the $\mathfrak{F}^\delta_\mu$ and K_μ are constructed simultaneously for all δ, by induction on μ. There is no superscript on the K_μ; but we arrange that for each δ, the system

$$\{A^\beta_{\alpha n}: \alpha < 2^\omega,\ 1 \leq n < \omega,\ \beta \in K_\mu\} \cup \{E_\gamma: \gamma \in K_\mu\}$$

is independent w.r.t. $\mathfrak{F}^\delta_\mu$. $K_\mu \setminus K_{\mu+1}$ is still finite; this presents no problem since at each stage μ, only one or two of the $\mathfrak{F}^{\delta}_{\mu}$ get extended properly. If $\mu \equiv 0 \pmod 3$, then for some $\delta = \delta_{\mu}$ and some $B = B_{\mu}$, $\mathfrak{F}^{\delta}_{\mu+1}$ contains either B or $\omega \setminus B$; the enumeration $\{\langle \delta_{\mu}, B_{\mu} \rangle : \mu < 2^{\omega}$ & $\mu \equiv 0 \pmod 3)\}$ is chosen to range over all of $2^{\omega} \times \mathscr{P}(\omega)$. If $\mu \equiv 1 \pmod 3$, then for some δ, $\mathfrak{F}^{\delta}_{\mu+1}$ is chosen to handle some sequence $\langle C_n : n \in \omega \rangle$ as in (7) of the proof of 3.1. Finally, if $\mu \equiv 2$ (mod 3), then for some distinct δ and ζ, and some $f: \omega \to \omega$, we choose $\mathfrak{F}^{\delta}_{\mu+1}$ and $\mathfrak{F}^{\zeta}_{\mu+1}$ to ensure that $p^{\delta} \neq f_*(p^{\zeta})$; this step is as in the proof in [2] of the existence of Rudin – Keisler incomparable points. ∎

REFERENCES

[1] C.C. Chang – H.J. Keisler, *Model theory,* 2-nd ed., North-Holland, Amsterdam, 1977.

[2] K. Kunen, Ultrafilters and independent sets, *Trans. Amer. Math. Soc.*, 172 (1972), 299-306.

[3] K. Kunen, Some points in βN, *Proc. Cambridge Phil. Soc.*, 80 (1976), 385-398.

[4] C. Mills, An easier proof of the Shelah *P*-point independence theorem, to appear.

[5] W. Rudin, Homogeneity problems in the theory of Cech compactifications, *Duke Math. J.*, 23 (1956), 409-420.

[6] E. Wimmers, The Shelah *P*-point independence theorem, to appear.

K. Kunen

University of Wisconsin – Madison, Department of Mathematics, Van Vleck Hall, 480 Lincoln Drive, Madison, Wisconsin 53706, USA.

Dimension Theory

E. Čech's work and the development of dimension theory

Miroslav Katětov

As is well known, there are two main kinds of dimension of topological spaces: The covering dimension dim, also called Lebesgue-Čech dimension, and the large inductive dimension Ind, also called Brouwer-Čech dimension. The small inductive dimension ind, which coincides with dim and Ind for separable metrizable spaces, seems to be of lesser importance, though it had been examined in a systematic way before the general definitions of dim and Ind were given. There are also other kinds of dimension of topological spaces, but we will concentrate on dim, Ind and ind.

The dimensions dim and Ind have been introduced in full generality by E. Čech [1931, 1932, 1933]. Let us add that the birthday of the large inductive dimension is known exactly: the pertinent note by E. Čech was presented to the Academy in Paris by Élie Cartan on November 16, 1931.

For the convenience of the reader, we are going to state the definitions of dim, Ind and ind.

If X is a topological space, then the *covering* or *Lebesgue-Čech dimension* $\dim X$ of X is defined as follows. If there is an $n \in \{-1, 0, 1, \ldots\}$ such that, for every finite open cover of a space X, there exists a finer finite open cover of order $\leq n$ (i.e., such that no point is in more than $n+1$ sets belonging to the cover), then $\dim X$ is equal to the least of such integers n. If there is no such n, then $\dim X = \infty$.

The *Čech-Brouwer* or *large inductive dimension* $\operatorname{Ind} X$ of a topological space X is defined inductively. If X is void, then $\operatorname{Ind} X = -1$. If $\operatorname{Ind} X = k$ is already defined for $k < n$, then $\operatorname{Ind} X = n$, if $\operatorname{Ind} X = k$ for no $k < n$ and, for every closed $F \subset X$ and every open $G \subset X$ containing F, there is an open U such that $F \subset U \subset G$ and $\operatorname{Ind}(\bar{U} \setminus U) < n$. If $\operatorname{Ind} X = k$ for no $k = -1, 0, 1, \ldots$, $\operatorname{Ind} X = \infty$.

The definition of the small inductive dimension differs from the definition of Ind only in that F is assumed to be a singleton (as a rule, it is required that all singletons should be closed).

In what follows, we shall not give the definitions of well-known standard concepts. On the other hand, as a rule, we will state the definitions of concepts of a more special kind, and of terms the usage of which is not unified. Let us add that, by a "topological space" or simply "space", we will always mean a topological space such that all singletons are closed. This convention does not extend to the articles reproduced in this chapter.

Since it was E. Čech who introduced dim and Ind in full generality, the present chapter should contain, strictly speaking, a complete survey of the development of dimension theory for topological spaces more general than the separable metrizable ones. This is impossible since it would mean writing a book or even books. Therefore, a short survey is given of only the theorems and examples connected directly, though sometimes only implicitly, with E. Čech's work. In addition, some other results are briefly mentioned.

The survey is not complete even for the selected topics we are concerned with. Its style rather resembles that of sight-seeing tours: some important or interesting objects are pointed out and some information is given making possible a more detailed acquaintance, whereas many other objects of equal interest are neglected.

In addition to this survey, we have inserted some pages of a different kind, namely concerning the origins of, or rather the preludes to, various kinds of dimension. The main reason lies in the fact that B. Bolzano's prelude to ind is still little known.

Many parts of dimension theory are completely neglected in our survey. No attention is paid to the dimension theory of proximity and uniform spaces. We do not consider results and problems which, by their nature, belong to the dimension theory of separable metrizable spaces, and we pay little attention to questions concerning general metrizable spaces. There are a lot of other topics not mentioned. To name only some of them: infinite-dimensional spaces; universal, in various sense, spaces; dimension of mappings and theorems on the change of dimension under mappings of a specific kind; metric-dependent dimension function, etc. Let us add that, in fact, one metric dependent function will be considered, namely the one introduced, almost explicitly, by Bolzano.

The chapter is organized as follows. The present survey consists of the following sections: Preludes to dimension theory, equality of dimensions, monotonicity, dimensions of Čech-Stone compactifications, sum theorem, and a section on some other questions. It is followed by a bibliography containing all articles mentioned in the survey. The bibliography also includes several books on dimension theory and some survey articles [ALEXANDROV 1951, 1960, 1964], [NAGATA 1967, 1971], [ALEXANDROV, FEDORČUK 1978], [PASYNKOV, FEDORČUK, FILIPPOV 1979], [FEDORČUK 1988]. References in the text are of the form e.g. [NISHIURA 1977], or of some other unambiguous form, e.g. "R. Engelking's book".

The survey is succeeded by the English translation of two papers [1932, 1933] by E. Čech and 7 articles (also in English) by other authors from the period 1949 – 1985. We endeavoured to select some of the most significant and interesting articles. However, due to space limitations, short notes have been preferred.

Preludes to dimension theory. Each of the three most important dimensions of topological spaces, the small and the large inductive dimensions and the covering dimension, came into being after a certain "prelude", partly of a mathematical kind, and partly rather intuitive.

We will not go into the mathematical prelude to Ind and dim. For these questions, we refer to books on dimension theory, in particular, to section 1.1 of R. Engelking's book which contains a short characteristic of the fundamental papers by L. E. J. Brouwer [1911, 1913] and H. Lebesgue [1911] as well as of some subsequent articles.

For Ind (but, as it seems, not for dim), there was also a prelude of a rather intuitive character. We mean, of course, some ideas of H. Poincaré. They are quoted in many books, but usually the references are not quite precise.

As is well known, in 1902 – 1913 there appeared four books by H. Poincaré in which he dealt mainly with various philosophical and methodological problems of science: *La Science et l'Hypothèse*, *La Valeur de la Science*, *La Science et la Méthode*, *Dernières Pensées* (prepared and published after the death of Poincaré). Two of these books contain passages presenting intuitive ideas which can lead, after some clarification, to a mathematical definition of dimension, namely of the large inductive one. However, H. Poincaré himself did not try give it an exact form.

One of these passages is contained in the book *La Valeur de la Science*, Paris, 1905, see, in particular, pp. 73 and 74 (we quote after the edition of 1911). This is the one quoted at the very beginning of the well-known book W. Hurewicz, H. Wallman, *Dimension Theory*, Princeton 1941. However, the authors of the book are mistaken when they say that the quotations are from an article [1912] by H. Poincaré in a philosophical journal (see below).

A part of the passage in question is as follows.

"... if to divide a continuum it suffices to consider a certain number of elements all distinguishable from one another, we say that the continuum is of one dimension;

if on the contrary, to divide a continuum it is necessary to consider as cuts a system of elements themselves forming one or several continua, we shall say that this continuum is of several dimensions.

If to divide a continuum C, cuts which form one or several continua of one dimension suffice, we shall say that C is a continuum of two dimensions; if cuts which form one or several continua of at most two dimensions suffice, we shall say that C is a continuum of three dimensions; and so on."

Another passage of a similar character is contained in the book *Dernières Pensées*, Paris, 1913, namely in the chapter entitled "Pourquoi l'Espace a trois dimensions", and also in the article (mentioned above) which appeared under the same title in *Révue de Métaphysique et de Morale* 20 (1912), 483–504. It is just this article to which many books and articles refer when speaking of the beginnings of dimension theory, although it was published seven years later and the pertinent passage is still more vague than that from which we have quoted.

The article mentioned above is also quoted by P. Urysohn in his first extensive publication [1925] on dimension theory, and it is quite possible that he was partly inspired by H. Poincaré's ideas.

L. E. J. Brouwer's paper [1913] is based, as the author states explicitly, on H. Poincaré's ideas, however, only the article from 1912 is quoted.

It is somewhat surprising that an article published in a journal having no direct connection with mathematics could influence mathematical work. This can be explained by the fact that Poincaré's books on general problems of science were widely read, and therefore no article by him, even if published in a philosophical journal, escaped the attention of scientists including mathematicians.

We now turn to the small inductive dimension. Its definition was published for the first time by P. Urysohn in 1922 and, independently, by K. Menger in 1923. We shall not consider the relations between results and publications of P. Urysohn and of K. Menger. For a pertinent information, we refer to R. Engelking's book (Section 1.1). We only want to recall that the small inductive dimension was investigated in those years primarily for compact metric spaces. In the full generality, it was introduced in P. Urysohn's paper [1925], though only in the form of a remark (Chapter I, section 11).

More than 80 years separate these dates from the prelude to the small inductive dimension. This prelude, contained in the work of B. Bolzano, was unknown to mathematicians for more than a century (the pertinent Bolzano's manuscript was published in 1948) and began to be investigated not earlier than about 15 years ago.

Bolzano's ideas concerning the concept of dimension are still not mentioned in the books on dimension theory or, at most, in a very vague way. Thus, in the well-known and very important book P. S. Alexandrov, B. Pasynkov [1973], B. Bolzano is mentioned only once: *"... this idea of Poincaré, which possibly stems already from Bolzano or even Lobachevskii ..."*. It seems that the first paper paying full attention to Bolzano's conception of dimension is that of D. M. Johnson [1977].

B. Bolzano's contribution to dimension theory is contained in his work "Uiber Haltung, Richtung, Krümmung und Schnörkelung bei Linien sowohl als Flächen sammt einigen verwandten Begriffen", written in 1843 and 1844 and first published in the book *Spisy Bernarda Bolzana*, Vol. 5: Geometrické práce, Ed. J. Vojtěch, Praha 1948.

In fact, this work could not have any influence on the origin and development of dimension theory. Nevertheless, an analysis of Bolzano's approach is quite interesting. His definitions are precise (up to minor changes) even from the modern standpoint. Extending them in a quite natural manner, we get, as already mentioned, a metric-dependent function defined for metric spaces. In a sense, this function is equivalent to the small inductive dimension (for metric spaces), as shown by M. Katětov in 1983.

We now quote some relevant passages from B. Bolzano's work "Uiber Haltung ...", as published in 1948, always adding the English translation and comments. Then we present a short analysis of these passages and perform a transition to an extended definition in modern terms.

(A) *"Wenn der Punct i in einem Raumdinge so liegt, dass keine auch so kleine Entfernung angeblich ist, von der behauptet werden könnte, für diese und für alle kleinere Entfernungen ... besitze i einen oder etliche Nachbarn: so sage ich, i stehe isolirt oder vereinzelt in diesem Raumdinge."*

Translation. If the point i so lies in a spatial object so that it is not possible to indicate a distance, however small, for which it can be asserted that for this distance and for all smaller distances ... i possesses one or several neighbours: then I say that i stands isolated or alone in this spatial object.

Comment. We translate "Raumding" as "spatial object". In fact, in Bolzano's terminology, "Raumding" means what we would call a non-void subset of the Euclidean space E_3. – "Nachbar", i.e., neighbour means a point at a given positive distance from the point under consideration.

(B) *"Ein Raumding, das keinen einzigen isolirt stehenden Punct hat, in welchem sonach jeder Punct für jede wenn auch nicht beliebig grosse, doch hinlänglich kleine*

Entfernung gewisse Nachbarn besitzt, nenne ich ein ausgedehntes, continuirliches Raumding, auch eine Ausdehnung. Ein Raumding dagegen, dessen jeder Punct vereinzelt stehet, nenne ich ein discontinuirliches Raumding."

Translation. I call a spatial object extended or continuous or else an extension if it contains no point standing isolated, in which therefore every point has some neighbours, not necessarily for an arbitrarily large distance, but certainly for all sufficiently small ones. I call a discontinuous spatial object an object whose every point stands alone.

(C) *"Ein Ausgedehntes, dessen jeder Punct für jede hinlänglich kleine Entfernung der Nachbarn nur so viele hat, dass ihr Inbegriff, für eine jede dieser Entfernungen für sich allein betrachtet, noch kein Ausgedehntes darstellt, nenne ich ein Raumding von einer einzigen oder einfachen Ausdehnung, auch eine Linie. Ein Raumding, dessen jeder Punct für jede hinlänglich kleine Entfernung der Nachbarn so viele hat, dass ihr Inbegriff für eine jede dieser Entfernungen für sich allein betrachtet selbst noch ein Raumding von einfacher Ausdehnung darstellt, nenne ich ein Raumding von zwiefacher oder doppelter Ausdehnung, auch eine Fläche. Ein Raumding endlich, dessen jeder Punct für jede hinlänglich kleine Entfernung der Nachbarn so viele hat, dass ihr Inbegriff für eine jede dieser Entfernungen für sich allein betrachtet schon ein Raumding von doppelter Ausdehnung darstellt, nenne ich ein Raumding von dreifacher Ausdehnung oder einen Körper."*

Translation. I call an extension whose every point has only so many neighbours for each sufficiently small distance that the set of these, considered in itself for each of the distances, still does not represent an extension, a "spatial object of a single or simple extension" or a "line". I call a spatial object whose every point has so many neighbours for each sufficiently small distance that the set of these, considered in itself for each of these distances, still represents a spatial object of a simple extension, a "spatial object of a twofold or double extension" or a "surface". Finally, I call a spatial object whose every point has so many neighbours for each sufficiently small distance that the set of these, considered in itself for each of these distances, already represents a spatial object of double extension, a "spatial object of a threefold extension" or a "solid".

(D) *"In einer Linie hat jeder Punct anzufangen von einer gewissen Entfernung für jede kleinere der Nachbarn insgemein nur zwei, zuweilen aber nur einen oder mehr als zwei, jedesmal doch nur so viele, dass ihr Inbegriff (wie wir schon wissen) ein discontinuirliches Raumding darstellt."*

Translation. In a line, every point possesses, beginning with a certain distance for every smaller one, usually only two neighbours, sometimes only one or several, but in any case only so many that their set (as we already know) represents a discontinuous spatial object.

This last sentence is remarkable; when confronted with the corresponding part of (C); it shows that, in Bolzano's usage, the expressions "noch kein Ausgedehntes" (... still no extension) and "ein discontinuirliches Raumding" (a discontinuous spatial object) have, or at least could have, the same meaning. This is not as strange as it may seem. B. Bolzano concentrated on cases, in which every set under consideration is of the same dimension at each of its points.

For an analysis of Bolzano's approach and his definitions, we refer to the note [KATĚTOV 1983], already mentioned. The definitions contained in (A) – (C) will now be transformed to a modern form. In addition, we replace, in (C) "noch kein Ausgedehntes" with "ein discontinuirliches Raumding". We use the current notation. If (P, ϱ) is a metric space, $T \subset P$, $x \in P$, $\varepsilon > 0$, $S_T(x, \varepsilon)$ denotes the set $\{y \in T : \varrho(x, y) = \varepsilon\}$. In this way, we get the following Bolzano definitions reformulated. Let $\emptyset \neq X \subset E_3$. Then

(I) if for any $x \in X$ and any $\varepsilon > 0$ there exists a positive $\delta < \varepsilon$ such that $S_X(x, \delta) = \emptyset$, we will say that X has the property $\mathcal{D}_0$.

(II) If $n = 1, 2, 3$ and for any $x \in X$ there exists an $\varepsilon > 0$ such that, for any positive $\delta \leq \varepsilon$ the set $S_X(x, \delta)$ has the property $\mathcal{D}_{n-1}$, we will say that X has the property $\mathcal{D}_n$.

These definitions correspond completely to Bolzano's original definitions except for the replacement mentioned above, which may be questionable since we cannot know whether it corresponds to Bolzano's intentions.

We now perform another transformation passing from E_3 and its subspaces to an arbitrary metric space. This transformation is obvious from the modern standpoint. On the other hand, we can take for certain that B. Bolzano never intended anything of this kind, since he never considered any metric space distinct from E_3 and its subspaces. We obtain the following definition of what we will call the Bolzano dimension (of metric space).

Definition. Let P be a metric space. We put $\text{B-ind}\, P = -1$ iff $P = \emptyset$, $\text{B-ind}\, P = 0$ iff $P \neq \emptyset$ and, for any $x \in P$ and any $\varepsilon > 0$, there exists a positive $\delta < \varepsilon$ such that $S_P(x, \delta) = \emptyset$. For $n = 1, 2, \ldots$, we put $\text{B-ind}\, P = n$ iff

(1) $\text{B-ind}\, P \leq n - 1$ does not hold,

(2) for every $x \in P$ there is an $\varepsilon > 0$ such that $\text{B-ind}\, S_P(x, \delta) \leq n - 1$ whenever $0 < \delta < \varepsilon$.

If B-ind $P = n$ for no $n = -1, 0, 1, \ldots$, we put B-ind $P = \infty$. We will call B-ind P Bolzano dimension of P.

For an examination of properties of the metric-dependent dimension function B-ind, we refer to [KATĚTOV 1983]. However, we will state the main theorem showing the relation of B-ind to the small inductive dimension.

Let X be a separable metrizable topological space; let $0 \leq n < \infty$. Then $\operatorname{ind} X = n$, if and only if

(1) there is a space $Y \subset \mathbb{R}^{2n+1}$ homeomorphic to X and satisfying B-ind $Y = n$,

(2) B-ind $Z \geq n$ for every $Z \subset \mathbb{R}^{2n+1}$ homeomorphic to X.

Equality of dimensions. The problem when the equality $\dim X = \operatorname{Ind} X$ takes place, emerged in Čech's fundamental paper "A contribution to the theory of dimension" (Czech), Čas. Pěst. Mat. Fys. 62 (1932), 277 – 279 (this paper is reprinted in this volume). To be precise, E. Čech conjectured that $\operatorname{Ind} X = \dim X$ for every perfectly normal space X.

It seems that the first deeper result concerning the equality in question was the proof that $\operatorname{Ind} X = \dim X$ for all metrizable spaces X. This theorem was proved independently by M. Katětov and K. Morita; see [KATĚTOV 1951, 1952], [MORITA 1954]. Later, a different proof was given by C. H. Dowker and W. Hurewicz [1956]; this paper is included in the present volume. Let us mention also a simple proof contained in [PRYMUSIŃSKI 1974].

The subsequent examination of conditions for Ind = dim went, as might be expected, in two directions: the search for wide classes of spaces in which the equality holds, and the search for spaces with Ind $\neq$ dim which have good properties in other respects. It is an interesting circumstance that several important examples possess certain good properties only under the assumption of (CH), i.e., of continuum hypothesis.

The question of the equalities $\operatorname{ind} X = \operatorname{Ind} X$ and $\operatorname{ind} X = \dim X$ was investigated on similar lines. We shall pay them less attention. However, it seems necessary to mention one of the most substantial results concerning the relation between ind and Ind, namely the P. Roy's construction of a metric space X with $\operatorname{ind} X = 0$, and $\operatorname{Ind} X = \dim X = 1$; see [ROY 1962] and a detailed exposition in [ROY 1968].

Relatively simple constructions of metric spaces X with $\operatorname{ind} X = 0$ and $\operatorname{Ind} X = 1$, and also some related results appear in [KULESZA 1990].

After the result on the equality of dim and Ind in arbitrary metric spaces, there has been found a number of classes of spaces for which the equality holds. In a prevalent majority, these classes of spaces are related in some appropriate sense to metrizable spaces. We shall not attempt here to give a complete survey of such classes or to present the latest results; we want only to point out some important and interesting theorems. The problem, formulated only vaguely, to find the widest natural class of spaces on which dim and Ind agree, is apparently still open, and it is not clear, whether it can have a solution.

One of the first results concerning wider classes of spaces on which the equality of dimensions holds was the following theorem proved by B. A. Pasynkov: If a normal space X admits a closed 0-dimensional continuous mapping onto a metric space, then $\operatorname{Ind} X = \dim X$; see [PASYNKOV 1964]. Later, I. M. Leĭbo [1974] proved a theorem which is, in a sense, a "reverse": If X is a Lashnev space, i.e., if there exists a closed continuous mapping of a metric space onto X, then $\operatorname{Ind} X = \dim X$. In the later paper by I. M. Leĭbo [1982], a theorem was proved which encompasses both these results. It reads as follows: If X is a normal space and if there are spaces Y, Z and continuous surjective mappings $f : X \to Y$, $g : Z \to Y$ such that Z is metrizable, f, g are closed and $\dim f^{-1}(y) = 0$ for every $y \in Y$, then $\operatorname{Ind} X = \dim X$.

In the years 1980 and 1981, there appears a number of papers containing proofs of $\dim X = \operatorname{Ind} X$ (and of some related results) for certain classes of spaces related to metric ones in some specified sense, different from that mentioned above in connection with B. Pasynkov's and I. Leĭbo's papers.

Among others, the following classes of spaces were examined. In 1970 and 1971, K. Nagami [1971, 1971a] introduced σ-metric spaces and μ-spaces. They are defined as follows: σ-metric spaces are unions of countably many metrizable closed subspaces; μ-spaces are subspaces of products of countably many paracompact σ-metric spaces.

Later on, K. Nagami [1980, 1980a, 1981] introduced L-spaces and weak L-spaces. As shown by M. Ito [1982], these two concepts coincide. We do not formulate the definition since it turned out [MIZOKAMI 1981] that they are contained in the class of μ-spaces, and many results (including dim = Ind) proved for them are valid for this wider class.

In 1980, the following classes were introduced by S. Oka. Patched spaces are defined as paracompact perfectly normal spaces which can be represented as a union of finitely many metrizable subspaces. Those paracompact spaces which can be expressed as a union of countably many closed patched subspaces are called σ-patched. Finally, free σ-patched spaces are subspaces of products of countably many σ-patched spaces.

The equality dim = Ind (and some related results) was proved for the classes mentioned above in the following papers. For L-spaces and weak L-spaces see [NAGAMI 1980, 1980a, 1981]. For free σ-patched spaces see [OKA 1980]; this paper also contains (as "added in proof") the assertion that free σ-patched spaces and μ-spaces coincide. For μ-spaces see [MIZOKAMI 1981]; in Mizokami's paper it is also proved that, for μ-spaces, $\dim X \le n$ is equivalent to $X = \bigcup_0^n X_i$, $\dim X_i \le 0$.

We have described this part of the development of dimension theory in a relatively detailed way. The reason is that it shows some features not uncommon in mathematics: it can happen that the search for wider and wider classes of objects with a certain desirable property proceeds from various directions and, nevertheless, leads to the same class, though characterized in different ways.

The first examples of spaces with Ind $\neq$ dim appeared somewhat earlier than the first positive results on dim = Ind in classes wider than that of separable metrizable spaces.

The first example of a compact space X with noncoinciding dimensions dim and Ind was given by A. L. Lunc [1949]. Somewhat later, a considerably simpler example was constructed by O. V. Lokucievskiĭ [1949]. This paper is reproduced in this volume. Lokucievskiĭ's example is a space X with $\operatorname{Ind} X = \operatorname{ind} X = 2$, $\dim X = 1$. Moreover, there is a partition into closed subspaces X_1, X_2 such that $\operatorname{Ind} X_1 = \operatorname{Ind} X_2 = 1$; hence in this space, the so-called sum theorem fails.

Almost 10 years later, the question on equality of dimensions dim and Ind (and of dim and ind) was fully solved, in the negative, by P. Vopěnka [1958]; this paper also appears in the present volume. Compact spaces X_{mn} and Y_{mn} satisfying $\dim X_{mn} = n$, $\operatorname{Ind} X_{mn} = m$, $\dim Y_{mn} = n$, $\operatorname{ind} Y_{mn} = m$ are constructed for arbitrary m and n with $m > n \ge 0$.

As for ind = Ind, it seems that the first example of a normal X with $\operatorname{ind} X \neq \operatorname{Ind} X$ appeared in [SMIRNOV 1951] whereas the first examples of compact spaces X with $\operatorname{ind} X \neq \operatorname{Ind} X$ were given by V. Filippov [1969, 1970, 1970a, 1970b]; the last of these notes is reprinted in this volume.

As for ind and dim, $\operatorname{ind} X > \dim X$ holds for the compact space constructed in [LOKUCIEVSKIĬ 1949]. Later on, V. V. Filippov [1970b] showed that, for any m, n with $1 \le m \le n$, there is a first-countable compact space X with $\dim X = m$, $\operatorname{ind} X = n$. In the same article, it is shown that, under (CH), there are perfectly normal compact spaces Q_n, $n = 1, 2, \ldots$, with $\dim Q_n = 1$, $\operatorname{ind} Q_n = \operatorname{Ind} Q_n = n$. Let us note that a normal space M with $\operatorname{ind} M = 0$, $\dim M = 1$ was constructed by C. H. Dowker [1955], and an example of a normal X with $\operatorname{ind} X = 0$, $\dim X = \infty$ was given by Smirnov [1958].

It is natural to ask whether there are compact spaces with additional good properties and non-coinciding dimensions. It seems that the first example of this kind was given by V. V. Fedorčuk [1968]. This author constructed a compact space X which was separable, first-countable and satisfied, still, $\dim X = 2$, $3 \leq \operatorname{ind} X \leq 4$. In another paper of the same author [1978], there are presented, though under (CH) only, examples of compact spaces X_n, $n = 1, 2 \ldots$, such that $\dim X_n = 1$, $\operatorname{Ind} X_n = n$ and, in addition, X_n is perfectly normal and hereditarily separable; there are also examples of hereditarily separable locally countable locally compact perfectly normal spaces X_{mn}, where $1 \leq m \leq n$, such that $\operatorname{ind} X_{mn} = 0$, $\dim X_{mn} = m$, $\operatorname{Ind} X_{mn} = n$.

As for the values which can be assumed simultaneously by dim, ind and Ind on arbitrary normal spaces, I. K. Lifanov [1973] proved that, for any k, m and n satisfying $0 \leq k \leq n$, $m \leq n$, there is a normal space X satisfying $\dim X = k$, $\operatorname{ind} X = m$, $\operatorname{Ind} X = n$.

In the following years, there were not many examples of spaces with good properties and non-coinciding dimensions. However, there were some important results. In [CHARALAMBOUS 1985] (this article is reprinted in this volume), the author presents examples showing that the assertions in [LIFANOV 1973] concerning possible values of dim, ind and Ind remain true with "normal" replaced with "separable normal" and, under (CH), even with "separable perfectly normal". There are also other examples of perfectly normal spaces not satisfying the equality $\dim X = \operatorname{Ind} X$, but all of them have been constructed under (CH) as well.

Monotonicity. We will say that a dimension function (usually Ind or dim) is monotone on a certain class of spaces, if for any X, Y belonging to the class in question, $Y \subset X$ implies that the dimension of Y is not greater than that of X.

It was already proved in Čech's papers that Ind and dim are monotone on the class of perfectly normal spaces. There appears a quite natural question as to whether dim or Ind are monotone on some suitable wider class of spaces. The question was for the first time explicitly posed by E. Čech, Problem 53, Colloq. Math. 1 (1948), 332; to be precise, he conjectured that dim is monotone on the class of all hereditarily normal spaces. For the dimension Ind, the problem of monotonicity was posed by C. H. Dowker [1953]; this article is included in the present volume.

Let us discuss now first the positive results concerning the monotonicity of Ind.

It seems that the first significant result in this direction was obtained by C. H. Dowker [1953]. He proved there that if X is a totally normal space and Y is its subspace, then Y is totally normal and $\operatorname{Ind} Y \leq \operatorname{Ind} X$. Recall that the total normality (this notion was introduced by C. H. Dowker, too) is defined as follows: a space X is called totally normal, if it is normal and if every open subset U of it is a union of a locally finite (in U) collection of open F_σ sets.

Later, B. A. Pasynkov [1967] introduced a larger class of spaces which he called Dowker spaces; the definition of these spaces is obtained if "locally finite" is replaced by "point-finite" in the definition of totally normal spaces. In that paper, among others, the statement on the monotonicity of Ind on the class of Dowker spaces was announced. The proofs were presented in the paper [PASYNKOV, LIFANOV 1970].

The next important step was the theorem proved by T. Nishiura [1977]. It asserts the monotonicity of dimension Ind for spaces which T. Nishiura called supernormal. They are the spaces such that the following holds: If A, B are disjoint closed sets, then there exist disjoint open sets $U \supset A$, $V \supset B$, which are a union of a locally finite (in U and V, respectively) collection of open F_σ-sets.

The results mentioned above originated in fact from modifications of considerations in Dowker's paper [1953]. By another modification and mainly by amalgamation of approaches contained in the papers by B. A. Pasynkov and T. Nishiura, R. Engelking then proved that Ind is monotone also on the class of all strongly hereditarily normal spaces. These spaces, introduced by R. Engelking, are defined as follows: If A, B are separated sets, then there exist disjoint open sets $U \supset A$, $V \supset B$, each of them being a union of a point-finite collection of open F_σ sets. The definition of strongly hereditarily normal spaces, the basic propositions on their properties and also the theorem on the monotonicity are contained R. Engelking's book; it seems that they were not published in a journal form. As for the monotonicity of dim, it also holds for the class of strongly hereditarily normal spaces; the proof is contained in Engelking's book. Earlier, this monotonicity had been proved for several classes included in the class of strongly hereditarily normal spaces. For totally normal spaces, the monotonicity of dim was proved by C. H. Dowker [1955]. It was extended to Dowker spaces and to supernormal spaces in the papers [PASYNKOV and LIFANOV 1970] and [NISHIURA 1977].

Let us note that the monotonicity of the dimension dim cannot be in general deduced from the monotonicity of Ind (and vice versa); namely, $\dim X$ may differ from $\operatorname{Ind} X$, as we have shown, even in perfectly normal spaces and they are of course strongly hereditarily normal.

There is also a number of theorems asserting that dim is monotone in some restricted sense, namely that $\dim Y \leq \dim X$, if Y is situated in X in a specific way or has certain specific properties.

The first theorem of this kind appeared in [SMIRNOV 1951a]: if X is normal, $Y \subset X$ and, for every open $G \supset Y$, there is an F_σ-set Z such that $Y \subset Z \subset G$, then $\dim Y \leq \dim X$. A significant result is contained in [ZOLOTAREV 1975]: $\dim Y \leq \dim X$ holds if X is normal and Y is fully paracompact. Substantial generalizations are contained in [FILIPPOV 1983].

It seems that the first example of a normal space, where the monotonicity of Ind (with respect to normal subspaces) is violated, is contained in Dowker's paper [1955]. It is a compact space Z satisfying $\operatorname{Ind} Z = 0$ and containing a normal subspace X with $\operatorname{Ind} X > 0$.

On the whole, there are fewer examples on the failure of monotonicity than examples on the failure of equality of dimensions. We shall mention here only one, which seems to be substantial. E. Pol and R. Pol investigated [1977, 1977a, 1979] the problem of normal spaces X with $\dim X = 0$ containing subspaces of positive dimension. In the article [1979], included in the present volume, a hereditarily normal space X is constructed such that $\dim X = 0$, but for $n = 1, 2, \ldots$, there exist subspaces Y with $\dim Y = \operatorname{Ind} Y = n$.

On dimensions of Čech-Stone compactification. E. Čech himself did not examine connections between the dimensions $\operatorname{Ind} X$ and $\dim X$ of a completely regular space X and the dimensions $\operatorname{Ind} \beta X$ and $\dim \beta X$ of the maximal compactification βX of the space X. However, it was found out relatively soon after the basic Čech's papers that $\dim \beta X = \dim X$ and $\operatorname{Ind} \beta X = \operatorname{Ind} X$ for every normal space X. The first of these equalities was proved by H. Wallman [1938], the second one by N. Vedenisov [1941].

Approximately 10 years later in [KATĚTOV 1950], the investigation of the dimension of completely regular spaces Y defined as $\dim \beta Y$ was suggested. Due to the equality $\dim \beta X = \dim X$ for normal spaces X, this dimension is equal to $\dim X$ of X normal, and therefore we can still denote this extended dimension by dim. In [KATĚTOV 1950] there are contained only the basic simple theorems concerning this dimension. Among these theorems there is the following statement: For every completely regular space Y, $\dim Y$ is equal to the least $n = -1, 0, 1, \ldots$ such that for every functionally open cover $\mathcal{U}$ of the space Y there is a finer functionally open cover $\mathcal{V}$, of the order at most $n+1$. Here a subset G of a topological space Z is called functionally open (or cozero-set) if there is a continuous function $f : Z \to \mathbb{R}$ such that $G = \{x \in X : f(x) > 0\}$; a cover is called functionally open, if it consists of functionally open sets. As it is clear from the proposition just mentioned, the dimension $\dim Y$ of a completely regular space Y is equal to the uniform dimension of the corresponding uniform space, but we shall not be concerned with this here.

All general theorems on the uniform covering dimension are of course valid (possibly after an appropriate reformulation) for the dimension dim in the sense of the previous paragraph. However, only relatively small attention has been paid to the investigation of its specific properties. For an explicit investigation of $\dim \beta X$ we can refer to [PASYNKOV 1980, 1981]. Implicitly, it occurs e.g. in [PASYNKOV 1971]. For a different approach to dim defined for completely regular spaces and various related concepts and results, see e.g. [ČIGOGIDZE 1977] and the subsequent articles by this author.

There are also some counterexamples. J. Terasawa [1977] constructed a completely regular space X such that $\dim X > 0$ and still there exists a functionally closed set $F \subset X$ such that $\dim F = 0$ and the subspace $X \setminus F$ is countable discrete. In the paper [1979] by E. Pol, preceded by a preliminary announcement [1976], an example was constructed of a completely regular space X such that $\dim X > 0$ and there are functionally closed subspaces F_1, F_2 such that $F_1 \cup F_2 = X$, $\dim F_1 = \dim F_2 = 0$. Thus, the finite sum theorem fails for dim defined on completely regular spaces.

Let us add that $\operatorname{ind} \beta X$ is not an extension of ind (defined for normal spaces) since, as shown already in [SMIRNOV 1951], there are normal spaces X with $\operatorname{ind} X < \operatorname{ind} \beta X$.

On the sum theorem. Already in the first years of the systematic development of dimension theory, it was proved that the (countable) sum theorem holds for ind and for all separable metrizable spaces. The theorem asserts that $\operatorname{ind} X = \sup(\operatorname{ind} X_n : n = 1, 2, \ldots)$, where X is separable metrizable and X_n are closed in X, $\bigcup X_n = X$. The equalities $\dim X = \operatorname{Ind} X = \operatorname{ind} X$, valid for these spaces, imply that the sum theorem holds also for dim and Ind on the class of separable metrizable spaces.

E. Čech proved the countable sum theorem for dim for all normal spaces, and for Ind for perfectly normal spaces. The theorem for dim cannot be generalized to the "extended" dimension dim defined for completely regular spaces; for an example, see [POL 1976, 1979]. As for Ind, the sum theorem has been proved for strongly hereditarily normal spaces; a proof is contained in R. Engelking's book on the dimension theory. Earlier, the sum theorem for Ind was proved successively for totally normal spaces, Dowker spaces and supernormal spaces. The proofs are contained in the same articles in which the monotonicity of Ind was proved for the corresponding classes.

Let us note that the sum theorem for Ind is valid for every normal space X satisfying the equality $\operatorname{Ind} X = \dim X$. This is an easy consequence of the inequality $\dim X \leq \operatorname{Ind} X$ valid for all normal spaces X.

The following problem, formulated only vaguely, seems to remain open: to find a possibly wide "natural" class of spaces for which the sum theorem for Ind holds. This problem is connected with the following question: which of various classes for which the equality Ind = dim is established are contained in the class of strongly hereditarily normal spaces?

As for examples of spaces for which the finite sum theorem for Ind fails, see the well-known article [LOKUCIEVSKIĬ 1949].

Let us add that the finite sum theorem for ind fails even for some metrizable non-separable spaces; see [PRZYMUSIŃSKI 1974].

Some further questions. For the sake of completeness, we are going to present some results on dim and Ind which are not directly connected with theorems and problems contained in E. Čech's work, but fall, in a sense, into the style of his papers. Namely, we will discuss some results concerning the addition and decomposition theorems on the dimension of the product of spaces.

It is well known that for separable metrizable spaces X the addition theorem, often called the Urysohn-Menger formula, holds for ind, Ind and dim. This means that, for arbitrary subspaces Y and Z of X, $\delta(Y \cup Z) \leq \delta Y + \delta Z + 1$, where δ means ind, Ind or dim. For these spaces, the so-called decomposition theorem also holds: if $\dim X \leq n = 0, 1, 2, \ldots$, then there are subspaces $X_0, \ldots, X_n$ such that $X = \bigcup_{i=0}^{n} X_i$ and $\dim X_i = 0$ for $i = 0, \ldots, n$.

In fact, the addition theorem holds for all three dimensions in the class of all hereditarily normal spaces. For ind and Ind, this is easily proved by slight modifications of the proof valid for separable metrizable spaces (see e.g. R. Engelking's book). For dim, the addition theorem was proved by Ju. V. Smirnov [1951] for hereditarily normal spaces and by V. Zarelua [1963] for arbitrary normal spaces. V. Zarelua's paper contains also the following theorem. Let X be a normal space, $X = Y \cup Z$. Let $m, n = 0, 1, 2, \ldots$. If Y is normal, $\dim Y \leq m$, and $\dim T \leq n$ whenever $T \subset Z$ is closed in X, then $\dim X \leq m+n+1$.

As for the decomposition theorem, it does hold (for Ind and dim) in all metrizable spaces, as has been proved in [KATĚTOV 1952] and, independently, in [MORITA 1954]. On the other hand, if X is a hereditarily normal space and $X = \bigcup_{i=0}^{n} X_i$, where $\dim X_i = 0$, then it is easy to see that $\operatorname{Ind} X \leq n$. Hence the decomposition theorem fails for hereditarily normal spaces X with $\dim X \neq \operatorname{Ind} X$. As stated in the section on the equality Ind = dim, there are various examples of such spaces.

The problem of validity of the inequalities $\dim(X \times Y) \leq \dim X + \dim Y$, $\operatorname{Ind}(X \times Y) \leq \operatorname{Ind} X + \operatorname{Ind} Y$ is rather complicated. Nevertheless, there are many important results, mainly for the case when X or Y is compact or metrizable. The results concerning dim and Ind are often similar except that the conditions for $\operatorname{Ind}(X \times Y) \leq \operatorname{Ind} X + \operatorname{Ind} Y$ are somewhat stronger.

For following two theorems and related results see [PASYNKOV 1969] and [FILIPPOV 1980] (preliminary announcement appeared in 1973). For the sake of simplicity, we state them in a weaker and slightly modified form, and we always assume $X \times Y$ non-void.

If $X \times Y$ is normal, X or Y is compact, then $\dim(X \times Y) \leq \dim X + \dim Y$; if, in addition, the finite sum theorem holds both for X and Y, then $\operatorname{Ind}(X \times Y) \leq \operatorname{Ind} X + \operatorname{Ind} Y$.

If $X \times Y$ is normal, X is metrizable and Y is countably paracompact (i.e., every countable open cover has a locally finite refinement), then $\dim(X \times Y) \leq \dim X + \dim Y$. If, in addition, the finite sum theorem is valid for Y, the $\operatorname{Ind}(X \times Y) \leq \operatorname{Ind} X + \operatorname{Ind} Y$.

Let us add that, by well known theorems, countable paracompactness can be replaced in the above theorem by the condition of normality of $Y \times \mathbb{R}$.

Since the finite sum theorem for Ind holds for every normal space X such that $\operatorname{Ind} Y = \dim Y$ whenever $Y \subset X$ is closed, we have, e.g., the following consequences. If $X \times Y$ is normal, X and Y are strongly hereditarily normal and X is compact, then $\operatorname{Ind}(X \times Y) \leq \operatorname{Ind} X + \operatorname{Ind} Y$. If $X \times Y$ is normal, X is metrizable, Y is strongly hereditarily normal countably paracompact, then $\dim(X \times Y) \leq \dim X + \dim Y$.

From the results stated above, the following theorems, proved earlier, easily follow.

If X is compact and Y is paracompact, then $\dim(X \times Y) \leq \dim X + \dim Y$. – This has been proved by K. Morita [1953].

If X is metrizable and Y is perfectly normal, then $\dim(X \times Y) \leq \dim X + \dim Y$. – See J. Nagata [1967a].

There are also various results which do not seem to be direct consequences of the general theorems stated above.

One of them is as follows: If $X \times Y$ is strongly paracompact, i.e., every open cover of X is refined by an open cover $\mathcal{U}$ such that no $U_0 \in \mathcal{U}$ meets infinitely many $U \in \mathcal{U}$, then $\dim(X \times Y) \leq \dim X + \dim Y$. This was proved by K. Morita [1953].

As for counterexamples, i.e., spaces X and Y violating the inequality $\dim(X \times Y) \leq \dim X + \dim Y$ or the corresponding inequality for Ind, we present only two of them.

Perhaps the first significant example was given by V. V. Filippov [1972]. He constructed compact spaces X and Y such that $\operatorname{Ind} X = \operatorname{ind} X = 1$, $\operatorname{Ind} Y = \operatorname{ind} Y = 2$ and, nevertheless, $\operatorname{ind}(X \times Y) \geq 4$. An important example is due to T. Przymusiński [1979]. He presents a separable Lindelöf first countable space X such that $\dim X = 0$, X^2 is normal and $\dim X^2 > 0$. This space also satisfies $\operatorname{Ind} X = 0$, $\operatorname{Ind} X^2 > 0$. The construction in [PRZYMUSIŃSKI 1979] is based to a considerable extent on the methods from [WAGE 1978]; however, in [WAGE 1978], a similar example is constructed only under the assumption of continuum hypothesis.

REFERENCES

P. S. ALEKSANDROV

[1951] *The present status of the theory of dimension* (Russian), Uspehi Mat. Nauk (N.S.) **6** (1951), 43 - 68. English translation: Amer. Math. Soc. Transl. (2) 1 (1955), 1 - 26.

[1960] *Some results in the theory of topological spaces obtained within the last twenty-five years* (Russian), Uspehi Mat. Nauk **15** (1960), 25 - 95. English translation: Russian Math. Surveys 15 (1960), 23 - 83.

[1964] *Some fundamental directions in general topology* (Russian), Uspehi Mat. Nauk **19** (1964), 3 - 46.

[1965] *Correction to the paper "Some fundamental directions in general topology"* (Russian), Uspehi Mat. Nauk **20** (1965), 253 - 254.

P. S. ALEKSANDROV and V. V. FEDORČUK

[1978] *Principal moments in the development of set-theoretic topology. With the collaboration of V. I. Zaĭcev* (Russian), Uspehi Mat. Nauk **33** (1978), 3 - 48. English translation: Russian Math. Surveys 33 (1978), 1 - 53.

P. S. ALEKSANDROV and B. A. PASYNKOV

[1973] *Introduction to Dimension Theory: An Introduction to the Theory of Topological Spaces and the General Theory of Dimension* (Russian), Moscow, 1973.

B. BOLZANO

[1843] *Uiber Haltung, Richtung, Krümmung und Schnörkelung bei Linien sowohl als Flächen sammt einigen verwandten Begriffen*, In: Spisy Bernarda Bolzana, Vol. 5: Geometrické práce, Ed. J. Vojtěch, Praha 1948; pp. 139 - 183. Not previously published. Written 1843 - 1844.

L. E. J. BROUWER

[1911] *Beweis der Invarianz der Dimensionenzahl*, Math. Ann. **70** (1911), 161 - 165.

[1913] *Über den natürlichen Dimensionsbegriff*, J. Reine Angew. Math. **142** (1913), 146 - 152.

E. ČECH

[1931] *Sur la théorie de la dimension*, C. R. Acad. Sci. Paris **193** (1931), 976 - 977.

[1932] *Sur la dimension des espaces parfaitement normaux*, Bull. Intern. Acad. Tchèque Sci. **33** (1932), 38 – 55.

[1933] *Contribution to dimension theory* (Czech), Časopis Pěst. Mat. Fys. **62** (1933), 277 - 291.

M. G. CHARALAMBOUS

[1985] *Spaces with noncoinciding dimensions*, Proc. Amer. Math. Soc. **94** (1985), 507 – 515.

A. Č. ČIGOGIDZE

[1977] *The relative dimensions of completely regular spaces* (Russian), Sakharth. SR Mecn. Akad. Moambe **85** (1977), 45 – 48.

C. H. DOWKER

[1953] *Inductive dimension of completely normal spaces*, Quart. J. Math. Oxford Ser. (2) **4** (1953), 267 - 281.

[1955] *Local dimension of normal spaces*, Quart. J. Math. Oxford Ser. (2) **6** (1955), 101 - 120.

C. H. DOWKER and W. HUREWICZ

[1956] *Dimension of metric spaces*, Fund. Math. **43** (1956), 83 - 88.

R. ENGELKING

[1978] *Dimension Theory*, Warsaw, 1978, - A revised and enlarged translation of Teoria wymiaru (Polish), Warsaw 1977.

V. V. FEDORČUK

[1968] *Bicompacta with noncoinciding dimensionalities* (Russian), Dokl. Akad. Nauk SSSR **182, no. 2** (1968), 275 - 277. English translation: Soviet. Math. Dokl. 9 (1968), 1148 - 1150.

[1978] *On the dimension of hereditarily normal spaces*, Proc. London Math. Soc. **36** (1978), 163 - 175.

[1988] *Principles of dimension theory* (Russian), In: Current problems in mathematics. Fundamental directions, Vol. **17**, 111 - 224, Moscow, 1988.

V. V. FILIPPOV

[1969] *A bicompactum with noncoinciding inductive dimensions* (Russian), Dokl. Akad. Nauk. SSSR **184** (1969), 1050 - 1053. English translation: Soviet. Math. Dokl. 10 (1969), 208 - 211.

[1970] *A solution of a certain problem of P. S. Aleksandrov (a bicompactum with noncoinciding inductive dimensions)* (Russian), Mat. Sb. (N.S.) **83 (125)** (1970), 42 - 46. English translation: Math. USSR – Sb. 12 (1970), 41 – 57.

[1970a] *Bicompacta with distinct inductive dimensions* (Russian), Dokl. Akad. Nauk SSSR **192** (1970), 289 – 292. English translation: Soviet Math. Dokl. 11 (1970), 635 – 638.

[1970b] *Bicompacta with distinct dimensions* ind *and* dim (Russian), Dokl. Akad. Nauk SSSR **192** (1970), 516 – 519. English translation: Soviet Math. Dokl. 11 (1970), 687 - 691.

[1972] *The inductive dimension of a product of bicompacta* (Russian), Dokl. Akad. Nauk SSSR **202** (1972), 1016 - 1019. English translation: Soviet Math. Dokl. 13 (1972), 250 - 254.

[1973] *The dimension of normal spaces* (Russian), Dokl. Akad. Nauk SSSR **209** (1973), 805 - 807. English translation: Soviet Math. Dokl. 14 (1973), 547 - 550.

[1980] *The dimension of products of topological spaces* (Russian), Fund. Math. **106** (1980), 181 - 212.

[1983] *Normally situated subspaces* (Russian), Trudy Mat. Inst. Steklov. **154** (1983), 239 - 251.

W. HUREWICZ and H. WALLMAN

[1941] *Dimension Theory*, Princeton, N. J., 1941.

M. ITO

[1982] *Weak L-spaces and free L-spaces*, J. Math. Soc. Japan **34** (1977), 262 - 295.

D. M. JOHNSON

[1977] *Prelude to dimension theory: the geometrical investigations of Bernard Bolzano*, Arch. History Exact Sci. **17** (1977), 262 - 295.

M. KATĚTOV

[1950] *A theorem on the Lebesgue dimension*, Čas. Pěst. Mat. Fys. **75** (1950), 79 - 87.

[1951] *On the dimension of metric spaces* (Russian), Dokl. Akad. Nauk SSSR **79** (1951), 189 - 191.

[1952] *On the dimension of non-separable spaces. I* (Russian), Czechoslovak. Math. J. **2** (1952), 333 - 368.

[1983] *On a dimension function based on Bolzano's ideas*, General Topology and its Relations to Modern Analysis and Algebra **V** (Prague 1981), 413 - 433, Berlin, 1983.

J. KULESZA

[1990] *Metrizable spaces where the inductive dimensions disagree*, Trans. Amer. Math. Soc. **318** (1990), 763 - 781.

H. LEBESGUE

[1911] *Sur la non applicabilité de deux domaines appartenant respectivement à des espaces, de n et $n+p$ dimensions*, Math. Ann. **70** (1911), 166 - 168.

I. M. LEĬBO

[1974] *The equality of dimensions for closed images of metric spaces* (Russian), Dokl. Akad. Nauk SSSR **216** (1974), 498 - 501. English translation: Soviet Math. Dokl. 15 (1974) 835 - 839.

[1982] *Dimensions of closed images of metric spaces* (Russian), Serdica **8, no. 4** (1982), 395 - 407.

I. K. LIFANOV

[1973] *The dimension of normal spaces* (Russian), Dokl. Akad. Nauk SSSR **209** (1973), 291 - 294. English translation: Soviet Math. Dokl. 14 (1973) 383 - 387.

O. V. LOKUCIEVSKIĬ

[1949] *On the dimension of bicompacta* (Russian), Dokl. Akad. Nauk SSSR (N.S.) **67** (1949), 217 - 219.

A. L. LUNC

[1949] *A bicompactum whose inductive dimension is greater than its dimension defined by means of coverings* (Russian), Dokl. Akad. Nauk SSSR (N.S.) **66** (1949), 801 - 803.

K. MENGER

[1923] *Über die Dimensionalität von Punktmengen I.*, Monatsh. für Math. und Phys. **33** (1923), 148 - 160.

T. MIZOKAMI

[1981] *On the dimension of μ-spaces*, Proc. Amer. Math. Soc. **83, no. 1** (1981), 195 - 200.

K. MORITA

[1953] *On the dimension of product spaces*, Amer. J. Math. **75** (1953), 205 - 223.

[1954] *Normal families and dimension theory for metric spaces*, Math. Ann. **128** (1954), 350 - 362.

K. NAGAMI

[1971] *Dimension for σ-metric spaces*, J. Math. Soc. Japan **23** (1971), 123 - 129.

[1971a] *Normality of products*, Actes de Congrès International des Mathématiciens (Nice, 1970), Tome **2**, 33 - 37, Gauthier-Villars, Paris, 1971.

[1980] *The equality of dimensions*, Fund. Math. **106** (1980), 239 - 246.

[1980a] *Dimension of free L-spaces*, Fund. Math. **108** (1980), 211 - 224.

[1981] *Weak L-structures and dimension*, Fund. Math. **112** (1981), 231 - 239.

J. NAGATA

[1965] *Modern Dimension Theory*, North-Holland, 1965.

[1967] *A survey of dimension theory*, General Topology and its Relations to Modern Analysis and Algebra II (Proc. Second Prague Topological Sympos., 1966), 259 - 270, Prague 1981.

[1967a] *Product theorems in dimension theory, I*, Bull. Acad. Polon. Sci. Sér. Sci. Math. Astronom. Phys. **15 (1967)**, 439 - 448.

[1971] *A survey of dimension theory, II*, General Topology and Appl. **1** (1971), 65 - 77.

[1983] *Modern Dimension Theory*, Berlin, 1983.

T. NISHIURA

[1977] *A subset theorem in dimension theory*, Fund. Math. **95** (1977), 105 - 109.

S. OKA

[1980] *Free patched spaces and the fundamental theorem of dimension theory*, Bull. Acad. Polon. Sci. Sér. Sci. Math. **28** (1980), 595 - 602.

B. A. PASYNKOV

[1964] *On a class of mappings and on the dimension of normal spaces* (Russian), Sibir. Math. J. **6** (1964), 356 - 376.

[1967] *Open mappings* (Russian), Dokl. Akad. Nauk SSSR **175** (1967), 292 - 295. English translation: Soviet. Math. Dokl. 8 (1967), 853 - 856.

[1969] *The inductive dimensions* (Russian), Dokl. Akad. Nauk SSSR **189** (1969), 254 - 257. English translation: Soviet. Math. Dokl. 10 (1969), 1402 - 1406.

[1971] *The dimensionality of normal spaces* (Russian), Dokl. Akad. Nauk SSSR **201** (1971), 1049 - 1052. English translation: Soviet Math. Dokl. 12 (1971) 1784 - 1787.

[1980] *On the dimension of topological products and limits of inverse sequences* (Russian), Dokl. Akad. Nauk SSSR **254** (1980), 1332 - 1336. English translation: Soviet Math. Dokl. 22 (1981) 596 - 601.

[1981] *Factorization theorems in dimension theory* (Russian), Uspehi Mat. Nauk **36 (219)** (1981), 147 - 175.

B. A. PASYNKOV and I. K. LIFANOV

[1970] *Examples of bicompacta with noncoinciding inductive dimensions* (Russian), Dokl. Akad. Nauk SSSR **192** (1970), 276 - 278. English translation: Soviet Math. Dokl. 11 (1970) 619 - 621.

B. A. PASYNKOV, V. V. FEDORČUK and V. V. FILIPPOV

[1979] *Dimension theory* (Russian), Algebra. Topology. Geometry, Vol. 17, 229 - 306, Moscow 1979.

A. R. PEARS

[1975] *Dimension Theory of General Spaces*, Cambridge Univ. Press, 1975.

H. POINCARÉ

[1905] *La Valeur de la Science*, Paris, 1905.

[1912] *Pourquoi l'espace a trois dimensions*, Révue de Métaph. et de Morale **20** (1912), 483 - 504.

[1913] *Dernières Pensées*, Paris, 1913.

E. POL

[1976] *Some examples in the dimension theory of Tychonoff spaces*, Bull. Acad. Polon. Sci. Sér. Math. **24** (1976), 893 - 897.

[1979] *Some examples in the dimension theory of Tychonoff spaces*, Fund. Math. **102** (1979), 29 - 43.

E. POL and R. POL

[1977] *A hereditarily normal strongly zero-dimensional space with a subspace of positive dimension and an N-compact space of positive dimension*, Fund. Math. **97** (1977), 43 - 50.

[1977a] *A hereditarily normal strongly zero-dimensional space containing subspaces of arbitrarily large dimension*, General Topology and its Relations to Modern Analysis and Algebra IV (Proc. Fourth Prague Topological Sympos., 1976), Part B, 357 - 360, Prague 1977.

[1979] *A hereditarily normal strongly zero-dimensional space containing subspaces of arbitrarily large dimension*, Fund. Math. **102** (1979), 137 - 142.

T. C. PRZYMUSIŃSKI
[1974] *A note on dimension theory of metric spaces*, Fund. Math. **85** (1974), 277 - 284.
[1979] *On the dimension of product spaces and an example of M. Wage*, Proc. Amer. Math. Soc. **76** (1979), 315 - 321.

P. ROY
[1962] *Failure of equivalence of dimension concepts for metric spaces*, Bull. Amer. Math. Soc. **68** (1962), 609 - 613.
[1968] *Nonequality of dimensions in metric spaces*, Trans. Amer. Math. Soc. **134** (1968), 117 - 132.

JU. M. SMIRNOV
[1951] *Some relations in the theory of dimension* (Russian), Mat. Sb. (N.S.) **29 (71)** (1951), 151 - 172.
[1951a] *On normally disposed sets of normal spaces* (Russian), Mat. Sb. (N.S.) **29** (1951), 173 - 176.
[1958] *An example of zero-dimensional normal space having infinite dimensions from the standpoint of covering* (Russian), Dokl. Akad. Nauk. SSSR **123** (1958), 40 - 42.

J. TERASAWA
[1977] *Spaces $N \cup R$ need not be strongly 0-dimensional*, Bull. Acad. Polon. Sci. Sér. Math. **25** (1977), 279 - 281.

P. URYSOHN
[1922] *Les multiplicités Cantoriennes*, C. R. Acad. Paris **175** (1922), 440 - 442.
[1923] *Mémoire sur les multiplicités Cantoriennes*, Fund. Math. **7** (1925), 30 - 137.

N. VEDENISOV
[1941] *On the dimension in the sense of E. Čech* (Russian), Izv. Akad. Nauk SSSR, Ser. Mat. **5** (1941), 211 - 216.

P. VOPĚNKA
[1958] *On the dimension of compact spaces* (Russian), Czechoslovak Math. J. **8** (1958), 319 - 327.

M. WAGE
[1978] *The dimension of product spaces*, Proc. Nat. Acad. Sci. U.S.A. **75** (1978), 4671 - 4672.

H. WALLMAN
[1938] *Lattices and topological spaces*, Ann. of Math. **39** (1938), 112 - 126.

V. ZARELUA
[1963] *On the equality of dimensions* (Russian), Mat. Sb. **62** (1963), 295 - 319.

V. P. ZOLOTAREV
[1975] *The dimension of subspaces* (Russian), Vestnik Moskov. Univ. ser. I. Mat. Meh. **30** (1975), 10 - 15.

ON THE DIMENSION OF PERFECTLY NORMAL SPACES

EDUARD ČECH

I modify slightly the recursive definition (Menger and Urysohn) of the dimension. In the case of separable spaces, up to now the only case studied, the modification is purely formal. I prove 1° the theorem on the dimension of a subset, 2° the theorem on the dimension of a sum (Summensatz), and 3° the theorem on the covering of a finite dimensional space (Zerlegungssatz) for very general spaces comprising as a particular case the metric spaces (metrische Räume).

The main theorems of this paper were announced without proofs in the note *Sur la théorie de la dimension* (C. R. nov. 1931).

I. AUXILIARY THEOREMS

1. A set R is called *topological space* (and the elements of R are called *points*) if there is given a family $\mathfrak{F}$ of subsets of R (called *closed* subsets *in* R) such that:

1.1. The empty set 0 and the space R are closed sets in R.

1.2. For arbitrary point x of R, the set (x) is closed in R.

1.3. The sum of a *finite* number of closed sets in R is closed in R.

1.4. The product (=common part) of an arbitrary number of closed sets in R is closed in R.

2. The set $A \subset R$ is called *open in* R if the set $R - A$ is closed in R.

3. If $S \subset R$, then each product AS, with A being closed in R, is called *closed in* R. Consequently, every subset S of the topological space R represents a topological space.

4. If $A \subset R$, I denote by $\overline{A}$ the *closure* of A (the least closed subset of R containing A).

5. If $A \subset S \subset R$, then the closure of A in the space S is $\overline{A} \cdot S$.

6. The sets $A, B \subset R$ are called *separated* if 1° $AB = 0$, 2° A and B are closed (or equivalently, open) in $A + B$.

6.1. If the sets A_i, B_j are separated for $1 \leqq i \leqq h$, $1 \leqq j \leqq k$, then the sets $\sum_{i=1}^{h} A_i$, $\sum_{j=1}^{k} B_j$ are also separated.

6.2. If the sets A, B are separated, and $A_1 \subset A$, $B_1 \subset B$, then the sets A_1 and B_1 are also separated.

7.1. The sets $A, B \subset R$ are closed in $A + B$ if and only if $A\overline{B} + \overline{A}B = AB$.

7.2. The sets A, B are separated if and only if $A\overline{B} = \overline{A}B = 0$.

8. If $U \subset R$ is open in R, then the set $H_R(U) = \overline{U} - U$ is called the *boundary* of U (in the space R).

8.1. $U \cdot H_R(U) = 0$.

8.2. $U + H_R(U) = \overline{U}$.

EDUARD ČECH

8.3. If U is open in R, then $H_R(U)$ is closed in R.

9.1.[1] Let U_i $(1 \leqq i \leqq m)$ be open sets in R. Then

$$H_R(\sum_{i=1}^{m} U_i) \subset \sum_{i=1}^{m} H_R(U_i).$$

9.2.[2] Let U, V be open sets in R. Then

$$H_R(U - \overline{V}) \subset H_R(U) + H_R(V).$$

10.[3] Let Q_ν, V_ν, $(\nu = 1, 2, 3, \dots)$ be open sets in R. Let $S = \prod_{\nu=1}^{\infty} Q_\nu$. Let $T \subset S$, $T \subset \sum_{\nu=1}^{\infty} V_\nu$. For $\nu = 1, 2, 3, \dots$, let $Q_\nu \supset Q_{\nu+1}$, $V_\nu \subset Q_\nu$. Then

$$H_R(\sum_{\nu=1}^{\infty} V_\nu) \subset \sum_{\nu=1}^{\infty} H_R(V_\nu) + M,$$

$$M = S \cdot H_R(\sum_{\nu=1}^{\infty} V_\nu) \subset S - T.$$

11.[4] Let $S \subset R$, and let U be an open subset of R. Then

$$H_S(SU) \subset S \cdot H_R(U).$$

12. A topological space R is called *normal*[5] if it satisfies the following condition: If $AB = 0$, the sets A, B being closed in R, then there exist open sets U, V in R such that $U \supset A$, $V \supset B$, $UV = 0$.

12.1.[6] Let R be a normal space. Let $A(U)$ be a closed (open) subset of R, and let $A \subset U$. Then there exists an open set V in R such that $A \subset V \subset \overline{V} \subset U$.

13. A topological space R is called *hereditarily normal*[7] if it satisfies the following condition: If the sets A, B are separated, then there exist open sets U, V in R such that $U \supset A$, $V \supset B$, $UV = 0$.

13.1.[8] A hereditarily normal space is normal.

13.2.[9] Every subset S of a hereditarily normal space R represents a hereditarily normal space.

13.3.[10] Let R be a hereditarily normal space, and let $S \subset R$. Let U_0 (U) be an open set in S (in R), and let $U_0 \subset U$. Then there exists a set V open in R such that

[1] K. Menger, *Dimensionstheorie, p. 36.*

[2] Menger, l. c., p. 36.

[3] Menger, l. c., p. 37. The assertion of Menger differs slightly from the one in the text.

[4] Menger, l. c., p. 35.

[5] P. Urysohn, *Über die Mächtigkeit zusammenhängender Mengen*, Math. Annalen, vol. 94, p. 265.

[6] Urysohn, l. c., p. 272.

[7] Urysohn, l. c., p. 265.

[8] Urysohn, l. c., p. 265.

[9] Urysohn, l. c., p.284.

[10] Menger, l. c., p. 36.

ON THE DIMENSION OF PERFECTLY NORMAL SPACES

$$V \subset U, \quad SV = U_0, \quad S \cdot H_R(V) = H_S(U_0).$$

14. A topological space R is called ***perfectly normal***[11] if it has the following two properties: 1° R is normal; 2° every open subset of R is a subset of the type F_σ in R (= a sum of a countable infinite system of closed sets in R). The property 2° can be formulated as follows: every closed subset of R is a subset of the type G_δ (= product of a countable infinite system of open sets in R).

14.1. Every perfectly normal space is hereditarily normal.[12]

14.2. Every subset of a perfectly normal space represents a perfectly normal space.[11]

14.3. Let R be a perfectly normal space, and let S be a closed subset of R. Then there exist closed sets Q_ν ($\nu = 1, 2, 3, \dots$) in R such that

$$S = \prod_{\nu=1}^{\infty} Q_\nu = \prod_{\nu=1}^{\infty} \overline{Q}_\nu; \quad Q_{\nu+1} \subset Q_\nu.$$

Proof. S being closed, it is of the type G_δ in R. Consequently, there exist open sets U_ν in R such that $S = \prod_{\nu=1}^{\infty} U_\nu$. According to 12.1, one can determine open sets V_ν in R such that $S \subset V_\nu \subset \overline{V}_\nu \subset U_\nu$. Therefore, it is sufficient to set $Q_\nu = \prod_{i=1}^{\nu} V_i$.

II. Definition of the dimension

15. Let R be a topological space. One says that the number of dimensions of R is equal to -1 (or is at most equal to -1), and one writes $\dim R = -1$ (or $\dim R \leqq -1$), if and only if $R = 0$. Let us suppose that one has already defined, for a certain value of n ($= 0, 1, 2, 3, \dots$), the topological spaces whose number of dimensions is at most equal to $n - 1$. Thus, let R be a topological space, and let A be a set closed in R. One says that the number of dimensions of R with respect to A is at most equal to n, and one writes $\dim_A R \leqq n$, if one can associate with each open set $U \supset A$ in R an open set V in R in such a way that $A \subset V \subset U$, $\dim H_R(V) \leqq n-1$. One says that the number of dimensions of R is at most equal to n, and one writes $\dim R \leqq n$, if $\dim_A R \leqq n$ for each closed subset A of R. One says that the number of dimensions of R (resp. the number of dimensions with respect to a closed subset A) is equal to n, and one writes $\dim R = n$ ($\dim_A R = n$), if $\dim R \leqq n$ ($\dim_A R \leqq n$), but $\dim R \leqq n - 1$ ($\dim_A R \leqq n - 1$) does not hold.

16.1. Let S be a closed set in R. Let $\dim R \leqq n$. Then $\dim S \leqq n$.

Proof. One can easily see that it is sufficient to deduce 16.1 for the dimension n, assuming the proposition 16.2 to be valid for the dimension $n - 1$. Thus, let $U_0 \supset A$ be an open subset of S. Then there exists an open subset U of R such that $U_0 = US$, and therefore $A \subset U$. Because $\dim_A R \leqq n$, there exists a set V open in R such that $A \subset V \subset U$, $\dim H_R(V) \leqq n - 1$. Let us set $V_0 = SV$. The set V_0

[11] By the way, this type of space has been considered (without a special name) by Urysohn, l. c., p. 286, remark[41] at the bottom of the page.

[12] Urysohn, l.c., sub[11].

is open in S, and one has $A \subset V_0 \subset U_0$. Moreover, the set $H_S(V_0)$ is closed in S, and consequently also in R, and according to 11 one has $H_S(V_0) \subset H_R(V_0)$. The proposition 16.2 being true for the dimension $n-1$, it follows $\dim H_S(V_0) \leqq n-1$.

17.1. Let S be a closed subset of R, and let A be a closed subset of S. Let us suppose that to each set $U \supset A$ one can attach an open set V in R such that $A \subset V \subset U$, $\dim S \cdot H_R(V) \leqq n-1$. Then $\dim_A S \leqq n$.

Proof. Let $U_0 \supset A$ be an open set in R. There exists a set U open in R such that $U_0 = SU$. Hence, it follows the existence of a set V open in R such that $A \subset V \subset U$, $\dim S \cdot H_R(V) \leqq n-1$. Let us set $V_0 = SV$. The set V_0 is open in S, and one has $A \subset V_0 \subset U_0$. According to 11, one has $H_S(V_0) \subset S \cdot H_R(V)$. The set $H_S(V_0)$ being closed in R, one concludes from 16.2 that $\dim H_S(V_0) \leqq n-1$.

17.2. Let R be a hereditarily normal space. Let S be a closed subset of R, and let A be a closed subset of S. Let $\dim_A S \leqq n$. Then to each set $U \supset A$ open in R one can attach a set V open in R such that $A \subset V \subset U$, $\dim S \cdot H_R(V) \leqq n-1$.

Proof. The set $U_0 = SU$ is open in S, and one has $U_0 \supset A$. Because $\dim_A S \leqq n$, there exists a set V_0 open in S such that $A \subset V_0 \subset U_0 \subset U$, $\dim H_S(V_0) \leqq n-1$. But, according to 13.3, there exists a set V open in R such that $V \subset U$, $SV = V_0$ (therefore $V \supset A$), $S \cdot H_R(V) = H_S(V_0)$.

18. Let R be a topological space. Let A, B be closed subsets of R, and let $C \subset R$. Let $R-C = P+Q$ with the sets P, Q being separated (see 6). Let $P \supset A$, $Q \supset B$. Then one says that *the set C separates A and B from each other in R.*

18.1. Let R be a normal space. Let A, B be two closed subsets of R with $AB = 0$. Let $\dim_A R \leqq n$. Then there exists a closed subset C of R separating A and B from each other in R and such that $\dim C \leqq n-1$.

Proof. According to 12.1, there exists a set U open in R such that $A \subset U \subset \overline{U} \subset R-B$. Because $\dim_A R \leqq n$, there exists a set V open in R such that $A \subset V \subset U$, $\dim H_R(V) \leqq n-1$. One can easily see that the set $C = H_R(V)$ has all the required properties.

18.2. Let R be a normal space, and let A be a closed subset of R. Let us suppose that to each set $B \subset R-A$ closed in R one can attach a set C closed in R, separating A and B from each other in R, with $\dim C \leqq n-1$. Then $\dim_A R \leqq n$.

Proof. Let $U \supset A$ be a set open in R. Setting $B = R-U$, one can see that there exist a set C closed in R and two separated sets P, Q such that $\dim C \leqq n-1$, $R-C = P+Q$, $P \supset A$, $Q \supset B$. One can easily see that the set P is open in R, that $A \subset P \subset U$, and that $H_R(P) \subset C$, whence, according to 16.2, $\dim H_R(P) \leqq n-1$.

18.3. Let R be a hereditarily normal space, and let $A, B, C \subset R$. Let us suppose that C separates A and B from each other in R. Then there exists a set $C^* \subset C$ closed in R and separating A and B from each other in R.

Proof. One has $R-C = P+Q$, where P, Q are separated. The space R being hereditarily normal, there exist sets U, V open in R such that $U \supset P$, $V \supset Q$, $UV = 0$. It is sufficient to set $C^* = R-(U+V)$.

ON THE DIMENSION OF PERFECTLY NORMAL SPACES

18.4. Let A be a closed subset of a hereditarily normal space R. Let us suppose that to each set $B \subset R - A$ closed in R one can attach a set C separating A and B from each other in R and such that $\dim C \leqq n - 1$. Then $\dim_A R \leqq n$.

Proof. By virtue of 13.1, 16.2, 18.2 and 18.3.

III. Theorem on the dimension of a sum

19. Let R be a perfectly normal space. Let S_i $(i = 1, 2, 3, \dots)$ be closed subsets of R, and let $\dim S_i \leqq n$ for $i = 1, 2, 3, \dots$. Then $\dim \sum_{i=1}^{\infty} S_i \leqq n$.

This theorem is trivial for $n = -1$. Therefore, one can proceed as follows: In the proofs of n° 20 one will make use of the theorem 19 for the dimension $n - 1$. In n° 21 one will prove the theorem 19, keeping the hypothesis that it is valid for the dimension $n - 1$ and making use of the proposition 20.3.

20.1. Let R be a perfectly normal space. Let S be a closed subset of R, and let A, B^* be closed subsets of S. Let C^* be a closed subset of S separating A and B^* from each other in S and such that $\dim C^* \leqq n - 1$. Let us suppose that $\dim_F R \leqq n$ for every subset $F \subset R - S$ closed in R. Let B be a closed subset of R such that $B^* = SB$. Then there exists a set C closed in R separating A and B from each other in R and such that $\dim C \leqq n$.

Proof. The set C^* separating A and B^* from each other in S, there exist two separated sets P, Q such that $S - C^* = P + Q$, $P \supset A$, $Q \supset B^*$. One can easily see that the set P is open in S, that $\overline{P} \cdot B = 0$, and that $H_S(P) \subset C^*$, whence, according to 16.2, $\dim H_S(P) \leqq n - 1$. The sets $\overline{P}$, B being closed in the normal space R, the relation $\overline{P}B = 0$ implies the existence of two sets U, T open in R such that $U \supset \overline{P}$, $T \supset B$, $UT = 0$, whence, according to 7.2, $\overline{U}T = 0$. According to 13.3 and 14.1, there exists a set V open in R such that $P = SV$, $V \subset U$, $SK = H_S(P)$, where $K = H_R(V)$. One can easily see that $A \subset V$, $BV = 0$, $BK = 0$. According to 14.3, there exist sets Q_ν open in R such that

$$Q_\nu \supset Q_{\nu+1}, \quad K = \prod_{\nu=1}^{\infty} Q_\nu = \prod_{\nu=1}^{\infty} \overline{Q}_\nu.$$

The set $K - S$ is open in K, and consequently is of the type F_σ in K. K being closed in R, $K - S$ is of the type F_σ in R. Therefore, there exist sets F_ν closed in R such that

$$K - S = \sum_{\nu=1}^{\infty} F_\nu.$$

According to 12.1, since $KB = 0$, there exist sets Z_ν open in R such that

$$F_\nu \subset Z_\nu \subset \overline{Z}_\nu \subset R - B.$$

Obviously, $F_\nu \subset R - S$, whence $\dim_{F_\nu} R \leqq n$. But $F_\nu \subset Q_\nu Z_\nu$, so that there exist sets W_ν open in R such that

$$F_\nu \subset W_\nu \subset Q_\nu Z_\nu, \quad \dim H_R(W_\nu) \leqq n - 1.$$

One has $K - S \subset \sum_{\nu=1}^{\infty} W_\nu$, so that, recalling that $SK = H_S(P)$, one deduces from the theorem 10 that

$$H_R(\sum_{\nu=1}^{\infty} W_\nu) \subset \sum_{\nu=1}^{\infty} H_R(W_\nu) + H_S(P).$$

Let us set

$$X = V + \sum_{\nu=1}^{\infty} W_\nu.$$

$C = H_R(X)$, so that, according to 9.1,

$$(*) \qquad C \subset \sum_{\nu=1}^{\infty} H_R(W_\nu) + H_S(P).$$

The theorem 19 being valid for the dimension $n-1$ according to the hypothesis, the number of dimensions of the second term of the relation (*) is at most equal to $n-1$, whence, according to 16.2, $\dim C \leqq n-1$. But one can easily verify that the set C separates A and B from each other in R, for

$$R - C = X + (R - \overline{X}), \quad X \supset A, \quad R - \overline{X} \supset B.$$

20.2. Let R be a perfectly normal space. Let S be a closed subset of R, and let T be an arbitrary subset of R. Let A be a closed subset of S. Let $\dim_A S \leqq n$ and $\dim T \leqq n$. Then $\dim_A (S+T) \leqq n$.

Proof. Without the loss of generality, one can suppose that $R = S + T$ (see 14.2). Therefore, from the relation $\dim T \leqq n$, one can easily deduce that $\dim_F R \leqq n$ for any choice of a set $F \subset R - S$ closed in R. Since $\dim_A S \leqq n$, according to 18.1, to each subset $B^* \subset S - A$ closed in S one can attach a set C^* closed in S separating A and B^* from each other in S and such that $\dim C^* \leqq n-1$. Now, let $B \subset R - A$ be a set closed in R. Setting $B^* = SB$, one concludes from 20.1 that there exists a set C closed in R separating A and B from each other in R and such that $\dim C \leqq n-1$. Therefore, according to 18.2, one has $\dim_A R = \dim_A (S+T) \leqq n$.

20.3. Let R be a perfectly normal space, and let S and T be sets closed in R. Let $\dim S \leqq n$, $\dim T \leqq n$. Then $\dim (S+T) \leqq n$.

Proof. One can again suppose that $R = S + T$. Let A be closed in R and $U \supset A$ open in R. It is necessary to construct a set V open in R such that $A \subset V \subset U$, $\dim H_R(V) \leqq n-1$. The set AS is closed in S, whence $\dim_{AS} S \leqq n$. Moreover, one has $\dim T \leqq n$, $R = S + T$. Therefore, one can deduce from 20.2 that $\dim_{AS} R \leqq n$. Consequently there exists a set V_1 open in R such that $AS \subset V_1 \subset U$, $\dim H_R(V_1) \leqq n-1$. For the reasons of symmetry, there exists a set V_2 open in R such that $AT \subset V_2 \subset U$, $\dim H_R(V_2) \leqq n-1$. Let us set $V = V_1 + V_2$. Obviously, V is an open subset of R and $A \subset V \subset U$. According to 9.1, one has

ON THE DIMENSION OF PERFECTLY NORMAL SPACES

$$(*) \qquad H_R(V) \subset H_R(V_1) + H_R(V_2).$$

But the theorem 20.3 represents only a special case of the theorem 19, which is supposed to be valid for the dimension $n-1$. Consequently, the theorem 20.3 is valid for the dimension $n-1$, so that the second term of (*) has the number of dimensions at most equal to $n-1$, whence dim $H_R(V \leqq n-1)$ according to 16.2.

21.1. Let us pass to the proof of the theorem 19 for the dimension n. The sets $\sum_{i=1}^{k} S_i$ are closed in R. From the theorem 23, one deduces recurrently that $\dim \sum_{k=1}^{k} S_i \leqq n$. Finally, $\sum_{i=1}^{\infty} S_k = \sum_{k=1}^{\infty}\sum_{i=1}^{k} S_i$. Hence, it follows that it is sufficient to prove the theorem 19 under the hypothesis

$$(1) \qquad S_k \subset S_{k+1} \quad \text{for} \quad k = 1, 2, 3, \ldots$$

Without loss of generality, one can also suppose (see 14.2) that

$$(2) \qquad \sum_{k=1}^{\infty} S_k = R.$$

Let us choose a set A closed in R and a set $Z \supset A$ open in R. The question is to construct a set U_ω open in R such that $A \subset U_\omega \subset Z$, dim $H_R(U_\omega) \leqq n-1$. Let us start by constructing recurrently, according to 12.1, sets Z_r open in R such that

$$(3) \qquad A \subset Z_r \subset Z, \quad \overline{Z}_r \subset Z_{r+1} \quad \text{for} \quad r = 1, 2, 3, \ldots$$

21.2. The important tool for the construction of the set U_ω under consideration will be an auxiliary construction of three sequences U_r, V_r, T_r ($r = 1, 2, 3, \ldots$) of open sets in R possessing, for $r = 1, 2, 3, \ldots$, the following ten properties:

$$(a_r) \qquad AS_r \subset U_r \subset Z_r,$$
$$(b_r) \qquad U_{r-1} \subset U_r,$$
$$(c_r) \qquad \dim S_r \cdot H_R(U_r) \leqq n-1,$$
$$(d_r) \qquad U_r - U_{r-1} \subset V_{r-1},$$
$$(e_r) \qquad H_R(U_r) - S_{r-1} \subset V_{r-1},$$
$$(f_r) \qquad H_R(U_{r-1}) - S_{r-1} \subset V_{r-1},$$
$$(g_r) \qquad A \subset V_{r-1},$$
$$(h_r) \qquad S_{r-1} - \overline{U}_{r-1} \subset T_{r-1},$$
$$(i_r) \qquad T_{r-2} \subset T_{r-1},$$
$$(j_r) \qquad T_{r-1} \cdot V_{r-1} = 0.$$

Further, we shall agree on setting

$$(4) \qquad U_0 = 0, \quad V_0 = R, \quad S_0 = 0, \quad T_0 = 0, \quad T_{-1} = 0.$$

We shall proceed in the following way: First (in 21.3) one will construct the set U_1 open in R in such a way that the conditions $(a_1) - (j_1)$ are satisfied. Then (in 21.4), supposing that for a certain value of k $(= 1, 2, 3, \dots)$ one has already constructed the sets U_r, V_s, T_s $(1 \leqq r \leqq k,\ 1 \leqq s \leqq k-1)$ in such a way that the conditions $(a_r) - (j_r)$ are satisfied for $1 \leqq r \leqq k$, one will construct the sets V_k, T_k open in R satisfying the conditions $(f_{k+1}) - (j_{k+1})$. Finally (in 21.5), supposing that for a given value of k $(= 1, 2, 3, \dots)$ one has already constructed the sets U_r, V_r, T_r $(1 \leqq r \leqq k)$ open in R satisfying the conditions $(a_r) - (e_r)$, $(f_s) - (j_s)$ $(1 \leqq r \leqq k$, $1 \leqq s \leqq k+1)$, one will construct the set U_{k+1} open in R satisfying the conditions $(a_{k+1}) - (e_{k+1})$. Then the proposed goal will be attained.

21.3. The set AS_1 is closed in S_1. The set Z_1 is open in R and $AS_1 \subset Z_1$ according to (3). The set S_1 is closed in R and $\dim S_1 \leqq n$. Therefore, it follows from 14.1 and 17.2 that there exists a set U_1 open in R such that $AS_1 \subset U_1 \subset Z_1$, $\dim S_1 \cdot H_R(U_1) \leqq n-1$. Taking into account (4), one can see that the conditions $(a_1) - (j_1)$ are realized.

21.4. Let $k = 1, 2, 3, \dots$ Let us suppose that the sets U_r, V_s, T_s $(1 \leqq r \leqq k$, $1 \leqq s \leqq k-1)$ satisfy the conditions $(a_r) - (j_r)$ for $1 \leqq r \leqq k$. First, let us show that

$$(\alpha_k) \qquad A,\ S_k - \overline{U}_k;$$

$$(\beta_k) \qquad H_R(U_k) - S_k, \quad S_k - \overline{U}_k;$$

$$(\gamma_k) \qquad A,\ T_{k-1};$$

$$(\delta_k) \qquad H_R(U_k) - S_k,\ T_{k-1}$$

are pairs of separated sets. The set $S_k - U_k$ is closed in R and contains $S_k - \overline{U}_k$. Therefore, according to (a_k),

$$\overline{A}(S_k - \overline{U}_k) + A \cdot \overline{S_k - \overline{U}_k} = A \cdot \overline{S_k - \overline{U}_k} \subset A(S_k - U_k) = AS_k - U_k = 0,$$

so that (see 7.2) the sets (α_k) are separated. Moreover,

$$\overline{H_R(U_k) - S_k} \cdot (S_k - \overline{U}_k) \subset H_R(\overline{U}_k) \cdot (R - U_k) = 0,$$
$$(H_R(U_k) - S_k) \cdot \overline{S_k - \overline{U}_k} \subset (R - S_k) S_k = 0,$$

so that the sets (β_k) are separated. The sets V_{k-1}, T_{k-1} being open in R, they are separated by virue of (j_k). Therefore, according to 6.2 and (g_k), the sets (γ_k) are separated, and, since $H_R(U_k) - S_k \subset H_R(U_k) - S_{k-1}$, according to (1) [for $k = 1$ according to (4)], it follows from 6.2 and (e_k) that the sets (δ_k) are separated. Consequently, one can see from 6.1 that the sets

$$A + [H_R(U_k) - S_k], \quad (S_k - \overline{U}_k) + T_{k-1}$$

are separated. Taking into account 14.1, one can deduce that there exist sets V_k, T_k open in R satisfying the conditions $(f_{k+1}) - (j_{k+1})$.

21.5. Let $k = 1, 2, 3, \ldots$ Let us suppose that the sets U_r, V_r, T_r $(1 \leqq r \leqq k)$ satisfy the conditions $(a_r) - (e_r)$, $(f_s) - (j_s)$ for $1 \leqq r \leqq k$, $1 \leqq s \leqq k+1$. According to 14.3, there exist sets Q_ν open in R such that

$$(5) \qquad Q_\nu \supset Q_{\nu+1} \quad \text{for} \quad \nu = 1, 2, 3, \ldots,$$

$$(6) \qquad [A + H_R(U_k)] \cdot S_{k+1} = \prod_{\nu=1}^{\infty} Q_\nu = \prod_{\nu=1}^{\infty} \overline{Q}_\nu.$$

The sets $[A + H_R(U_k)] \cdot S_{k+1}$, S_k being closed in the perfectly normal space R, their difference is of the type F_σ in R. Therefore, there exist sets F_ν closed in R such that

$$(7) \qquad [A + H_R(U_k)] \cdot S_{k+1} - S_k = \sum_{\nu=1}^{\infty} F_\nu.$$

According to (7), (f_{k+1}), and (g_{k+1}) one has $F_\nu \subset V_k$. Therefore, by virtue of 12.1, there exist sets P_ν open in R such that

$$(8) \qquad \overline{P}_\nu \subset V_k$$

and $F_\nu \subset P_\nu$. According to (6) and (7), $F_\nu \subset Q_\nu$. From (3), (7), and (a_k), one can easily deduce that $F_\nu \subset Z_{k+1}$. According to (7), one has $F_\nu \subset R - S_k$. Therefore, the set F_ν closed in S_{k+1} [according to (7)] constitutes a part of the set $P_\nu Q_\nu Z_{k+1} \cdot (R - S_k)$. But S_{k+1} is closed in R and $\dim S_{k+1} \leqq n$. Therefore, it results from 14.1 and 17.2 that there exist sets W_ν open in R such that

$$(9) \qquad F_\nu \subset W_\nu \quad \text{for} \quad \nu = 1, 2, 3, \ldots,$$

$$(10) \qquad W_\nu \subset P_\nu Q_k Z_{k+1} - S_k \quad \text{for} \quad \nu = 1, 2, 3, \ldots,$$

$$(11) \qquad \dim S_{k+1} \cdot H_R(W_\nu) \leqq n - 1 \quad \text{for} \quad \nu = 1, 2, 3, \ldots$$

EDUARD ČECH

According to (5), (6), (7), (9), and (10), the hypotheses of the theorem 10 are satisfied if one replaces S, T, Q_ν, V_ν by (6), (7), Q_ν, W_ν. Therefore,

$$H_R(\sum_{\nu=1}^{\infty} W_\nu) \subset \sum_{\nu=1}^{\infty} H_R(W_\nu) + S_k \cdot (A + H_R(U_k)). \tag{12}$$

Let us set

$$U_{k+1} = U_k + \sum_{\nu=1}^{\infty} W_\nu \tag{13}$$

so that U_{k+1} is an open subset of R.

From (a_k), (3), (7), (9), (13), one deduces easily that the condition (a_{k+1}) is realized. The validity of (b_{k+1}) is obvious. According to (a_k), (12), (13), and 9.1, one has

$$H_R(U_{k+1}) \subset H_R(U_k) + \sum_{\nu=1}^{\infty} H_R(W_\nu). \tag{14}$$

According to (7), (9), and (13), one has $H_R(U_{k+1}) \cdot [S_{k+1} \cdot H_R(U_k) - S_k] = 0$, so that (14) gives

$$S_{k+1} \cdot H_R(U_{k+1}) \subset S_k \cdot H_R(U_k) + \sum_{\nu=1}^{\infty} S_{k+1} \cdot H_R(W_k). \tag{15}$$

But we suppose the validity of the theorem 19 for the dimension $n-1$. Therefore, from (c_k), (11), (15), and 16.2, it results (c_{k+1}). According to (8) and (10), $W_\nu \subset V_k$, so that (13) gives (d_{k+1}). According to (8) and (10), $H_R(W_\nu) \subset V_k$. According to (f_{k+1}), $H_R(U_k) - S_k \subset V_k$. Therefore, (14) gives (e_{k+1}).

21.6. The construction of sets U_r, V_r, T_r open in R possessing the properties (a_r) – (j_r) for $r = 1, 2, 3, \ldots$ is thus accomplished. We must (see 21.1) construct a set U_ω open in R such that

$$A \subset U_\omega \subset Z, \tag{16}$$

$$\dim H_R(U_\omega) \leqq n - 1. \tag{17}$$

For this purpose let us set

$$U_\omega = \sum_{r=1}^{\infty} U_r,$$

ON THE DIMENSION OF PERFECTLY NORMAL SPACES

so that U_ω is an open subset of R. The condition (16) is satisfied by virtue of (2), (3), and (a_r). Let us choose an arbitrary value of k $(= 1, 2, 3, \dots)$. According to (b_r) and (d_r), one has

$$U_\omega \subset U_k + \sum_{r=k}^{\infty} V_r. \tag{18}$$

According to (i_r) and (j_r), one has $T_k \cdot \sum_{r=k}^{\infty} V_k = 0$. The sets T_k and $\sum_{r=k}^{\infty} V_r$ being open in R, they are separated, whence (see 7.2)

$$T_k \cdot \overline{\sum_{r=k}^{\infty} V_r} = 0.$$

But according to (18),

$$H^R(U_\omega) \subset \overline{U}_\omega \subset \overline{U}_k + \overline{\sum_{r=k}^{\infty} V_r},$$

so that

$$T_k \cdot H_R(U_\omega) \subset \overline{U}_k.$$

Therefore, by virtue of (h_{k+1}), $(S_k - \overline{U}_k \cdot H_R(U_\omega)) = 0$, i. e. $S_k \cdot H_R(U_\omega) \subset \overline{U}_k$. But $U_k \subset U_\omega \subset R - H_R(U_\omega)$, $\overline{U}_k - U_k = H_R(U_k)$. Consequently, $S_k \cdot H_R(U_\omega) \subset S_k \cdot H_R(U_k)$. This being true for $k = 1, 2, 3 \dots$, one deduces from (2) that

$$H_R(U_\omega) \subset \sum_{k=1}^{\infty} S_k \cdot H_R(U_k)\,. \tag{19}$$

Since we suppose the validity of the theorem 19 for the dimension $n-1$, the relation (17) results from (19) and (c_r) by virtue of 16. 2.

IV. Some consequences

22. Let R be a perfectly normal space, and let $\dim R \leqq n$. Let A be an arbitrary subset of R. Let $U \supset A$ be an open subset of R. Then there exists a set V open in R such that

$$A \subset V \subset U\,, \qquad H_R(V) = \Phi_1 + \Phi_2\,,$$
$$\dim \Phi_1 \leqq n-1\,, \quad \Phi_2 = \overline{A} \cdot H_R(V) = (\overline{A} - A) \cdot H_R(V)\,.$$

Proof. The space R being perfectly normal, the set $\overline{A} \cdot U$ is of the type F_σ in R. Therefore, there exist sets F_ν closed in R such that

$$\overline{A} \cdot U = \sum_{\nu=1}^{\infty} F_\nu\,. \tag{1}$$

According to 14.3, there exist sets Q_ν open in R such that

$$Q_\nu \supset Q_{\nu+1}\ , \quad \overline{A} = \prod_{\nu=1}^{\infty} Q_\nu = \prod_{\nu=1}^{\infty} \overline{Q}_\nu\ . \tag{2}$$

According to (1) and (2), $F_\nu \subset UQ_\nu$. Because $\dim R \leqq n$, there exist sets W_ν open in R such that

$$F_\nu \subset W_\nu \subset UQ_\nu\ , \tag{3}$$

$$\dim H_R(W_\nu) \leqq n-1\ . \tag{4}$$

According to 10,

$$H_R(\sum_{\nu=1}^{\infty} W_\nu) \subset \sum_{\nu=1}^{\infty} H_R(W_\nu) + (\overline{A} - U)\ . \tag{5}$$

Let us set

$$V = \sum_{\nu=1}^{\infty} W_\nu\ ; \quad \Phi_1 = H_R(V) \cdot \sum_{\nu=1}^{\infty} H_R(W_\nu)\ ; \quad \Phi_2 = \overline{A} \cdot H_R(V)\ .$$

The set V is open in R. Since $A \subset U$, according to (1) and (3), one gets $A \subset \overline{A} \cdot U \subset V$. Therefore, $A \cdot H_R(V) = 0$, whence $\Phi_2 = (\overline{A} - A) \cdot H_R(V)$. According to (5), $H_R(V) = \Phi_1 + \Phi_2$. According to (4), 19, and 16. 2, $\dim \Phi_1 \leqq n-1$.

23. Let R be a perfectly normal space, and let $\dim R \leqq n$. Let S be an arbitrary subset of R. Then $\dim S \leqq n$.

Proof. The theorem being trivial for $n = -1$, let us suppose it is true for the dimension $n-1$. Let A be a set closed in S. Let $U_0 \supset A$ be a set open in S, so that there exists a set U open in R such that $U_0 = SU$, whence $U \supset A$. According to 22, there exists a set V open in R such that $A \subset V \subset U$, $H_R(V) = \Phi_1 + \Phi_2$, $\dim \Phi_1 \leqq n-1$, $\Phi_2 \subset \overline{A} - A$. Let us set $V_0 = SV$. The set V_0 is open in S, and one has $A \subset V_0 \subset U_0$. According to 5, $A = S\overline{A}$, whence $S\Phi_2 = 0$. Therefore, $S \cdot H_R(V) \subset \Phi_1$, and consequently $H_S(V_0) \subset \Phi_1$ by virtue of 11. But we suppose the validity of the theorem under consideration for the dimension $n-1$, which implies $H_S(V_0) \leqq n-1$. Consequently, $\dim S \leqq n$.

24. 1. Let R be a perfectly normal space, and let $\dim R \leqq n$. Let S be a closed subset of R. Let U_0 be a set open in S, and let $U \supset U_0$ be a set open in R. Let $\dim H_S(U_0) \leqq n-1$. Then there exists a set V open in R such that $U_0 \subset V \subset U$, $SV = U_0$, $S \cdot H_R(V) = H_S(U_0)$, $\dim H_R(V) \leqq n-1$.

Proof. According to 13.3 and 14.1, there exists a set W open in R such that $SW = U_0$, $U_0 \subset W \subset U$, $S \cdot H_R(W) = H_S(U_0)$. According to the theorem 22 (where one replaces A, U by U_0, W), there exists a set V open in R such that $U_0 \subset V \subset W \subset U$ (therefore, $SV = U_0$), $H_R(V) = \Phi_1 + \Phi_2$, dim $\Phi_1 \leqq n-1$, $\Phi_2 = \overline{U}_0 \cdot H_R(V) = (\overline{U}_0 - U)H_R(V)$. The set Φ_2 is closed in R. One has $H_R(V) \subset \overline{V} \subset \overline{W}$. But $SW = U_0 \subset V \subset R - H_R(V)$, so that $S \cdot H_R(V) \subset S \cdot H_R(W) = H_S(U_0)$, whence, according to 11, $S \cdot H_R(V) = H_S(U_0)$. But $H_S(U_0) = S \cdot \overline{U}_0 - U_0 = \overline{U}_0 - U_0$ (for $U_0 \subset S$ implies $\overline{U}_0 \subset S$, S being closed in R), and consequently $\Phi_2 = (\overline{U}_0 - U_0)H_R(V) = H_S(V_0)$. Therefore, dim $\Phi_2 \leqq n-1$. The set Φ_2 being closed in R, the set $H_R(V) - \Phi_2 \subset \Phi_1$ is (the space R being perfectly normal) of the type F_σ in R, whence $H_R(V) = \Phi_2 + \sum_{\nu=1}^{\infty} F_\nu$, with F_ν being subsets of Φ_1 closed in R. Since dim $\Phi_1 \leqq n-1$, according to 23 (or according to 16.2), one has dim $F_\nu \leqq n-1$. Therefore, according to 19, dim$H_R(V) \leqq n-1$.

24.2. Let R be a perfectly normal space. Let $S_1, S_2, \ldots, S_k$ be closed subsets in R. Let $R = S_1 \supset S_2 \supset \cdots \supset S_k$. Let U_0 be a set open in S_k. Let $U \supset U_0$ be a set open in R. Let dim $S_\nu \leqq n_\nu$ for $1 \leqq \nu \leqq k$. Let dim $H_{S_k}(U_0) \leqq n_k - 1$. Then there exists a set V open in R such that $U_0 \subset V \subset U$, $S_k \cdot V = U_0$, $S_k \cdot H_R(V) = H_{S_k}(U_0)$, dim $S_\nu \cdot H_R(V) \leqq n_\nu - 1$ for $1 \leqq \nu \leqq k$.

Proof. The theorem being trivial for $k=1$, let us suppose it holds for $k-1$. Then there exists (see 14.2) a set V_0 open in S_2 such that $U_0 \subset V_0 \subset S_2 \cdot U$, $S_k \cdot V_0 = U_0$, $S_k \cdot H_{S_2}(V_0) = H_{S_k}(U_0)$, dim $S_\nu \cdot H_{S_2}(V_0) \leqq n_\nu - 1$ for $2 \leqq \nu \leqq k$. By virtue of 24.1 (where one replaces n, S, U_0 by n_1, S_2, V_0), there exists a set V open in R such that $U_0 \subset V_0 \subset V \subset U$, $S_2 V = V_0$ (and consequently $S_k V = U_0$), $S_2 \cdot H_R(V) = H_{S_2}(V_0)$, dim $H_R(V) \leqq n_1 - 1$. Since $S_k \subset S_2$, one has $S_k \cdot H_R(V) = S_k \cdot H_{S_2}(V_0) = H_{S_k}(U_0)$. Since $S_1 = R$, the relation dim $S_\nu \cdot H_R(V) \leqq n_\nu - 1$ is true for $\nu = 1$. For $2 \leqq \nu \leqq k$, one has $S_\nu \subset S_2$. Therefore, $S_\nu \cdot H_R(V) = S_\nu \cdot H_{S_2}(V_0)$, whence again dim $S_\nu \cdot H_R(V) \leqq n_\nu - 1$.

V. Theorem on the covering of a finite dimensional space

25.1. Let R be a normal space. Let $U_1, \ldots, U_m$ be open subsets of R, and let $\sum_{\nu=1}^{m} U_\nu = R$. Then there exists a set V_1 open in R such that $\overline{V}_1 \subset U_1$, $V_1 + \sum_{\nu=2}^{m} U_\nu = R$.

Proof. [13] The set $R - \sum_{\nu=2}^{m} U_\nu$ is closed in R, and is contained in U_1. Therefore, according to 12.1, there exists a set V_1 open in R such that

$$R - \sum_{\nu=2}^{m} U_\nu \subset V_1 \subset \overline{V}_1 \subset U_1 .$$

Obviously $V_1 + \sum_{\nu=2}^{m} U_\nu = R$.

25.2. Let R be a normal space. Let $U_1, \ldots, U_m$ be open subsets of R, and let $\sum_{\nu=1}^{m} U_\nu = R$. Then there exist sets $V_1, \ldots, V_m$ open in R such that $\overline{V}_1 \subset$

[13] Menger, l. c. p. 159–160 (Bemerkung).

$U_1, \ldots, \overline{V}_m \subset U_m$, $\sum_{\nu=1}^{m} U_\nu = R$.

Proof. [13] The sets V_ν can be obtained by applying n-times the proposition 25.1.

25.3. Let R be a perfectly normal space, and let $\dim R \leqq n$. Let $U_1, \ldots, U_m$ be open subsets of R, and let $\sum_{\nu=1}^{m} U_\nu = R$. Then there exist sets $V_1, \ldots, V_m$ open in R such that: $\overline{V}_\nu \subset U_\nu$ for $1 \leqq \nu \leqq m$; $\sum_{\nu=1}^{m} \overline{V}_\nu = R$; $V_\mu V_\nu = 0$ for $1 \leqq \mu < \nu \leqq m$; $\dim H_R(V_\nu) \leqq n-1$ for $1 \leqq \nu \leqq m$.

Proof. According to 25.2. there exist sets $F_1, \ldots, F_m$ closed in R such that $F_\nu \subset U_\nu$ for $1 \leqq \nu \leqq m$, $\sum_{\nu=1}^{m} F_\nu = R$. According to 12.1., there exist sets $W_1, \ldots, W_m$ open in R such that $F_\nu \subset W_\nu \subset \overline{W}_\nu \subset U_\nu$ for $1 \leqq \nu \leqq m$. Since $\dim R \leqq n$, there exist sets $Z_1, \ldots, Z_m$ open in R such that $F_\nu \subset Z_\nu \subset W_\nu$, $\dim H_R(Z_\nu) \leqq n-1$ for $1 \leqq \nu \leqq m$. Let us set $V_1 = Z_1$. For $2 \leqq \nu \leqq m$, let us set $V_\nu = Z_\nu - \sum_{\mu=1}^{\nu-1} \overline{Z}_\mu$. The sets V_ν are open in R, and one has $\overline{V}_\nu \subset \overline{Z}_\nu \subset \overline{W}_\nu \subset U_\nu$. Let p be an arbitrary point of R. Since $R = \sum_{\nu=1}^{m} F_\nu \subset \sum_{\nu=1}^{m} \overline{Z}_\nu$, let ν be the least index such that $(p) \subset \overline{Z}_\nu$. One can easily see that $(p) \subset \overline{V}_\nu$. Therefore, $\sum_{\nu=1}^{m} \overline{V}_\nu = R$. For $1 \leqq \mu < \nu \leqq m$, one has $V_\nu \subset R - \sum_{i=1}^{\nu-1} \overline{Z}_i \subset R - \overline{Z}_\mu \subset R - \overline{V}_\mu \subset R - V_\mu$, whence $V_\mu V_\nu = 0$. Since $V_1 = Z_1$, one has $\dim H_R(V_1) \leqq n-1$. For $2 \leqq \nu \leqq m$, one has

$$V_\nu = Z_\nu - \sum_{\mu=1}^{\nu-1} \overline{Z}_\mu = Z_\nu - \overline{\sum_{\mu=1}^{\nu-1} Z_\mu},$$

and therefore, according to 9.1 and 9.2,

$$H_R(V_\nu) \subset \sum_{\mu=1}^{\nu} H_R(Z_\mu),$$

whence, according to 19 and 23 (or 16.2) $\dim H_R(V_\nu) \leqq n-1$.

26. Let R be a perfectly normal space, and let $\dim R \leqq n$. Let $U_1, \ldots, U_m$ be open subsets of R, and let $\sum_{\nu=1}^{m} U_\nu = R$. Then there exist sets V_i $(1 \leqq i \leqq (n+1)m)$ open in R possessing the following properties:

1° $\overline{V}_i \subset U_\nu$ for $1 \leqq \nu \leqq m$, $(n+1)(\nu-1)+1 \leqq i \leqq (n+1)\nu$;

2° $\sum_{i=1}^{(n+1)m} \overline{V}_i = R$;

3° $V_i \cdot V_j = 0$ for $1 \leqq i < j \leqq (n+1)m$;

4° $\dim H_R(V_i) \leqq n-1$ for $1 \leqq i \leqq (n+1)m$;

ON THE DIMENSION OF PERFECTLY NORMAL SPACES

5° let $i_1, i_2, \dots, i_r$ $(2 \leqq r \leqq n+2)$ be a combination of the indices $1, 2, \dots, (n+1)m$: then $\dim \prod_{s=1}^{r} \overline{V}_{i_s} \leqq n - r + 1$.

Proof. For $k = 1$, it follows easily from 25.3 that there exist sets $V_i^{(k)}$ $(1 \leqq i \leqq km)$ open in R such that

$$(a_k) \qquad \overline{V}_i^{(k)} \subset U_\nu \quad \text{for} \quad 1 \leqq \nu \leqq m, \quad k(\nu - 1) + 1 \leqq i \leqq k\nu;$$

$$(b_k) \qquad \sum_{i=1}^{km} \overline{V}_i^{(k)} = R;$$

$$(c_k) \qquad V_i^{(k)} \cdot V_j^{(k)} = 0 \quad \text{for} \quad 1 \leqq i < j \leqq km;$$

$$(d_k) \qquad \dim H_R(V_i^{(k)}) \leqq n - 1 \quad \text{for} \quad 1 \leqq qi \leqq km;$$

(e_k) for each combination $i_1, i_2, \dots, i_r$ of the indices $1, 2, \dots, km$ such that $2 \leqq r \leqq k+1$ one has $\dim \prod_{s=1}^{r} \overline{V}_{i_s}^{(k)} \leqq n - r + 1$.[14]

One must prove that such sets exist also for $k = n + 2$. Thus, let us suppose that, for a certain value of k $(1 \leqq k \leqq n+1)$, one has already constructed sets $V_i^{(k)}$ $(1 \leqq i \leqq km)$ open in R satisfying the conditions (a_k) – (e_k). The question is only to deduce from them sets $V_i^{(k+1)}$ $(1 \leqq i \leqq (k+1)m)$ open in R satisfying (a_{k+1}) – (e_{k+1}).

26.1. For $1 \leqq r \leqq k$, let S_r be the set of all points p of the space R such that $(p) \subset V_i^{(k)}$ for at least r distinct values of the index i $(1 \leqq i \leqq km)$. According to (b_k), one has $S_1 = R$. Obviously, $S_1 \supset S_2 \supset \cdots \supset S_k$, and the sets S_r $(1 \leqq r \leqq k)$ are closed in R. According to 19 and (e_k), one has $\dim S_r \leqq n - r + 1$ for $1 \leqq r \leqq k$. The sets $S_k U_\nu$ $(1 \leqq \nu \leqq m)$ are open in S_k, and one has $\sum_{\nu=1}^{m} S_k U_\nu = S_k$. Therefore, by virtue of 14.2 and 25.3, there exist sets T_ν $(1 \leqq \nu \leqq m)$ open in S_k such that

$$(1) \qquad \overline{T}_\nu \subset U_\nu \quad \text{for} \quad 1 \leqq \nu \leqq m;$$

$$(2) \qquad \sum_{\nu=1}^{m} \overline{T}_\nu = S_k;$$

$$(3) \qquad T_\mu T_\nu = 0 \quad \text{for} \quad 1 \leqq \mu < \nu \leqq m;$$

$$(4) \qquad \dim H_{S_k}(T_\nu) \leqq n - k \quad \text{for} \quad 1 \leqq \nu \leqq m.$$

[14] For $k = 1$, there is $r = 2$, and one has $P = \prod_{s=1}^{r} \overline{V}_{i_s}^{(k)} = \overline{V}_{i_1}^{(1)} \cdot \overline{V}_{i_2}^{(1)}$. But, according to (c_i), $V_{i_1}^{(1)} \cdot V_{i_2}^{(1)} = 0$, whence, according to 7.2, $V_{i_1}^{(1)} \overline{V}_{i_2}^{(1)} = 0$. Therefore, $P \subset H_R(V_{i_1}^{(1)})$, whence, according to 16.2 and (d_1), $\dim P \leqq n - r + 1$.

26.2. We are going to construct sets W_ν $(1 \leqq \nu \leqq m)$ open in R such that:

$$(\alpha_\nu) \qquad \overline{W}_\nu \subset U_\nu \quad \text{for} \quad 1 \leqq \nu \leqq m;$$

$$(\beta_\nu) \qquad S_k \cdot W_\nu = T_\nu \quad \text{for} \quad 1 \leqq \nu \leqq m;$$

$$(\gamma_\nu) \qquad S_k \cdot H_R(W_\nu) = H_{S_k}(T_\nu) \quad \text{for} \quad 1 \leqq \nu \leqq m;$$

$$(\delta_\nu) \qquad \dim S_k \cdot H_R(W_\nu) \leqq n - r \quad \text{for} \quad 1 \leqq r \leqq k,\ 1 \leqq \nu \leqq m;$$

$$(\varepsilon_\nu) \qquad \overline{W}_\mu W_\nu = 0 \quad \text{for} \quad 1 \leqq \mu < \nu \leqq m;$$

$$(\zeta_\nu) \qquad \overline{W}_\mu \cdot \overline{W}_\nu \subset S_k \quad \text{for} \quad 1 \leqq \mu < \nu \leqq m.$$

The conditions (ε_1) and (ζ_1) can be regarded as obvious. According to (1) and 12.1 one constructs first sets Z_ν open in R such that

$$(5) \qquad \overline{T}_\nu \subset Z_\nu \subset \overline{Z}_\nu \subset U_\nu \quad \text{for} \quad 1 \leqq \nu \leqq m.$$

According to 26.1, the hypotheses of the theorem 24.2 are satisfied if one replaces U_0, U, n_r $(1 \leqq r \leqq k)$ by T_1, Z_1, $n - r$. Therefore, there exists a set W_1 open in R satisfying (α_1) – (ζ_1). Thus, let us suppose that for a certain value of ν $(2 \leqq \nu \leqq m)$ one has already constructed sets $W_1, \dots, W_{\nu-1}$ open in R such that the conditions (α_μ) – (ζ_μ) are satisfied for $1 \leqq \mu < \nu$. One must construct a set W_ν open in R satisfying the conditions (α_ν) – (ζ_ν). For $1 \leqq \mu < \nu$, one has according to (β_μ) – (γ_μ)

$$(\overline{W}_\mu - S_k)\overline{T}_\nu \subset (\overline{W}_\mu - S_k)S_k = 0, \quad \overline{\overline{W}_\mu - S_k} \cdot T_\nu \subset \overline{W}_\mu \cdot S_k T_\nu \subset \overline{T}_\mu T_\nu.$$

The sets T_μ and T_ν being open in S_k, they are, according to (3), separated, whence $\overline{T}_\mu \cdot T_\nu = 0$. Therefore (see 7.2), the sets $\overline{W}_\mu - S_k$ and T_ν are separated. The sets $\sum\limits_{\mu=1}^{\nu-1} (\overline{W}_\mu - S_k)$, T_ν are also separated by virtue of 6.1. Therefore (see 14.1), there exist sets P_ν, Q_ν open in R such that

$$(6) \qquad P_\nu \supset \sum_{\mu=1}^{\nu-1} \overline{W}_\mu - S_k, \quad Q_\nu \supset T_\nu, \quad P_\nu Q_\nu = 0.$$

But one can apply the theorem 24.2, replacing there U_0, U, n_r $(1 \leqq r \leqq k)$ by T_ν, $Z_\nu Q_\nu$, $n - r$. Hence, it follows the existence of a set W_ν open in R such that

$$(7) \qquad T_\nu \subset W_\nu \subset Z_\nu Q_\nu .$$

with the conditions (β_ν), (γ_ν), (δ_ν) being realized. According to (7), one has $\overline{W}_\nu \subset \overline{Z}_\nu$, whence, according to (5), it results (α_ν). The sets P_ν, Q_ν open in R are separated according to (6), so that (see 7.2) $P_\nu \overline{Q}_\nu = 0$, and therefore $\overline{W}_\mu \overline{Q}_\nu \subset S_k$ for $1 \leqq \mu < \nu$. But one has $\overline{W}_\nu \subset \overline{Q}_\nu$ by virtue of (7), and therefore one has (ζ_ν). Consequently, for $1 \leqq \mu < \nu$, one has $\overline{W}_\mu \cdot W_\nu = [W_\mu + H_R(W_\mu)] \cdot W_\nu \subset S_k$, whence, taking into account (β_μ), (γ_μ), and (β_ν),

$$\overline{W}_\mu \cdot W_\nu = [S_k W_\mu + S_k H_R(W_\mu)] \cdot S_k W_\nu = [T_\mu + H_{S_k}(T_\mu)] T_\nu = \overline{T}_\mu T_\nu = 0$$

according to (3) and 7.2. Hence, it results (ε_ν).

26.3. Let us set

$$(8)\qquad W_{i+r}^{(k+1)} = V_i^{(k)} - \sum_{\nu=1}^{m} \overline{W}_\nu \quad \text{for} \quad 0 \leqq r \leqq m-1,\ rk+1 \leqq i \leqq (r+1)k,$$

$$(9)\qquad V_{r(k+1)}^{(k+1)} = W_\nu \quad \text{for} \quad 1 \leqq \nu \leqq m,$$

so that $V_i^{(k+1)}$ $(1 \leqq i \leqq (k+1)m)$ are open sets of R. According to (a_k) and (α_ν), one has (a_{k+1}). For $0 \leqq r \leqq m-1$, $rk+1 \leqq i \leqq (r+1)k$, one has, according to (8), $V_i^{(k)} \subset V_{i+r}^{(k+1)} + \sum_{\nu=1}^{m} \overline{W}_\nu$, whence $\overline{V}_i^{(k)} \subset \overline{V}_{i+r}^{(k+1)} + \sum_{\nu=1}^{m} \overline{W}_\nu$, and therefore (b_{k+1}) by virtue of (9) and (b_k). According to (8), (9), (c_k) and (ε_ν), one has (c_{k+1}). According to 9.1 and 9.2, one has for $0 \leqq r \leqq m-1$, $rk+1 \leqq i \leqq (r+1)k$

$$H_R(V_{i+r}^{(k+1)}) \subset H_R(V_i^{(k)}) + \sum_{\nu=1}^{m} H_R(W_\nu),$$

whence (d_{k+1}) according to (δ_ν), (d_k), 16.2, and 20.3.

26.4. It remains to prove (e_{k+1}). Therefore, let $j_1, j_2, \ldots, j_r$ $(2 \leqq r \leqq k+1)$ be a combination of the indices $1, 2, \ldots, (k+1)m$. Let

$$Q = \prod_{s=1}^{r} \overline{V}_{j_s}^{(k+1)}.$$

One must prove that $\dim Q \leqq n - r + 1$. There are four cases to be distinguished:

F i r s t c a s e. At least two of the sets $V_{j_s}^{(k+1)}$ $(1 \leqq s \leqq r)$ are given by (9). For fixing the ideas, let

$$W_{j_1}^{(k+1)} = W_\mu, \quad V_{j_2}^{(k+1)} = W_\nu$$

with $\mu, \nu = 1, 2, \ldots, m$; $\mu < \nu$. One has $Q \subset \overline{W}_\mu \cdot \overline{W}_\nu$, whence, according to (ζ_ν), $Q \subset S_k$. Besides, according to (ε_ν), $\overline{W}_\mu W_\nu = 0$, whence $Q \subset H_R(W_\nu)$. Therefore, according to 16.2 and (δ_ν), $\dim Q \leqq n - k \leqq n - r + 1$, for $r \leqq k+1$.

S e c o n d c a s e. Among the r sets $V_{j_s}^{(k+1)}$, only one is given by (9). There exists an index ν $(1 \leqq \nu \leqq m)$ and a combination $i_1, i_2, \ldots, i_{r-1}$ of the indices $1, 2, \ldots, km$ so that

$$Q = \overline{W}_\nu \cdot \overline{\prod_{s=1}^{r-1} V_{i_s}^{(k)} - \sum_{\mu=1}^{m} \overline{W}_\mu} \subset \overline{W}_\nu \cdot \prod_{s=1}^{r-1} \overline{V}_{i_s}^{(k)} \subset \overline{W}_\nu \cdot S_{r-1}.$$

The sets W_ν, $V_{i_s}^{(k)} - \sum_{\mu=1}^{m} \overline{W}_\mu$ without a common point being open in R, they are separated. Therefore, according to 7.2, $W_\nu \cdot \overline{\prod_{s=1}^{r-1} V_{i_s}^{(k)} - \sum_{\mu=1}^{m} \overline{W}_\mu} = 0$, whence $Q \subset S_{r-1} \cdot H_R(W_\nu)$, and, according to 16.2 and (δ_ν), $\dim Q \leqq n - r + 1$.

EDUARD ČECH

Third case. All the sets $V_{j_s}^{(k+1)}$ are given by (8), and $2 \leqq r \leqq k$. There exists a combination $i_1, i_2, \ldots, i_r$ of the indices $1, 2, \ldots, km$ such that

$$(*) \qquad Q = \overline{\prod_{s=1}^{r} V_{i_s}^{(k)} - \sum_{\mu=1}^{m} \overline{W}_\mu} \subset \prod_{s=1}^{r} \overline{V}_{i_s}^{(k)},$$

and therefore, according to 16.2 and (e_k), $\dim Q \leqq n - r + 1$.

Fourth case. All the sets $V_{j_s}^{(k+1)}$ are given by (8), and $r = k+1$. One has again the formula (*). Since $r = k+1$, one has $Q \subset S_k$ by virtue of the definition of S_k, whence, according to (2) and (β_ν), $Q \subset \sum_{\nu=1}^{m} \overline{W}_\nu$. But $(V_{i_s}^{(k)} - \sum_{\mu=1}^{m} \overline{W}_\mu) \cdot W_\nu = 0$, whence, according to 7.2, $\overline{V_{i_s}^{(k)} - \sum_{\mu=1}^{m} \overline{W}_\mu} \cdot W_\nu = 0$, so that $Q \cdot W_\nu = 0$ for $1 \leqq \nu \leqq m$. Therefore, $Q \subset \sum_{\nu=1}^{m} (\overline{W}_\nu - W_\nu)$, $Q \subset S_k$, and consequently $Q \subset \sum_{\nu=1}^{m} S_k \cdot H_R(W_\nu)$. Therefore, according to 16.2 and (δ_ν), $\dim Q \leqq n - k = n - r + 1$.

27. Let R be a perfectly normal space. Let S be a closed subset of R, and let $\dim S \leqq n$. Let $U_1, \ldots, U_m$ be open subsets of R, and let $\sum_{\nu=1}^{m} U_\nu \supset S$. Then there exist sets V_i $(1 \leqq i \leqq (n+1)m)$ open in R such that:

1° $\overline{V}_i \subset U_\nu$ for $1 \leqq \nu \leqq m$, $(n+1)(\nu - 1) + 1 \leqq i \leqq (n+1)\nu$;

2° $\sum_{i=1}^{(n+1)m} V_i \supset S$;

3° $V_i V_j = 0$ for $1 \leqq i < j \leqq (n+1)m$;

4° $\dim S \cdot H_R(V_i) \leqq n - 1$ for $1 \leqq i \leqq (n+1)m$;

5° for each combination $i_1, i_2, \ldots, i_r$ $(2 \leqq r \leqq n+2)$ of the indices $1, 2, \ldots, (n+1)m$ one has $\prod_{s=1}^{r} \overline{V}_{i_s} \subset S$, $\dim \prod_{s=1}^{r} \overline{V}_{i_s} \leqq n - r + 1$.

Proof. According to 26 and 14.2, there exist sets T_i open in S such that:

6° $\overline{T}_i \subset U_\nu$ for $1 \leqq \nu \leqq m$, $(n+1)(\nu - 1) + 1 \leqq i \leqq (n+1)\nu$;

7° $\sum_{i=1}^{(n+1)m} \overline{T}_i = S$;

8° $T_i \cdot T_j = 0$ for $1 \leqq i < j \leqq (n+1)m$;

9° $\dim H_S(T_i) \leqq n - 1$ for $1 \leqq i \leqq (n+1)m$;

10° for every combination $i_1, i_2, \ldots, i_r$ $(2 \leqq r \leqq n+2)$ of the indices $1, 2, \ldots, (n+1)m$, one has $\dim \prod_{s=1}^{r} \overline{T}_{i_s} \leqq n - r + 1$.

According to 6° and 12.1, there exist sets Z_i open in R such that

$$\overline{T}_i \subset Z_i \subset \overline{Z}_i \subset U_\nu \quad \text{for} \quad 1 \leqq \nu \leqq m, \quad (n+1)(\nu - 1) + 1 \leqq i \leqq (n+1)\nu.$$

Obviously, it is sufficient to construct sets V_i $(1 \leqq i \leqq (n+1)m)$ open in R, so that

$$(a_i) \qquad T_i \subset V_i \subset Z_i, \quad T_i = SV_i \quad \text{for} \quad 1 \leqq i \leqq (n+1)m;$$

$$(b_i) \qquad V_i \cdot V_j = 0 \quad \text{for} \quad 1 \leqq j < i \leqq (n+1)m;$$

$$(c_i) \qquad S \cdot H_R(V_i) = H_S(T_i) \quad \text{for} \quad 1 \leqq i \leqq (n+1)m;$$

$$(d_i) \qquad \overline{V}_i \cdot \overline{V}_j = \overline{T}_i \cdot \overline{T}_j \quad \text{for} \quad 1 \leqq j < i \leqq (n+1)m.$$

ON THE DIMENSION OF PERFECTLY NORMAL SPACES

According to 13.3 and 14.1, there exists a set V_1 open in R satisfying (a_1) and (c_1), for (b_1) and (d_1) require nothing. Thus, let us suppose that, for a certain value of i $(2 \leqq i \leqq (n+1)m)$, one has already constructed sets V_j $(1 \leqq j < i)$ open in R such that the conditions (a_j) – (d_j) are satisfied for $1 \leqq j < i$. One has to construct a set V_i open in R satisfying (a_i) – (d_i). For $1 \leqq j < i$, one has

$$(\overline{V}_j - S) \cdot \overline{T}_i \subset (R - S)S = 0,$$
$$\overline{V_j - S} \cdot T_i \subset \overline{V}_j \cdot S \cdot T_i = [S \cdot V_j + S \cdot H_R(V_j)]T_i =$$
$$= [T_j + H_S(T_j)]T_i = \overline{T}_j \cdot T_i,$$

and, according to 8° and 7.2, $\overline{T}_j \cdot T_i = 0$. Consequently, by virtue of 6.1 and 7.2, the sets $\sum_{j=1}^{i-1} V_j - S$ and T_i are separated. Consequently, there exist (see 14.1) sets P_i, Q_i open in R such that

$$P_i \supset \sum_{j=1}^{i-1} \overline{V}_j - S, \quad Q_i \supset T_i, \quad P_i Q_i = 0.$$

According to 7.2, $P_i \overline{Q}_i = 0$. According to 13.3 and 14.1, there exists a set V_i open in R such that

$$T_i \subset V_i \subset Z_i Q_i, \quad T_i = SV_i, \quad H_S(T_i) = S \cdot H_R(V_i).$$

The properties (a_i) and (c_i) are obvious. Since $V_i \subset Q_i$, $V_j \subset S + P_i$ $(1 \leqq j < i)$, one has $\overline{V}_i \overline{V}_j \subset P_i \overline{Q}_i + S = S$. Therefore, $V_i V_j = S$, whence $V_i V_j = SV_i \cdot SV_j = T_i T_j = 0$, which gives (b_i). Moreover, $S\overline{V}_i = S \cdot [V_i + H_R(V_i)] = T_i + H_S(T_i) = \overline{T}_i$. Since $\overline{V}_i \overline{V}_j \subset S$, one has $\overline{V}_i \overline{V}_j = S\overline{V}_i \cdot S\overline{V}_j = \overline{T}_i \overline{T}_j$, whence (d_i).

Contribution to Dimension Theory

Eduard Čech

(Received March 14, 1933)

Introduction. Let $n = 1, 2, 3, \ldots$. Denote by C_n the set of those "points", for which $0 \leqq x_i \leqq 1$ $(1 \leqq i \leqq n)$. The set C_n is the simplest example of a point set *of dimension n*. In particular, *for* $m < n$, *the dimension of the set* C_m *is less than the dimension of the set* C_n. The question on the precise mathematical meaning of the italicized statement is, however, not quite easy. On the first sight it may seem that C_n has *more points* than C_m. But as soon as in 1877, G. Cantor showed[1]) that this is not the case, that there exists a *one-to-one* mapping of C_m onto C_n. Try then otherwise! It is clear that (still for $m < n$) C_m is a continuous image of the set C_n; in comparison with the opinion that C_n is not a continuous image of the set C_m. But even this way does not lead to the aim, because in 1890, G. Peano[2]) showed in turn that there exists a continous mapping of the set C_m onto the set C_n, too.

In 1910, H. Lebesgue stated[3]) a theorem (*Lebesgue lemma*): C_n *can be covered by a finite number of arbitrarily small closed sets so that every point belongs to at most* $n+1$ *of them, but not so that every point would belong to at most* n *of them.* It is apparent that on the basis of Lebesgue lemma one can formulate a general definition of dimension. Lebesgue however did not do so, but used his lemma only for giving reasons of the fact that C_m and C_n $(m \neq n)$ have a distinct topological structure.[4])

The first general definition of dimension was stated by L. E. J. Brouwer[5]) in 1913. Brouwer's paper remained however isolated and it is necessary to regard K. Menger and P. Urysohn[6]) as founders of the general dimension theory. The origin of these papers is to be placed approximately to the period 1921 - 1923. From the other authors, who contributed to the dimension theory, it is necessary to mention mainly W. Hurewicz.

The theory of dimension is built up in separable[7]) spaces. However, numer-

[1]) Journal f. Math., vol. 84 (1877), p. 242.

[2]) Math. Ann., vol. 36 (1890), p. 257.

[3]) Math. Ann., vol. 70 (1911), p. 166. The proof indicated here is not complete and Lebesgue gave a precise exposition only in Fund. Math., vol. 2 (1921), p. 256. The first precise proof of Lebesgue's lemma was given by Brouwer in the paper quoted in the remark [5]). Lebesgue's proof was later substantially simplified by W. Hurewicz [Math. Ann., vol. 101 (1929), p. 210] and by E. Sperner [Hamb. Abh., vol. 6 (1928), p. 265].

[4]) The first precise proof of this fact was given by Brouwer in Math. Ann., vol. 70 (1911), p. 161.

[5]) Journ. f. Math., vol. 142 (1913), p. 210 [see also vol. 153 (1924), p. 253].

[6]) A systematic exposition of the theory is given by K. Menger in his book *Dimensionstheorie* (1928). See also posthumously published Urysohn's treatise in Fund. Math., vol. 7 (1925), pp. 30–137 and vol. 8 (1926), pp. 225–351. The paper by V. Jarník in Časopis, vol. 58 (1929), p. 367, can help well for the first information.

[7]) A space R is separable, if it is metric (see head 5) and if it is possible to choose from every system of open sets covering R (see head 1) at most countable system, which covers R.

ous theorems hold only in compact[8]) or semicompact[9]) spaces. The restriction to separable spaces is, as it seems, justified by the nature of a matter at several problems[10]). But things are different with the other problems. As I have showed,[11]) the Menger-Uryson definition of dimension can be modified so that the validity of some theorems extend to much more general perfectly normal spaces,[12]) which contain as a special case all metric spaces.[13])

Urysohn proved[14]) that for compact spaces, dimension can be defined by the property formulated in Lebesgue lemma; but he notes that this definition is retricted to compact spaces. However later Hurewicz stated Lebesgue lemma in such a form, that it gives a characteristic property of dimension even in arbitrary separable spaces. Hence an attempt occurs readily to deduce some theorems of the dimension theory so that the indicated property will be chosen for the definition. I show in the present paper, how one can this way deduce so called *sum theorem.*[15]) The proof is not only easier than the one used until now, but it has the advantage that it holds in every normal[16]) space.

The definition chosen has two drawbacks. First, it is not obvious that the dimension of a part cannot exceed the dimension of the total; but I shall show that at least in perfectly normal spaces the theorem on the dimension of a part is a corollary of sum theorem. The second drawback is more substantial: In this way it is possible to define the dimension of a space as of *total*, not the dimension at single points or parts of it. In connection with it, I conjecture that in perfectly normal spaces, the definition of dimension chosen here is equivalent with that one I used in the paper cited in the remark [11]). By my opinion, the proof of this conjecture could provide an important progress in the dimension theory.

There are two advantages against the mentioned drawbacks. First, the definition chosen here is most suitable for the proof of the fact that Euclidean space and its elementary parts have, in the sense of the general theory, the dimension ajudged to them since long ago.[17]) Second, there are important theorems, the proof of which needs *only* the property of dimension that was chosen for the definition here. This is the case e.g. of the theorem due to P. Alexandroff; *A compact space of dimension n can be transformed by an arbitrarily small continuous deformation into a polyhedron of dimension n, but not into a polyhedron of dimension $< n$.*[18])

[8]) A space R is compact, if it is metric and if from every system of open subsets in R covering R it is possible to choose a finite system, covering R.

[9]) A space R is semicompact, if it is metric and if it is a sum of at most countable system of compact parts.

[10]) E.g., at problems studied in Chap. 4. in Menger's book.

[11]) Rozpr. II. tř. čes. Akad., vol. 42, 13 (1932). See also Comptes Rendus Paris, vol. 193 (1931), pp. 976–977.

[12]) See head 25.

[13]) See head 26.

[14]) Fund. Math., vol. 8, p 301.

[15]) See heads 23 and 24.

[16]) See head 9.

[17]) The beginner reader will perhaps lack the proof of this fact. I shall have an opportunity to present such a proof in another paper, which will soon appear in this Journal.

[18]) Annals of Math., vol 30 (1929), p. 120 (Überführungsatz). An easy proof will appear in the paper by C. Kuratowski *Sur un théorème fondamental* etc. in vol. 20, Fund. Math.

I remark yet that *no mathematical knowledge is necessary for the complete understanding of the text*; it suffices to read few first paragraphs (1 – 6) of my paper *Sets irreducibly connected between n points.*[19]) It is not incidental that it was possible to choose so elementary form of the exposition. Indeed, general topology in the course of its penetrating analysis of the space by no means gathers from the finely branching supply of knowledge of classical mathematics, still conquers its problems with success by a quite simple weapon, that is, by elementary rules of logical reasoning, applied directly to simple notions obtained from the intuition and axiomatically precised. Let this modest paper contribute somewhat to this beautiful mathematical discipline to enjoy even in our country such a favour, which it deserves for its inner grace and its significance in the totality of mathematical sciences.

1. Let A be a part of a topological space[20]) R. Let $\mathfrak{S}$ be a system of parts of a space R. We shall say that $\mathfrak{S}$ *covers* A, if every point $a \in A$[21]) is a part of some $S \in \mathfrak{S}$.

2. Let $n = -1, 0, 1, 2, \ldots$. Let $\mathfrak{S}$ be a system of parts of a space R. We shall say that *order of* $\mathfrak{S}$ is $\leqq n$, if no point $a \in R$ is contained in more than $n+1$ elements of the system $\mathfrak{S}$. We shall say that *order of* $\mathfrak{S}$ is n, if 1° order of $\mathfrak{S}$ is $\leqq n$, 2° either $n = -1$ or the order of $\mathfrak{S}$ is not $\leqq n-1$.

3. Let $n = -1, 0, 1, 2, \ldots$. Let R be a topological space. We shall say that *dimension* of the space R is $\leqq n$ and write $\dim R \leqq n$, if it is possible to assign to every finite[22]) system $\mathfrak{U}$ of open[20]) sets covering R a finite system $\mathfrak{V}$ of sets open in R such that 1° $\mathfrak{V}$ covers R; 2° every $V \in \mathfrak{V}$ is a part of some $U \in \mathfrak{U}$; 3° order of $\mathfrak{V}$ is $\leqq n$. We shall say that *the dimension of R is n* and write $\dim R = n$, if 1° $\dim R \leqq n$; 2° either $n = -1$ or $\dim R \leqq n-1$ does not hold.

Clearly, $\dim R = -1$ if and only if $R = 0$.[23])

By M, 1,2 the dimension of a one-point space equals to 0.

Clearly $\dim R = n$ implies that $\dim R^* = n$ for every space R^* homeomorphic[24]) to the space R.

4. *Let R be a topological space. Let $S \subset R$.*[25]) *Let S be closed in R. Let $\dim R \leqq n$. Then $\dim S \leqq n$.*[26])

P r o o f. Let $\mathfrak{U}'$ be a finite system of open sets in S, let $\mathfrak{U}'$ cover S. We have to show that there exists a finite system $\mathfrak{V}'$ consisting of open sets in S such that 1° $\mathfrak{V}'$ covers S; 2° every $V' \in \mathfrak{V}'$ is a part of some $U' \in \mathfrak{U}'$; 3° order of V' is $\leqq n$. By the definition of sets open in S[27]) one can assign to each $U' \in \mathfrak{U}'$ a set U open in R such that $U' = U \cdot S$.[28]) Let us add to the sets just determined moreover a set

19) Časopis, vol. 61 (1932), p. 109. Quoted under an abbr. M.

20) M, I.

21) $a \in A$ means that a is an element of a set A.

22) It means that the number of elements is finite.

23) M, [2]).

24) M, 3.

25) M, [6]).

26) S is a topological space by M, 5.

27) M, 5.

28) M, [8]).

$R - A$ open in R.[29]) Thus one gets a finite system $\mathfrak{U}$ of sets open in R; clearly $\mathfrak{U}$ covers R. Since $\dim R \leqq n$, there is a finite system $\mathfrak{V}'$ of sets open in R such that 4° $\mathfrak{V}$ covers R; 5° each $V \in \mathfrak{V}$ is a part of some $U \in \mathfrak{U}$; 6° order of $\mathfrak{V}$ is $\leqq n$. Let us assign to each $V \in \mathfrak{V}$ a set $V' = S \cdot V$. These V' form a system $\mathfrak{V}'$. Clearly $\mathfrak{V}'$ is a finite system of sets open in S. From the properties 4°, 5°, 6° of the system $\mathfrak{V}$ follow the properties 1°, 2°, 3° of the system $\mathfrak{V}'$.

5. Let us call a set R (composed from arbitrary elements) a *metric space* (and call its elements *points*), if to every pair a, b of points of R a real number $\varrho(a,b)$ has been assigned (the *distance* of points a, b in the space R). At the same time, the following three axioms must be satisfied:

5.1. If $a \in R$, then $\varrho(a,a) = 0$.

5.2. If $a \in R$, $b \in R$, $\varrho(a,b) = 0$, then $a = b$.

5.3. If $a \in R$, $b \in R$, $c \in R$, then $\varrho(a,b) + \varrho(c,b) \geqq \varrho(a,c)$.

The following theorems 5.4 and 5.5 then hold.[30])

5.4. If $a \in R$, $b \in R$, $a \neq b$, then $\varrho(a,b) > 0$. Proof. By 5.3, $2\varrho(a,b) \geqq \varrho(a,a)$, hence $\varrho(a,b) \geqq 0$ by 5.1, consequently $\varrho(a,b) > 0$ by 5.2.

5.5. If $a \in R$, $b \in R$, then $\varrho(a,b) = \varrho(b,a)$. Proof. By 5.3, $\varrho(a,a) + \varrho(b,a) \geqq \varrho(a,b)$, hence $\varrho(b,a) \geqq \varrho(a,b)$ by 5.1; and one shows in the same way that $\varrho(a,b) \geqq \varrho(b,a)$.

6. Let R be a metric space. We define then: The set $A \subset R$ is called *open in* R, if for each $a \in A$ there is a positive number η such that $x \in R$, $\varrho(a,x) < \eta$ implies $x \in A$. One can easily observe that as a consequence of this definition, theorems M, 1,5–1,8 hold true, i.e. *metric space is* (on the basis of our definition of an open set) *a topological space*.

7. Let R be a metric space. Let $A \subset R$; let $A \neq 0$. Let $r > 0$. Call a ball with centre A and radius r and denote by $K(A,r)$ the set of those $x \in R$, which satisfy $\varrho(a,x) < r$ for a suitable $a \in A$. If $A = \{a\}$ is a one-point set, we shall write $K(A,r) = K(a,r)$.[31])

8. *Let R be a metric space. Let $A \subset R$; let $A \neq 0$. Let $r > 0$. Then the set $K(A,r)$ is open in R.*

Proof. We have to show that it is possible to assign to every $x \in K(A,r)$ a real $\eta > 0$ so that $y \in R$, $\varrho(x,y) < \eta$ implies $y \in K(A,r)$. So let $x \in K(A,r)$. Then there is a point $a \in A$ such that $\vartheta = \varrho(a,x) < r$. Thus the number $\eta = r - \vartheta$ is positive. Let $y \in R$, $\varrho(x,y) < \eta$. By 5.5, we have $\varrho(y,x) < \eta$. By 5.3, $\varrho(a,y) \leqq \varrho(a,x) + \varrho(y,x) < \vartheta + \eta = r$. Hence $\varrho(a,y) < r$, consequently $y \in K(A,r)$.

9. A topological space R is called *normal*,[32]) if it has the following property: *If A, B is a pair of closed sets in R and if $A \cdot B = 0$, then there exist sets U, V open in R such that $A \subset U$, $B \subset V$, $U \cdot V = 0$.*

---

[29]) M, [7]).

[30]) The properties 5.4 and 5.5 are usually included into the definition of metric space. Deducing them from the properties is due to A. Lindenbaum in Fund. Math., vol 8 (1926), p. 211.

[31]) Neither the set A nor r need to be uniquely determined by the set $K(A,r)$. An example: Let the space R contain only two points a, b and let $\varrho(a,b) = 1$. Then $K(a,2) = K(b,3)$.

[32]) This notion appears for the first time in H. Tietze [Math. Ann., vol. 88 (1923), p. 301]; the name *normal* in P. Urysohn [Math. Ann., vol. 94 (1925), p. 265].

10. *Let R be a normal space. Let $A \subset U \subset R$. Let A be closed in R; let U be open in R. Then there exists a set V open in R such that $A \subset V \subset \overline{V} \subset U$.*[33])

Proof. The sets A, $R—U$ are closed in R and $A \cdot (R—U) = 0$. Since R is normal, there are sets V, W open in R such that $A \subset V$, $R—U \subset W$, $V \cdot W = 0$. Since $V \cdot W = 0$, one has $V \subset R—U$; since $R—W$ is closed in R, one has[34]) $\overline{V} \subset R—W$, i.e. $\overline{V} \cdot W = 0$. Since $R—U \subset W$, it is $\overline{V} \cdot (R—U) = 0$, i.e. $\overline{V} \subset U$.

11. *Let a topological space R have the following property: If $A \subset U \subset R$, if A is closed in R and if U is open in R, then there is a set V open in R such that $A \subset V \subset \overline{V} \subset U$. Then R is a normal space.*

Proof. Let $A \subset R$, $B \subset R$. Let A, B be closed in R. Let $A \cdot B = 0$. We have to prove that there exist open sets V, W such that $A \subset V$, $B \subset W$, $V \cdot W = 0$. Set $U = R—B$. Since B is closed in R, the set U is open in R. Since $A \cdot B = 0$, one has $A \subset U$. So there exists an open set V such that $A \subset V \subset \overline{V} \subset U$. Put $W = R—\overline{V}$. Since $\overline{V}$ is closed in R,[33]) the set W is open in R. Since $V \subset \overline{V}$, we have $V \cdot W = 0$. Since $U = R—B$, $\overline{V} \subset U$, one has $\overline{V} \cdot B = 0$ and consequently $B \subset R—\overline{V}$, i.e. $B \subset W$.

12. *Let R be a normal space. Let $F_1, F_2, \ldots, F_m$ be a finite number of sets closed in R. Then one can find to each F_i $(1 \leqq i \leqq m)$ an open set $U_i \supset F_i$ such that if for some combination $(i_1, i_2, \ldots, i_k)$ $(1 \leqq k \leqq m)$ of indices it is true that $\prod_{r=1}^{k} \overline{U_{i_r}} \neq 0$, then also $\prod_{r=1}^{k} F_{i_r} \neq 0$.*[34])

Proof. Let $\Phi_1, \Phi_2, \ldots, \Phi_{2^m-1}$ be all products of the form $\prod_{r=1}^{k} F_{i_r}$, where $(i_1, i_2, \ldots, i_k)$ runs through all combinations of indices $1, 2, \ldots, m$.[35]) Let $S = \sum \Phi_j$, where j runs through all such values $(1 \leqq j \leqq 2^m - 1)$ for which $F_1 \cdot \Phi_j = 0$. The set S as well as the set F_1 is closed in R[36]) and $F_1 \cdot S = 0$. Since R is normal, there exist open sets U_1, V_1 such that $F_1 \subset U_1$, $S \subset V_1$, $U_1 \cdot V_1 = 0$. We have $U_1 \subset R—V_1$; since $R—V_1$ is closed in R, one has[33]) $\overline{U_1} \subset R—V_1$; i.e. $\overline{U_1} \cdot V_1 = 0$, hence $\overline{U_1} \cdot S = 0$, consequently $\overline{U_1} \cdot \Phi_j \neq 0$ $(1 \leqq j \leqq 2^m - 1)$ implies $F_1 \cdot \Phi_j \neq 0$. In this manner we obtained from the system $F_1, F_2, \ldots, F_m$ *by the first modification* the system $\overline{U_1}, F_2, \cdots, F_m$, from which, using the same method (starting with F_2 instead of F_1), we get *by the second modification* the system $\overline{U_1}, \overline{U_2}, F_3, \ldots, F_m$. After m modifications like this we obtain the required system $\overline{U_1}, \overline{U_2}, \ldots, \overline{U_m}$.

13. It is obvious that theorem 12 cannot hold even in the special case $m = 2$, if the space R is not normal.

14. *Let R be a normal space. Let the sets $U_1, U_2, \ldots, U_m$ be open in R and cover R, the number of them be finite. Then it is possible to find for every U_i $(1 \leqq i \leqq m)$ a set V_i open in R such that 1° $\overline{V_i} \subset U_i$ for $1 \leqq i \leqq m$; 2° the sets $V_1, V_2, \ldots, V_m$*

[33]) M, 4.

[34]) W. Hurewicz and K. Menger, Math. Ann., vol. 100 (1928), remark [22]).

[35]) The number of all combinations of indices 1, 2, ..., m is $\binom{m}{1} + \binom{m}{2} + \cdots + \binom{m}{m} = 2^m - 1$. After all, for our purposes it is essential only that this number is finite.

[36]) M, 1,3.

cover R.[37])

Proof.[38]) Put $F_i = R - U_i$ $(1 \leqq i \leqq m)$. The the sets F_i are closed in R and $\prod_{i=1}^{m} F_i = R - \sum_{i=1}^{m} U_i = 0$.[39]) By head 12, there exist open $W_i \supset F_i$ such that $\prod_{i=1}^{m} \overline{W_i} = 0$. Set $V_i = R - \overline{W_i}$, thus $V_i \subset R - W_i$. Since $R - W_i$ is closed in R, one has $\overline{V_i} \subset R - W_i$, therefore $\overline{V_i} \cdot W_i = 0$; since $W_i \supset F_i$, one has $\overline{V_i} \cdot F_i = 0$, so $\overline{V_i} \subset R - F_i$, i.e. $\overline{V_i} \subset U_i$. In addition, $\sum_{i=1}^{m} V_i = \sum_{i=1}^{m} (R - \overline{W_i}) = R - \prod_{i=1}^{m} \overline{W_i} = R - 0 = R$.

15. Also the theorem from head 14 holds only under the assumption that R is a normal space. Even the next theorem holds: *Let a topological space R posses the following property: For every finite system $\mathfrak{U}$ of open sets which covers R, one can find a finite system $\mathfrak{F}$ of sets closed in R so that 1° $\mathfrak{F}$ covers R; 2° every $F \in \mathfrak{F}$ is a part of some $U \in \mathfrak{U}$. Then R is a normal space.*

Proof. Let $A \subset U \subset R$. Let A be closed in R; let U be open in R. By 11 it suffices to show that there is a set V open in R such that $A \subset V \subset \overline{V} \subset U$. Two sets U and $R - A$ obviously constitute a finite system $\mathfrak{U}$ covering R and consisting of open sets. Hence there exists a finite system $\mathfrak{F}$ of sets closed in R, which covers R and such that for every $F \in \mathfrak{F}$ the relation $A \cdot F = 0$ implies $F \subset U$. Let $F'_1, F'_2, \cdots, F'_h$ be the elements $F \in \mathfrak{F}$ for which $A \cdot F = 0$; let $F''_1, F''_2, \ldots, F''_k$ be the elements $F \in \mathfrak{F}$, for which $A \cdot F \neq 0$; thus $F \subset U$. Set $\Phi' = \sum_{i=1}^{h} F'_i$, $\Phi'' = \sum_{j=1}^{k} F''_j$. The sets Φ' and Φ'' are closed in R and $A \cdot \Phi' = 0$, $\Phi'' \subset U$ and $\Phi' + \Phi'' = R$, for $\mathfrak{F}$ covers R. Let $V = R - \Phi'$; then V is open in R. As $A \cdot \Phi' = 0$, we have $A \subset V$. As $\Phi' + \Phi'' = R$, we have $V \subset \Phi''$. As Φ'' is closed in R, $\overline{V} \subset \Phi''$. But $\Phi'' \subset U$. So $A \subset V \subset \overline{V} \subset U$.

16. *Let R be a metric space. Then R is a normal space.*

Proof. Let A, B be two closed sets in R; let $A \cdot B = 0$. We have to show that there exist two open sets U, V in R such that $A \subset U$, $B \subset V$, $U \cdot V = 0$. Since the set $R - B$ is open in R and since $A \subset R - B$, according to 6 one can assing to each $a \in A$ a real $\eta_a > 0$ such that $K(a, 2\eta_a) \subset R - B$ (see 7). Similarly, for each $b \in B$ there is $\zeta_b > 0$ with $K(b, 2\zeta_b) \subset R - A$. Set

$$U = \sum_{a \in A} K(a, \eta_a); \qquad V = \sum_{b \in B} K(b, \zeta_b).$$

By 5.1 one has $a \in K(a, \eta_a)$, hence $A \subset U$; similarly $B \subset V$. It remains to show that $U \cdot V = 0$. Aiming for a contradiction, assume that $c \in U \cdot V$. Then there exist points $a \in A$, $b \in B$ so that $c \in K(a, \eta_a)$, $c \in K(b, \zeta_b)$, i.e. $\varrho(a, c) < \eta_a$, $\varrho(b, c) < \zeta_b$. We may suppose that $\eta_a \geqq \zeta_b$. By 5.3 we get $\varrho(a, b) \leqq \varrho(a, c) + \varrho(b, c) < \eta_a + \zeta_b \leqq 2\eta_a$, i.e. $b \in K(a, 2\eta_a)$, which is a contradiction, because $b \in B$, $K(a, 2\eta_a) \subset R - B$.

---

[37]) K. Menger, *Dimensionstheorie*, p. 159–160 (Bemerkung).

[38]) Following the manuscript of *Topologie* by Kuratowski. (It will appear in the series *Monografie Matematyczne* in vol. 3.)

[39]) M, [4])

17. Let R be a topological space. Let $A \subset R$. We shall say that A is F_σ in R (G_δ in R), if there exist sets A_ν closed in R (open in R) such that $A = \sum_{\nu=1}^{\infty} A_\nu$ ($A = \prod_{\nu=1}^{\infty} A_\nu$).

A is G_δ in R (F_σ in R) if and only if $R - A$ is F_σ in R (G_δ in R). Indeed, $A = \prod_{\nu=1}^{\infty} A_\nu$ means the same as $R - A = \sum_{\nu=1}^{\infty}(R - A_\nu)$.

If A is open in R (closed in R), then A is G_δ in R (F_σ in R). Indeed, if $A_\nu = A$ ($\nu = 1, 2, 3, \ldots$), then $\prod_{\nu=1}^{\infty} A_\nu = \sum_{\nu=1}^{\infty} A_\nu = A$.

18. *Let R be a normal space. Let $A \subset R$, $B \subset R$. Let A, B be F_σ in R. Let $A \cdot \overline{B} = \overline{A} \cdot B = 0$. Then there exist open sets U, V in R such that $A \subset U$, $B \subset V$, $U \cdot V = 0$.*[40])

Proof. Since A, B are F_σ in R, there exist A_ν, B_ν closed in R such that $A = \sum_{\nu=1}^{\infty} A_\nu$, $B = \sum_{\nu=1}^{\infty} B_\nu$, hence $A_\nu \subset A$, $B_\nu \subset B$. Since $A \cdot B \subset A \cdot \overline{B} = 0$, $A_\nu \cdot B_\nu = 0$. Since A_ν, B_ν are closed in a normal space R and since $A_\nu \cdot B_\nu = 0$, there exist sets G_ν, H_ν open in R and such that $A_\nu \subset G_\nu$, $B_\nu \subset H_\nu$, $G_\nu \cdot H_\nu = 0$. Since $A_\nu \subset A$, $A \cdot \overline{B} = 0$, one has $A_\nu \cdot \overline{B} = 0$, similarly $B_\nu \cdot \overline{A} = 0$. Thus $A_\nu \subset G_\nu - \overline{B}$, $B_\nu \subset H_\nu - \overline{A}$. The set $G_\nu - \overline{B} = G_\nu \cdot (R - \overline{B})$ is by M, 1,7 open in R; since $A_\nu \subset G_\nu - \overline{B}$, by 10 there is an open P_ν in R such that $A_\nu \subset P_\nu \subset \overline{P_\nu} \subset G_\nu - \overline{B}$. Similarly, there exists an open Q_ν in R such that $B_\nu \subset Q_\nu \subset \overline{Q_\nu} \subset H_\nu - \overline{A}$. Let us put

$$U_1 = P_1\,, \quad V_1 = Q_1 - \overline{P_1}\,; \quad U_\nu = P_\nu - \sum_{\mu=1}^{\nu-1} \overline{Q_\mu}\,,$$

$$V_\nu = Q_\nu - \sum_{\mu=1}^{\nu} \overline{P_\mu} \quad (\nu = 2, 3, \ldots); \quad U = \sum_{\nu=1}^{\infty} U_\nu\,, \quad V = \sum_{\nu=1}^{\infty} V_\nu\,.$$

Since P_ν, Q_ν are open in R, by M, 1,3 and by M, 1,7 the sets U_ν, V_ν are open in R, too; thus by M, 1,8 also the sets U, V are open in R. Since $\overline{Q_\mu} \subset H_\mu - \overline{A} \subset R - \overline{A}$, $A_\nu \subset P_\nu$, $A_\nu \subset A \subset \overline{A}$, we have $A_\nu \subset U_\nu$; similarly, $B_\nu \subset V_\nu$. Hence $A = \sum_{\nu=1}^{\infty} A_\nu \subset U$ and similarly $B \subset V$. It remains to show that $U \cdot V = 0$, i.e. that $U_\mu \cdot V_\nu = 0$ holds for $\mu, \nu = 1, 2, 3, \ldots$. Suppose first $\mu \leqq \nu$. Then $U_\mu \subset P_\mu$, $V_\nu = Q_\nu - \sum_{\mu=1}^{\nu} \overline{P_\mu} \subset R - \overline{P_\mu} \subset P_\mu$, thus $U_\mu \cdot V_\nu = 0$. Second, suppose $\mu > \nu$. Then $V_\nu \subset Q_\nu$, $U_\mu = P_\mu - \sum_{\nu=1}^{\mu-1} \overline{Q_\nu} \subset R - \overline{Q_\nu} \subset -Q_\nu$, thus $U_\mu \cdot V_\nu = 0$.

19. *Let R be a normal space. Let $S \subset R$. Let S be F_σ in R. Then S is a normal space.*[40])

[40]) Theorems 18, 19 and 27 are due to Urysohn; quoted in [32]), pp. 285–288.

Proof. Since S is F_σ in R, there exist closed sets S_ν in R such that $S = \sum_{\nu=1}^{\infty} S_\nu$. Let A, B be closed in S; let $A \cdot B = 0$. We have to show that in S, there exist open U_0, V_0 such that $A \subset U_0$, $B \subset V_0$, $U_0 \cdot V_0 = 0$. Since A is closed in S, by M, 6 we have $A = \overline{A} \cdot S$. Since $S = \sum_{\nu=1}^{\infty} S_\nu$, $A = \sum_{\nu=1}^{\infty} \overline{A} \cdot S$. Since S_ν are closed in R, by M, 1,4 the sets $\overline{A} \cdot S_\nu$ are closed in R. Thus A is F_σ in R; similarly, B is F_σ in R. Since $A = \overline{A} \cdot S$, $A \cdot B = 0$, $B \subset S$, we have $\overline{A} \cdot B = 0$; similarly, $A \cdot \overline{B} = 0$. Therefore by 18 there exist sets U,V open in R such that $A \subset U$, $B \subset V$, $U \cdot V = 0$. Let us put $U_0 = U \cdot S$, $V_0 = V \cdot S$, then U_0, V_0 are[27]) open in S and $U_0 \cdot V_0 = 0$. Since $A \subset S$, $B \subset S$, we have $A \subset U_0$, $B \subset V_0$.

20. *Let R be a normal space. Let $\dim R \leqq n$. Let finitely many sets $U_1, U_2, \dots, U_m$ be open in R and cover R. Then it is possible to find for every U_i $(1 \leqq i \leqq m)$ a set V_i open in R such that 1° $\overline{V_i} \subset U_i$ for $1 \leqq i \leqq m$; 2° the sets $V_1, V_2, \dots, V_m$ cover R; 3° order of the system of sets $\overline{V_1}, \overline{V_2}, \dots, \overline{V_m}$ is $\leqq n$.*

Proof. According to 3 there exists a finite system $\mathfrak{W}$ of open sets in R such that 1° $\mathfrak{W}$ covers R; 2° every $W \in \mathfrak{W}$ is a part of some U_i $(1 \leqq i \leqq m)$; 3° order of the system $\mathfrak{W}$ is $\leqq n$. Let $W_1, W_2, \dots, W_k$ be elements of the system $\mathfrak{W}$. Assign to each index j $(1 \leqq j \leqq k)$ an index $\varphi(j) = i$ $(1 \leqq i \leqq m)$ so that $W_j \subset U_i$. Let us set $Z_i = \sum_j W_j$ for $1 \leqq i \leqq m$, where j runs through those values $(1 \leqq j \leqq k)$ for which $\varphi(j) = i$. Obviously the sets $Z_1, Z_2, \dots, Z_m$ are open in R, they constitute a system of order $\leqq n$, cover R and $Z_i \subset U_i$ holds for $1 \leqq i \leqq m$. Using 14, let us find sets V_i open in R $(1 \leqq i \leqq m)$ covering R and such that $\overline{V_i} \subset Z_i$ for $1 \leqq i \leqq m$. Clearly the sets $V_1, V_2, \dots, V_m$ have the required properties.

21. *Let $n = -1, 0, 1, 2, \dots$. Let a topological space R have the following property: For every finite system $\mathfrak{U}$ of sets open in R and covering R one can find a finite system $\mathfrak{F}$ of sets closed in R such that: 1° $\mathfrak{F}$ covers R; 2° every $F \in \mathfrak{F}$ is a part of some $U \in \mathfrak{U}$; 3° order of $\mathfrak{F}$ is $\leqq n$. Then R is a normal space and $\dim R \leqq n$.*

Proof. R is a normal space by 15. It remains to show that for every $\mathfrak{U}$ one can find a finite system $\mathfrak{V}$ of open sets in R that: 1° $\mathfrak{V}$ covers R; 2° every $V \in \mathfrak{V}$ is a part of some $U \in \mathfrak{U}$; 3° order of $\mathfrak{V}$ is $\leqq n$. By the assumption, for $\mathfrak{U}$ given, we can find an $\mathfrak{F}$. Let $F_1, F_2, \dots, F_m$ be all elements of the system $\mathfrak{F}$. We can assign to every F_i $(1 \leqq i \leqq m)$ a set $U_i \in \mathfrak{U}$ so that $F_i \subset U_i$. By 12, one can find for every F_i $(1 \leqq i \leqq m)$ an open set $W_i \supset F_i$ such that $\prod_{r=1}^{k} \overline{W_{i_r}} \neq 0$ implies $\prod_{r=1}^{k} F_{i_r} \neq 0$ for every combination $(i_1, i_2, \dots, i_k)$ $(1 \leqq k \leqq m)$ of indices $1, 2, \dots, m$. It is easy to see that the sets $V_i = U_i \cdot W_i$ form the required system $\mathfrak{V}$.

22. *Let R be a normal space. Let $A \subset R$; let A be closed in R. Let $\dim A \leqq n$. Let finitely many sets $U_1, U_2, \dots, U_m$ be open in R and cover A. The it is possible to assign to every U_i $(1 \leqq i \leqq m)$ a set V_i open in R so that: 1° $\overline{V_i} \subset U_i$; 2° the sets $V_1, V_2, \dots, V_m$ cover A; 3° order of the system of sets $\overline{V_1}, \overline{V_2}, \dots, \overline{V_m}$ is $\leqq n$.*

Proof. Let $U_i' = A \cdot U_i$ $(1 \leqq i \leqq m)$. The sets $U_1', U_2', \dots, U_m'$ are open in A and cover A. By 19, A is a normal space. Since $\dim A \leqq n$, by 20 we can for every U_i' $(1 \leqq i \leqq m)$ find a set F_i closed in A so that 1° $F_i \subset U_i'$ for $1 \leqq i \leqq m$; 2° the sets $F_1, F_2, \dots, F_m$ cover A; 3° order of the system $F_1, F_2, \dots, F_m$ is $\leqq n$. Since F_i are closed in A and since A is closed in R, the sets F_i are[27]) closed in R. Since

$F_i \subset U_i^! \subset U_i$, by 10 there exist sets Z_i open in R that $F_i \subset Z_i \subset \overline{Z_i} \subset U_i$. Since the order of the system $F_1, F_2, \dots, F_m$ is $\leqq n$, then according to 12 there exist open sets W_i $(1 \leqq i \leqq m)$ such that 1° $F_i \subset W_i$, 2° order of the system $\overline{W_1}, \overline{W_2}, \dots, \overline{W_m}$ is $\leqq n$. Obviously the sets $V_i = Z_i \cdot W_i$ $(1 \leqq i \leqq m)$ have the required properties.

23. *Let R be a normal space. Let $R = \sum_{\nu=1}^{\infty} A_\nu$. Let the set A_ν be closed in R and have dimension $\leqq n$ for $\nu = 1,2,3,\dots$. Then* $\dim R \leqq n$.

Proof. Let U_i $(1 \leqq i \leqq m)$ be open sets in R covering R. It suffices to construct sets V_i $(1 \leqq i \leqq m)$ open in R and such that 1° the system $V_1, V_2, \dots, V_m$ covers R and its order is $\leqq n$, 2° $V_i \subset U_i$.

By 22 there exists a system $\mathfrak{V}^1$ consisting of sets V_i^1 $(1 \leqq i \leqq m)$ open in R such that 1° $\overline{V_i^1} \subset U_i$; 2° $\mathfrak{V}^i$ covers A_1; 3° the system $\overline{\mathfrak{V}^1}$ of sets $\overline{V_i^1}$[41]) is of order $\leqq n$. More generally, let us assume that for a certain ν $(= 1, 2, \dots)$, a system $\mathfrak{V}^\nu$ of sets V_i^ν open in R $(1 \leqq i \leqq m)$ has been already constructed, such that 1° $\mathfrak{V}^\nu$ covers $\sum_{\lambda=1}^{\nu} \overline{A_\lambda}$; 2° $\overline{V_i^\nu} \subset U_i$; 3° order of $\overline{\mathfrak{V}^\nu}$ is $\leqq n$. We shall show that then it is possible to construct an analogous system $\mathfrak{V}^{\nu+1}$.

Since $\overline{\mathfrak{V}^\nu}$ is a finite system of order $\leqq n$ of sets closed in R, by 12 there exist sets S_i open in R $(1 \leqq i \leqq m)$ such that 1° $\overline{V_i^\nu} \subset U_i$; 2° order of the system $S_1, S_2, \dots, S_m$ is $\leqq n$. Since $\overline{V_i^\nu} \subset U_i$, by 10 there exist sets T_i open in R such that $\overline{V_i^\nu} \subset T_i \subset \overline{T_i} \subset U_i$. Let us set $W_i = S_i \cdot T_i$. The system $\mathfrak{W}$ of sets $W_1, W_2, \dots, W_m$ is a finite system of order $\leqq n$ consisting of sets open in R and $\overline{V_i^\nu} \subset W_i \subset \overline{W_i} \subset U_i$ holds. By 10, there exist sets P_i $(1 \leqq i \leqq m)$ open in R such that $\overline{V_i^\nu} \subset P_i \subset \overline{P_i} \subset W_i$.

For $1 \leqq i \leqq m$ let: 1° $\mathfrak{M}_i$ denote the system consisting of two sets P_i and $R - \overline{V_i^\nu}$; 2° $\mathfrak{N}_i$ denote the system consisting of two sets W_i and $R - \overline{P_i}$. Clearly each of $2m$ systems $\mathfrak{M}_i$, $\mathfrak{N}_i$ covers R. Let us denote by $\mathfrak{H}$ the system consisting of $m \cdot 4^m$ open sets of form $U_i \cdot \prod_{j=1}^{m} M_j \cdot \prod_{k=1}^{m} N_k$ $(1 \leqq i, j, k \leqq m;\ M_j \in \mathfrak{M}_j;\ N_k \in \mathfrak{N}_k)$. Then $\mathfrak{H}$ is a finite system covering R of sets open in R. Since $A_{\nu+1}$ is closed in R and since $\dim A_{\nu+1} \leqq n$, by 22 there exists a finite system $\mathfrak{Z}$ of sets Z_r open in R $(1 \leqq r \leqq t = m \cdot 4^m)$ such that 1° every $\overline{Z_r}$ is a part of some element of system $\mathfrak{H}$; 2° $\mathfrak{Z}$ covers $A_{\nu+1}$; 3° order of $\mathfrak{Z}$ is $\leqq m$. The next items follow from the property 1° by the definition of systems $\mathfrak{M}_i$, $\mathfrak{N}_i$, $\mathfrak{H}$: 4° $\overline{Z_r} \cdot \overline{V_i^\nu} \neq 0$ implies $\overline{Z_r} \subset P_i$; 5° $\overline{Z_r} \cdot \overline{P_i} \neq 0$ implies $\overline{Z_r} \subset W_i$.

Let us divide the sets Z_r $(1 \leqq r \leqq t)$ into three classes: Z_r is of the *first* class, if there exists some i $(1 \leqq i \leqq m)$ such that $\overline{Z_r} \cdot \overline{V_i^\nu} \neq 0$, Z_r is of the *second* class, if it is not of the first class and if there exists some i $(1 \leqq i \leqq m)$ such that $\overline{Z_r} \cdot \overline{P_i} \neq 0$, Z_r is of the *third* class, if it is neither of the first nor of the second class.

Let us assign every Z_r of the first class to *every* index i such that $Z_r \cdot \overline{V_i^\nu} \neq 0$. Let us assign every Z_r of the second class to a *unique* index i chosen so that $Z_r \cdot \overline{P_i} \neq 0$. Let us assign every Z_r of the third class to a *unique* index i chosen so

[41]) In what follows, the symbol $\overline{\mathfrak{V}^\nu}$ has a similar meaning

that $\overline{Z_r} \subset U_i$.[42])

Let us set for $1 \leqq i \leqq m$, $W_i' = V_i^\nu + \sum Z_r$, where Z_r runs through all elements from $\mathfrak{Z}$ of the first and of the second class assigned to the index i. Clearly $V_i^\nu \subset W_i'$ and by the property 5° of system $\mathfrak{Z}$ one has $\overline{W_i'} \subset W_i$. Futher let us set $V_i^{\nu+1} = W_i' + \sum Z_r$, where Z_r runs through all elements from $\mathfrak{Z}$ of the third class assigned to the index i. Clearly $V_i^{\nu+1}$ are open sets in R and $V_i^\nu \subset V_i^{\nu+1} \subset \overline{V_i^{\nu+1}} \subset U_i$ holds. Since $\mathfrak{V}^\nu$ covers $\sum_{\lambda=1}^{\nu} A_\lambda$ and since $\mathfrak{Z}$ covers $A_{\nu+1}$, the system $\mathfrak{V}^{\nu+1}$ of sets $V_1^{\nu+1}, V_2^{\nu+1}, \ldots, V_m^{\nu+1}$ covers $\sum_{\lambda=1}^{\nu+1} A_\lambda$.

It is still to be shown that order of $\mathfrak{V}^{\nu+1}$ is $\leqq n$, i.e. that every point $a \in R$ is contained in at most $n+1$ elements of the system $\mathfrak{V}^{\nu+1}$. Let us distinguish two cases. *First*, let there be an index j $(1 \leqq j \leqq m)$ such that $a \in \overline{P_j}$. Then $a \in Z_r$ implies that $Z_r \cdot \overline{P_j} \neq 0$, i.e. that Z_r is not of the third class. Thus $a \in V_i^{\nu+1}$ implies $a \in W_i' \subset W_i$. Thus the point a is contained in at most as many $V_i^{\nu+1}$'s as in W_i's; since the order of the system $\mathfrak{W}$ is $\leqq n$, the point a is in at most $n+1$ sets $V_i^{\nu 1}$. *Second*, let $a \in \overline{P_i}$ hold for no index i. By the property 4° of the system $\mathfrak{Z}$, the relation $a \in Z_r$ implies $Z_r \cdot \overline{V_i^\nu} = 0$, consequently the point a neither belongs to any V_i^ν nor it belongs to any Z_r of the first class. Thus $a \in W_i'$ implies $a \in Z_r$, where Z_r is of the second class and is assigned to the index i. Since every Z_r of the second class is assigned to unique i, the point a belongs to at most as many W_i''s as to Z_r's of the second class. Similarly, a belongs to at most as many $V_i^{\nu+1}$'s as to Z_r of the third class. Thus the point a belongs to at most as many sets $V_i^{\nu+1} - W_i'$ as to sets Z_r; since the system $\mathfrak{Z}$ is of order $\leqq n$, the point a belongs to at most $n+1$ sets $V_i^{\nu+1}$.

We have proved that one can recurrently define systems $\mathfrak{V}^\nu$ $(\nu = 1, 2, 3, \ldots)$ of sets V_i^ν $(1 \leqq i \leqq m)$ open in R so that: 1° $V_i^\nu \subset V_i^{\nu+1}$; 2° $V_i^\nu \subset U_i$; 3° $\mathfrak{V}^\nu$ covers A_ν; 4° order of V^ν is $\leqq n$.

Let us put $V_i = \sum_{\nu=1}^{\infty} V_i^\nu$ $(1 \leqq i \leqq m)$. Then V_i are sets open in R. By 2°, $V_i \subset U_i$. Since $R = \sum_{\nu=1}^{\infty} A_\nu$, by 3°, the system $\mathfrak{V}$ of sets $V_1, V_2, \ldots, V_m$ covers R. It remains to show that order of $\mathfrak{V}$ is $\leqq n$. In the opposite case it would be possible to find $n+2$ distinct indices $i_1, i_2, \ldots, i_{n+2}$ and a point $a \in R$ such that $a \in V_{i_s}$ $(1 \leqq s \leqq n+2)$. Then by 1°, there would exist an index ν such that $a \in V_{i_s}^\nu$ $(1 \leqq s \leqq n+2)$. This contradicts to 4°.

24. *Let R be a normal space. Let for $\nu = 1, 2, 3, \ldots$ the set A_ν be F_σ in R and its dimension be $\leqq n$. Then* $\dim \sum_{\nu=1}^{\infty} A_\nu \leqq n$.

Proof. Let $S = \sum_{\nu=1}^{\infty} A_\nu$. Since A_ν are F_σ in R, there exist $B_{\nu\mu}$ closed in R $(\mu = 1, 2, 3, \ldots)$ such that $A_\nu = \sum_{\mu=1}^{\infty} B_{\nu\mu}$. Since $B_{\nu\mu} \subset A_\nu$ and since $B_{\nu\mu}$ is closed

[42]) This is possible by the definiton of $\mathfrak{H}$ and by the property 1° of the system $\mathfrak{Z}$.

in R, $B_{\nu\mu}$ is closed in A_ν. Since $\dim A_\nu \leqq n$, we have $\dim B_{\nu\mu} \leqq n$ by 4. Let us order[43]) the double sequence $B_{\nu\mu}$ (ν, μ = 1, 2, 3, ...) into a single sequence B_λ ($\lambda = 1, 2, 3, \ldots$). Then $S = \sum_{\lambda=1}^{\infty} B_\lambda$ holds, the sets B_λ are closed in R, hence in S, and $\dim B_\lambda \leqq n$. The set S is F_σ in R, so S is a normal space by 19. Therefore $\dim S \leqq n$ by 23.

25. Let R be a topological space. We shall say that R is *perfectly normal*, if: 1° R is normal; 2° every set open in R is F_σ in R. The condition 2° can be stated also like this: Every set closed in R is G_δ in R.

26. *Let R be a metric spoace. Then R is a perfectly normal space.*

Proof. By 16, it suffices to show that every closed set in R is G_δ in R. So let A be a closed set in R; we may obviously assume that $A \neq 0$. Let (see 7) $U_\nu = K(A, \frac{1}{\nu})$ ($\nu = 1, 2, 3, \ldots$). The sets U_ν are open in R by 8, hence it is enough to show that $A = \prod_{\nu=1}^{\infty} U_\nu$. Clearly $A \subset \prod_{\nu=1}^{\infty} U_\nu$, thus it remains to show that for every $x \in R - A$ there is a ν such that $x \in R - U_\nu$. Since $R - A$ is open in R, there exists an $\eta > 0$ such that $y \in R$, $\varrho(x, y) < \eta$ implies $y \in R - A$. For a suitable ν, it will be $\frac{1}{\nu} < \eta$, hence $x \in R - U_\nu$ by the definition of U_ν.

27. *Let R be a perfectly normal space. Let $S \subset R$. Then S is a perfectly normal space.*[40])

Proof. Let Z be open in S; then there exists a G open in R such that $Z = S \cdot G$. Since R is perfectly normal, there exist sets F_ν closed in R such that $G = \sum_{\nu=1}^{\infty} F_\nu$. If we put $\Phi_\nu = S \cdot F_\nu$, we have $Z = \sum_{\nu=1}^{\infty} \Phi_\nu$ and the sets Φ_ν are closed in S. Thus every set open in S is F_σ in S. It remains to show that S is a normal space. Let A, B be closed in S and let $A \cdot B = 0$. We have to show that there exist sets U', V' open in S such that $A \subset U'$, $B \subset V'$, $U' \cdot V' = 0$. Since A, B are closed in S, by M, 6 one has $A = \overline{A} \cdot S$, $B = \overline{B} \cdot S$. Since $A \cdot B = 0$, one has $\overline{A} \cdot \overline{B} \cdot S = 0$. If we put $A_0 = \overline{A} - \overline{B}$, $B_0 = \overline{B} - \overline{A}$, we can see easily that $A \subset A_0$, $B \subset B_0$. Since $A_0 \subset \overline{A}$, by M, 4,2 $\overline{A_0} \subset \overline{A}$, thus $\overline{A_0} \cdot B_0 - 0$; similarly, $A_0 \cdot \overline{B_0} = 0$. Since R is perfectly normal, $R - \overline{B}$ is F_σ in R; from which it follows easily that also $A_0 = \overline{A} \cdot (R - \overline{B})$ is F_σ in R; similarly B_0 is F_σ in R. Since $A_0 \cdot \overline{B_0} = \overline{A_0} \cdot B_0 = 0$, by 18 there exist sets U, V open in R such that $A_0 \subset U$, $B_0 \subset V$ (therefore $A \subset U$, $B \subset V$), $U \cdot V = 0$. It suffices to set $U' = S \cdot U$, $V' = S \cdot V$.

28. *Let R be a perfectly normal space. Let $\dim R \leqq n$. Let $S \subset R$. Then $\dim S \leqq n$.*

Proof. Let $\mathfrak{U}'$ be a finite system of sets open in S, which covers S. We have to show that there exists a finite system $\mathfrak{V}'$ of sets open in S such that 1° $\mathfrak{V}'$ covers S; 2° every $V' \in \mathfrak{V}'$ is a part of some $U' \in \mathfrak{U}'$; $\mathfrak{V}'$ is of order $\leqq n$. Let U_i' ($1 \leqq i \leqq m$) be elements of $\mathfrak{U}'$. Then there exist sets U_i open in R that $U_i' = S \cdot U_i$. Let us set $A = \sum_{i=1}^{m} U_i$. The set A is open in R; since R is perfectly normal, A is F_σ in

43) e.g. diagonally: $B_{11} = B_1$; $B_{21} = B_2$; $B_{12} = B_3$; $B_{31} = B_4$; $B_{22} = B_5$; $B_{13} = B_6$; $B_{41} = B_7$, ...

R. Thus there exist sets A_ν ($\nu = 1, 2, 3, \dots$) closed in R such that $A = \sum_{\nu=1}^{\infty} A_\nu$. Since $\dim R \leqq n$, one has $\dim A_\nu \leqq n$ by 4. Thus by 24, $\dim A \leqq n$. Since the sets $U_1, U_2, \dots, U_m$ cover A, there is a finite system $\mathfrak{V}$ of closed sets such that 1° $\mathfrak{V}$ covers A; 2° each $V \in \mathfrak{V}$ is a part of some U_i; 3° order of $\mathfrak{V}$ is $\leqq n$. We obtain the required system $\mathfrak{V}'$, if we replace every $V \in \mathfrak{V}$ by the set $V' = S \cdot V$.

Доклады Академии наук СССР
1949, Том LXVII, № 2, 217 – 219

MATHEMATICS

O. V. LOKUCIEVSKIJ

ON THE DIMENSION OF BICOMPACTA*

(Presented by academician A. N. Kolmogorov 19. V. 1949)

As widely known, the following theorem takes place.

Theorem. *If a normal space X can be represented as a union of not more than countably many of its closed subspaces X_n with* $\dim X_n \leqslant r$, *then* $\dim X \leqslant r$**.

In § 1 we shall construct a compact space $S = S_1 \cup S_2$, where S_1, S_2 are closed in S, $\operatorname{ind} S_1 = \operatorname{ind} S_2 = 1$, but $\operatorname{ind} S = 2$. From the inequality $\dim R \leqslant \operatorname{ind} R$, proved for compact spaces by P. S. Alexandrov [1], and from the equivalence of statements $\operatorname{ind} R = 0$, $\dim R = 0$ it follows that $\dim S_1 = \dim S_2 = 1$. Therefore, according to the above theorem, $\dim S = 1$. Hence, the forthcoming implies (via essentially easier construction) the result due to A. L. Lunc [2] on the existence of compact spaces in which ind does not coincide with dim. In § 2, this result applies to the case of dyadic compacta.

§ **1.1.** Let us assign to each transfinite $\alpha < \omega_1$ a copy I_α of a segment $0 \leqslant x \leqslant 1$ of coordinate plane, identifying its left end with α and the right one with $\alpha + 1$. We shall topologize the set, obtained by adding a number ω_1 to these and denoted by P, by introducing the relation of an order***. For distinct $\xi_1\, \xi_2 \in P \setminus (\omega_1)$ we shall set $\xi_1 < \xi_2$ if one of two conditions holds: 1° $\xi_1 \in I_{\alpha_1}$, $\xi_2 \in I_{\alpha_2}$, where $\alpha_1 < \alpha_2$; 2° ξ_1, $\xi_2 \in I_\alpha$, and $\xi_1 < \xi_2$ on this segment. In addition $\omega_1 > \xi$ for all $\xi \in P \setminus (\omega_1)$, by definition. It is easy to see that P is compact Hausdorff.

2. Let C be a Cantor discontinuum (it is convenient to consider it as an ordered set) and $R = P \times C$. Then every point of R is a pair (ξ, x), where $\xi \in P$, $x \in C$.

Let us introduce some notation: $C_{x_1}^{x_2}(\xi_0) = \mathcal{E}(\xi = \xi_o, x_1 \leqslant x \leqslant x_2)$, $P_{\xi_1}^{\xi_2}(x_0) = \mathcal{E}(\xi_1 \leqslant \xi \leqslant \xi_2, x = x_0)$. Let us consider an open set $G \subseteq R$, satisfying $G_1 = G \cap C_0^1(\omega_1) \neq \emptyset$****, while the upper bound of G_1 is a two-sided point $(\omega_1, x_0) \in C_0^1(\omega_1)$. The following fact is true.

A. *If* $P_\xi^{\omega_1}(x_0) \nsubseteq \bar{G} \setminus G$ *for an arbitrary* $\xi < \omega_1$, *then* $C_{x_0}^{x'_0}(\omega_1) \subseteq \bar{G} \setminus G$ *for some* $x'_0 > x_0$.

Proof. There is a sequence $\{(\omega_1, x_n)\} \to (\omega_1, x_0)$, where $(\omega_1, x_n) \in G$, $n = 1, 2, \ldots$. As G is open, for some $\xi_n < \omega_1$ one has $P_{\xi_n}^{\omega_1}(x_n) \subseteq G$. But ω_1 is a

* The research was done under the guidance of P. S. Alexandrov.

**dim denotes the dimension, defined with the aid of coverings, and ind stands for the inductive dimension.

***Open sets are defined as order intervals and arbitrary unions of them.

****$\emptyset$ denotes empty set; 0, 1 are the endpoints of C.

regular number. Therefore all ξ_n do not exceed some $\xi_0 < \omega_1$. It is easy to see that $P_{\xi_0}^{\omega_1} \subseteq \bar{G}$. The assumptions of the statement imply the existence of an uncountable set $D = \{(\xi_\nu, x_0)\} \subseteq G$ having the point (ω_1, x_0) as a unique condensation point. Since G is open, for every ν there is an $x_\nu > x_0$ such that $C_{x_0}^{x_\nu}(\xi_\nu) \subseteq G$. From the uncountability of D we conclude that for some $x_0' > x_0$ the inequality $x_\nu > x_0'$ holds for uncountably many values of ν. It is easy to see that $C_{x_0}^{x_0'}(\omega_1) \subseteq \bar{G}$, and this proves the proposition, because x_0 is an upper bound of G_1.

3. Let R' be a compact space obtained from R by pairwise identification of end-points of every deleted interval of $C_0^1(\omega_1)$. By this identification the set $C_0^1(\omega_1)$ will turn into a segment, which we denote by F, and the set of all one-sided points of $C_0^1(\omega_1)$ will come into some M. As M is dense in F, it is not difficult to show that $\operatorname{ind} R' = 1$. Nevertheless, A implies

A'. *If G' is an open subset in R' such that $G_1' = G' \cap F \neq \emptyset$ and the upper bound of G' does not belong to M, then* $\operatorname{ind}(\bar{G} \setminus B) \geqslant 1$.

4. Let us consider a compact space $Q = R_1' \cup R_2'$, where R_1', R_2' are disjoint copies of the space R'. Denote by F_i, M_i the sets in R_i' ($i = 1, 2$), corresponding to sets F, M, as defined in R'. Let us choose in F_2 an arbitrary set N, which is dense in F_2, an order type of which is that of M_1 and which does not have common points with M_2 except the endpoints of F_2. By a well-known fact [3], there is a similar mapping g of an interval F_2 onto F_1 which maps N onto M_1. It is possible to consider g as a continuous mapping from Q onto some compact space $S = g(Q)$. Obviously, $g(R_1') = S_1$, $g(R_2') = S_2$ are homeomorphisms, $S = S_1 \cup S_2$. If $E = g(F_1) = g(F_2)$, then $S_1 \cap S_2 = E$.

We shall show that $\operatorname{ind} S = 2$. Let y be an arbitrary inner point of E and let U be some neighborhood of it. Denote by U_1 the intersection $U \cap E$ and let y_1 be an upper bound of U_1. We can assume without loss of generality that y_0 is an inner point of E. Then y_0 cannot belong simultaneously to both of $g(M_1)$ and $g(M_2)$. Suppose, for instance, that $y_0 \notin g(M_1)$. Setting $O = U \cap S_1$, we get by A' that $\operatorname{ind}(\bar{O} \setminus O) \geqslant 1$. But since $\bar{O} \setminus O \subseteq \bar{U} \setminus U$, we have $\operatorname{ind}(\bar{U} \setminus U) \geqslant 1$, consequently, $\operatorname{ind} S \geqslant 2$. But it is not difficult to show that $\operatorname{ind} S \leqslant 2$, and since $\operatorname{ind} S_1 = \operatorname{ind} S_2 = 1$, the construction of an example is finished.

§ **2.** Let S be an arbitrary compact space of weight τ. We shall construct a dyadic compact space of the same weight, which contains S and which will be called a dyadic envelope of S in the sequel.

By the well-known theorem due to P. S. Alexandrov [4], there exists a closed subset $\tilde{S}$ in D^τ and a continuous mapping f such that $f(\tilde{S}) = S$. We shall denote by Ω the continuous decomposition [5], generated by f, and shall define a decomposition Ω' of the space D^τ by the following: 1° the elements of Ω' agree with elements of Ω on $\tilde{S}$; 2° the elements of Ω' are points on $D^\tau \setminus \tilde{S}$. The continuity of Ω' can be proved easily. Let us denote by g a mapping, generated by this decomposition: $g(D^\tau) = R$. As known, g coincides with f on the set $\tilde{S}$ and consequently $S \subseteq R$. It is obvious that R is a dyadic compact space of wieight τ. The following take place.

Theorem 1. *If* $\dim S < +\infty$, *then* $\dim R = \dim S$.

Theorem 2. *If* $\operatorname{ind} S < +\infty$, *then* $\operatorname{ind} R \leqslant \operatorname{ind} S + 1$.

In particular, if S is a compact space constructed in § 1, then the theorems immediately imply for its dyadic envelope R the following: $\dim R = 1$, $2 \leqslant \operatorname{ind} R \leqslant 3$.

Let us remark that Theorem 2 cannot be strengthened in the obvious direction: There exist a compact space S and its dyadic envelope R such that $\operatorname{ind} S = 1$, but $\operatorname{ind} R = 2$. The existence of a compact space, whose inductive dimension is less than the inductive dimension of any dyadic envelope of it, presents an open problem.

Proof of Theorem 1. Let $\omega = \{U_1, U_2, \dots, U_p\}$ be an arbitrary open cover of R; $\tilde{U}_\nu = g^{-1}(U_\nu)$, $U_\nu^S = S \cap U_\nu$. Then $\tilde{\omega} = \{\tilde{U}_1, \tilde{U}_2, \dots, \tilde{U}_p\}$ and $\omega^S = \{U_1^S, U_2^S, \dots, U_p^S\}$ are open covers of D^τ and S respectively. Let us consider a closed cover $\alpha^S = \{F_1^S, F_2^S, \dots, F_k^S\}$ of a space S of order $n+1$, which refines ω^S. In R, we shall construct a system of open sets $\{H_1, H_2, \dots, H_k\}$, $H_\nu \supseteq F_\nu^S$, such that for $F_\nu = \bar{H}_\nu$, the system $\alpha = \{F_1, F_2, \dots, F_k\}$ is similar to α^S and refines ω, which is always possible. Let $\tilde{F}_\nu = g^{-1}(F_\nu)$, $\tilde{H}_\nu = g^{-1}(H_\nu)$. Then $\tilde{H} = \bigcup_{\nu=1}^k \tilde{H}_\nu$ is a set open in D^τ and $\tilde{S} \subseteq \tilde{H}$. Since $\dim D^\tau = 0$, there exists a clopen $\tilde{A} \subseteq D^\tau$ such that $\tilde{S} \subseteq \tilde{A} \subseteq \tilde{H}$. Let $\tilde{B} = D^\tau \setminus \tilde{A}$, $\tilde{F}_\nu^A = \tilde{A} \cap \tilde{F}_\nu$, $\tilde{U}_\nu^B = \tilde{B} \cap \tilde{U}_\nu$. By the zero-dimensionality of $\tilde{B}$ there is a system of closed disjoint sets $\{\tilde{F}_1^B, \tilde{F}_2^B, \dots, \tilde{F}_m^B\}$ refining $\tilde{\omega}^B = \{\tilde{U}_1^B, \dots, \tilde{U}_p^B\}$ and such that $\bigcup_{\nu=1}^m \tilde{F}_\nu^B = \tilde{B}$. If $F_\nu^A = g\tilde{F}_\nu^A)$, $F_\nu^B = g(\tilde{F}_\nu^B)$, then the system $\beta = \{F_1^A, F_2^A, \dots, F_k^A, F_1^B, F_2^B, \dots, F_m^B\}$ represents a closed cover R, which refines ω. Since the mapping is one-to-one on $\tilde{B}$, one has $F_\mu^A \cap F_\nu^B = \emptyset$ for arbitrary μ, ν and $F_\mu \cap F_\nu^B = \emptyset$ for $\mu \neq \nu$. Moreover, $F_\nu^A \subseteq F_\nu$. Therefore the order of β does not exceed $n+1$, which means $\dim R \leqslant \dim S$. From the fact that $S \subseteq R$ and from the monotonicity of dim with respect to closed subsets of a compact space we conclude that $\dim R = \dim S$.

Proof of Theorem 2. Let x be a point in R and let U be an arbitrary neighborhood of it. Clearly, the theorem will be proved if we find a neighborhood V of the point that $V \subseteq U$ and $\bar{V} \setminus V \subseteq S$. Let U' be a neighborhood of a point x contained in U together with its closure: $\bar{U}' \subseteq U$. Denote $\bar{U}'$ by F, and let $g^{-1}(F) = \tilde{F}$, $g^{-1}(U) = \tilde{U}$. One can find a clopen set $\tilde{A} \subseteq D^\tau$ such that $\tilde{F} \subseteq \tilde{A} \subseteq \tilde{U}$. If $A = g(\tilde{A})$, then apparently $F \subseteq A \subseteq U$. If V is an interior of A, then the system of inclusion holds: $U' \subseteq V \subseteq A \subseteq U$, which implies that V is a neighborhood of x contained in U. It remains to show that $\bar{V} \setminus V \subseteq S$. Let $y \in \bar{V} \setminus S$. Then also $y \in A \setminus S$. Since $y \notin S$, then $g^{-1}(y)$ is a unique point $\tilde{y}$ and $\tilde{y} \in \tilde{A} \setminus \tilde{S}$. If $\tilde{H}$ is an arbitrary neighborhood of $\tilde{y}$ contained in $\tilde{A} \setminus \tilde{S}$, then $y \in g(\tilde{H}) = H \subseteq A \setminus S$. It is easy to see that H is an open set. Therefore y is in A together with its neighborhood, and so $y \in V$, because V is the interior of A. From this follows that $\bar{V} \setminus V \subseteq S$, which proves the theorem.

Recieved
19 V 1949

References

[1] P. S. Alexandroff, *The sum theorem in the dimension theory of bicompact spaces*, Soobšč. Akad. Nauk Gruz.SSR **2, 1** (1941), 1 – 6.

[2] A. L. Lunc, *A bicompactum whose inductive dimension is greater than its dimension defined*

by means of coverings, Dokl. Akad. Nauk SSSR (N.S.) **66,5** (1949), 801 - 803.
[3] F. Hausdorff, Teorija množestv (1937), ONTI, Moscow.
[4] P. S. Alexandroff, *On the notion of a space in topology*, Uspechi Mat. Nauk **1 (17)** (1947), 5 - 57.
[5] P. S. Alexandroff, H. Hopf, *Topologie I*, Springer Verlag, Berlin, 1935.

INDUCTIVE DIMENSION OF COMPLETELY NORMAL SPACES

By C. H. DOWKER (*London*)

[Received 13 March 1953]

USING the dimension defined inductively in terms of closed sets, I show that in a completely normal space the dimension of the union of two disjoint sets, one of which is closed, is at most equal to the greatest of their dimensions. A corresponding theorem is proved for a countable union of disjoint sets. It follows that in completely normal spaces the subset theorem implies the sum theorem. The subset theorem and the open subset theorem are shown to be equivalent. Therefore (Theorem 1) the sum and subset theorems hold for any completely normal space in which the dimension of a set A is never less than the dimension of a relatively open subset of A.

E. Čech (**3**) extended the sum and subset theorems from separable metric spaces to perfectly normal spaces. I introduce a new class of normal spaces, intermediate between completely normal and perfectly normal, which I call *totally normal.* A normal space X is *totally normal* if each open set G of X has a locally finite covering by open subsets each of which is an F_σ set of X. The totally normal spaces include the hereditarily paracompact Hausdorff spaces as well as the perfectly normal spaces. It is shown that the open subset theorem and hence the subset theorem and sum theorem hold for totally normal spaces. The covering theorem of Čech also holds for totally normal spaces.

The open subset theorem does not hold for all normal Haurdorff spaces nor for all completely normal spaces. The question of whether it holds for all completely normal Hausdorff spaces is still undecided.

1. Definitions and known theorems.

A space X is called *normal* if for each pair of disjoint closed sets E and F of X there exist disjoint open sets U and V with $E \subset U$ and $F \subset V$. It is no restriction on the space to require also that the closures $\overline{U}$ and $\overline{V}$ of U and V should be disjoint or that the open sets U and V should be F_σ sets, i.e. countable unions of closed sets. A space X is called *completely normal* if every subset of X is a normal space. Clearly every subset of a completely normal space is completely normal.

[1.1] *If every open subset of a space X is a normal space, X is completely normal.*

Proof. Let every open subset of X be normal and let A be an arbitrary subset of X. Let E and F be any two disjoint sets closed in A. Let $G = X - \overline{E} \cap \overline{F}$, $E_1 = \overline{E} \cap G$, and $F_1 = \overline{F} \cap G$. Then $A \subset G$, G is open and hence normal, and E_1 and F_1 are disjoint closed sets of G. Hence there exist disjoint open sets U_1 and V_1

Quart. J. Math. Oxford (2), 4 (1953), 267-81.

C. H. DOWKER

of G with $E_1 \subset U_1$ and $F_1 \subset V_1$. Let $U = U_1 \cap A$ and $V = V_1 \cap A$; then U and V are open in A, $E \subset U$, $F \subset V$ and $U \cap V = 0$. Therefore A is normal. Therefore X is completely normal.

The inductive dimension of a space X, $\operatorname{Ind} X$, is defined inductively as follows. If X is the empty set, $\operatorname{Ind} X = -1$. For $n = 0, 1, \ldots$, $\operatorname{Ind} X \leqslant n$ means that for every open set G containing E there is an open set U with $E \subset U \subset G$ and $\operatorname{Ind}(\overline{U} - U) \leqslant n-1$. $\operatorname{Ind} X = \infty$ means that there is no n for which $\operatorname{Ind} X \leqslant n$.

[1.2] *If A is any closed subset of a space X,* $\operatorname{Ind} A \leqslant \operatorname{Ind} X$.

Proof.† It is sufficient to show that, if $\operatorname{Ind} X \leqslant n$, then $\operatorname{Ind} A \leqslant n$. This is trivially true for dimension -1 and we assume it true for dimension $n-1$. Let $\operatorname{Ind} X \leqslant n$, let A be closed in X, and let $E \subset G \subset A$ with E closed in A and G open in A. Then E is closed in X and there exists G_1 open in X with $G_1 \cap A = G$, and hence with $E \subset G_1$. Hence there exists U_1 open in X with $E \subset U_1 \subset G_1$ and $\operatorname{Ind}(\overline{U_1} - U_1) \leqslant n-1$. Let $U = U_1 \cap A$, then U is open in A and $E \subset U \subset G$. Then $\overline{U} \subset \overline{U_1}$, and hence

$$\overline{U} \cap A - U \subset \overline{U_1} \cap A - U = \overline{U_1} \cap A - U_1 \cap A = (\overline{U_1} - U_1) \cap A \subset \overline{U_1} - U_1.$$

And $\overline{U} \cap A - U$ is closed in A, hence in X and in $\overline{U_1} - U_1$. Hence, by the induction hypothesis, $\operatorname{Ind}(\overline{U} \cap A - U) \leqslant n-1$. Therefore $\operatorname{Ind} A \leqslant n$, as was to be shown.

[1.3] $\operatorname{Ind} X \leqslant n$ *is equivalent to the following condition on X:*

(α) *If $E \subset G \subset X$ with E closed and G open, then X is the union of three disjoint sets U, V, and C with U and V open, $E \subset U \subset G$ and* $\operatorname{Ind} C \leqslant n-1$.

Proof. If there exists U with $E \subset U \subset G$ and $\operatorname{Ind}(\overline{U} - U) \leqslant n-1$, we set $C = \overline{U} - U$ and $V = X - \overline{U}$. Then U, V, and C are disjoint, $X = U \cup V \cup C$, U and V are open, and $\operatorname{Ind} C \leqslant n-1$.

If, on the other hand, X is the union of disjoint sets U, V, and C with U and V open, $E \subset U \subset G$ and $\operatorname{Ind} C \leqslant n-1$, then U is contained in the closed set $U \cup C$ and hence $\overline{U} - U \subset C$. Therefore, since $\overline{U} - U$ is closed, $\operatorname{Ind}(\overline{U} - U) \leqslant \operatorname{Ind} C \leqslant n-1$.

[1.4] *If X is normal,* $\operatorname{Ind} X \leqslant n$ *is equivalent to the following condition:*

(β) *If E and F are disjoint closed sets of X, then X is the union of disjoint sets U, V, and C with U and V open, $E \subset U$, $F \subset V$, and*

$$\operatorname{Ind} C \leqslant n-1$$

Proof.†† It is sufficient to show that the conditions (α) and (β) are equivalent when X is normal. First let (α) be satisfied and let E and F be disjoint closed sets of X. Since X is normal, there exists an open set G such that $E \subset G \subset \overline{G} \subset X - F$. Then, by ($\alpha$), X is the union of disjoint sets U, V, and C with U and V open, $E \subset U \subset G$, and $\operatorname{Ind} C \leqslant n-1$. Since $U \subset G$, $\overline{U} \subset \overline{G} \subset X - F$ and hence $F \subset X - \overline{U} = V$. Thus ($\beta$) is satisfied.

Conversely let (β) be satisfied and let $E \subset G \subset X$ with E closed and G open. Let $F = X - G$; then E and F are closed and disjoint. Hence, by (β), X is the union of disjoint sets U, V, and C with U and V open, $E \subset U$, $F \subset V$ and $\operatorname{Ind} C \leqslant n-1$. Since $X - G = F \subset V \subset X - U$, therefore $U \subset G$. Thus (α) is satisfied.

† See Čech (**3**), Proposition 16.2.

†† See Čech (**3**), § 18. Condition (β) is closely related to the original definition of dimension by L. E. J. Brouwer (**2**).

ON INDUCTIVE DIMENSION OF NORMAL SPACES

2. Sum theorem for disjoint sets.

In the following lemma I show that, if a completely normal space Y is the union of a sequence $\{D_i\}$ of disjoint sets such that the partial unions $\bigcup_{j\leqslant i} D_j$ are closed in Y, then $\operatorname{Ind} Y \leqslant \sup \operatorname{Ind} D_i$. In particular, if $Y = D_1 \cup D_2$ with $D_1 \cap D_2 = 0$ and D_1 closed, then

$$\operatorname{Ind} Y \leqslant \max(\operatorname{Ind} D_1, \operatorname{Ind} D_2).$$

This lemma does not extend to arbitrary normal spaces. O. V. Lokutzievski (**5**) has given an example of a space S and a closed subset S_1 such that S, S_1, and $S - S_1$ are normal, $\operatorname{Ind} S_1 = 1$, $\operatorname{Ind}(S - S_1) = 1$ but $\operatorname{Ind} S = 2$.

[2.1] *Let Y_i ($i = 1, 2, \dots$) be open sets in a completely normal space Y such that $Y = Y_1 \supset Y_2 \supset \dots$, $\bigcap_{i=1}^{\infty} Y_i = 0$ and, for each i, $\operatorname{Ind}(Y_i - Y_{i+1}) \leqslant n$. Then $\operatorname{Ind} Y \leqslant n$.*

Proof. This is trivially true for dimension -1 and we assume it true for dimension $n - 1$. Let E and F be any two disjoint closed sets of Y. Since Y is normal, there exist open sets U_0 and V_0 with $E \subset U_0$, $F \subset V_0$, and $\overline{U_0} \cap \overline{V_0} = 0$.

Let $D_i = Y_i - Y_{i+1}$. We construct disjoint sets U_i, V_i, and C_i for $i = 1, 2, \dots$, with U_i and V_i open in Y_i and hence open in Y and with $C_i \subset D_i \subset U_i \cup V_i \cup C_i$, $\operatorname{Ind} C_i \leqslant n - 1$, $\overline{U_i} \cap \overline{V_i} \cap Y_{i+1} = 0$, $U_i \supset \overline{U_{i-1}} \cap Y_i$, and $V_i \supset \overline{V_{i-1}} \cap Y_i$.

We already have the sets U_0 and V_0, and, assuming U_{i-1} and V_{i-1} have been constructed so that $\overline{U_{i-1}} \cap \overline{V_{i-1}} \cap Y_i = 0$, we construct the sets U_i, V_i, and C_i as follows.

The sets $\overline{U_{i-1}} \cap D_i$ and $\overline{V_{i-1}} \cap D_i$ are disjoint and closed in D_i and $\operatorname{Ind} D_i \leqslant n$. Hence D_i is the union of disjoint sets G_i, H_i, and C_i, with G_i and H_i open in D_i and with $\overline{U_{i-1}} \cap D_i \subset G_i$, $\overline{V_{i-1}} \cap D_i \subset H_i$ and $\operatorname{Ind} C_i \leqslant n - 1$. Then C_i is closed in $D_i = Y_i - Y_{i+1}$ which is closed in Y_i; hence $Y_i - C_i$ is open in Y_i and hence open in Y.

The sets G_i and H_i are closed in $D_i - C_i$ and hence closed in $Y_i - C_i$. Let $E_i = (\overline{U_{i-1}} \cup G_i) \cap (Y_i - C_i)$ and $F_i = (\overline{V_{i-1}} \cup H_i) \cap (Y_i - C_i)$, then E_i and F_i are closed sets of $Y_i - C_i$. Since $\overline{V_{i-1}} \cap D_i \subset H_i$ and $G_i \subset D_i - H_i$, therefore $\overline{V_{i-1}} \cap G_i = 0$, and similarly, $\overline{U_{i-1}} \cap H_i = 0$. Hence, since

$$\overline{U_{i-1}} \cap \overline{V_{i-1}} \cap (Y_i - C_i) \subset \overline{U_{i-1}} \cap \overline{V_{i-1}} \cap Y_i = 0 \quad \text{and} \quad G_i \cap H_i = 0,$$

E_i and F_i are disjoint.

Since Y is completely normal, $Y_i - C_i$ is normal. Hence there exist sets U_i and V_i open in $Y_i - C_i$ and hence open in Y such that $E_i \subset U_i$, $F_i \subset V_i$, and $U_i \cap V_i = 0$ and such that, moreover, $\overline{U_i} \cap \overline{V_i} \cap (Y_i - C_i) = 0$. Then $\overline{U_i} \cap \overline{V_i} \cap Y_{i+1} = 0$, and, since U_i and V_i are disjoint and contained in $Y_i - C_i$, the sets U_i, V_i, and C_i are disjoint.

Since $D_i = G_i \cup H_i \cup C_i$ and $G_i \subset E_i \subset U_i$ and $H_i \subset F_i \subset V_i$, therefore $C_i \subset D_i \subset U_i \cup V_i \cup C_i$. Since $\overline{U_{i-1}} \cap D_i \subset G_i$, therefore $\overline{U_{i-1}} \cap C_i = 0$ and hence $\overline{U_{i-1}} \cap Y_i = \overline{U_{i-1}} \cap (Y_i - C_i) \subset E_i \subset U_i$. Similarly $V_i \supset \overline{V_{i-1}} \cap Y_i$. Thus the sets U_i, V_i, and C_i have the required properties.

C. H. DOWKER

Let $U = \bigcup_{i=0}^{\infty} U_i$, $V = \bigcup_{i=0}^{\infty} V_i$, $Z_i = \bigcup_{j=i}^{\infty} C_j$ and $C = Z_1 = \bigcup_{j=1}^{\infty} C_j$. Then the sets U and V are unions of open sets and hence open; and

$$E \subset U_0 \subset U \quad \text{and} \quad F \subset V_0 \subset V.$$

Every point of Y is in some D_i, hence in U_i, V_i, or C_i and hence in U, V, or C; thus $Y \subset U \cup V \cup C$.

If $i \leqslant j$, $U_i \cap Y_j \subset U_j$ and $V_i \cap Y_j \subset V_j$. Therefore $U_i \cap V_j \subset U_j \cap V_j = 0$ and $U_j \cap V_i \subset U_j \cap V_j = 0$, and hence $U \cap V = 0$. If $i \leqslant j$,

$$U_i \cap V_j \subset U_j \cap C_j = 0$$

and, if $i > j$, since $U_i \subset Y_i$, $U_i \cap C_j \subset Y_i \cap C_j = 0$. Hence $U \cap C = 0$ and similarly $V \cap C = 0$. Thus the sets U, V, and C are disjoint.

As a subset of a completely normal space, C is completely normal. Each $Z_i = C_i \cap Y_i$ is open in C, $Z_i \supset Z_{i-1}$ and

$$\bigcap_{i=1}^{\infty} Z_i \subset \bigcap_{i=1}^{\infty} Y_i = 0.$$

We have $Z_i = \bigcup_{j=i}^{\infty} C_j = C_i \cup Z_{i+1}$ and, for $i < j$, $C_i \cap C_j \subset C_i \cap Y_j = 0$ and hence $C_i \cap Z_{i+1} = 0$. Therefore $C_i = Z_i - Z_{i+1}$ and

$$\operatorname{Ind}(Z_i - Z_{i+1}) = \operatorname{Ind} C_i \leqslant n - 1.$$

Therefore, by the induction hypothesis, $\operatorname{Ind} C \leqslant n-1$. Thus condition ($\beta$) is satisfied and $\operatorname{Ind} Y \leqslant n$, as was to be shown.

[2.2] *If A is a closed subset of a completely normal space Y and if* $\operatorname{Ind} A \leqslant n$ *and* $\operatorname{Ind}(Y - A) \leqslant n$, *then* $\operatorname{Ind} Y \leqslant n$.

Proof. Let $Y_1 = Y$, $Y_2 = Y - A$, and $Y_3 = Y_4 = \cdots = 0$. Then

$$Y = Y_1 \supset Y_2 \supset Y_3 \supset \ldots, \qquad \bigcap_{i=1}^{\infty} Y_i = 0,$$

and $\operatorname{Ind}(Y_1 - Y_2) = \operatorname{Ind} A \leqslant n$, $\operatorname{Ind}(Y_2 - Y_3) = \operatorname{Ind}(Y - A) \leqslant n$, and, for $i \geqslant 3$, $\operatorname{Ind}(Y_i - Y_{i+1}) = -1 \leqslant n$. Therefore, by [2.1], $\operatorname{Ind} Y \leqslant n$.

3. Consequences of the open subset theorem.

In order to discuss the relations between the subset theorem, the open subset theorem, and the sum theorem, we consider the following conditions which a space X may satisfy.

(a_n) If $B \subset A \subset X$ and $\operatorname{Ind} A \leqslant n$, then $\operatorname{Ind} B \leqslant n$.

(b_n) If $G \subset A \subset X$ with G open in A and $\operatorname{Ind} A \leqslant n$, then $\operatorname{Ind} G \leqslant n$.

(c_n) If $A = B \cup C \subset X$ with B closed in A, $\operatorname{Ind} B \leqslant n$ and $\operatorname{Ind} C \leqslant n$, then $\operatorname{Ind} A \leqslant n$.

ON INDUCTIVE DIMENSION OF NORMAL SPACES

(d_n) If $A = \bigcup_{i=1}^{\infty} A_i \subset X$ with each A_i closed in A and $\operatorname{Ind} A_i \leqslant n$, then $\operatorname{Ind} A \leqslant n$.

If $Y \subset X$ and X satisfies condition (a_n) [or (b_n), (c_n), (d_n)], then Y also satisfies (a_n) [or (b_n), (c_n), (d_n)].

Clearly, (a_n) implies (b_n).

[3.1] *If a space X satisfies (a_{n-1}) and (b_n), then it satisfies (a_n).*

Proof. Let X satisfy conditions (a_{n-1}) and (b_n), let $B \subset A \subset X$, and let $\operatorname{Ind} A \leqslant n$. Let $E \subset G \subset B$ with E closed in B and G open in B. Then there exist E_1 closed in A and G_1 open in A with $E_1 \cap B = E$ and $G_1 \cap B = G$. Let $H = (A - E_1) \cup G_1$, then $H \supset (B - E) \cup G = B$. By (b_n), since H is open in A, $\operatorname{Ind} H \leqslant n$. We have $E_1 \cap H$ closed in H, G_1 open in H, and

$$E_1 \cap H = E_1 \cap [(A - E_1) \cup G_1] = E_1 \cap G_1 \subset G_1 .$$

Therefore, since $\operatorname{Ind} H \leqslant n$, there exists U_1 open in H with

$$E_1 \cap H \subset U_1 \subset G_1 \qquad \text{and} \qquad \operatorname{Ind}(\overline{U_1} \cap H - U_1) \leqslant n - 1.$$

Let $U = U_1 \cap B$. The $E = E_1 \cap B \subset E_1 \cap H \subset U_1$ and $E \subset B$; hence $E \subset U$. And $U = U_1 \cap B \subset G_1 \cap B = G$; thus $E \subset U \subset G$. The boundary of U in B is

$$\overline{U} \cap B - U = \overline{U} \cap B - U_1 \cap B \subset \overline{U_1} \cap B - U_1 \cap B = (\overline{U_1} - U_1) \cap B ,$$

and, since $B \subset H$,

$$(\overline{U_1} - U_1) \cap B = (\overline{U_1} \cap H - U_1) \cap B \subset \overline{U_1} \cap H - U_1 .$$

Therefore, by (a_{n-1}), $\operatorname{Ind}(\overline{U} \cap B - U) \leqslant n - 1$. Thus $\operatorname{Ind} B \leqslant n$, and condition (a_n) is satisfied. This completes the proof.

Thus we have the implications: $(a_{n-1}) + (b_n) \to (a_n) \to (b_n)$. The condition (a_{-1}) is trivially satisfied. Hence, if (b_n) is satisfied for every n, (a_n) is also satisfied for every n. Thus the open subset theorem implies the subset theorem.

[3.2] *If X is completely normal, (b_n) implies (c_n).*

Proof. Let X be a completely normal space satifsying (b_n) and let $A = B \cup C \subset X$ with B closed in A, and with $\operatorname{Ind} B \leqslant n$ and $\operatorname{Ind} C \leqslant n$. Since B is closed in A, $A - B$ is open in A, hence open in C, and hence, by (b_n), $\operatorname{Ind}(A - B) \leqslant n$. Since X is completely normal, so is A; and B is closed in A, $\operatorname{Ind} B \leqslant n$ and $\operatorname{Ind}(A - B) \leqslant n$. Hence, by [2.2], $\operatorname{Ind} A \leqslant n$. Thus condition (c_n) is satisfied.

[3.3] *If X is completely normal, (b_n) implies (d_n).*

Proof. Let X be a completely normal space satisfying (b_n) and let $A = \bigcup_{i=1}^{\infty} A_i \subset X$ with A_i closed in A and $\operatorname{Ind} A_i \leqslant n$. Let $D_i = A_i - \bigcup_{j<i} A_j$; then

$$A = \bigcup_{i=1}^{\infty} A_i = \bigcup_{i=1}^{\infty} D_i .$$

C. H. DOWKER

Let

$$Y_i = \bigcup_{j=i}^{\infty} D_j = A - \bigcup_{j<i} A_j\,;$$

then $Y_i \supset Y_{i+1}$ and

$$\bigcap_{i=1}^{\infty} Y_i = A - \bigcup_{j=1}^{\infty} A_j = 0\,.$$

Since $\bigcup_{j<i} A_j$ is closed in A, D_i is open in A_i and Y_i is open in A. Then, by (b_n), $\operatorname{Ind} D_i \leqslant n$. The sets D_i are disjoint and

$$Y_i = \bigcup_{j=i}^{\infty} D_j = D_i \cup Y_{i+1}\,.$$

Hence $D_i = Y_i - Y_{i+1}$. Therefore $\operatorname{Ind}(Y_i - Y_{i+1}) \leqslant n$. Since $A \subset X$, A is completely normal. Therefore, by [2.1], $\operatorname{Ind} A \leqslant n$. Thus condition (d_n) is satisfied.

Thus the open subset theorem implies the sum theorem for completely normal spaces.

THEOREM 1. *If X is completely normal space satisfying condition (b_n) for all n, then X also satisfies (a_n), (c_n), and (d_n) for all n.*

Proof. This follows from [3.1], [3.2], and [3.3].

4. Totally normal spaces.

I introduce a class of normal spaces for which the subset theorem and sum theorem of inductive dimension can be shown to hold.

Definition. A space X is called *totally normal* if it is normal and if each open set G of X is the union of a collection $\{G_\alpha\}$, locally finite in G, of open F_σ sets of X.

It will be shown that perfectly normal spaces are totally normal and totally normal spaces are completely normal.

[4.1] *Perfectly normal spaces are totally normal.*

Proof. Each open set G of a perfectly normal space X is an F_σ set of X. Thus the collection $\{G_\alpha\}$ of open F_σ sets covering G may consist of the one set G.

[4.2] *Hereditarily paracompact Hausdorff spaces are totally normal.*

Proof. Let X be an hereditarily paracompact Hausdorff space and let G be an open set in X. Then X and G are normal Hausdorff spaces (**4**). Then X is regular and hence each point x of G is contained in an open set U_x whose closure $\overline{U_x}$ is contained in G. The sets U_x form a covering of the paracompact space G; hence there is a locally finite refinement $\{V_\alpha\}$ of the covering by the sets U_x. Then (**4**) there is a covering $\{F_\alpha\}$ of G by closed sets of G such that $F_\alpha \subset V_\alpha$. Hence, since G is normal, there exist open F_σ sets G_α of G with $F_\alpha \subset G_\alpha \subset V_\alpha$. Since $G = \bigcup F_\alpha$ and $F_\alpha \subset G_\alpha \subset G$, therefore $G = \bigcup G_\alpha$. Since $G_\alpha \subset V_\alpha$ and $\{V_\alpha\}$ is locally finite in G, therefore $\{G_\alpha\}$ is locally finite in G. Since G_α is open in the open set G, it is open in X. Since G_α is an F_σ set in G and, for some x, $G_\alpha \subset \overline{V_\alpha} \subset \overline{U_x} \subset G$, therefore G_α is an F_σ set of the closed set $\overline{V_\alpha}$ of X and hence is an F_σ set of X. Thus the space X is totally normal.

ON INDUCTIVE DIMENSION OF NORMAL SPACES

Let us consider the following two conditions on an open set G of a space X.

(h) G is the union of a collection, locally finite in G, of open F_σ sets of X.

(k) For each $i = 1,2,\ldots,$ there is a collection $\{W_{i\alpha}\}$, locally finite in G, of disjoint open sets and a corresponding collection $\{F_{i\alpha}\}$ of closed sets of X such that $F_{i\alpha} \subset W_{i\alpha} \subset G$ and such that $\bigcup_{i=1}^{\infty} \bigcup_{\alpha} F_{i\alpha} = G$.

The condition (h) is satisfied by open sets of totally normal spaces.

[4.3] *For an open set G of a normal space X the conditions (h) and (k) are equivalent.*

Proof. (h) $\to$ (k). Let G be an open set of a normal space X and let $G = \bigcup G_\alpha$, where $\{G_\alpha\}$ is locally finite in G and each G_α is an open F_σ set of X. Then there is a continuous function f_α $(0 \leqslant f_\alpha(x) \leqslant 1)$ such that $f_\alpha(x) > 0$ if and only if $x \in G_\alpha$.

Let the indices α be well ordered and let

$$W_{i\alpha} = \{x \mid f_\alpha(x) > (i+1)^{-1}, f_\beta(x) < (i+1)^{-1} \text{for all } \beta < \alpha\}.$$

Then from the local finiteness of $\{G_\alpha\}$ in G it follows that $W_{i\alpha}$ is open in G and hence in X. Since $W_{i\alpha} \subset G_\alpha$ and $\{G_\alpha\}$ is locally finite in G, therefore, for each i, $\{W_{i\alpha}\}$ is locally finite in G. Clearly, if $\beta < \alpha$, then

$$W_{i\alpha} \cap W_{i\beta} = 0.$$

Let $F_{i\alpha} = \{x \mid f_\alpha(x) \geqslant 1/i, x \notin G_\beta \text{ for } \beta < \alpha\}$. Then $F_{i\alpha}$ is closed in X and $F_{i\alpha} \subset W_{i\alpha} \subset G$. If $x \in G$, let G_α be the first set of the covering $\{G_\alpha\}$ of G which contains x. Then $f_\alpha(x) > 0$ and hence, for some i, $f_\alpha(x) \geqslant 1/i$ while $x \notin G_\beta$ for $\beta < \alpha$; hence $x \in F_{i\alpha}$. Thus $\bigcup_{i=1}^{\infty} \bigcup_{\alpha} F_{i\alpha} = G$. Thus condition ($k$) is satisfied.

(k) $\to$ (h). Let G be an open set of a normal space X and, for $i = 1,2,\ldots,$ let $\{W_{i\alpha}\}$ be a collection, locally finite in G, of disjoint open sets. Let $F_{i\alpha}$ be closed in X, $F_{i\alpha} \subset W_{i\alpha} \subset G$ and let $\bigcup_{i=1}^{\infty} \bigcup_{\alpha} F_{i\alpha} = G$. Since X is normal, there is a continuous function $g_{i\alpha}$ $(0 \leqslant g_{i\alpha}(x) \leqslant 1)$ such that $g_{i\alpha}(x) = 0$ for $x \in X - W_{i\alpha}$ and $g_{i\alpha}(x) = 1$ for $x \in F_{i\alpha}$. Let

$$g_i(x) = \sum_\alpha g_{i\alpha}(x) = \sup_\alpha g_{i\alpha}(x).$$

Then, since $\{W_{i\alpha}\}$ is locally finite in G, $g_i(x)$ exists and is continuous in G. Let

$$G_{i\alpha} = \{x \mid g_{i\alpha}(x) > 0, g_j(x) < 1/i \text{ for } j < i\},$$

and let $H_{i\alpha} = \{x \mid g_{i\alpha}(x) > 0\}$. Then $H_{i\alpha}$ is an open F_σ set of X, $H_{i\alpha} \subset G$ and $G_{i\alpha}$ is an open F_σ set in $H_{i\alpha}$ and hence in X.

If $x \in G$, then, for some j and β, $x \in F_{j\beta}$ and hence $g_{j\beta}(x) = 1$ and $g_j(x) = 1$. If i is the least number such that, for some α, $g_{i\alpha}(x) > 0$, then, for $j < i$, $g_j(x) = 0$ and hence $x \in G_{i\alpha}$. Thus $G \subset \bigcup_{i,\alpha} G_{i\alpha}$. Hence, since $G_{i\alpha} \subset H_{i\alpha} \subset G$, $G = \bigcup_{i,\alpha} G_{i\alpha}$.

Since $g_{j\beta}(x) = 1$, there is a neighbourhood N of x in G such that $g_j(y) > \frac{1}{2}$ for $y \in N$. Hence, for $i > j$ and $i > 2$, $N \cap G_{i\alpha} = 0$. Thus N can meet $G_{i\alpha}$

C. H. DOWKER

only if $i \leqslant i_0 = \max(2, j)$. For each i, $\{W_{i\alpha}\}$ is locally finite in G and hence some neighbourhood N_i of x meets only a finite number of the sets $W_{i\alpha}$. Since $G_{i\alpha} \subset H_{i\alpha} \subset W_{i\alpha}$, $N_i \cap G_{i\alpha} \neq 0$ for at most a finite number of values of α. Then the intersection $N \cap \bigcap_{i \leqslant i_0} N_i$ is a neighbourhood of x which meets only a finite number of the sets $G_{i\alpha}$. Thus the collection $\{G_{i\alpha}\}$, for all i and all α, is locally finite in G. Thus condition (h) is satisfied.

[4.4] *Let a space X be the union of disjoint sets G_α each of which is open and closed in X. If each G_α is normal, then X is normal.*

Proof. Let E and F be closed and disjoint in X. Then $E \cap G_\alpha$ and $F \cap G_\alpha$ are closed and disjoint in the normal space G_α. Hence there exist disjoint sets U_α and V_α open in G_α, and hence open in X, such that $E \cap G_\alpha \subset U_\alpha$ and $F \cap G_\alpha \subset V_\alpha$. Let $U = \bigcup U_\alpha$ and $V = \bigcup V_\alpha$, then U and V are open and disjoint and $E = \bigcup E \cap G_\alpha \subset \bigcup U_\alpha = U$ and $F \subset V$. Therefore X is normal.

[4.5] *Let X be a space and let $\{C_i\}$ be a sequence of closed sets whose interiors cover X. If each C_i is normal, then X is normal.*

Proof. Let E_0 and F_0 be any two disjoint closed sets of X. We construct increasing sequences $\{E_i\}$ and $\{F_i\}$ of disjoint closed sets of X such that, for $i = 1, 2 \ldots$,

$$E_{i-1} \cap C_i \subset G_i \subset E_i \qquad \text{and} \qquad F_{i-1} \cap C_i \subset H_i \subset F_i\,,$$

where G_i and H_i are open and disjoint in C_i.

Assume that we already have E_{i-1} and F_{i-1}. Then $E_{i-1} \cap C_i$ and $F_{i-1} \cap C_i$ are disjoint closed sets of the normal space C_i. Hence there exist disjoint open sets G_i and H_i of C_i such that $E_{i-1} \cap C_i \subset G_i$, $F_{i-1} \cap C_i \subset H_i$, and $\overline{G_i} \cap \overline{H_i} = 0$. Let $E_i = E_{i-1} \cup \overline{G_i}$ and $F_i = F_{i-1} \cup \overline{H_i}$. Then E_i and F_i are disjoint closed sets of X and $G_i \subset E_i$ and $H_i \subset F_i$.

Let $\operatorname{int} C_i$ be the interior of C_i and let

$$U = \bigcup_{i=1}^{\infty} G_i \cap \operatorname{int} C_i \qquad \text{and} \qquad V = \bigcup_{i=1}^{\infty} H_i \cap \operatorname{int} C_i\,.$$

Since G_i is open in C_i, $G_i \cap \operatorname{int} C_i$ is open in $\operatorname{int} C_i$, and hence is open in X. Therefore U is open in X and similarly V is open in X. Since

$$G_i \cap \operatorname{int} C_i \subset \overline{G_i} \subset E_i\,,$$

therefore $U \subset \bigcup E_i$ and similarly $V \subset \bigcup F_i$. And, if $k = \max(i, j)$,

$$E_i \cap F_j \subset E_k \cap F_k = 0\,.$$

Hence $U \cap V = 0$. Since $X = \bigcup_{i=1}^{\infty} \operatorname{int} C_i$,

$$E_0 = \bigcup_{i=1}^{\infty} E_0 \cap \operatorname{int} C_i \subset \bigcup_{i=1}^{\infty} E_{i-1} \cap \operatorname{int} C_i \subset \bigcup_{i=1}^{\infty} G_i \cap \operatorname{int} C_i = U\,.$$

Similarly $F_0 \subset V$. Therefore X is normal.

ON INDUCTIVE DIMENSION OF NORMAL SPACES

[4.6] *Totally normal spaces are completely normal.*

Proof. Let G be any open set in a totally normal space X. Then, by [4.3], for each $i = 1, 2, \ldots$, there is a collection $\{W_{i\alpha}\}$, locally finite in G, of disjoint open sets and a corresponding collection $\{F_{i\alpha}\}$ of closed sets of X with $F_{i\alpha} \subset W_{i\alpha} \subset G$ and $\bigcup_{i=1}^{\infty} \bigcup_{\alpha} F_{i\alpha} = G$. Then, since X is normal, there exists a closed set $C_{i\alpha}$ with $F_{i\alpha} \subset \operatorname{int} C_{i\alpha} \subset C_{i\alpha} \subset W_{i\alpha}$. Since X is normal and $C_{i\alpha}$ is closed in X, $C_{i\alpha}$ is normal. Let $C_i = \bigcup_{\alpha} C_{i\alpha}$. Since $C_{i\alpha} \subset W_{i\alpha}$, then, for each i, $\{C_{i\alpha}\}$ is a collection of disjoint sets locally finite in G and hence locally finite in C_i. Hence each $C_{i\alpha}$ is open as well as closed in C_i. Hence, by [4.4], C_i is normal. Also $F_{i\alpha} \subset \operatorname{int} C_{i\alpha} \subset \operatorname{int} C_i$ and hence $G = \bigcup_{i,\alpha} F_{i\alpha} \subset \bigcup_{i} \operatorname{int} C_i$. Hence, by [4.5], G is normal. Thus every open set of X is normal and hence, by [1.1], X is completely normal.

[4.7] *Every subset of a totally normal space is totally normal.*

Proof. Let $A \subset X$ with X totally normal. Then, by [4.6], A is normal. Let G be any open set in A. Then there is an open set H of X such that $H \cap A = G$. Since X is totally normal, H is the union of a collection $\{H_\alpha\}$, locally finite in H, of open F_σ sets of X. Then, if $G_\alpha = H_\alpha \cap A$,

$$G = H \cap A = \bigcup H_\alpha \cap A = \bigcup G_\alpha ,$$

and G_α is an open F_σ set of A. Each point x of $G \subset H$ has a neighborhood N in H which meets only a finite number of the sets H_α and hence has the neighborhood $N \cap G$ in G which meets only a finite number of the sets G_α. Hence $\{G_\alpha\}$ is locally finite in G. Therefore A is totally normal.

Examples. Let Z be a space consisting of a non-countable set of points one of which is distinguished and called z_0. A subset of Z is called *open* (i) if it does not contain z_0 or (ii) if its complement is finite. Then Z is an hereditarily paracompact Hausdorff space which is not perfectly normal. Bing's example H [(**1**), 185] is perfectly normal but not paracompact and so not hereditarily paracompact. His example G [(**1**), 184] is totally normal but neither perfectly normal nor paracompact. The space consisting of ordinal numbers $\leqslant \omega_1$ with the usual topology is completely normal but not totally normal.

5. Dimension in totally normal spaces.

Returning to the theory of inductive dimension, I show (Theorem 2) that the subset theorem, and consequently the sum theorem, holds for totally normal spaces.

[5.1] *Let a space be the union of disjoint sets G_α each of which is open and closed in X. If each $\operatorname{Ind} G_\alpha \leqslant n$, then $\operatorname{Ind} X \leqslant n$.*

Proof. This is trivially true for dimension -1, and we assume it true for dimension $n-1$. Let $E \subset W \subset X$ with E closed and W open. Then $E \cap G_\alpha \subset W \cap G_\alpha \subset G_\alpha$ with $E \cap G_\alpha$ closed in G_α and $W \cap G_\alpha$ open in G_α. Hence, since $\operatorname{Ind} G_\alpha \leqslant n$, G_α is the union of disjoint sets U_α, V_α, and C_α with U_α and V_α open in G_α and hence open in X, $E \cap G_\alpha \subset U_\alpha \subset W \cap G_\alpha$ and $\operatorname{Ind} C_\alpha \leqslant n-1$. Let

$$U = \bigcup U_\alpha \qquad V = \bigcup V_\alpha, \qquad \text{and} \qquad C = \bigcup C_\alpha .$$

C. H. DOWKER

Then U, V, and C are disjoint, their union is X, and U and V are open sets. Also

$$E = \bigcup_\alpha E \cap G_\alpha \subset \bigcup_\alpha U_\alpha = U$$

and

$$U = \bigcup_\alpha U_\alpha \subset \bigcup_\alpha W \cap G_\alpha = W .$$

Each set $C_\alpha = C \cap G_\alpha$ is open and closed in C and $\operatorname{Ind} C_\alpha \leqslant n-1$; hence, by the induction hypothesis, $\operatorname{Ind} C \leqslant n-1$. Thus $\operatorname{Ind} X \leqslant n$, as was to be shown.

[5.2] *Let X be a normal space satisfying the condition (d_{n-1}) of § 3. Let $\{C_i\}$ and $\{F_i\}$ be sequences of closed sets of X such that $F_i \subset \operatorname{int} C_i$, $X = \bigcup_{i=1}^{\infty} F_i$, and $\operatorname{Ind} C_i \leqslant n$. Then $\operatorname{Ind} X \leqslant n$.*

Proof. Let $E \subset G \subset X$ with E closed and G open. Then, since X is normal, there is a closed set K and a sequence $\{W_i\}$ of open sets such that

$$E \subset K \subset \overline{W_{i+1}} \subset W_i \subset G$$

and

$$K = \bigcap_{i=1}^{\infty} W_i .$$

Then $F_i \cap K \subset (\operatorname{int} C_i) \cap W_i$, $F_i \cap K$ is closed and $(\operatorname{int} C_i) \cap W_i$ is open and is contained in C_i. Hence, since $\operatorname{Ind} C_i \leqslant n$, there exists U_i open in C_i with $F_i \cap K \subset U_i \subset (\operatorname{int} C_i) \cap W_i$ and $\operatorname{Ind}(\overline{U_i} \cap C_i - U_i) \leqslant n-1$. Since U_i is open in C_i and $U_i \subset \operatorname{int} C_i$, therefore U_i is open in $\operatorname{int} C_i$ and hence open in X. And, since $U_i \subset C_i$ and C_i is closed, $\overline{U_i} \cap C_i = \overline{U_i}$; hence

$$\operatorname{Ind}(\overline{U_i} - U_i) \leqslant n-1 .$$

Let $U = \bigcup_{i=1}^{\infty} U_i$, then U is open in X and

$$E \subset K = \bigcup_i (F_i \cap K) \subset \bigcup_i U_i = U$$

and

$$U \subset \bigcup_i W_i \subset G ;$$

thus $E \subset U \subset G$.

Let $x \notin K$; then, for some j, $x \notin \overline{W_j}$, and hence x has a neighbourhood $X - \overline{W_j}$ which meets at most a finite number of the sets U_i ($\subset W_i$) and hence $x \in \overline{U}$ if and only if $x \in \overline{U_i}$ for some i. Hence, since $K \subset U \subset \overline{U}$ and $K \subset \bigcup U_i \subset \bigcup \overline{U_i}$, we have $\overline{U} = \bigcup_{i=1}^{\infty} \overline{U_i}$. Hence

$$\overline{U} - U = \bigcup \overline{U_i} - \bigcup U_i \subset \bigcup (\overline{U_i} - U_i) .$$

ON INDUCTIVE DIMENSION OF NORMAL SPACES

Since U_i is open, $\overline{U_i}-U_i$ is closed, and we have $\operatorname{Ind}(\overline{U_i}-U_i) \leqslant n-1$; hence, by (d_{n-1}), $\operatorname{Ind}[\bigcup(\overline{U_i}-U_i)] \leqslant n-1$. Thus $\operatorname{Ind} X \leqslant n$, as was to be shown.

[5.3] *If X is a totally normal space, condition (d_{n-1}) implies (b_n).*

Proof. Let X be a totally normal space satisfying condition (d_{n-1}) and let $G \subset A \subset X$ with G open in A and $\operatorname{Ind} A \leqslant n$. By [4.7], A is totally normal and hence, by [4.3], there is, for each $i = 1, 2, \ldots$, a collection $\{W_{i\alpha}\}$, locally finite in G, of disjoint open sets of A and a collection $\{F_{i\alpha}\}$ of closed sets of A with

$$F_{i\alpha} \subset W_{i\alpha} \subset G$$

and

$$\bigcup_{i=1}^{\infty} \bigcup_{\alpha} F_{i\alpha} = G.$$

Since A is normal, there exist $V_{i\alpha}$ open in A and $C_{i\alpha}$ closed in A with $F_{i\alpha} \subset V_{i\alpha} \subset C_{i\alpha} \subset W_{i\alpha}$. Then, for each i, the sets $C_{i\alpha}$ are disjoint, $\{C_{i\alpha}\}$ is locally finite in G and hence, if $C_i = \bigcup_\alpha C_{i\alpha}$, $\{C_{i\alpha}\}$ is a locally finite collection of disjoint closed sets of C_i. Therefore $C_{i\alpha}$ is open and closed in C_i. Since $C_{i\alpha}$ is closed in A, $\operatorname{Ind} C_{i\alpha} \leqslant n$. Hence, by [5.1], $\operatorname{Ind} C_i \leqslant n$.

Let $F_i = \bigcup_\alpha F_{i\alpha}$ and $V_i = \bigcup_\alpha V_{i\alpha}$ as well as $C_i = \bigcup_\alpha C_{i\alpha}$. Then F_i and C_i are unions of locally finite collections of closed sets of G, and hence are closed in G, while V_i is a union of open sets in G and hence is open in G. Clearly $F_i \subset V_i \subset C_i$; hence G_i is contained in the interior of C_i with respect to G.

Since X is totally normal, G is normal, and, since X satisfies condition (d_{n-1}), so does G. Hence, by [5.2], since

$$G = \bigcup_{i,\alpha} F_{i\alpha} = \bigcup_{i=1}^{\infty} F_i,$$

$\operatorname{Ind} G \leqslant n$. Thus X satisfies condition (b_n). This completes the proof.

THEOREM 2. *Let $A \subset X$ with X totally normal and* $\operatorname{Ind} X \leqslant n$. *Then* $\operatorname{Ind} A \leqslant n$.

THEOREM 3. *Let a totally normal space X be the union of two sets A and B with A closed and* $\operatorname{Ind} A \leqslant n$ *and* $\operatorname{Ind} B \leqslant n$. *Then* $\operatorname{Ind} X \leqslant n$.

THEOREM 4. *Let $\{A_i\}$ be a sequence of closed sets in a totally normal space and let each* $\operatorname{Ind} A_i \leqslant n$. *Then* $\operatorname{Ind} \bigcup_{i=1}^{\infty} A_i \leqslant n$.

Proof. Let X be a totally normal space. Then X is completely normal and hence, by [3.3], (b_n) implies (d_n), and, by [5.3], (d_{n-1}) implies (b_n). Hence, since (b_{-1}) and (d_{-1}) are trivially satisfied, (b_n) and (d_n) hold for all n. Hence, by Theorem 1, (a_n) and (c_n) also hold for all n. Then Theorems 2, 3, 4 follow respectively from (a_n), (c_n), (d_n).

Examples. Let I be the segment $0 \leqslant x \leqslant 1$, and let J be a space consisting of the points of I and a special point j_0; a set of J is open (i) if it is the whole space J or (ii) if it is an open set of I. The space J is trivially normal; there are no disjoint

C. H. DOWKER

non-empty closed sets. Any subset either is a subset of I and hence is normal or it contains j_0 and hence is trivially normal. Thus J is completely normal. If A is non-empty and closed in J, then $j_0 \in A$, the only open set containing A is J, and $\overline{J}—J = 0$. Hence $\operatorname{Ind} J = 0$. But I is an open set in J and $\operatorname{Ind} I = 1$. Thus the open subset theorem does not hold in the completely normal space J.

Of course J is not a Hausdorff space. An example is given elsewhere† of a normal Hausdorff space X and an open subset A of X such that A is normal and $\operatorname{Ind} A = 1$ but $\operatorname{Ind} X = 0$.

Problem. Does the open subset theorem hold for every completely normal Hausdorff space?

6. Čech's covering theorem for finite dimensional spaces.

Čech ends his paper (**3**) with a proof that a finite-dimensional perfectly normal space admits a certain kind of covering by the closures of disjoint open sets. Theorem 5 below is an extension of Čech's result to totally normal spaces.

[6.1] *Let R be a completely normal space satisfying (b_n) and let $\operatorname{Ind} R \leqslant n$. Let U be open, S closed, and U_0 open in S with $U_0 \subset U$ and $\operatorname{Ind}(\overline{U_0}—U_0) \leqslant n-1$. Then there exists V open in R with*

$$U_0 \subset V \subset U, \qquad V \cap S = U_0, \qquad (\overline{V}—V) \cap S = \overline{U_0}—U_0$$

and

$$\operatorname{Ind}(\overline{V}—V) \leqslant n-1.$$

Proof. Let $W = R—(S—U_0)$ and $Y = R—(\overline{U_0}—U_0)$; then W and Y are open and $U_0 \subset W \subset Y$. Since U_0 is closed in the normal space Y and $U_0 \subset W \cap U \subset Y$, there exists H open in Y with

$$U_0 \subset H \subset \overline{H} \cap Y \subset W \cap U.$$

By (b_n), $\operatorname{Ind} Y \leqslant n$, and hence there exists V open in Y and hence open in R with $U_0 \subset V \subset H$ and $\operatorname{Ind}(\overline{V} \cap Y—V) \leqslant n-1$.

Then $V \subset H \subset U$ and $U_0 \subset V \cap S \subset W \cap S = U_0$. Hence $V \cap S = U_0$. And

$$\overline{V} \cap S \cap Y \subset \overline{H} \cap S \cap Y \subset W \cap U \cap S = U_0;$$

hence

$$\overline{U_0} \subset \overline{V} \cap S \subset U_0 \cup (R—Y) = \overline{U_0}.$$

Thus $\overline{V} \cap S = \overline{U_0}$ and

$$(\overline{V}—V) \cap S = \overline{V} \cap S—V \cap S = \overline{U_0}—U_0.$$

The completely normal space $\overline{V}—V$ is the union of the disjoint sets $\overline{U_0}—U_0$ and $\overline{V} \cap Y—V$, where $\overline{U_0}—U_0$ is closed and both sets have $\operatorname{Ind} \leqslant n-1$. Hence, by [2.2], $\operatorname{Ind}(\overline{V}—V) \leqslant n-1$. This completes the proof.

† See 'Local dimension of normal spaces' by the author, to appear. [Editor's note: Quart. J. Math. Oxford Ser. (2), 6, 1955, 101 – 120]

ON INDUCTIVE DIMENSION OF NORMAL SPACES

THEOREM 5. *Let R be a totally normal space, or more generally any completely normal space satisfying condition* (b_N) *for all* $N = 0, 1, \ldots$. *Let S be closed in R with* $\operatorname{Ind} S \leqslant n$. *Let* $U_1, \ldots, U_m$ *be open in R with* $S \subset \bigcup_{\nu=1}^{m} U_\nu$. *Then there exist* V_i $(1 \leqslant i \leqslant (n+1)m)$ *open in R with the properties:*

(i) $\overline{V_i} \subset U_\nu$ *for* $1 \leqslant \nu \leqslant m$, $(n+1)(\nu-1)+1 \leqslant i \leqslant (n+1)\nu$;

(ii) $\bigcup_{i=1}^{(n+1)m} \overline{V_i} \supset S$;

(iii) $V_i \cap V_j = 0$ *for* $1 \leqslant i < j \leqslant (n+1)m$;

(iv) $\operatorname{Ind}((\overline{V_i} - V_i) \cap S) \leqslant n-1$ *for* $1 \leqslant i \leqslant (n+1)m$;

(v) *if* $2 \leqslant r \leqslant n+2$ *and if* $i_1, \ldots, i_r$ *is any combination (without repetition) of indices* $1, 2, \ldots, (n+1)m$, *then* $\bigcap_{s=1}^{r} \overline{V_{i_s}} \subset S$ *and* $\operatorname{ind} \bigcup_{s=1}^{r} \overline{V_{i_s}} \leqslant n - r + 1$.

Proof. Using [3.1], [3.3], and [6.1] above in place of Čech's propositions 23, 19, and 24.1, Čech's proof applies with trivial modifications.

REFERENCES

1. R. H. Bing, 'Metrization of topological spaces', *Canadian J. Math.* **3** (1951), 175-86.
2. L. E. J. Brouwer, 'Über den natürlichen Dimensionsbegriff', *J. f. Math.* **142** (1913), 146-52.
3. E. Čech, 'Dimense dokonale normálních prostorů', *Rozpravy České Akad.* II **42** (1932), no. 13; 'Sur la dimension des espaces parfaitement normaux ', *Bull. int. Acad. Prague* 33 (1932), 38-55.
4. J. Dieudonné, 'Une généralization des espaces compacts', *J. de Math.* **23** (1944), 65-76.
5. O. V. Lokutzievski, 'On the dimension of bicompacta', *Doklady Akad. Nauk SSSR* **67** (1949), 217-19.

Dimension of metric spaces

by

C. H. Dowker (London) and W. Hurewicz (Cambridge, Mass.)

1. It is to be shown that a metric space has dimension $\leqslant n$ if and only if there exists a sequence $\{\alpha_i\}$ of locally finite open coverings, each of order $\leqslant n$, with mesh tending to zero as $i \to \infty$, such that

(a) the closure of each member of α_{i+1} is contained in some member of α_i.

For a compact metric space, every sequence of coverings of order $\leqslant n$ with mesh tending to zero contains a subsequence satisfying condition (a). But condition (a) can not in general be omitted, as is shown by K. Sitnikov's example [8] of a two-dimensional metric separable space which has a sequence of coverings, each of order one, with mesh tending to zero.

In the course of proving the above proposition, we incidentally give a new proof of the theorem of M. Katětov (see [4]; also [5], theorem 3.4 and also K. Morita [7], theorem 8.6) that for an arbitrary metric space X the covering dimension ($\dim X$) is equal to the dimension ($\operatorname{Ind} X$) defined inductively in terms of the separation of closed sets.

2. By a *covering* of a topological space X we mean a collection of open sets of X whose union is X. A covering β is called a *refinement* of a covering α if each member of β is contained in some member of α.

The *order* of a collection of subsets of X is the largest integer n such that some point of X is contained in $n+1$ members of the collection, or is ∞ if there is no such largest integer.

Definition 1. The *dimension of a space* X ($\dim X$) is the least integer n such that every finite covering of X has a refinement of order $\leqslant n$, or the dimension is ∞ if there is no such integer.

A collection of subsets of X is called *locally finite* if every point of X has a neighborhood meeting at most a finite number of members of the collection. If X is a metric space, it is known ([9], corollary 1, and [3], theorem 3.5) that $\dim X \leqslant n$ if and only if every covering of X has a locally finite refinement of order $\leqslant n$.

The *mesh* of a collection of subsets of a metric space is the upper bound of the diameters of the members of the collection.

Definition 2. The *sequential dimension of a metric space* X (ds X) is the least integer n such that there exists a sequence $\{a_i\}$ of locally finite coverings, each of order $\leqslant n$, with mesh $a_i \to 0$ as $i \to \infty$, such that

(a) the closure of each member of a_{i+1} is contained in some member of a_i.

If there is no such integer, ds $X = \infty$.

LEMMA 1. *If X is a metric space,* ds $X \leqslant$ dim X.

Proof. It is sufficient to show that if dim $X \leqslant n$ then ds $X \leqslant n$. Let dim $X \leqslant n$ and suppose that the locally finite coverings $a_1, \dots, a_{i-1}$ of order $\leqslant n$ have been constructed so that mesh $a_k \leqslant 2^{-k}$ and, for $1 < k < i$, the closure of each member of a_k is contained in some member of a_{k-1}. We now construct the covering a_i.

It follows from [9], corollary 1, that a_{i-1} has a locally finite refinement β_i of mesh $\leqslant 2^{-i}$. By [3], theorem 3.5, since dim $X \leqslant n$, β_i has a locally finite refinement $\gamma_i = \{U_{i\lambda}\}$ of order $\leqslant n$. By [6], p. 26, (33.4), the covering γ_i can be shrunk to a covering $a_i = \{V_{i\lambda}\}$ such that each $\overline{V}_{i\lambda} \subset U_{i\lambda}$. Then a_i is locally finite and of order $\leqslant n$, and mesh $a_i \leqslant 2^{-k}$. And, since γ_i is a refinement of a_{i-1}, each $\overline{V}_{i\lambda}$ is contained in some member of a_{i-1}. Thus the required sequence $\{a_i\}$ (see definition 2) can be constructed, and hence ds $X \leqslant n$ as was to be shown.

Definition 3. The *inductive dimension of a space* X (Ind X) is defined inductively as follows: If X is empty, Ind $X = -1$. For $n = 0, 1, \dots$, Ind $X \leqslant n$ means that for each closed set E and open set G with $E \subset G$ there exists an open set U with $E \subset U \subset G$ and Ind$(\overline{U} - U) \leqslant n-1$.

Ind $X = \infty$ means that there is no integer n for which Ind $X \leqslant n$.

It is known ([1], § 18) that, if X is a normal space, Ind $X \leqslant n$ if and only if, for each pair E, F of disjoint closed sets, X is the union of three disjoint sets U, V and K with U and V open, $E \subset U$, $F \subset V$ and Ind $K \leqslant n-1$.

LEMMA 2. *If X is a metric space,* Ind $X \leqslant$ ds X.

Proof. It is sufficient to show that if ds $X \leqslant n$ then Ind $X \leqslant n$. The proof is by induction. It is clear that if ds $X = -1$ then X is empty and hence Ind $X = -1$. We assume it proved that ds $X \leqslant n-1$ implies Ind $X \leqslant n-1$.

Let X be a metric space for which ds $X \leqslant n$. That is, let there exist a sequence $\{a_i\}$ of locally finite coverings as in definition 2 above. We are to prove that Ind $X \leqslant n$. Let E and F be an arbitrary pair of disjoint closed sets of X.

For each $i=0,1,\ldots$ we define a decomposition of X into the union of three disjoint sets M_i, N_i and K_i, of which M_i and N_i are closed and hence K_i is open. Let $M_0=N_0=0$; for $i\geqslant 1$ the decompositions (M_i, N_i, K_i) are defined inductively as follows.

Let the members of a_i be put in the following three classes: a_{i1} consists of those members of a_i whose closures do not meet $F\cup N_{i-1}$, a_{i2} consists of those members of a_i whose closures meet $F\cup N_{i-1}$ but do not meet $E\cup M_{i-1}$, and a_{i3} consists of those members of a_i whose closures meet both $F\cup N_{i-1}$ and $E\cup M_{i-1}$. Let G_i be the union of the open sets which are elements of a_{i1}, let H_i be the union of a_{i2} and let J_i be the union of a_{i3}. Then G_i, H_i and J_i are open sets and their union is X. Let $M_i=X-H_i-J_i$, $N_i=X-G_i-J_i$ and $K_i=X-M_i-N_i=(G_i\cap H_i)\cup J_i$. Then M_i and N_i are closed and K_i is open. Since (G_i, H_i, J_i) covers X, therefore $M_i\cap N_i=0$ and hence (M_i, N_i, K_i) is a decomposition of X into disjoint sets. Thus the sequence of decompositions is defined.

If $U\in a_{i+1}$ then, for some $V\in a_i$, $\overline{U}\subset V$. We will verify that

(1) $\quad V\in a_{i1}\Rightarrow \overline{U}\cap N_i=0, \quad \overline{U}\cap F=0,$

(2) $\quad V\in a_{i2}\Rightarrow \overline{U}\cap M_i=0, \quad \overline{U}\cap E=0,$

(3) $\quad V\in a_{i3}\Rightarrow \overline{U}\cap M_i=0, \quad \overline{U}\cap N_i=0.$

For, if $V\in a_{i1}$, then $\overline{V}\cap(F\cup N_{i-1})=0$ and hence $\overline{V}\cap F=0$. Also $V\subset G_i=\bigcup a_{i1}$ and hence $V\cap N_i=0$. Since $\overline{U}\subset V$, therefore $\overline{U}\cap N_i=0$ and $\overline{U}\cap F=0$. Similarly, if $V\in a_{i2}$, then $V\cap E=0$ and $V\subset H_i$ and hence $V\cap M_i=0$, from which (2) follows. And, if $V\in a_{i3}$, then $V\subset J_i$ and hence $V\cap M_i=0$ and $V\cap N_i=0$, from which (3) follows.

By (1) and (3), if $\overline{U}\cap N_i\neq 0$ then $V\in a_{i2}$ and hence, by (2), $\overline{U}\cap M_i=0$ and $\overline{U}\cap E=0$. Similarly, if $\overline{U}\cap M_i\neq 0$ then $V\in a_{i1}$ and hence $\overline{U}\cap N_i=0$ and $\overline{U}\cap F=0$. Hence, if $U\in a_{i+1,3}$, that is if $\overline{U}\cap(E\cup M_i)\neq 0$ and $\overline{U}\cap(F\cup N_i)\neq 0$, then $\overline{U}\cap N_i=0$ and $\overline{U}\cap M_i=0$. Thus

$$(4)\qquad \begin{gathered} U\in a_{i+1,3}\Rightarrow \overline{U}\cap N_i=0, \\ \overline{U}\cap M_i=0, \quad \overline{U}\cap E\neq 0, \quad \overline{U}\cap F\neq 0. \end{gathered}$$

Since the closure of the union of a locally finite collection of sets is the union of the closures of the sets, $\overline{J}_{i+1}$ is the union of the sets $\overline{U}$ with $U\in a_{i+1,3}$. Hence, by (4), $\overline{J}_{i+1}\cap M_i=0$ and $\overline{J}_{i+1}\cap N_i=0$. Also $\overline{G}_{i+1}$ is the union of all $\overline{U}$ with $U\in a_{i+1,1}$; hence $\overline{G}_{i+1}\cap N_i=0$. Similarly $\overline{H}_{i+1}$ is the union of all $\overline{U}$ with $U\in a_{i+1,2}$; hence $\overline{H}_{i+1}\cap M_i=0$. Therefore

$$(5)\qquad M_i\subset X-\overline{H}_{i+1}-\overline{J}_{i+1}=\operatorname{Int} M_{i+1},$$

$$(6)\qquad N_i\subset X-\overline{G}_{i+1}-\overline{J}_{i+1}=\operatorname{Int} N_{i+1}.$$

Let $M=\bigcup_{i=1}^{\infty} M_i$ and $N=\bigcup_{i=1}^{\infty} N_i$. It follows from (5) and (6) that M and N are open sets. And, since, for each i, $M_i \frown N_i=0$, it follows that $M\frown N=0$. Let $K=X-M-N=\bigcap_{i=1}^{\infty} K_i$.

If $x \in E$ then the distance $\varrho(x,F)>0$ and hence, for some i, $\varrho(x,F)>\operatorname{mesh} a_i$. For any $U \in a_i$ with $x \in U$, we have $\bar{U}\frown E\neq 0$ and $\bar{U}\frown F=0$. Thus, since $\bar{U}\frown E\neq 0$, $U \notin a_{i2}$. And, since $\bar{U}\frown F=0$, therefore (see (4)) $U \notin a_{i3}$. Hence $x \notin H_i$ and $x \notin J_i$. Therefore $x \in M_i$. Thus $E\subset M$ and similarly $F\subset N$.

Thus X is decomposed into three disjoint sets M, N and K with M and N open and $E\subset M$ and $F\subset N$. To show that $\operatorname{Ind} X\leqslant n$ it is sufficient to show that $\operatorname{Ind} K\leqslant n-1$.

Let $C_i=K-J_i$; then, since J_i is open, C_i is closed. If $U \in a_{i+1,3}$ then $\bar{U}\subset V$ with $V \in a_i$. It follows from (4) that $V\frown E\neq 0$ and $V\frown F\neq 0$, and hence that $V \in a_{i3}$. Therefore $J_{i+1}\subset J_i$ and hence $C_i\subset C_{i+1}$. Thus $\{C_i\}$ is an ascending sequence of closed sets.

For each point $x \in X$, either $\varrho(x,E)>0$ or $\varrho(x,F)>0$. Hence, for sufficiently large i, if $x \in U \in a_i$ then either $\bar{U}\frown E=0$ or $\bar{U}\frown F=0$. Hence, by (4), $U \notin a_{i3}$ and hence $x \notin J_i$. Thus $\bigcap_{i=1}^{\infty} J_i=0$ and therefore $\bigcup_{i=1}^{\infty} C_i=K$.

We now show that $\operatorname{ds} C_i\leqslant n-1$ for each $i=1,2,\ldots$ Let β_{ij} be the family of open subsets $U\frown C_i$ of C_i with $U \in a_{i+j,2}$. Since $C_i\subset K\subset K_{i+j}$ $=(G_{i+j}\frown H_{i+j})\smile J_{i+j}$ and $C_i\subset C_{i+j}=K-J_{i+j}$, therefore $C_i\subset G_{i+j}\frown H_{i+j}$. Thus each point x of C_i is contained in some element of $a_{i+j,1}$ and also in some element of $a_{i+j,2}$ and, since β_{i+j} is of order $\leqslant n$, x is in at most n elements of $a_{i+j,2}$. Hence β_{ij} is a covering of C_i and is of order $\leqslant n-1$. Since a_{i+j} is locally finite, so is β_{ij}. Also $\operatorname{mesh}\beta_{ij}\leqslant \operatorname{mesh} a_{i+j}$ and hence $\operatorname{mesh}\beta_{ij}\to 0$ as $j\to\infty$.

Let $U \in a_{i+j+1,2}$ and $U\frown C_i\neq 0$ so that $U\frown C_i$ is a non-empty member of the covering $\beta_{i,j+1}$. Then $\bar{U}\subset V$ for some $V \in a_{i+j}$. Since $V\frown C_i\neq 0$, $V \operatorname{non}\subset J_i$ and hence $V \notin a_{i+j,3}$. Also, if V were an element of $a_{i+j,1}$ then, by (1), $\bar{U}\frown(F\smile N_{i+j})=0$ and hence $U \in a_{i+j+1,1}$ contrary to assumption. Therefore $V \in a_{i+j,2}$ and hence

$$\overline{U\frown C_i}\subset \bar{U}\frown C_i\subset V\frown C_i \in \beta_{ij}\,.$$

Thus we have $\operatorname{ds} C_i\leqslant n-1$. Hence, by the induction hypothesis, $\operatorname{Ind} C_i\leqslant n-1$ and hence, by the sum theorem ([1], § 19) for inductive dimension, $\operatorname{Ind} K\leqslant n-1$. Therefore $\operatorname{Ind} X\leqslant n$ as was to be shown. This completes the proof of Lemma 2.

The inequality $\dim X\leqslant \operatorname{Ind} X$ was proved by E. Čech ([1], § 26) for perfectly normal spaces and later by N. Vedenissoff [10] for arbitrary normal spaces. For completeness we include a proof of this result.

LEMMA 3. (Vedenissoff) *If X is a normal space,* $\dim X \leqslant \operatorname{Ind} X$.

Proof. Let $\operatorname{Ind} X \leqslant n$; it is to be shown that $\dim X \leqslant n$. The proof is by induction, the case $n=-1$ being trivial.

Let $\{U_1, \ldots, U_k\}$ be a finite covering of X. Since X is normal there exists a covering $\{V_1, \ldots, V_k\}$ of X with $\overline{V}_i \subset U_i$. Since $\operatorname{Ind} X \leqslant n$ there exist open sets W_i with boundaries $B_i = \overline{W}_i - W_i$ such that $\overline{V}_i \subset W_i \subset U_i$ and $\operatorname{Ind} B_i \leqslant n-1$. Let $Y_i = W_i - \bigcup_{j<i} \overline{W}_j$; then $\{Y_i\}$ is a collection of disjoint open sets. Each point $x \in X$ is in some W_i, hence in a first W_i and hence, unless $x \in B_j$ for some $j<i$, we have $x \in Y_i$. Thus, if $B = \bigcup_{j=1}^{k} B_j$ and $Y = \bigcup_{i=1}^{k} Y_i$, we have $X = B \cup Y$.

By the induction hypothesis, since $\operatorname{Ind} B_i \leqslant n-1$, $\dim B \leqslant n-1$. The closed set B is normal and hence by the sum theorem ([2], § 23), since each B_i is closed, $\dim B = \dim \bigcup_j B_j \leqslant n-1$. Hence the covering $\{B \cap U_i\}$ of B has a refinement $\{G_j\}$ of order $\leqslant n-1$, where the sets G_j are open in B. Let each G_j be associated with one of the sets U_i containing it, and let H_i be the union of the sets G_j associated with U_i. Then $\{H_i\}$ is a covering of B of order $\leqslant n-1$ and $H_i \subset U_i$.

The covering $\{H_i\}$ of the normal space B can be shrunk ([6], p. 26, (33.4)) to a covering K_i with $\overline{K}_i \subset H_i$. The family $\{\overline{K}_i\}$ of closed sets of X can be extended to a system $\{L_i\}$ of open sets of X similar to $\{\overline{K}_i\}$ and hence of order $\leqslant n-1$ ([2], § 12). If $M_i = L_i \cap U_i$ then M_i is open, $\{M_i\}$ is of order $\leqslant n-1$, $M_i \subset U_i$ and, since $\overline{K}_i \subset M_i$, $\{M_i\}$ covers B.

Adding the collection $\{Y_i\}$ of disjoint open sets, we get a covering $\{M_i, Y_j\}$ of X which is a refinement of $\{U_i\}$ and which is of order $\leqslant n$. Thus $\dim X \leqslant n$ as was to be shown.

THEOREM 1. *If X is a metric space,* $\dim X = \operatorname{ds} X = \operatorname{Ind} X$.

Proof. This is an immediate consequence of Lemmas 1, 2 and 3.

References

[1] E. Čech, *Dimense doskonale normálních prostorů*, Rozpravy České Akad. II 42 (1932), no 13; *Sur la dimension des espaces parfaitement normaux*, Bull. Int. Acad. Prague 33 (1932), p. 38-55.

[2] — *Contribution à la théorie de la dimension*, Čas. Mat. Fys. 62 (1933), p. 277-291.

[3] C. H. Dowker, *Mapping theorems for non-compact spaces*, Amer. J. Math. 69 (1947), p. 200-242.

[4] М. Катетов, *О размерности метрических пространств*, Доклады АН СССР 79 (1951), p. 189-191.

[5] — *О размерности несепарабельных пространств I*, Чехословацкий Мат. Журнал 2 (1952), p. 333-368.

[6] S. Lefschetz, *Algebraic topology*, New York 1942.

[7] K. Morita, *Normal families and dimension theory for metric spaces*, Math. Ann. 128 (1954), p. 350-362.

[8] К. Ситников, *Пример двумерного множества в трехмерном евклидовом пространстве*, Доклады АН СССР 88 (1953), p. 21-24.

[9] A. H. Stone, *Paracompactness and product spaces*, Bull. Amer. Math. Soc. 54 (1948), p. 977-982.

[10] N. Vedenissoff, *Sur la dimension au sens de E. Čech*, Bull. Acad. Sci. URSS, sér. math., 5 (1951), p. 211-216.

Reçu par la Rédaction le 5.8.1955

Czechoslovak Mathematical Journal v. 8 (83), 1958, 319 – 327.

ON THE DIMENSION OF COMPACT SPACES

PETR VOPĚNKA

(Received 30/I 1958)

In the present paper, for arbitrary m, n, $1 \leqq m \leqq n \leqq \infty$ the existence of compact spaces X, Y such that $\dim X = \dim Y = m$, $\operatorname{ind} X = \operatorname{Ind} Y = n$ is proved.

In 1935, P. S. Alexandroff [1][1] posed a question whether for all compact spaces the equality $\dim X = \operatorname{ind} X$ holds. In 1949, A. Lunc [4] and later O. Lokucievskij [3] constructed a compact space X with $\dim X = 1$, $\operatorname{ind} X = 2$. In the present paper, it is proved that for m, n satisfying the inequalities $0 < m \leqq n \leqq \infty$, there exist compact X, Y such that $\dim X = \dim Y = m$, $\operatorname{ind} X = n$, $\operatorname{Ind} Y = n$.

The existence of these spaces was proved in the first version of the paper[2] with the aid of a construction given for a certain particular case. The generalization of the construction (see 2.1, 2.5) as well as adopting of a notion of "nearly open" mapping were pointed out by M. Katětov.

The paper is divided into two subsections: § 1 contains several (partly known) auxiliary theorems, § 2 is devoted to the actual realization of a construction.

1

Throughout the whole paper, a space is understood to be a topological space (with the axiom $\overline{\overline{A}} = \overline{A}$), in which the Hausdorff separation axiom is satisfied. The boundary of a set A in the space X, i.e. the set $\overline{A} \cap \overline{X \setminus A}$, will be denoted by a symbol $\operatorname{Fr} A$. If X is a space, then symbols $\dim X$, $\operatorname{ind} X$ and $\operatorname{Ind} X$ denote its dimension, defined by means of covers ("combinatorial" dimension), small inductive dimension and large inductive dimension, respectively.[3]

It is well known that $\operatorname{ind} X \leqq \operatorname{Ind} X$ for an arbitrary space X; $\dim X \leqq \operatorname{Ind} X$ in the case of a normal space X; $\dim X \leqq \operatorname{ind} X$ in the case of a compact X; if (for an arbitrary space X) $\dim X = 0$, then also $\operatorname{Ind} X = \operatorname{ind} X = 0$. If $Y \subset X$, then $\operatorname{ind} Y \leqq \operatorname{ind} X$; if Y is closed in X, then $\dim Y \leqq \dim X$, $\operatorname{Ind} Y \leqq \operatorname{Ind} X$.

[1] Numbers in square brackets denote a reference to a list of literature at the end of the paper.

[2] The first version of the manuscript has been received 15/III 1957.

[3] Let us briefly remind their definitions: $\dim X \leqq n$ means that every finite open cover of a space X has a finite open refinement $\{G_i\}$ of order $\leqq n+1$, i.e. such that any $n+2$ distinct G_j's have an empty intersection; $\operatorname{ind} X$ and $\operatorname{Ind} X$ are defined inductively as follows: $\operatorname{ind} \emptyset = \operatorname{Ind} \emptyset = -1$; $\operatorname{ind} X \leqq n$ means that for every point $x \in X$ and its neighborhood U there is a neighborhood $V \subset U$ such that $\operatorname{ind} \operatorname{Fr} V \leqq n-1$; $\operatorname{Ind} X \leqq n$ means that for an arbitrary closed set $F \subset X$ and its neighborhood U there exists a neighborhood $V \subset U$ of the set F with $\operatorname{Ind} \operatorname{Fr} V \leqq n-1$.

PETR VOPĚNKA

1.1. *Let spaces* X, Y *be compact,* $\dim X = 0$. *Then* $\dim(X \times Y) = \dim Y$, $\operatorname{Ind}(X \times Y) = \operatorname{Ind} Y$.

Proof. Clearly, it suffices to show assuming $\dim Y = m < \infty$, $\operatorname{Ind} Y = n < \infty$ that $\dim(X \times Y) \leqq m$, $\operatorname{Ind}(X \times Y) \leqq n$.

We shall show the first inequality. Let $\{G_i\}$ be a finite open cover of the space $X \times Y$. The compactness of the space Y easily implies that for every point $x \in X$ there exists a neighborhood U_x such that for suitable open $V_{x,i} \subset Y$ one has $U_x \times V_{x,i} \subset G_i$, $\bigcup V_{x,i} = Y$; moreover, according to the inequality $\dim Y \leqq m$, one can assume that the order of $\{V_{x,i}\}$ is $\leqq m+1$. Since X is compact, there are x_j, $j = 1, \ldots, p$ such that $\bigcup_1^p U_{x_j} = X$. By the equality $\dim X = 0$ it follows that there are open $U_j^* \subset U_{x_j}$ such that $\bigcup_1^p U_j^* = X$ and the system $\{U_j^*\}$ is disjoint. In such a case the collection of all $\{U_j^* \times V_{x_j,i}\}$ obviously refines $\{G_i\}$ and its order is $\leqq m+1$.

Now we shall prove the second inequality by induction on n, showing only the step from n to $n+1$, because the proof for $n = 0$ is completely analogous. An easy consequence of compactness of Y is that whenever a closed F and an open $U \supset F$ are given, then for an arbitrary point $x \in X$ one can find its neighborhood U_x such that for suitable open sets $V_x \subset Y$, $W_x \subset Y$ the following will hold:

$$F \cap (U_x \times Y) \subset U_x \times V_x\,, \quad \overline{V_x} \subset W_x\,, \quad U_x \times W_x \subset U\,.$$

Since X is compact, there are x_j, $j = 1, \ldots, p$, such that $\bigcup_1^p U_{x_j} = X$; from the equality $\dim X = 0$ one has that there are clopen sets $U_j^* \subset U_{x_j}$ such that $\bigcup U_j^* = X$. Select now for every j a set V_j^* open in Y in such a way that

$$\overline{V_{x_j}} \subset V_j^*\,, \quad \overline{V_j^*} \subset W_{x_j}\,, \quad \operatorname{Ind} \operatorname{Fr} V_j^* \leqq n-1$$

(this is possible, because Y is normal, $\operatorname{Ind} Y = n$), and set $V = \bigcup(U_j^* \times V_j^*)$. Obviously, $F \subset V \subset U$. It is easy to check that

$$\operatorname{Fr} V = \bigcup_1^p (U_j^* \times \operatorname{Fr} V_j^*)\,.$$

By the induction assumption we get $\operatorname{Ind} \operatorname{Fr} V \leqq n-1$.

1.2. *Let* X *be a normal space, let a set* $F \subset X$ *be closed and* $\dim F \leqq n$. *Suppose* $\dim(X \setminus U) \leqq m$ *for every open* $U \supset F$. *Then* $\dim X \leqq \max(m, n)$.

Remark. This result is due to C. H. Dowker [2].

Proof. Let $\{G_i\}$ be a finite open cover of the space X. By $\dim F \leqq n$ and by the fact that F is normal, one gets immediately that there are closed $A_i \subset G_i \cap F$ such that $\bigcup A_i = F$, and the order of the system $\{A_i\}$ does not exceed $n+1$. Now, according to known theorems, we can claim that there exist sets $U_i \supset A_i$, open in X and such that order of $\{U_i\}$ is not more than $n+1$. Set $U = \bigcup(U_i \cap G_i)$. As $F \subset U$, there exist open sets V, W such that $F \subset V$, $\overline{V} \subset W$, $\overline{W} \subset U$.

Let us put $G_i' = (G_i \setminus \overline{W}) \cup (U_i \cap G_i)$; obviously, $\bigcup G_i' = X$. We have $\dim(X \setminus V) \leqq m$, consequently, there are open sets in $X \setminus V$, say B_i, such that

$$B_i \subset (X \setminus V) \cap G_i'\,, \quad \bigcup B_i = X \setminus V\,,$$

the order of $\{B_i\}$ does not exceed $m+1$. Set

$$H_i = (W \cap U_i \cap G_i) \cup B_i .$$

Then H_i's are open,

$$H_i \subset G_i' \subset G_i , \bigcup H_i = X .$$

If $x \in \overline{W}$, then x can belong to at most $n+1$ of the sets $U_i \cap G_i$, consequently, to at most $n+1$ sets G_i' and henceforth it cannot be a member of more than $n+1$ sets H_i; if $x \in X \setminus \overline{W}$, then x cannot belong to more than $m+1$ of the sets H_i, because in the opposite case it would be contained in more than $m+1$ many sets B_i, which is impossible. Thus, order of $\{H_i\}$ is at most $\max(m,n)+1$.

1.3. *Let P be a compact space, $S \subset P$ a closed set, and suppose that for every neighborhood U of the set S there exists a clopen neighborhood $V \subset U$. Let* $\operatorname{Ind} S = m \geqq 0$ *and suppose that for every closed $Q \subset P$, disjoint with S, one has* $\operatorname{Ind} Q \leqq n$. *Further, suppose that there exists a continuous mapping φ of a space P onto S such that $\varphi(x) = x$ for $x \in S$. Then* $\operatorname{Ind} P \leqq m+n$.

Proof. We shall prove the theorem by induction on m, showing only the step from $m-1$ to m, because the proof of the theorem for $m=0$ is quite analogous. Let $F \subset U \subset P$, where F is closed and U an open set. There is a set $G \subset S$ open in S such that

$$S \cap F \subset G , \overline{G} \subset S \cap U , \operatorname{Ind} \operatorname{Fr} G \leqq m-1 .$$

Set $G^* = \varphi^{-1}(G)$; then $\operatorname{Fr} G^* \subset \varphi^{-1}(\operatorname{Fr} G)$. Since the subspace $\varphi^{-1}(\operatorname{Fr} G)$ and its subset $\operatorname{Fr} G$ satisfy the assumptions of the theorem (with $m-1$ instead of m), which is easy to verify, the inductive assumption implies that $\operatorname{Ind} \varphi^{-1}(\operatorname{Fr} G) \leqq m-1+n$. From the relations

$$S \cap \overline{G^*} \subset S \cap \varphi^{-1}(\overline{G}) \subset U , S \cap F \subset G^* ,$$

we easily conclude that there exists a clopen subset H of P such that

$$S \subset H , H \cap \overline{G^*} \subset U , H \cap F \subset G^* .$$

We have $\operatorname{Ind}(P \setminus H) \leqq n$; consequently, there is a set G_1, open in $P \setminus H$ (hence, also in P), such that

$$F \setminus H \subset G_1 \subset U \setminus H , \operatorname{Ind} \operatorname{Fr} G_1 \leqq n-1 .$$

Now, let us set $V = (G^* \cap H) \cup G_1$. Then $F \subset V$, $V \subset U$; the set $\operatorname{Fr} V$ is a disjoint union of $\operatorname{Fr}(G^* \cap H) \subset \operatorname{Fr} G^*$ and $\operatorname{Fr} G_1$, so we immediately obtain the inequality $\operatorname{Ind} \operatorname{Fr} V \leqq m-1+n$. From this it follows that $\operatorname{Ind} P \leqq m+n$.

1.4. *Let R be a space, let $R_2 \subset R_1 \subset R$ with R_2 nonempty connected. Let $G \subset R$ be an open set satisfying $G \cap R_2 \neq \emptyset$, $R_2 \setminus \overline{G} \neq \emptyset$. Then either* (a) *there is a set U open in R_1 such that $U \cap R_2 \neq \emptyset$, $U \subset \operatorname{Fr} G$, or* (b) $\overline{G \cap R_1} \cap \overline{R_1 \setminus \overline{G}} \cap R_2 \neq \emptyset$.

PETR VOPĚNKA

Proof. Let $H = R \setminus G$. Let $A = \overline{G \cap R_1}$, $B = \overline{H \cap R_1}$, $U = R_1 \setminus A \setminus B$. If $R_2 \setminus A \setminus B \neq \emptyset$, then clearly the case (a) takes place, since

$$U \subset R_1 \setminus G \setminus H \subset R_1 \cap \operatorname{Fr} G .$$

If $R_2 \subset A \cup B$, then the relations

$$R_2 \cap A \supset R_2 \cap G \neq \emptyset , \; R_2 \cap B \supset R_2 \cap H \neq \emptyset$$

together with the connectedness of R_2 imply that $R_2 \cap A \cap B \neq \emptyset$, thus the case (b) holds.

1.5. Definition. A mapping f from a space P into a space Q will be called nearly open,[4] if for an arbitrary $y \in f(P)$ there is an $x \in P$ such that $f(x) = y$ and for an arbitrary neighborhood U of a point x the set $f(U)$ is a neighborhood of y in the space $f(P)$.

1.6. *Let X be a compact space. Then there is a compact subspace Y such that $\dim Y = 0$ and a nearly open continuous mapping f of the space Y onto X.*

Proof. It is clear that there is a system $\{\mathfrak{G}_\mu\}$, where the index μ runs through some set M, such that: (1) $\mathfrak{G}_\mu = \{G_{\mu,i}\}$, where $i = 1, \ldots, n(\mu)$, are finite open covers of the space X, (2) the collection of all $G_{\mu,i}$ is an open base of the space X, (3) if $x_1 \in X$, $x_2 \in X$, $x_1 \neq x_2$, then there exists a μ such that for no $i = 1, \ldots, n(\mu)$ can be $x_1 \in \overline{G_{\mu,i}}$ and $x_2 \in \overline{G_{\mu,i}}$ simultaneously. For an arbitrary natural p let us denote by K_p the space consisting of p points $1, \ldots, p$. Set $R = \mathfrak{P}_\mu K_{n(\mu)}$. It follows by the property (3) of the system $\{\mathfrak{G}_\mu\}$ that for an arbitrary $y = \{y_\mu\} \in R$ the set $\Phi(y) = \bigcap_\mu \overline{G_{\mu,y\mu)}}$ is either one-point or void. The compactness of X implies that in the case of $\Phi(y) = \emptyset$ there is a finite set $M' \subset M$ with $\bigcap_{\mu \in M'} \overline{G_{\mu,y(\mu)}} = \emptyset$; but from this it follows immediately that the set Y of all points $y \in R$ with the property $\Phi(y) \neq \emptyset$ is closed in R; in particular, the space Y is compact. Now, put $\Phi(y) = (f(y))$ for $y \in Y$. Obviously, f is a mapping of Y onto X. If $y = \{y_\mu\} \in Y$ and if V is a neighborhood of the point $f(y)$ in X, then $\bigcap_\mu \overline{G_{\mu,y(\mu)}} = (f(y)) \subset V$; next, by compactness of X there exists a finite set $M' \subset M$ with $\bigcap_{\mu \in M'} \overline{G_{\mu,y(\mu)}} \subset V$. We shall denote by U the set of all $z \in Y$ such that $z(\mu) = y(\mu)$ for $\mu \in M'$; U is a neighborhood of y in Y and $f(U) \subset V$ then. The mapping f is, as a consequence, continuous.

We shall show now that f is a nearly open mapping. Let $x \in X$. For every μ choose $y(\mu)$ in such a manner that $x \in G_{\mu,y(\mu)}$; set $y = \{y_\mu\}$, thus $y \in Y$, $f(y) = x$. Let U be a neighborhood of y in Y. Then there is a finite set $M' \subset M$ such that $U' \subset U$, where U' stands for the set of those $z \in Y$, which satisfy $z(\mu) = y(\mu)$ for $\mu \in M'$. Finally, let us set

$$V' = \bigcap_{\mu \in M'} G_{\mu,y(\mu)} .$$

Then V' is a neighborhood of the point x in X and it is easy to check that $V' \subset f(U')$, thus $f(U)$ is a neighborhood of the point x.

[4] Let us remark that the term "nearly open mapping" has already appeared in the literature, but in an essentially different sense; see e.g. V. Pták, On complete topological linear spaces, Czech. Math. J. 1953, 3 (78), 301 – 364.

ON THE DIMENSION OF COMPACT SPACES

2

2.1. Let X, Y, Z be spaces, f a continuous mapping of the space Y onto X, N an infinite discrete space. Let us set

$$T = X \cup (Y \times N \times Z);$$

and let an open base of the space T be the system of all open subsets of the topological product $Y \times N \times Z$ and of all sets of the form

$$U \cup f^{-1}(U) \times (N \setminus K) \times Z),$$

where $U \subset X$ is open in X and $K \subset N$ is a finite set. It is easy to verify that T with the topology defined that way is actually a Hausdorff topological space. We shall denote it by the symbol $T(X, Y, Z, f, N)$.

2.2. Let $T = T(X, Y, Z, f, N)$. If we put for $t \in T$, $\varphi(t) = t$ in the case $t \in T$, $\varphi(t) = f(y)$ in the case $t = (y, \nu, z) \in Y \times N \times Z$, then, as may be easily observed, φ is a continuous mapping of T *onto* X.

2.3. *If X, Y, Z are compact spaces, then $T = T(X, Y, Z, f, N)$ is compact, too, and for an arbitrary neighborhood U of the set $X \subset T$ there exists a clopen neighborhood $V \subset U$. If moreover* $\dim Y = 0$, *then* $\dim = \max(\dim X, \dim Z)$, $\operatorname{Ind} T \leqq \operatorname{Ind} X + \operatorname{Ind} Z$.

Proof. I. Let $\mathfrak{G}$ be an open cover of P. We shall show that it is possible to find a finite subcover of $\mathfrak{G}$; it is enough to prove this statement in the case when $\mathfrak{G}$ consists of members of the open base described in 2.1. But then it is possible to choose from $\mathfrak{G}$ a finite number of sets

$$V_i = U_i \cup (f^{-1}(U_i) \times (N \setminus K_i) \times Z),$$

where $U_i \subset X$ are open in X and $K_i \subset N$ are finite sets, in such a manner that $\bigcup_1^n U_i = X$. Then, obviously,

$$T \setminus \bigcup_1^n V_i \subset Y \times \bigcup_1^n K_i \times Z;$$

consequently, $T \setminus \bigcup_1^n V_i$ is compact. But from this it follows that one can select a finite subcover from $\mathfrak{G}$. Thus T is a compact space. It is also easy to verify that the sets $X \cup (Y \times (N \setminus K) \times Z)$ with finite $K \subset N$ form a neighborhood base of X in T; these sets are obviously clopen. The relation

$$\dim T \leqq \max(\dim X, \dim Z)$$

follows now immediately by 1.1 and 1.2, and the inequality

$$\operatorname{Ind} T \leqq \operatorname{Ind} X + \operatorname{Ind} Z$$

is a consequence of 2.2, 1.1 and 1.3.

PETR VOPĚNKA

2.4. *Let X, Y, Z be compact spaces,* $\dim Y = 0$, *let f be a nearly open continuous mapping from Y onto X and let N be a discrete space of cardinality greater than the cardinality of X. Finally, let $T = T(X, Y, Z, f, N)$. Then the following is true: If U, V are open in T and $\overline{U \cap X} \cap \overline{V \cap X} \neq \emptyset$, then $\overline{U} \cap \overline{V}$ contains a part homeomorphic to the space Z.*

Proof. We set $U_1 = U \cap X$, $V_1 = V \cap X$. Let $x_0 \in \overline{U_1} \cap \overline{V_1}$. There is a $y_0 \in Y$ such that $f(y_0) = x_0$ and for every neighborhood G of the point y_0 in Y the set $f(G)$ is a neighborhood of the point x_0 in X; then clearly $y_0 \in \overline{f^{-1}(U_1)}$, $y_0 \in \overline{f^{-1}(V_1)}$. Since the sets U, V are open, for every $x \in U_1$ there is a finite set $K_x \subset N$ such that

$$f^{-1}(x) \times (N \setminus K_x) \times Z \subset U,$$

and for every $x \in V_1$ there is a finite set $K'_x \subset N$ such that

$$f^{-1}(x) \times (N \setminus K'_x) \times Z) \subset V.$$

Let us set

$$M = N \setminus \bigcup_{x \in U_1} K_x \setminus \bigcup_{x \in V_1} K'_x;$$

obviously, $M \neq \emptyset$. One has

$$f^{-1}(U_1) \times M \times Z \subset U, \quad f^{-1} \times M \times Z \subset V;$$

from which it follows that

$$(y_0) \times M \times Z \subset \overline{U}, \quad (y_0) \times M \times Z \subset \overline{V}.$$

2.5. *If the space X is compact,* $\dim X > 0$, *then there exists a compact space $T \supset X$ having the following properties:* (1) $\dim T = \dim X$, $\operatorname{Ind} T \leqq 2 \operatorname{Ind} X + 1$; (2) *there is a point $t \in T$ and its neighborhood U in T such that for an arbitrary neighborhood $G \subset U$ of the point $t \in T$ the set* $\operatorname{Fr} G$ *contains a part which is homeomorphic to X.*

Proof. Let $I =< 0, 1 >$. Let g be a continuous mapping of a Cantor set C onto I; let M be a discrete space of cardinality $\aleph_0$. Put $T' = T(I, C, X, g, M)$. According to 2.3 and 1.6 there exists a compact space Y and a mapping f of Y onto T' such that $\dim Y = 0$ and f is a nearly open continuous mapping. Let N be a discrete space whose cardinality is greater than the cardinality of the space X and greater than $2^{\aleph_0}$. Let us set $T = T(T', Y, X, f, N)$. We shall prove that T has the required properties.

2.3 implies that

$$\dim T = \dim T' = \dim X, \quad \operatorname{Ind} T \leqq \operatorname{Ind} T' + \operatorname{Ind} X,$$
$$\operatorname{Ind} T' \leqq \operatorname{Ind} I + \operatorname{Ind} X, \quad \operatorname{Ind} T \leqq 2 \operatorname{Ind} X + 1.$$

Let us denote by t the point $0 \in I \subset T' \subset T$; choose its neighborhood U in T so that the point $1 \in I$ does not belong to $\overline{U}$. Let $G \subset U$ be a neighborhood of t in T. Then by 1.4 (setting $R = T$, $R_1 = T'$, $R_2 = I$) we obtain either (a) there is an open set V in T' such that $V \cap I \neq \emptyset$, $V \subset \operatorname{Fr} G$, or (2) $\overline{G \cap T'} \cap \overline{T' \setminus \overline{G}} \neq \emptyset$. In the case (a), one can see immediately from the construction of T' that V contains a part homeomorphic to X. In the case (b), 2.4 implies that the intersection $\overline{G} \cap \overline{T \setminus \overline{G}} \subset \operatorname{Fr} G$ contains a part, homeomorphic to X.

2.6. *Let* $1 \leqq n < \infty$. *Then there exists a compact space* X *such that* $\dim X = 1$, $\operatorname{ind} X = n$, $\operatorname{Ind} X < \infty$ *and a compact space* Y *such that* $\dim Y = 1$, $\operatorname{Ind} Y = n$.

Proof. We shall denote by A (by B, resp.) the set of all natural numbers n for which there is X (Y, resp) with the properties as stated. According to 2.5, the following is true: If $n \in A$ (or $n \in B$), then there is a $k > n$ such that $k \in A$ (or $k \in B$). From the inductive definition of dimension (and by the fact that $\dim X = 0$ cannot hold provided $\operatorname{ind} X > 0$) one has a straightforward conclusion that once $n > 1$ belongs to the set A (to the set B, resp.), then $n - 1$ belongs to A (B, resp.) as well. This implies our proposition.

2.7. Theorem. *Let* $1 \leqq m \leqq n \leqq \infty$. *Then there exists a compact space* X *such that* $\dim X = m$, $\operatorname{ind} X = n$ *and a compact space* Y *such that* $\dim Y = m$, $\operatorname{Ind} Y = n$.

Proof. The theorem has been already proved for $1 = m \leqq n < \infty$ (2.6). Let $m = 1$, $n = \infty$. Let X_k, $k = 1, 2, \dots$, be compact spaces, $\dim X_k = 1$, $\operatorname{ind} X_k = k$. We can of course assume that X_k are pairwise disjoint; choose an element ξ not belonging to any X_k and set $X = (\xi) \cup \bigcup_1^\infty X_k$, finally, for an open base in X we take a system of all sets, which are open in some X_k, and of all sets of the form $(\xi) \cup \bigcup_{k=p}^\infty X_k$. It is easy to check that X is compact. Clearly, $\operatorname{Ind} X = \operatorname{ind} X = \infty$; by the known theorem on the dimension of a union of countable number of closed sets we obtain $\dim X = 1$.

If $m > 1$, then it suffices to choose compact spaces X', Y' such that $\dim X' = 1$, $\operatorname{ind} X' = n$, $\operatorname{Ind} Y' = n$, and let X or Y be the disjoint union of X' or Y' with an m-dimensional cube I^m.

References

[1] P. S. Alexandroff, *Nekotorye problemy teoretiko-množestvennoj topologii*, Mat. Sb. **1 (43)** (1936), 619 – 634.

[2] C. H. Dowker, *Local dimension of normal spaces*, Quart. J. Math. Oxford (2) **6** (1955), 101 – 120.

[3] O. Lokucievskij, *On the dimension of bicompacta*, Dokl. Akad. Nauk SSSR **67, 2** (1949), 217 – 219.

[4] A. Lunc, *A bicompactum, the inductive dimension of which is greater that the dimension, defined with aid of coverings*, Dokl. Akad. Nauk SSSR **66, 5** (1949), 801 – 803.

Доклады Академии наук СССР
1970, Том 192, № 3, 516 – 519

MATHEMATICS

V. V. FILIPPOV

BICOMPACTA WITH DISTINCT DIMENSIONS ind AND dim

(Presented by academician P. S. Alexandrov 20. X. 1969)

There are numerous examples of compact spaces with noncoinciding dimensions ind and dim recently (see [1-4]). The present note shows a number of examples of that kind. They are in certain respects better than the previous ones.

§ 1*. A compact space P with $\dim P = 1$, $\operatorname{ind} P = \operatorname{Ind} P = 2$. The ordinary square $I^2 = I \times I$, where $I = [0,1]$, is of Euclidean structure (in which the factors are orthogonal and of length 1); which enables us to measure angles. By the metric we shall understand the Euclidean metric.

Let us equip the product $P = I^2 \times S$, where S is an ordinary circle, by a somewhat nonstandard topology. Let $I_0 \subset I$ be the set of all rational points of the segment, $I_0^2 \subset I^2$ the corresponding set of points of the square, both coordinates of which are rational. We identify the points of the circle S with the directions of the rays emanating from a point, taking as the zero direction the direction of the axis of abscissas. Let V be a set which is open in the circle $\{x\} \times S$. We use the symbol V^* to denote: 1) the union of all rays lying in I^2 (without the point x) radiating from the point x at angles, which correspond to the points of the set V, provided $x \in I_0^2$, 2) the union of all circles lying in I^2 (or, more precisely, of their intersections with I^2) with centers at the point x and radii r such that $1/r$ is the measure of an angle lying in V, if $x \in I^2 \setminus I_0^2$.

Let $\pi : P \to I_2$ be the projection onto a factor, U the neighborhood of a point x in I^2. For an arbitrary point of the set $V \subset \{x\} \times S$, we declare the set $V \cup \pi^{-1}(V^* \cap U)$ to be its neighborhood. It is easy te see that the topology is specified correctly and that the resulting space P is compact.

The estimate $\dim P \geqslant 1$ is obvious. The estimate $\dim P \leqslant 1$ can be obtained as follows. Consider an arbitrary cover ω of the space P. For $x \in I^2 \setminus I_0^2$, contract to points those sets of the form $\{x\} \times S$, which are as a whole covered by at least one member of the cover (it is easy to see that all sets, with the possible exception of finitely many, are such). This mapping is the ω-mapping of the compact space P onto a metric compact space, whose one-dimensionality is clear. Therefore, $\dim P \leqslant 1$.

Let $V \subset P$ be an arbitrary open set, the complement of which has a nonempty interior. Obviously, the boundary of the set $\operatorname{Int} \pi(V)$ contains some connected nonsingleton set F. Since such a set cannot be countable, there is a point $x \in F \setminus I_0^2$. Then, the connectedness of the set F, the definition of a topology of P and the

*It is useful to compare the construction of §§ 1–4 with V. V. Fedočuk's example.

closedness of the boundary of the set V imply that the boundary of the set V contains the whole one-dimensional set $\{x\} \times S$. The inequality $\operatorname{Ind} P \geqslant \operatorname{ind} P \geqslant 2$ follows. The estimate $\operatorname{Ind} P \leqslant 2$ can be obtained from the self-evident geometric considerations.

§ **2.** A compact space P_i with $\dim P_i = 1$, $\operatorname{ind} P_i = \operatorname{Ind} P_i = i$. We got the compact space P out of a square by replacing points by circles and "distributing" the neighborhoods "over sectors" — case 1) or "over rings" — case 2). But in the latter case, it is possible to insert the compact space just constructed instead of circles (we shall show below how). As before, an arbitrary boundary contains the whole copy of an inserted space, which raises the inductive dimensions by at least one; dim, as before, remains equal to 1. The exact upper estimates of the inductive dimensions, however, are very difficult to obtain. Therefore we modify a bit the geometry of our example.

As our P_1 we take a square, where all points have been replaced by circles with the convergence over sectors. Suppose that in this way we have already constructed the compact space P_{i-1} with $\dim P_{i-1} = 1$, $\operatorname{ind} P_{i-1} = \operatorname{Ind} P_{i-1} = i-1$. Let π_{i-1} be the natural projection onto a square. Let $f : (0,1] \to I^2$ be at most two-to-one mapping of the half-interval $(0,1]$ into the square I^2, which is a linear parametrization of some piecewise linear curve, and such that $\overline{f((0,r))} = I^2$, if $0 < r \leqslant 1$.

We replace the point $x = (r_1, r_2)$ of the square I^2 by a circle with convergence over sectors, provided $r_1 \neq 0$ and by the compact space P_{i-1}, if $r_1 = 0$. In the latter case, we define the neighborhoods as follows. Let V be an open subset of the copy of P_{i-1}, $V^* = \operatorname{Int} V$ and let V^{**} be the union of all circles with the center at the point x and radii from $f^{-1}(V^*)$, let U be a neighborhood of the point x in I^2, and let $\pi_i : P_i \to I^2$ be the projection of the set P_i onto a square, obtained as a result of the insertions, which collapses the inserted spaces into points. We define a neighborhood of any point from V as a set $V \cup \pi_i^{-1}(V^{**} \cap U)$.

The estimates $\dim P_i = 1$, $\operatorname{Ind} P_i \geqslant \operatorname{ind} P_i \geqslant i$ can be obtained almost in the same manner as for the compact space P. We show that $\operatorname{ind} P_i \leqslant i$. It is clearly possible to find neighborhoods with zero-dimensional boundaries for points contained in the sets $\pi_i^{-1}((r_1, r_2))$, where $r_1 \neq 0$. Let $\pi(y) = (0, r_2) = x$, let V_1 be a neighborhood of the point y in the subspace $\pi_i^{-1}(x)$ with an $(i-2)$-dimensional boundary, and let $V = V_1 \cup \pi_i^{-1}(V_1^{**} \cap U)$ be a neighborhood of a point y in the space P_i, constructed with the aid of V_i. Let y' be a boundary point of the set V_1. We can find an arbitrarily small neighborhood V_2 of the point y' in the subspace $\pi_i^{-1}(x)$ which is a union of countably many open sets G_n, $n = 1, 2, \ldots$, $\overline{G_n} \subset G_{n+1}$. Let V_2^{**} and $\overline{G_n}^{**}$ be sets constructed starting with V_2 and $\overline{G_n}$, respectively, as indicated before. These sets consist of concentric rings. Let L be a ring, which is a component of the open set V_2^{**} and the inner radius of which is larger that $1/n$. Let L^* be a ring, $\overline{G_n}^{**} \cap L \subset L^* \subset L$, whose boundary does not meet the boundary of the set V_1^{**} — it is easy to see that such a ring exists. Let Λ be the union of all such rings L^* over all components L of the set V_2^{**}. It is easy to see that the set $V_2 \cup \Lambda$ is open in P_{i-1}, and the intersection of its boundary with the boundary of the set V is contained in the boundary of the set $V_1 \subset \pi_i^{-1}(x)$, and, consequently, its dimension is $\leqslant i-2$. For the remaining boundary points of the set V, arbitrarily small neigborhoods with boundaries of dimension $\leqslant i-2$ can be found in self-evident fashion. Thus, the ind of the boundary of the set V equals to $i-1$; consequently, $\operatorname{ind} P_i \leqslant i$. Proof of the inequality $\operatorname{Ind} P_i \leqslant i$ is analogous.

§ **3.** A perfectly normal compact space Q with $\dim Q = 1$, $\operatorname{ind} Q = \operatorname{Ind} Q = 2$. We shall construct this compact space under the assumption $2^{\aleph_0} = \aleph_1$.

We show that if $2^{\aleph_0} = \aleph_1$, then there is a subset L of the square I^2 having the following properties: a) the intersection of the set L with an arbitrary zero-dimensional compact set is at most countable, b) every connected compact set contains at least one point of the set L, c) $L \cap L_0^2 = \emptyset$.

To this end we enumerate by transfinite ordinals less than ω_1 all zero-dimensional compact sets contained in I^2; and also enumerate by transfinite ordinals less than ω_1 all connected compact sets contained in I^2. It is obvious that every zero-dimensional compact set is nowhere dense in an arbitrary connected compact set containing it, thus, if we omit from a connected compact set with a number $\alpha < \omega_1$ all points lying in all zero-dimensional compact sets with smaller numbers and points from the set I_0^2, the remaining set will be non-void. Let us choose an arbitrary point in it and denote it by x_α. It is clear that the set $L = \{x_\alpha\}_{\alpha<\omega_1}$ satisfies the required conditions.

Clearly, as a result of decomposition of the space P from § 1, nonsingleton elements of which are the sets $\{x\} \times S$, where $x \in I^2 \setminus (I_0^2 \cup L)$, one gets a (Hausdorff) compact space Q. Let $\pi' : Q \to I^2$ be the mapping, generated by the projection π.

We shall show that the compact space Q is perfectly normal. To do this it is enough to prove that any family W consisting of sets open in Q has at most countable refinement with the same union. Let $\{u_1, u_2 \ldots\}$ be the set of those elements of some countable base of the square I^2 whose complete preimages under π' are contained in elements of the family W. It is easy to see that the sets $\{\pi'^{-1}(u_1), \pi'^{-1}(u_2), \ldots\}$ do not cover only the set M consisting of those points, covered by the family W, which are mapped by the mapping π' into the boundary of the set $\bigcup_{i=1}^{\infty} u_i$. We shall show that the set $M_0 = \pi'(M)$ is at most countable. For every point $m \in M_0 \setminus I_0^2 \subset L$ we choose a point $x(m) \in \pi'^{-1}(m)$ covered by the family W. Some member of the family W contains a neighborhood of a point $x(m)$ of the form $V \cup (\pi'^{-1}(V^* \cap U)$ (see the construction of the space P), where U is an ε-neighborhood of the point m. Let H_n be the set of those points $m \in M_0$, for which $\varepsilon > 1/n$. It is not difficult to verify that the set $\overline{H_n}$ is zero-dimensional, so that $H_n \subset L$ is at most countable; this in turn implies that the set $M_0 = \bigcup_{n=1}^{\infty} H_n \cup (M_0 \cap I_0^2)$ is countable, too. It is not difficult to find for every point $m \in M_0$ a countable family $V(m)$ refining W, consisting of open sets in Q and such that it covers in $\pi'^{-1}(m)$ everything, what is covered by the family W. Then the family $\{\pi'^{-1}(u_1), \pi'^{-1}(u_2), \ldots\} \cup (\bigcup_{m\in M_0} V(m))$ is clearly countable, it refines the family W and its union is the same as that of W. We have thus shown that the compact space Q is perfectly normal.

§ **4.** A perfectly normal compact space Q_i with $\dim Q_i = 1$, $\operatorname{ind} Q_i = \operatorname{Ind} Q_i = i$. These compact spaces are constructed under the assumption that

$2^{\aleph_0} = \aleph_1$.

We have already constructed the compact space $Q_2 = Q$. The inductive argument is the same as in § 2. Let us suppose that we have already constructed the compact space Q_{i-1} and the mapping $\pi'_{i-1} : Q_{i-1} \to I^2$ of the compact space onto a square. Let $f' : (0,1] \to I^2$ be at most two-to-one mapping of the half-interval $(0,1]$ into the square I^2, which is a linear parametrization of some piecewise linear curve and such that $1/2n$-neighborhood of the set $f'((1/(n+1), 1/n))$ covers I^2. The distinction between this and the corresponding passage in § 2 is necessary for the perfect normality of the compact space we are about to construct.

We do not replace a point $x \in I^2$ by anything, if $x \in I^2 \setminus (I_0^2 \cup L)$; we replace it by a circle with convergence over sectors provided $x \in I_0^2$; and we replace it by the compact space Q_{i-1} if $x \in L$. In the latter case, we shall define neighborhoods as follows. Let V be an open set in this copy of a compact space Q_{i-1}, let $V^* = \operatorname{Int} \pi'_{i-1}(V)$, let V^{**} be a union of all circles with a center in a point x and with radii from ${f'}^{-1}(V^*)$, let U be a neighborhood of a point $x \in I^2$ and finally, let $\pi'_i : Q_i \to I^2$ be the projection of the set Q_i onto the square resulting from the insertions, which collapses inserted sets into points. We define as a neighborhood of an arbitrary point from V the set $V \cup {\pi'_i}^{-1}(V^{**} \cap U)$.

The proof of perfect normality of this compact space proceeds along similar lines as in § 3.

Estimates of the dimensions can be obtained similarly as above.

§ **5.** Remarks. It is easy to see that the series of examples constructed in §§ 2 and 4 can be extended to transfinite dimensions.

Taking a direct sum of a suitable number of compact spaces and compactifying this locally compact space by a single point, we can construct compact spaces P_∞, Q_∞ with $\dim P_\infty = 1$, $\operatorname{ind} P_\infty = \infty$, $\dim Q_\infty = 1$, $\operatorname{ind} Q_\infty = \infty$. Let $m \geqslant n$, $P_{m,n} = P_m \cup I^n$, $Q_{m,n} = Q_m \cup I^n$. It is easy to check that $\dim P_{m,n} = n$, $\operatorname{ind} P_{m,n} = m$, $\dim Q_{m,n} = n$, $\operatorname{ind} Q_{m,n} = m$.

§ **6.** The construction of a compact space R^* with $\dim R^* = l$, $\operatorname{ind} R^* = m+1$, $\operatorname{Ind} R^* = n+1$ from a compact space R with $\dim R = l$, $\operatorname{ind} R = m$, $\operatorname{Ind} R = n$.

Let us introduce the following lexicographic order in the set $C_0 = I \times I \times D$, where $D = \{0,1\}$: $(r_1, r_2, r_3) > (r'_1, r'_2, r'_3)$ if $r_1 > r'_1$ or $r_1 = r'_1$, $r_2 > r'_2$ or $r_1 = r'_1$, $r_2 = r'_2$, $r_3 > r'_3$.

Let $\varphi : \theta \to R$ be a continuous mapping of some zero-dimensional compact space θ onto our compact space R. There is at least one compact space θ like that — every compact space is an image of some closed subset of a compact space D^τ.

Let us introduce a topology in the set $C = C_0 \cup (I \times I \times \theta)$ as follows. Every set $\{r_1\}, \times \{r_2\} \times \theta$ will be a clopen subspace with the topology inherited from θ. Convergence to points of C_0 is given by intervals, and we assume that the set $\{r_1\} \times \{r_2\} \times \theta$ lies between the points $(r_1, r_2, 0)$ and $(r_1, r_2, 1)$.

Let z be a point not belonging to R. We equip the set $I^* = I \times (\{z\} \cup R)$ by a topology as follows. The sets $\{r\} \times R$, $r \in I$, will be clopen subspaces with the topology inherited from R. A standard neighborhood of a point $r \in I \times \{z\}$ is of the form $(l \times (\{z\} \cup R)) \setminus (\{r\} \times R)$, where l is an interval in I.

Let us define the following gluings in the Tychonoff product $R_1 = I^* \times C$. We make the identification φ in the set $\{(r, z)\} \times (\{r_1\} \times \{r_2\} \times \theta)$. Let R^* be the compact space resulting from these gluings.

The estimates $\dim R^* = l$, $\operatorname{ind} R^* \leqslant m+1$, $\operatorname{Ind} R^* \leqslant n+1$ are straightforward. Standard reasonings, an example of which can be found e.g. in [4], lead to the conclusion, that an arbitrary partition between the points $(0, z) \times (0, 0, 1)$ and $(1, z) \times (0, 0, 1)$ contains a copy of the space R and, consequently, $\operatorname{ind} R^* \geqslant m+1$, $\operatorname{Ind} R^* \geqslant n+1$.

In order to satisfy the first countability axiom in the compact space R^* it is necessary and sufficient that it is satisfied in the compact spaces I and θ.

§ 7. A compact space S_i with $\dim S_i = 1$, $\operatorname{ind} S_i = \operatorname{Ind} S_i = i$, $S_i = \bigcup_{j=1} S_i^j$, where S_i^j is a compact space one-dimensional in all senses.

Suppose we have already constructed the compact space S_{i-1}. Carrying out the construction from § 6, we obtain the compact space S_i. The representation of this compact space as a union of a finite number of compact spaces which are one-dimensional in all senses is self-evident.

I am deeply grateful to my scientific supervisor A. V. Arhangel'skij.

Mechanics-Mathematics Faculty
Moscow State University

Recieved
15 X 1969

REFERENCES

[1] O. V. Lokucievskij, *On the dimension of bicompacta*, (Russian), Dokl. Akad. Nauk SSSR **67** (1949), 217 - 219.

[2] A. Lunc, *A bicompactum whose inductive dimension is greater than its dimension defined by means of coverings*, (Russian), Dokl. Akad. Nauk SSSR **66** (1949), 801 - 803.

[3] V. Fedorčuk, *Bicompacta with noncoinciding dimensions*, (Russian), Dokl. Akad. Nauk SSSR **182** (1968), 275 - 277.

[4] V. V. Filippov, *A bicompactum satisfying the first axiom of countability with noncoinciding* ind *and* dim *dimensions*, (Russian), Dokl. Akad. Nauk SSSR **186** (1969), 1020 - 1022.

FUNDAMENTA
MATHEMATICAE
XCVII (1977)

A hereditarily normal strongly zero-dimensional space with a subspace of positive dimension and an N-compact space of positive dimension

by

Elżbieta Pol and **Roman Pol** (Warszawa)

Abstract. In this paper we give a solution of an old Čech's problem on dimension by constructing a hereditarily normal strongly zero-dimensional space containing a subspace of positive dimension. We give also an example of an N-compact space of positive dimension.

The aim of this paper is to construct spaces with the properties mentioned in the title.

The problem of existence of a hereditarily normal space X containing a subspace with the covering dimension greater than the covering dimension of X is an old problem of Čech (see [2]; compare also [7] Appendix, [3], [11] Problem 11-14, [1] VII, Introduction). Recently, V. V. Filippov [6] showed that the existence of a Souslin Tree yields a space of this kind. Further examples, with many additional properties, were constructed by V. V. Fedorčuk [5]; he used, however, some additional set theoretic assumptions, too. The example we shall construct needs only the usual axioms for the set theory. It solves at the same time a problem on the local dimension raised by C. H. Dowker in [3].

The problem of existence of a closed subspace with the positive covering dimension in a product of countable discrete spaces appears in the natural way in the theory of N-compactness (see [12]). It was solved recently by S. Mrówka [10] (see also [13]). We give another example of this kind (it seems to us that it is simpler than the Mrówka's one).

1. Notation and terminology. Our terminology will follow [4]. We shall use the following notation: I denotes the closed real unit interval, Q stands for rationals of I, P — for irrationals of I and N — for natural numbers. For an ordinal α we shall denote by $D(\alpha)$ the set of all ordinals less than α with the discrete topology and by $W(\alpha)$ the same set with the order topology. The word "dimension" will denote the covering dimension dim (see [4], § 7.1); a space X with $\dim X = 0$ is called *strongly zero-dimensional.* We say that the *local dimension* of a space X is at most n (abbreviated $\operatorname{locdim} X \leqslant n$) if each point $x \in X$ has an open neighbourhood U with $\dim \overline{U} \leqslant n$ (see [3] and [11] Definition 11-6). All spaces under discussion are assumed completely regular.

2. Auxiliary construction. The construction of the Broom due to Knaster and Kuratowski (see [8] and [4] P. 6.3.23) is a source of the following observation which play the key role in the sequel.

Let X be a topological space, A a subspace of X and let $Q_0 \supset \{0, 1\}$ be a subset of Q such that the set $Q_1 = Q \setminus Q_0$ is dense in Q. Let

$$B(X, A) = (X \times Q_0) \cup (A \times P) \cup [(X \setminus A) \times Q_1]$$

be the subspace of the Cartesian product $X \times I$. For $Y \subset X$ put

$$C(Y) = (Y \times I) \cap B(X, A) = B(Y, A \cap Y).$$

We have the following

LEMMA 1. *If A is not an F_σ-set in X, then for arbitrary $q, q' \in Q_0$ the sets $X \times \{q\}$ and $X \times \{q'\}$ cannot be separated in $B(X, A)$ by the empty set. In particular,* $\dim B(X, A) > 0$.

Proof. Suppose that $B(X, A)$ is the union of two disjoint open-and-closed subsets U and U' such that $U \supset X \times \{q\}$ and $U' \supset X \times \{q'\}$. The set $F = \overline{U} \cap \overline{U}'$, where bar denotes the closure in $X \times I$, separates the sets $X \times \{q\}$ and $X \times \{q'\}$ in $X \times I$ and $F \cap B(X, A) = \emptyset$. For each $s \in Q_1$ the set $F(s) = \{x \in X\colon (x, s) \in F\}$ is closed in X. We shall show that $A = \bigcup_{s \in Q_1} F(s)$, i.e. that A is an F_σ-set in X. Indeed, if $x \in F(s)$ for some $s \in Q_1$ then $(x, s) \in F = F \setminus B(X, A)$, hence $x \in A$; for every $x \in X$ there exists a $t \in I$ such that $(x, t) \in F = F \setminus B(X, A)$ and if $x \in A$, then $t \in Q_1$ so that $x \in F(t)$.

3. A hereditarily normal strongly zero-dimensional space with a subspace of positive dimension. C. H. Dowker [3] showed that the existence of such a space is equivalent to the existence of a hereditarily normal space L with $\operatorname{locdim} L = 0 < \dim L$ (see [11] Remark 11-18); for the construction of L we shall need the following

LEMMA 2. *There exists a perfectly normal and locally second-countable space K with* $\operatorname{locdim} K = 0$ *which contains a locally countable subset A which is not an F_σ-set in K.*

We take the space X defined in Example of [14] as the space K; we recall the construction below. Let $B(\aleph_1) = D(\omega_1)^N$ be the Baire space of weight $\aleph_1$ (see [4] Example 4.2.12). For each $x \in B(\aleph_1)$ let $\varkappa(x) = \min\{\alpha\colon x(i) < \alpha \text{ for } i \in N\}$ and let K be the graph $\{(x, \varkappa(x))\colon x \in B(\aleph_1)\} \subset B(\aleph_1) \times W(\omega_1)$ of the function $\varkappa$. The space K is perfectly normal (see [14] Proposition 1) and, since $K \cap (B(\aleph_1) \times W(\xi)) = K \cap (D(\xi)^N \times W(\xi))$ for every $\xi < \omega_1$, K is locally second-countable and $\operatorname{locdim} K$

$= 0$. Finally, if we choose for each $\xi<\omega_1$ a point $x_\xi \in \varkappa^{-1}(\xi)$ then the set $A = \{(x_\xi, \xi): \xi<\omega_1\}$ has the required property, by [14] Remark 3, Proposition 2′.

EXAMPLE 1. *There exists a perfectly normal locally second-countable space L such that* $\operatorname{locdim} L = 0 < \dim L$.

Let us put $Q_0 = \{0, 1\}$ and let $L = B(K, A)$, where K and A are as in Lemma 2. By Morita's theorem L is perfectly normal (see for example [4] P. 4.5.16) and it is locally second-countable. By Lemma 1 we have $\dim L > 0$. It remains to show that $\operatorname{locdim} L = 0$. Take an arbitrary point $(x, t) \in L$, where $x \in K$, $t \in I$. There exists an open-and-closed neighbourhood U of x such that $\dim U = 0$ and $|U \cap A| \leqslant \aleph_0$. The set

$$C(U) = (U \times Q_0) \cup [(U \cap A) \times (P \cup Q_0)] \cup [(U \setminus A) \times Q_1]$$

is the countable union of its closed strongly zero-dimensional subsets $U \times \{t\}$ for $t \in Q_0$, $(U \setminus A) \times \{t\}$ for $t \in Q_1$ and $\{y\} \times (P \cup Q_0)$ for $y \in U \cap A$. Hence by the Sum Theorem $\dim C(U) = 0$. It follows that the point (x, t) has an open-and-closed strongly zero-dimensional neighbourhood.

We shall use Dowker's construction (see [11] Theorem 11-17) to obtain the following

EXAMPLE 2. *There exists a hereditarily normal strongly zero-dimensional Lindelöf space containing a subspace of positive dimension.*

Let $L^* = L \cup \{p\}$ where L is the space from Example 1 and p is a point which does not belong to L. The topology of L^* consists of all open subsets of L and the sets V such that $p \in V$ and $L \setminus V$ is a second-countable closed subspace of L. It is easy to see that L^* is Lindelöf. By the construction it follows that each separable subset of K is contained in an open-and-closed second-countable and strongly zero-dimensional subspace of K; the same holds in L. Thus the space L^* is hereditarily normal, because for any separated sets A, B either $A \cup B \subset L$ or one of the sets is second-countable. Finally, it is not hard to verify that $\dim L^* = 0 < \dim L$.

Remark 1. The space L we have constructed is collectionwise normal. Indeed, the space K is perfect and collectionwise normal (see [14] Remark 2) and thus the same is true for the product $K \times I$ which contains L (cf. [4], P. 4.5.16, P. 5.5.19 and P. 5.5.1). Let us notice that L cannot be paracompact because in the class of paracompact spaces locdim = dim (see [11], Corollary 11-8).

Remark 2. As proved by C. H. Dowker [3], for the function locdim the Finite Sum Theorem holds in the class of normal spaces. It is easy to show that the space $Z = (N \times L) \cup \{a\}$, where the sets of the form $\{a\} \cup \bigcup_{k \geqslant n} [\{k\} \times L]$ form a base of neighbourhoods at the point a, is the union of countably many closed subsets with $\operatorname{locdim} = 0$, whereas $\operatorname{locdim} Z > 0$. Thus the Countable Sum Theorem fails for locdim in the class of perfectly normal spaces.

4. An N-compact space of positive dimension. A space X is N-compact if it can be embedded as a closed subspace in a product of copies of N (see [16]).

Let S be a set of cardinality $\aleph_1$. For every $T\subset S$ by $p_T\colon N^S\to N^T$ we shall denote the projection; if $|T|\leqslant\aleph_0$ then the set $p_T^{-1}(x)$, where $x\in N^T$, will be called an $\aleph_0$-cube.

The following lemma was proved by the authors in [15] (Example 2).

LEMMA 3. *There exists a subset E of N^S which has the following properties:*

(i) *E is locally an F_σ-set in N^S* (¹),

(ii) *E is not an F_σ-set in N^S,*

(iii) *E is the union of $\aleph_0$-cubes.*

Notice that $N^S\setminus E$ is also the union of $\aleph_0$-cubes, because by (i) it is the union of G_δ-sets and every G_δ-set in N^S is the union of $\aleph_0$-cubes.

EXAMPLE 3 (cf. [10]). *An N-compact space M which is not strongly zero-dimensional.*

Let Q_0 be a dense subset of Q such that $Q_1 = Q\setminus Q_0$ is also dense in Q. Define $M = B(N^S, E)$, where E is as in Lemma 3. By Lemma 1 and (ii) it follows that $\dim M>0$. It remains to show that M is N-compact.

First we shall prove that

(1) M is realcompact.

Indeed, the space $N^S\times I$ is realcompact and the complement $(N^S\times I)\setminus M = (E\times Q_1)\cup[(N^S\setminus E)\times P]$ is the union of $\aleph_0$-cubes and thus, as each $\aleph_0$-cube is a G_δ-set in $N^S\times I$, (1) follows by Mrówka's theorem (see [4] P. 3.12.25).

Let U be an open-and-closed subset of N^S. Then

(2) $\dim C(U) = 0$ if and only if $U\cap E$ is an F_σ-set in U.

If $U\cap E$ is not an F_δ-set in U then by Lemma 1 $\dim C(U)>0$. Conversely, let $U\cap E$ be an F_σ-set in N^S. Since U is open-and-closed it depends on countably many coordinates, i.e. $U = p_{T_1}(U)\times N^{S\setminus T_1}$ for some countable set $T_1\subset S$ (see [4], P. 2.7.12). Because $U\cap E$ is an F_σ-set which is the union of $\aleph_0$-cubes,

$$U\cap E = p_{T_2}(U\cap E)\times N^{S\setminus T_2}, \quad \text{where } T_2\subset S \text{ is countable},$$

by Theorem 2 of [15]. By Remark 2 of [15] there exists $T\subset S$, $T\supset T_1\cup T_2$, $|T|\leqslant\aleph_0$ such that $p_T(U\cap E)$ is an F_σ-set in N^T. Thus we have

$$U = U'\times N^{S\setminus T}, \quad \text{where } U' \text{ is open in } N^T,$$

and

$$U\cap E = E'\times N^{S\setminus T}, \quad \text{where} \quad E' = p_T(U\cap E) = \bigcup_{i=1}^{\infty} F_i,$$

(¹) This means that for each $x\in N^S$ there exists a neighbourhood V of x such that $V\cap E$ is an F_σ-set in V.

where F_i are closed in N^T. It follows that

$$\begin{aligned} C(U) &= (U\times Q_0)\cup[(U\cap E)\times P]\cup[(U\setminus E)\times Q_1] \\ &= (U'\times N^{S\setminus T}\times Q_0)\cup(E'\times N^{S\setminus T}\times P)\cup[(U'\setminus E')\times N^{S\setminus T}\times Q_1] \\ &\overset{\text{top}}{=} \{(U'\times Q_0)\cup(E'\times P)\cup[(U'\setminus E')\times Q_1]\}\times N^{S\setminus T}. \end{aligned}$$

The space $Z=(U'\times Q_0)\cup(E'\times P)\cup[(U'\setminus E')\times Q_1]\subset N^T\times I$ is a metrizable separable space and it is the union of its closed zero-dimensional subsets $U'\times\{t\}$, for $t\in Q_0$, $(U'\setminus E')\times\{t\}$, for $t\in Q_1$ and $F_i\times(P\cup Q_0)$, for $i=1,2,\ldots$ Hence $\dim Z=0$ by the Sum Theorem. Thus $C(U)$ is the product of zero-dimensional second-countable spaces and hence $\dim C(U)=0$ by Morita's theorem ([9], Theorem 3). The proof of (2) is completed.

Let U be an open-and-closed subset of M. For $q\in Q_0$ put

$$U(q)=\{x\in N^S\colon (x,q)\in U\};$$

clearly $U(q)$ is open-and-closed in N^S.

We shall verify that

(3) $(U(q)\setminus U(q'))\cap E$ is an F_σ-set in $U(q)\setminus U(q')$ for every $q,q'\in Q_0$.

Indeed, the set $V=U\cap C(U(q)\setminus U(q'))$ is open-and-closed in $C(U(q)\setminus U(q'))$ and $(U(q)\setminus U(q'))\times\{q\}\subset V\subset C(U(q)\setminus U(q'))\setminus(U(q)\setminus U(q'))\times\{q'\}$. Hence from Lemma 1 it follows that $(U(q)\setminus U(q'))\cap E$ is an F_σ-set in $U(q)\setminus U(q')$.

For an open-and-closed set $U\subset M$ we define

$$J_U=\bigcap_{q\in Q_0}U(q)\subset N^S.$$

The sets J_U satisfy

(4) $$C(J_U)\subset U$$

and

(5) $U\setminus C(J_U)$ is the countable union of strongly zero-dimensional open-and-closed subsets of M.

Consider an $(x,t)\in C(J_U)$. Then $x\in\bigcap_{q\in Q_0}U(q)$ and because Q_0 is dense in Q and U is closed, we have $U\supset C(\{x\})\ni(x,t)$. Thus (4) holds. To establish (5) let us assume that $(x,t)\in U\setminus C(J_U)$. Then $(x,q')\notin U$ for some $q'\in Q_0$. Since U is open and Q_0 is dense in Q there exists $q\in Q_0$ such that $(x,q)\in U$. Thus $x\in U(q)\setminus U(q')$ and $(x,t)\in C(U(q)\setminus U(q'))$. We have obtained the equality

$$U\setminus C(J_U)=\bigcup_{q,q'\in Q_0}C(U(q)\setminus U(q'))\cap U$$

which proves (5) by (3) and (2).

We shall prove now that M is N-compact. Since by (i) and (2) it follows that

(6) $$\operatorname{locdim} M = 0,$$

it suffices only to verify that every open-and-closed ultrafilter in M with the countable intersection property has nonempty intersection (see [16], p. 478). Let $\mathcal{U}$ be such an ultrafilter. We shall show that

(7) there exists $U \in \mathcal{U}$ with $\dim U = 0$.

Suppose on the contrary that $\dim U > 0$ for each $U \in \mathcal{U}$. Fix an arbitrary $q_0 \in Q_0$ and let

$$\mathcal{V} = \{U(q_0) \colon U \in \mathcal{U}\}.$$

We shall prove that $\mathcal{V}$ is an open-and-closed ultrafilter in N^S and has the countable intersection property. $\mathcal{V}$ is a filter because for $U_1, U_2 \in \mathcal{U}$ the intersection $U_1(q_0) \cap U_2(q_0) = (U_1 \cap U_2)(q_0)$ belongs to $\mathcal{V}$ and if an open-and-closed set $A \subset N^S$ contains $U_1(q_0)$ then $A = (C(A) \cup U_1)(q_0)$ also belongs to $\mathcal{V}$ (because $\mathcal{U}$ is a filter). Now let U be an open-and-closed subset of N^S. Then either $C(U) \in \mathcal{U}$ or $M \setminus C(U) = C(N^S \setminus U) \in \mathcal{U}$, hence either $U \in \mathcal{V}$ or $N^S \setminus U \in \mathcal{V}$; thus $\mathcal{V}$ is an ultrafilter. Let $U_i \in \mathcal{U}$ for $i = 1, 2, \ldots$, we shall show that $\bigcap_{i=1}^{\infty} U_i(q_0) \neq \emptyset$. As shown in (5), we have $\bigcup_{i=1}^{\infty} (U_i \setminus C(J_{U_i})) = \bigcup_{j=1}^{\infty} V_j$, where the sets V_j are strongly zero-dimensional and open-and-closed in M. Since $\mathcal{U}$ is an ultrafilter it follows from the negation of (7) that $M \setminus V_j \in \mathcal{U}$ for $j = 1, 2, \ldots$ By the countable intersection property of $\mathcal{U}$ there exists a point

$$(x_0, t_0) \in \bigcap_{i=1}^{\infty} U_i \cap \bigcap_{j=1}^{\infty} (M \setminus V_j) = \bigcap_{i=1}^{\infty} U_i \setminus \bigcup_{j=1}^{\infty} V_j = \bigcap_{i=1}^{\infty} U_i \setminus \bigcup_{i=1}^{\infty} (U_i \setminus C(J_{U_i})) = \bigcap_{i=1}^{\infty} C(J_{U_i}).$$

We obtain

$$x_0 \in \bigcap_{i=1}^{\infty} J_{U_i} \subset \bigcap_{i=1}^{\infty} U_i(q_0).$$

Now $\mathcal{V}$, being an open-and-closed ultrafilter in N^S with the countable intersection property, has the nonempty intersection, and thus there exists an $x \in \bigcap \mathcal{U}$. By (6) there exists an open-and-closed strongly zero-dimensional neighbourhood U of x. We have $U \in \mathcal{U}$ contrary to our assumption that $\mathcal{U}$ does not contain strongly zero-dimensional sets. This completes the proof of (7).

Let us take an open-and-closed set $U_0 \in \mathcal{U}$ with $\dim U_0 = 0$. The set U_0 is realcompact by (1) and thus it is N-compact (see [16], p. 478). The family $\mathcal{W}$

$= \{U_0 \cap U\colon U \in \mathscr{U}\}$ is an open-and-closed ultrafilter in U_0 with the countable intersection property and hence $\emptyset \neq \bigcap \mathscr{W} \subset \bigcap \mathscr{U}$. This completes the proof that M is N-compact.

Remark 3. If we take in the above construction K instead of $N^{\aleph_1}$ and A instead of E, where K and A are as in Lemma 2, then we obtain a space M' which is perfectly normal, locally second-countable, N-compact, and satisfies $\dim M' > \operatorname{loc}\dim M' = 0$ (the space M' is a slight modification of Example 1). The proof of N-compactness of M' is analogous to the proof in Example 3 and reduces to the proof of N-compactness of K (²) (which follows by Mrówka's result [10] from the fact that K can be mapped continuously in a one-to-one way into the metrizable strongly zero-dimensional space $B(\aleph_1)$) and of realcompactness of M' (which follows from the fact that M' can be mapped in a one-to-one way into the space $B(\aleph_1) \times I$ (see [4] Exercise 3.11.B)). The remaining properties of M' can be proved in the same way as the properties of the space L in Example 1.

We are grateful to Professor R. Engelking for valuable discussions about the subject of this paper.

Added in proof.

(a) In the paper *A hereditarily normal strongly zero-dimensional space containing subspaces of arbitrarily large dimensional,* Fund. Math. (to appear) the authors have developed essentially the idea described in Section 3.

(b) E. Pol, Bull. Acad. Polon. Sci. 24 (1976), pp. 749–752 gave under CH an example of a locally compact perfectly normal space X_n with $\operatorname{loc}\dim X_n = 0$ and $\dim X_n > n$, where $n = 1, 2, \ldots$; some very strong examples of this kind, also under CH, were constructed recently by V. V. Fedorčuk, *On the dimension of hereditarily normal spaces* (to appear).

References

[1] P. S. Alexandroff and B. A. Pasynkov, *Introduction to Dimension Theory* (in Russian), Moskva 1973.

[2] E. Čech, *Problem* 53, Colloq. Math. 1 (1948), p. 332.

[3] C. H. Dowker, *Local dimension of normal spaces,* Quart. J. Math. Oxford 6 (1955), pp. 101–120.

[4] R. Engelking, *General Topology,* Warszawa 1977.

[5] V. V. Fedorčuk, *Compatibility of some theorems of the general topology with axioms of the theory of sets* (in Russian), DAN SSSR 220 (1975), pp. 786–788.

[6] V. V. Filippov, *On the dimension of normal spaces* (in Russian) DAN SSSR 209 (1973), pp. 805–807.

[7] W. Hurewicz and H. Wallman, *Dimension Theory,* Princeton 1941.

[8] B. Knaster et K. Kuratowski, *Sur les ensembles connexes,* Fund. Math. 2 (1921), pp. 206–255.

[9] K. Morita, *On the dimension of the product of Tychonoff spaces,* Gen. Top. and its Appl. 3 (1973), pp. 125–133.

(²) In fact, one can prove that K is strongly zero-dimensional.

[10] S. Mrówka, *Recent results on E-compact spaces*, TOPO 72, Proc. of Second Pittsburgh International Conference, Lecture Notes 378, Springer-Verlag 1974.
[11] K. Nagami, *Dimension Theory*, New York 1970.
[12] P. Nyikos, *Strongly zero-dimensional spaces*, Proc. of Third Prague Top. Symp. 1971, Prague 1972, pp. 341–344.
[13] — *Prabir Roy's space Δ is not N-compact*, Gen. Top. and its Appl. 3 (1973), pp. 197–210.
[14] R. Pol, *A perfectly normal locally metrizable not paracompact space*, Fund. Math. 97 (1977), pp. 37–42.
[15] — and E. Pol, *Remarks on Cartesian products*, Fund. Math. 93 (1976), pp. 57–69.
[16] J. van der Slot, *A survey of realcompactness*, Theory of Sets and Topology (in honour of Felix Hausdorff), Berlin 1972, pp. 473–494.

DEPARTMENT OF MATHEMATICS AND MECHANICS, WARSAW UNIVERSITY
WYDZIAŁ MATEMATYKI I MECHANIKI UNIWERSYTETU WARSZAWSKIEGO

Accepté par la Rédaction le 18. 8. 1975

PROCEEDINGS OF THE
AMERICAN MATHEMATICAL SOCIETY
Volume 94, Number 3, July 1985

SPACES WITH NONCOINCIDING DIMENSIONS

M. G. CHARALAMBOUS

ABSTRACT. For any given nonnegative integers l, m, n with $\max\{l, m\} \leqslant n$ and $n = 0$ if $m = 0$, we construct a normal, Hausdorff and separable space X with $\operatorname{ind} X = l$, $\dim X = m$ and $\operatorname{Ind} X = n$. We also construct a space X_n with $\dim X_n = 1$ and $\operatorname{ind} X_n = \operatorname{Ind} X_n = n$ which is the limit space of an inverse limit sequence of compact, Hausdorff and separable spaces all of whose dimensions are one.

1. Introduction. In §2, for any given nonnegative integers l, m, n with $\max\{l, m\} \leqslant n$ and $n = 0$ if $m = 0$, we construct a normal, Hausdorff and separable space X with $\operatorname{ind} X = l$, $\dim X = m$ and $\operatorname{Ind} X = n$. In particular, we establish for $m \leqslant n$ the existence of a normal Hausdorff space X with $\operatorname{ind} X = 0$, $\dim X = m$ and $\operatorname{Ind} X = n$. Apparently, the only existing example of this sort is that of Nagami [**11**] for $m = 1$ and $n = 2$. Furthermore, there is no example of a space X in the literature known to satisfy, for example, $\dim X = 1$, $\operatorname{ind} X = 2$ and $\operatorname{Ind} X = 4$, although Filippov [**8**] constructs a compact Hausdorff space X with $\dim X = 1$, $\operatorname{ind} X = i$ and $\operatorname{Ind} X = 2i - 1$ for any given positive integer i.

Assuming CH, the Continuum Hypothesis, our space X can be constructed to be additionally perfectly normal. Several examples of perfectly normal separable spaces with noncoinciding dimensions have been constructed using CH (see e.g. [**7, 9, 15**]). However, the only known perfectly normal spaces with noncoinciding dimensions that were constructed using no set-theoretical assumptions beyond ZFC are not separable [**12, 14**].

In §3, given a positive integer n, we construct an inverse limit sequence of compact, Hausdorff and separable spaces each of which has all dimensions equal to one while the limit space X_n of the sequence satisfies $\dim X_n = 1$ and $\operatorname{ind} X_n = \operatorname{Ind} X_n = n$. For $n = 2$, we constructed such an example in [**2**]. Again assuming CH, all the spaces of the sequence as well as the limit space X_n can be constructed to be additionally perfectly normal. This is in sharp contrast with the inverse limit theorem for covering dimension which holds not only for arbitrary inverse systems of compact spaces, but also for sequences of perfectly normal spaces [**1**].

As regards method of construction, in both cases we construct new topologies from old using a mixture of the techniques of replacing a point by a space (cf. [**5, 7, 9**]) and the techniques of assigning limit points to enough sequences so as not to deviate too much from the original topology (see [**4, 10, 13, 15**]).

Received by the editors November 14, 1983 and, in revised form, July 2, 1984.
1980 *Mathematics Subject Classification*. Primary 54F45, 54G20.
Key words and phrases. Covering and inductive dimensions, compact, Hausdorff, separable, normal and perfectly normal spaces.

In this paper, all spaces are at least Tychonoff. N denotes the set of positive integers, C the Cantor set, $I = [0, 1]$ the unit interval, I^n the n-dimensional unit cube, $|X|$ the cardinality of a set X, c the cardinality of the continuum and $\omega(c)$ the first ordinal of cardinality c.

For the standard results in Dimension Theory we refer to [**16, 17**].

2. Construction of normal spaces with dimensions given integers.

THEOREM 1. *Let X be a normal, Hausdorff and separable space X with* $\dim X \geq 1$. *Then there exists a normal, Hausdorff and separable space $Y = Y(X)$ such that*

(i) $\operatorname{ind} Y = \operatorname{ind} X$,

(ii) $\dim Y = \dim X$, *and*

(iii) $\operatorname{Ind} Y = \operatorname{Ind} X + 1$.

PROOF. Let D be a separable completely metrizable one-dimensional space that contains two disjoint closed subsets E_0, F_0 which cannot be separated by countable closed set. An example of such a space is $C \times I$. Let Q, P be countable dense subsets of D, X, respectively. Let $\{(S_\alpha, T_\alpha): \alpha < \omega(c)\}$ be the collection of all sequences of D with $\bar{S}_\alpha \cap \bar{T}_\alpha$ uncountable. Note that, in fact, $|\bar{S}_\alpha \cap \bar{T}_\alpha| = c$ since D is separable and completely metrizable. For each $\alpha < \omega(c)$, pick by transfinite induction a point x_α and a sequence $\{x_{\alpha n}\}$ coverging to x_α so that $x_\alpha \neq x_\beta$ for $\alpha \neq \beta$, $\{x_{\alpha n}\}$ contains infinitely many points from each of S_α, T_α and Q and $x_{\alpha n} \lhd x_\alpha$ for each n in N, where $\lhd$ is some well ordering on D of the same type as $\omega(c)$ such that $q \lhd x$ whenever $q \in Q$ and $x \notin Q$. Let $f_\alpha: \{x_{\alpha n}\} \to P$ be any function such that, for every x in P, $f_\alpha^{-1}(x)$ consists of infinitely many points from each of S_α, T_α and Q. Let $A = \{x_\alpha: \alpha < \omega(c)\}$. For each $x \notin A$, we also fix a sequence $\{x_n\}$ in Q converging to x.

We define a new topology on D by transfinite induction as follows. We let every point of Q be isolated and, assuming we have defined basic open neighbourhoods of all points $\lhd x$, we declare basic open neighbourhoods of x sets of the form $\{x\} \cup W$ where W is an open neighbourhood of all but finitely many points of $\{x_n\}$. D^* will denote D with this new topology, which is clearly separable and finer than the standard topology on D. Moreover, each point of D^* has countable basic open neighbourhoods which are closed in D and hence in D^*. Thus $\operatorname{ind} D^* = 0$ and D^* is Tychonoff.

We now let

$$Y = D \times \{p\} \cup A \times X,$$

where p is a fixed point of P, with topology defined as follows.

(a) If $x \notin A$, (x, p) has basic open neighbourhoods sets of the form $\pi^{-1}(U)$ where $\pi: Y \to D$ is the coordinate projection and U is an open neighbourhood of x in D^*.

(b) $(x_\alpha, y) \in A \times X$ has basic open neighbourhoods sets of the form

$$O(x_\alpha, U, V) = \{x_\alpha\} \times V \cup \pi^{-1}(U),$$

where V is an open neighbourhood of y in X and U is an open neighbourhood in D^* of all but finitely many points of $f_\alpha^{-1}(V)$.

Clearly Y is a T_1 separable space and π is continuous both as a function onto D and D^*. Furthermore,

(1) $\{x_\alpha\} \times X \subset \overline{Z}$ whenever $\pi(Z)$ contains S_α or T_α.

(2) If E, F are disjoint closed sets of Y, then the set $\overline{\pi(E)} \cap \overline{\pi(F)}$ is countable in D and hence in D^*. Otherwise, for some $\alpha < \omega(c)$, $S_\alpha \subset \pi(E)$ and $T_\alpha \subset \pi(F)$ and hence, by (1), $\{x_\alpha\} \times X \subset E \cap F$.

Lemma 1. *Y is normal.*

Proof. Let E, F be disjoint closed sets of Y. The countable set $\overline{\pi(E)} \cap \overline{\pi(F)}$ is contained in a countable cozero set Z of the locally countable space D^*. Further, since ind $D^* = 0$, Z is Lindelöf with dim $Z = 0$, and we can insert a clopen set of D^* between $\overline{\pi(E)} \cap \overline{\pi(F)}$ and Z. To prove the lemma, it clearly suffices to show that $\pi^{-1}(Z)$ is normal for all basic neighbourhoods of D^*, which are countable and clopen. If Z is such a neighbourhood of $x \notin A$, then $\pi^{-1}(Z)$ is homeomorphic with Z which is normal. Let us assume, therefore, that Z is a basic neighbourhood of x_α and E, F are disjoint closed sets of $\pi^{-1}(Z)$. We may assume further that $x \triangleleft x_\alpha$ for $x \in Z - \{x_\alpha\}$. Let V_1, V_2 be disjoint open sets of X with $(\{x_\alpha\} \times X \cap E) \subset \{x_\alpha\} \times V_1$ and $(\{x_\alpha\} \times X \cap F) \subset \{x_\alpha\} \times V_2$. If $K = \overline{\pi(E)} \cap \overline{\pi(F)} - \{x_\alpha\}$, there exist disjoint clopen sets Z_1, Z_2 of $Z - \{x_\alpha\}$ containing K and $\{x_{\alpha n}\} - K$, respectively. Then $\pi^{-1}(\{x_\alpha\} \cup Z_2)$ is in fact clopen in $\pi^{-1}(Z)$ since no point of $\{x_\alpha\} \times X$ is a limit point of $\pi^{-1}(K)$. Since $x \triangleleft x_\alpha$ for $x \in Z_1$, by transfinite induction, it suffices to prove normality in the case $Z = \{x_\alpha\} \cup Z_2$, i.e. when $\overline{\pi(E)} \cap \overline{\pi(F)} = \{x_\alpha\}$. Let U_1, U_2 be disjoint clopen sets of Z_2 containing, respectively, $(f_\alpha^{-1}(V_1) - \pi(F)) \cup (\pi(E) - \{x_\alpha\})$ and the union of its complement in $\{x_{\alpha n}\}$ with $\pi(F) - \{x_\alpha\}$. Then $O(x_\alpha, U_1, V_1), O(x_\alpha, U_2, V_2)$ are disjoint open sets of Y containing, respectively, E, F, which concludes the proof of Lemma 1.

We note that, in fact, bd $O(x_\alpha, U_1, V_1) = \{x_\alpha\} \times$ bd V_1, which readily implies the following two results.

Lemma 2. ind Y = ind X.

Lemma 3. *If Z is a countable closed or open subspace of D^*, then* Ind $\pi^{-1}(Z) \leqslant$ Ind X.

Lemma 4. dim Y = dim X.

Proof. Set $n = \dim X \geqslant 1$. Let $(E_1, F_1), \ldots, (E_{n+1}, F_{n+1})$ be pairs of disjoint closed subsets of Y. As in Lemma 1, there exists a countable clopen set Z of D^* which contains the closed sets $\overline{\pi(E_i)} \cap \overline{\pi(F_i)}$ of D, $i = 1, \ldots, n+1$. Since $n \geqslant 1$ and $\dim D - Z \leqslant \dim D \leqslant 1$, $\overline{\pi(E_i)} - Z, \overline{\pi(F_i)} - Z$, $i = 1, \ldots, n+1$, can be separated in $D - Z$ by closed sets K_i such that $\bigcap_{i=1}^{n+1} K_i = \varnothing$. Further, the normal space $\pi^{-1}(Z)$ is the union of a countable collection of closed subsets each of which is a singleton or a copy of X. Hence $\dim \pi^{-1}(Z) \leqslant n$ and $E_i \cap \pi^{-1}(Z)$, $F_i \cap \pi^{-1}(Z)$ can be separated in $\pi^{-1}(Z)$ by disjoint closed sets L_i, $i = 1, \ldots, n+1$, with $\bigcap_{i=1}^{n+1} L_i = \varnothing$. Let $M_i = \pi^{-1}(K_i) \cup L_i$. This is clearly a closed partition of the pair E_i, F_i in Y and $\bigcap_{i=1}^{n+1} M_i = \varnothing$. Hence $\dim Y \leqslant \dim X$. That $\dim Y \geqslant \dim X$ follows from the fact that Y contains a copy of X.

LEMMA 5. *Let L be a closed subset of Y with $\pi(L)$ a zero-dimensional subset of D. Then* $\operatorname{Ind} L \leqslant \operatorname{Ind} X$.

PROOF. Let E, F be disjoint closed sets of L. Then the closed set $\overline{\pi(E)} \cap \overline{\pi(F)}$ of D is contained in a countable clopen subset Z of D^*. Now $\pi(E) - Z$, $\pi(F) - Z$ are separated by the empty set in the zero-dimensional space $\pi(L) - Z$. Hence E, F can be separated by the empty set in $L - \pi^{-1}(Z)$ and so any partition of these sets in $\pi^{-1}(Z)$ separates them in L as well. By Lemma 3, there is such a partition with $\operatorname{Ind} \leqslant \operatorname{Ind} X - 1$. Hence $\operatorname{Ind} L \leqslant \operatorname{Ind} X$.

LEMMA 6. $\operatorname{Ind} Y \leqslant \operatorname{Ind} X + 1$.

PROOF. Any two disjoint closed sets E, F can be separated by a set of the form $K = \pi^{-1}(M) \cup L$ where $L \subset \pi^{-1}(Z)$ for some countable clopen set Z of D^*, $\operatorname{Ind} L \leqslant \operatorname{Ind} X - 1$, and M is a zero-dimensional closed subset of the one-dimensional space $D - Z$. By Lemma 5, $\operatorname{Ind} \pi^{-1}(M) \leqslant \operatorname{Ind} X$. Hence, since $\pi^{-1}(M)$, L are clopen in K, $\operatorname{Ind} K \leqslant \operatorname{Ind} X$ and so $\operatorname{Ind} Y \leqslant \operatorname{Ind} X + 1$.

LEMMA 7. $\operatorname{Ind} Y > \operatorname{Ind} X$.

PROOF. Let E_0, F_0 be disjoint closed sets of D which cannot be separated by a countable set. Let U, V be open sets of D with $E_0 \subset U$, $F_0 \subset V$, and $\overline{U} \cap \overline{V} = \varnothing$. Let E, F be closed sets of Y with $Y = E \cup F$, $\pi^{-1}(\overline{U}) \subset E$ and $\pi^{-1}(\overline{V}) \subset F$. It suffices to show $\operatorname{Ind}(E \cap F) \geqslant \operatorname{Ind} X$. But $\overline{\pi(E)} \cap \overline{\pi(F)}$ is a partition in D between E_0 and F_0. Hence for some $\alpha < \omega(c)$, S_α, T_α are dense subsets of $\pi(E)$, $\pi(F)$, respectively. Hence, by (1), $E \cap F$ contains $\{x_\alpha\} \times X$ and so $\operatorname{Ind}(E \cap F) \geqslant \operatorname{Ind} X$.

Lemmas 6 and 7 imply that $\operatorname{Ind} Y = \operatorname{Ind} X + 1$, which concludes the proof of Theorem 1.

REMARK 1 (CH). If X is perfectly normal, Y can be made perfectly normal.

PROOF. Let $\{R_\alpha: \alpha < \omega(c)\}$ be the set of all sequences of D with $R_0 = Q$. Then assuming CH, $L_x = \{\alpha: x_\alpha \triangleleft x\}$ is countable for each x in D. Thus we can choose the sequence $\{x_n\}$ with the additional property that it contains infinitely many points from each R_α with $\alpha \in L_x$, $y \triangleleft x$ for each y in R_α and $x \in \overline{R}_\alpha$. Now for each closed set E of Y,

(3) $\overline{\pi(E)} - \pi(E)$ in D is countable.

PROOF. Otherwise, let R_α be a dense countable subset of $\pi(E)$ and choose x in $\overline{\pi(E)} - \pi(E)$ such that $x_\alpha \triangleleft x$ and $y \triangleleft x$ for each $y \in R_\alpha$. Then $\pi^{-1}(x)$ is contained in the closed set E, contradicting the fact that $x \notin \pi(E)$.

Also, if G is open in Y,

(4) $\pi(G) - \pi^{\#}(G)$ is countable, where $\pi^{\#}(G) = D - \pi(Y - G)$.

PROOF. If not, for some $\alpha < \omega(c)$, R_α is a dense subset of $\pi(G) - \pi^{\#}(G)$. Let x be a point of this set such that $x_\alpha \triangleleft x$ and $y \triangleleft x$ for each $y \in R_\alpha$. The definition of the topology of Y assures that an open set $O(x, U, V)$ inside G contains $\pi^{-1}(y)$ for infinitely many points y of R_α, contradicting the fact that R_α is a subset of $\pi(G) - \pi^{\#}(G)$.

It follows from (3) and (4) that

$$G = \pi^{-1}\big(D - \overline{\pi(Y - G)}\big) \cup (G \cap \pi^{-1} Z)$$

where Z is a countable subset of D. The conclusion that Y is perfectly normal if X is is now immediate.

COROLLARY 1. *Let l, m, n be nonnegative integers with $l \leqslant m \leqslant n$ and $m \geqslant 1$. Then there exists a separable, Hausdorff and normal space Z with* $\operatorname{ind} Z = l$, $\dim Z = m$ *and* $\operatorname{Ind} Z = n$.

PROOF. Let X be the disjoint union of I^l with a normal T_2 and separable space with ind = 0 and dim = Ind = m (e.g. [3]). Clearly, $\operatorname{ind} X = l$ and $\dim X = \operatorname{Ind} X = m$. Finally, apply Theorem 1 to get spaces $X_1 = Y(X)$, $X_2 = Y(X_1), \ldots, Z = X_{n-m} = Y(X_{n-m-1})$. Clearly, Z has the stated properties.

COROLLARY 2 (CH). *Let l, m, n be nonnegative integers with $l \leqslant m \leqslant n$ and $m \geqslant 1$. Then there exists a perfectly normal, separable and Hausdorff space Z with* $\operatorname{ind} Z = l$, $\dim Z = m$ *and* $\operatorname{Ind} Z = n$.

PROOF. A modification of the construction in [3] as in Remark 1 will produce a perfectly normal, separable and Hausdorff space T with $\operatorname{ind} T = 0$ and $\dim T = \operatorname{Ind} T = m$.

We note that Fedorčuk [7] also constructs spaces with properties like those of T as well as spaces with the properties of Z of Corollary 2. His construction uses CH too.

COROLLARY 3. *Let l, m, n be positive integers with $l \leqslant m \leqslant n$. Then there exists a normal, separable and Hausdorff space with* $\dim Z = l$, $\operatorname{ind} Z = m$ *and* $\operatorname{Ind} Z = n$.

PROOF. Let X be the disjoint union of I^l and a compact T_2 and separable space with dim = 1 and ind = Ind = m (see [9] or §3 of this paper). If $X_1 = Y(X)$, $X_2 = Y(X_2), \ldots, Z = X_{n-m} = Y(X_{n-m-1})$, then Z has the stated properties.

COROLLARY 4 (CH). *Let l, m, n be positive integers with $l \leqslant m \leqslant n$. Then there is a perfectly normal, Hausdorff and separable space Z with* $\dim Z = l$, $\operatorname{ind} Z = m$ *and* $\operatorname{Ind} Z = n$.

PROOF. There exist, under CH, compact T_2, perfectly normal and separable spaces with dim = 1 and ind = Ind = m (see [7, 9] or §3 of this paper).

3. Dimension and inverse limit sequences.

THEOREM 2. *For each integer $n \geqslant 1$, there is an inverse limit sequence of compact, Hausdorff and separable spaces each of which has* dim = ind = Ind = 1 *with limit space X_n satisfying* $\dim X_n = 1$ *while* $\operatorname{ind} X_n = \operatorname{Ind} X_n = n$.

PROOF. We perform a construction similar to that of the previous section. Here D will be the unit interval I together with the union of a collection $\{I_n: n \in N\}$ of copies of I such that I_{2n-1}, I_{2n} have endpoints $(0, 1/(2n-1)), (1, 1/2n)$ and $(1, 1/2n), (0, 1/(2n+1))$, respectively. $\{(S_\alpha, T_\alpha): \alpha < \omega(c)\}$ is the collection of all sequences of I with $|\bar{S}_\alpha \cap \bar{T}_\alpha| = c$ and x_α in $\bar{S}_\alpha \cap \bar{T}_\alpha$ is chosen so that $x_\alpha \neq x_\beta$ for $\alpha \neq \beta$. For each x in I, we choose a sequence $\{x_n\}$ converging to x and such that if $x = x_\alpha$ for some $\alpha < \omega(c)$, then $x_{\alpha n} \in S_\alpha$ if n is odd and $x_{\alpha n} \in T_\alpha$ if n is even. Let $Q_1, Q_2, Q_3, \ldots$ be countable dense subsets of I contained respectively in $Q(\sqrt{2}) - Q$,

$Q(\sqrt{2},\sqrt{3}) - Q(\sqrt{2})$, $Q(\sqrt{2},\sqrt{3},\sqrt{5}) - Q(\sqrt{2},\sqrt{3}),\ldots$, where $Q(a_1, a_2,\ldots,a_k)$ denotes the field extension of Q, the field of rational numbers, generated by $a_1, a_2,\ldots,a_k$. It is important that if $q \in Q_m$, $r \in Q_n$ and $m \neq n$, the lines $x + y = q$ and $x - y = r$ do not meet on D. For $0 < x < 1$, let h_x: $I - \{x\} \to (x,1]$ be a continuous extension of the identity on $(x,1]$ that maps $[0, x)$ homeomorphically onto $(x,1]$ and takes $Q_n - \{x\}$ to $Q_n \cap (x,1]$ for each n in N; let g_x: $(x,1] \to D - I$ be a homeomorphism and define f_x to be $g_x \circ h_x$. For $x = 0, 1$, we let f_x: $I - \{x\} \to D - I$ be any homeomorphism. For all x in I, we can assume that, for each n in N, f_x maps only points of Q_n to points of intersection of lines of the form $x + y = q$, $x - y = r$ with q and r in Q_n, and that the distance between $f_x(x_n)$ and $f_x(x_{n+1})$ tends to zero. The following properties of f_x are immediate:

(5) $f_x^{-1}(V)$ contains infinitely many points of $\{x_n\}$ and, if $x = x_\alpha$, it contains infinitely many points from both S_α and T_α, whenever V is an open set of D with $V \cap I \neq \emptyset$.

(6) $\{x\} \cup f_x^{-1}(V)$ is open in I whenever V is an open set of D that contains all but finitely many I_n's.

Next we consider the category $\mathscr{C}$ whose objects are pairs (X, π) where X is a compact, T_2 and separable space and π: $X \to I$ is a continuous surjection such that, for each x in I,

(7) For every finite open cover $\{V_1, V_2,\ldots, V_k\}$ of $\pi^{-1}(x)$, $\{x\} \cup \bigcup_{i=1}^{k} \pi^{\#}(V_i)$ is open in I, where $\pi^{\#}(V) = I - \pi(X - V)$, i.e. π is fully closed [6].

(8) Every point $\pi^{-1}(x)$ has basic neighbourhoods V such that $\pi(V) - \{x\}$ is open in I and its closure contains x. For each such neighbourhood which in the sequel is referred to as special, V^* denotes the set consisting of the intersection of D with the perimeters of all squares one diameter of which has endpoints y, $h_x(y)$ with either y or $h_x(y)$ in $\pi(V) - \{x\}$, where $h_0(y) = -y$ and $h_1(y) = 1 + y$.

(9) Any two distinct points of $\pi^{-1}(x)$ have special neighbourhoods V_1, V_2 such that $V_1^* \cap V_2^* = \emptyset$.

A morphism $(X, \pi) \to (Y, \sigma)$ in $\mathscr{C}$ is a map τ: $X \to Y$ such that $\pi = \sigma \circ \tau$. It will prove convenient to occasionally denote the object (X, π) of $\mathscr{C}$ more simply by X.

Given (X, π) in $\mathscr{C}$ and $A \subset I$, we define a topology on

$$Y(A) = Y(A, (X, \pi)) = I \times \{q\} \cup A \times X$$

where q is a fixed point of X, as follows;

(a) For $x \notin A$, (x, q) has basic open neighbourhoods of the form $\sigma^{-1}(U)$ where U is an open neighbourhood of x in I and σ: $Y(A) \to I$ is the natural projection.

(b) For $x \in A$, (x, y) has basic open neighbourhoods of the form

$$O(x, U, V) = \{x\} \times V \cup \sigma^{-1}\big(f_x^{-1}(V^*) \cap U\big)$$

where U is an open neighbourhood of x in I and V is a special neighbourhood of y in X.

It can be verified that $(Y(A), \sigma)$ is in $\mathscr{C}$ and $\{x_\alpha\} \times X$ is contained in the closure of any subset Z of $Y(A)$ with S_α or $T_\alpha \subset \sigma(Z)$. The fact that σ is fully closed, for instance, follows from (6) and the fact that π is fully closed. Furthermore, if to a morphism $\pi: X \to Z$ in $\mathscr{C}$, we associate $\tau^* = \tau^*(A)$: $Y(A, X) \to Y(A, Z)$ defined by $\tau^*(x, y) = (x, \tau(y))$, we see that we have a functor $Y: \mathscr{C} \to \mathscr{C}$.

LEMMA 8. $\operatorname{ind} Y(I) \geqslant \operatorname{ind} X + 1$ *and* $\operatorname{Ind} Y(I) \geqslant \operatorname{Ind} X + 1$.

PROOF. Let E, F be nontrivial regular closed subsets of $Y(I)$ with $Y(I) = E \cup F$. If $\sigma(E) \cap \sigma(F)$ is uncountable, then for some $\alpha < \omega(c)$, $S_\alpha \subset \sigma(E)$, $T_\alpha \subset \sigma(F)$ and $\{x_\alpha\} \times X \subset E \cap F$. If $\sigma(E) \cap \sigma(F)$ is countable, it contains an isolated point x. Then the regularity of E and F assures that $\{x\} \times X$ is contained in $E \cap F$. This proves the result.

To compute the inductive dimensions of $Y(A)$, it is convenient to assume that X satisfies the following:

(10) For n in N sufficiently large, every point of $\pi^{-1}(x)$ has arbitrarily small special neighbourhoods V with $\pi(\operatorname{bd} V) - \{x\} \subset Q_n$, $\operatorname{ind} \operatorname{bd} V < \operatorname{ind} X$ and $\operatorname{Ind} \operatorname{bd} V < \operatorname{Ind} X$.

And

(11) If B is a subset of I of the form $Q_1 \cup \cdots \cup Q_n \cup \{x_1, \ldots, x_n\}$, then $\operatorname{ind} \pi^{-1}(B) < \operatorname{ind} X$ and $\operatorname{Ind} \pi^{-1}(B) < \operatorname{Ind} X$.

LEMMA 9. *If X satisfies* (10) *and* (11), *then* $\operatorname{ind} Y(I) = \operatorname{ind} X + 1$, $\operatorname{Ind} Y(I) = \operatorname{Ind} X + 1$ *and $Y(I)$ satisfies* (10) *and* (11).

PROOF. If V is as in (10) and U is an open interval around x with endpoints in Q_n, then the construction of f_x assures that $O(x, U, V)$ has boundary in $\sigma^{-1}(Q_n \cup \{x\})$. Now consider a subset $B = Q_1 \cup \cdots \cup Q_n \cup \{x_1, x_2, \ldots, x_n\}$ of I. Let (x, y) be in $\sigma^{-1}(B)$ and V a special neighbourhood of y in X with $\operatorname{bd} \pi(V) - \{\pi(y)\} \subset Q_{n+1}$ and U on open interval with endpoints in Q_{n+1} containing x but no point from $\{x_1, x_2, \ldots, x_n\} - \{x\}$. Then the boundary of $O(x, U, V)$ in $\sigma^{-1}(B)$ is $\{x\} \times \operatorname{bd} V$. Hence, $\operatorname{ind} \sigma^{-1}(B) \leqslant \operatorname{ind} X$ and $\operatorname{ind} Y(I) \leqslant \operatorname{ind} X + 1$. Thus, in view of Lemma 8, $\operatorname{ind} Y(I) = \operatorname{ind} X + 1$ and $Y(I)$ satisfies (10). The rest of the lemma follows from the fact that X satisfied (11) and for every closed set F inside an open set G of X there are special neighbourhoods $V_1, \ldots, V_k$ satisfying (10) and $F \subset \bigcup_{i=1}^{k} V_i \subset G$.

In the sequel, for each $n \in N$, A_n denotes the set $I - \bigcup_{i=n}^{\infty} Q_i$.

LEMMA 10. *If X satisfies* (10) *and* $\operatorname{ind} X = 1$, *then the same holds for* $Y(A_n)$.

PROOF. Let V be a special neighbourhood of y in X with $\operatorname{ind} \operatorname{bd} V \leqslant 0$ and $\pi(\operatorname{bd} V - \{y\})$ a subset of Q_{n+1}. Let U be an open interval containing x with endpoints in Q_{n+1}. Then $\operatorname{bd} O(x, U, V)$ consists of $\{x\} \times \operatorname{bd} V$ together with a subset of $\sigma^{-1}(Q_{n+1})$, which is homeomorphic with its projection under σ into I. Now the sum theorem for covering dimension assures that $\operatorname{bd} O(x, U, V)$ is zero-dimensional in every sense.

To conclude the proof of Theorem 2, two remarks are needed. First, if $A \subset B \subset I$ then there is a map $Y(B, X) \to Y(A, X)$ in $\mathscr{C}$ that sends (x, y) to (x, q) for

$x \in B - A$ and leaves all other points fixed. Moreover, if $X \to Z$ is a map in $\mathscr{C}$, we have a map $Y(B, X) \to Y(A, Z)$ in $\mathscr{C}$ which is the composite of $Y(B, X) \to Y(A, X)$ with the map $Y(A, X) \to Y(A, Z)$ considered before. Second, if

$$Z \to \cdots \to Z_n \to Z_{n-1} \to \cdots \to Z_1$$

is an inverse limit sequence in $\mathscr{C}$ with limit Z, then

$$Y(I, Z) \to \cdots \to Y(A_n, Z_n) \to Y(A_{n-1}, Z_{n-1}) \to \cdots \to Y(A_1, Z_1)$$

is an inverse limit sequence in $\mathscr{C}$ with limit $Y(I, Z)$.

Thus by repeated application of the functor Y we obtain the following sequences and limit spaces in $\mathscr{C}$:

$$X_2 = Y(I, I) \to \cdots \to Y(A_m, I) \to Y(A_{m-1}, I) \to \cdots \to Y(A_1, I),$$

$$X_3 = Y(I, X_2) \to \cdots \to Y(A_m, Z_{m,2}) \to Y(A_{m-1}, Z_{m-1,2}) \to \cdots \to Y(A_1, Z_{1,2}),$$

$$X_n = Y(I, X_{n-1}) \to \cdots \to Y(A_m, Z_{m,n-1}) \to Y(A_{m-1}, Z_{m-1,n-1}) \to \cdots \to Y(A_1, Z_{1,n-1}),$$

where $Z_{m,k} = Y(A_m, Z_{m,k-1})$ for $k > 1$ and $Z_{m,1} = I$.

Repeated application of Lemma 9 shows that $\operatorname{ind} X_n = \operatorname{Ind} X_n = n$, while Lemma 10 assures that $\operatorname{ind} Z_{n,k} = \operatorname{Ind} Z_{n,k} = 1$. That $\dim X_n = 1$ follows from the fact that X_n is the inverse limit of compact spaces with covering dimension 1.

REMARK 2. As in Theorem 1, assuming CH, the spaces of Theorem 2 can be constructed to be additionally perfectly normal.

REMARK 3. It would be interesting to know whether Remarks 1 and 2 as well as Corollaries 2 and 4 remain valid without CH. Indeed the following open problems pertaining to spaces not all of whose dimensions coincide present themselves. The last three of these are open even assuming CH. Roy's example of a metric space with ind $= 0$ and dim $= 1$ is the only known partial solution of (6). As for (4), apart from Filippov's examples in **[8]** various other examples from the same period exist that provide partial solutions, but in its general form, we believe, the problem is as yet unresolved.

1. Given nonnegative integers l, m, n with $l \leqslant m \leqslant n$ and $m \geqslant 1$, construct a normal Hausdorff space X such that $\operatorname{ind} X = l$, $\dim X = m$, $\operatorname{Ind} X = n$ and X is additionally perfectly normal or completely normal.

2. Given positive integers l, m, n with $l \leqslant m \leqslant n$, construct a normal Hausdorff space X such that $\dim X = l$, $\operatorname{ind} X = m$, $\operatorname{Ind} X = n$ and X is additionally perfectly normal or completely normal.

3. Construct a separable perfectly normal or ever completely normal space X with $\operatorname{ind} X < \dim X$.

4. Given positive integers l, m, n with $l \leqslant m \leqslant n$, construct a compact Hausdorff space X with $\dim X = l$, $\operatorname{ind} X = m$ and $\operatorname{Ind} X = n$.

5. The same as (4) with X being additionally perfectly normal or completely normal.

6. Given nonnegative integers l and m with $l \leqslant m$, construct a metric space X with $\operatorname{ind} X = l$ and $\dim X = n$.

References

1. M. G. Charalambous, *The dimension of inverse limits*, Proc. Amer. Math. Soc. **58** (1976), 289–295.

2. ______, *Inductive dimension and inverse sequences of compact spaces*, Proc. Amer. Math. Soc. **81** (1981), 482–484.

3. ______, *The dimension of inverse limit and N-compact spaces*, Proc. Amer. Math. Soc. **85** (1982), 648–652.

4. E. Van Douwen, *A technique for constructing honest locally compact submetrizable spaces*, Preprint.

5. V. V. Fedorčuk, *Bicompacta with non-coinciding dimensionalities*, Soviet Math. Dokl. **9** (1968), 1148–1150.

6. ______, *Mappings that do not reduce dimension*, Soviet Math. Dokl. **10** (1969), 314–317.

7. ______, *On the dimension of hereditarily normal spaces*, Proc. London Math. Soc. (3) **36** (1978), 163–175.

8. V. V. Filippov, *On bicompacts with non-coinciding inductive dimensions*, Soviet Math. Dokl. **11** (1970), 635–638.

9. V. V. Filippov, *On compacta with unequal dimensions* ind *and* dim, Soviet Math. Dokl. **11** (1970), 687–691.

10. I. Juhász, K. Kunen and M. E. Rudin, *Two more hereditarily separable non-Lindelöf spaces*, Canad. J. Math. **28** (1976), 998–1005.

11. K. Nagami, *A normal space Z with* ind $Z = 0$, dim $Z = 1$, Ind $Z = 2$, J. Math. Soc. Japan **18** (1966), 158–165.

12. E. Pol and R. Pol, *A hereditarily normal strongly zero-dimensional space with a subspace of positive dimension and an N-compact space of positive dimension*, Fund. Math. **97** (1977), 43–50.

13. T. Przymusiński, *On the dimension of product spaces and an example of M. Wage*, Proc. Amer. Math. Soc. **76** (1979), 315–321.

14. P. Roy, *Failure of equivalence of dimension concepts for metric spaces*, Bull. Amer. Math. Soc. **68** (1962), 609–613.

15. M. Wage, *The dimension of product spaces*, Proc. Nat. Acad. Sci. U.S.A. **75** (1978), 4671–4672.

16. R. Engelking, *Dimension theory*, North-Holland, Amsterdam, 1978.

17. A. R. Pears, *Dimension theory of general spaces*, Cambridge University Press, Cambridge, 1975.

Department of Mathematics, University of Nairobi, P.O. Box 30197, Nairobi, Kenya

Algebraic Topology

E. G. Sklyarenko

On the paper by E. Čech "Théorie générale de l'homologie dans un espace quelconque", Fund. Math. 1932, 19, 149 – 183.

E. Čech's papers on algebraic topology appeared in the period of intensive development of homology theory. They concern both the classical part, polyhedra, and generalizations.

By that time, foundations of the classical part of the theory had been completed. Soon after 1925, in a large extent thanks to the influence of E. Noether, homology groups became the main object of the research. They replaced the purely integer invariants such as Betti and torsion numbers. It became possible to use, besides the integers, other groups of coefficients in the definition of homology. Moreover, it became possible to investigate the homology of more complicated objects than just finite complexes (the homology of which cannot be described in the terms of the integer invariants). The topological invariance problem for homology of complexes was finally solved as a corollary of the homotopical invariance by J. W. Alexander ([1], 1926). The homology $H_n(K, L)$ of pairs of complexes (K, L) (the homology K *mod* L, or the relative homology) was introduced (S. Lefschetz [21],1927). The homomorphism f_* corresponding to a continuous mapping f, and the connecting homomorphism $\delta : H_n(K, L) \to H_{n-1}(L)$ were defined. Some of the most typical applications were outlined.

All that does not mean, however, that the homology theory of polyhedra was completely finished. Thus, the exactness of the homological sequence of a pair was fully established as late as in 1942 (W. Hurewicz [16]). In 1935, as an independent topic, cohomology was introduced (J. W. Alexander [2], A. N. Kolmogorov [17]). Soon, many new applications appeared (in particular, the famous Lefschetz's fixed point theorem [24]).

Simultaneously, with the development of the classical theory, we encounter intensive attempts to investigate more general homology constructions applied to the topological spaces which either have not the structure of a complex or, if they have, this structure is too complicated to be used for the computation of homology groups (note that for a triangulation of such a simple figure as the torus one needs at least 42 simplexes, and the n–simplex itself has $2^n - 1$ faces).

The first definition of the homology groups which used a limit process was given by E. Vietoris ([32], 1927). His construction, however, was applied only to compact metric spaces, and used the concrete metric of such spaces for the description of cycles. At the same time, P. S. Alexandroff ([4]) proposed a method of approximation of metric compacta by inverse sequences of finite complexes (the theory of projection spectra). These complexes are the nerves of arbitrarily small finite closed coverings. Due to the fact that the natural projections of refinement covering nerves are not unique, some specific multi-valued projections of the complexes in the spectra are considered instead of the usual simplicial mappings. As a particular case, sequences of complexes with the simplicial projections were also studied (one can find such sequences also in chapter VII of Lefschetz's book [22]). Using the projection spectra, P. S. Alexandroff defined Betti numbers of any metric compact. L. S. Pontrjagin ([26], 1931) was the first to consider the direct and inverse group sequences. He defined the notion of direct limit and using the results of his own character theory (instead of the inverse limit) he obtained homology groups of metric compacta. About this time, the Čech's first paper on algebraic topology was published.

In this paper (in Chapter II), using the nerve system of the finite open coverings of a topological space, E. Čech virtually builds a homological apparatus of the chain complex inverse spectra which is much more suitable for the purposes of homology theory than Alexandroff's projection spectra. In this theory the projections of nerves of refinements of coverings are the usual simplicial mappings and although there are many such projections for a pair of coverings, it is shown that they all are mutually homological and hence determine the same homomorphism of the homology groups. In modern terminology, Čech's homology group $\check{H}_n(R)$ of a topological space R is simply the inverse limit of the homology groups of all finite open coverings of R.

Following tradition, E. Čech himself defined his homology groups in terms of the homological cycle classes which correspond to the coverings and are mapped by the projections of covering nerves in homological cycles. The notion of inverse limit in the manifest shape was first employed in the literature later (for the compact group sequences in [27], for the systems of topological spaces in [31], etc). In spite of it, all the principal constructions used in the Čech's paper are in fact constructions of inverse spectrum theory. The more general theory created by S. Lefschetz for the abstract complexes some time later (including the spectra with multiple projections) was considered by the author himself as an extension of the Čech's homology theory of topological spaces (see Chapter VI in [25]).

The homology groups in Čech's paper are constructed for pairs (R, A) of topological spaces (homology R *mod* A) where A is a closed subspace of R, (see Chapter III). It is shown (see § 30) that for the definition of homology it suffices to use any confinal part of the system of all finite open coverings of R (and this, in particular, permits us to contemplate the dimension, the influence of the topological weight of R, etc). It is also shown that without any trouble the open coverings can be replaced by the closed ones. E. Čech does not restrict himself to the description of the homology groups and their most standard properties. He studies the relations between the homology groups of a space R and a covering α of R. The existence of a natural homomorphism $\check{H}_n(R) \to H_n(\alpha)$ is shown (§ 28) and in case the coefficient group is a field, a description of the image is obtained. By this description, the image is generated by certain cycles, the so called essential ones: such cycles of α are images of a cycle of any refinement of α under the corresponding projection. The essential cycles are studied in detail. It is shown that the projections in the inverse spectra formed by the linear spaces of the essential cycles are epimorphic. If the n-th Betti number is finite, it is shown that there is a covering α for which one has the inclusion $\check{H}_n(R) \subset H_n(\alpha)$ (see § 29).

In Chapter IV, the problem of the dependence of Betti numbers of the union $R = R_1 \cup R_2$ of closed subspaces on Betti numbers of R_1, R_2 and the intersection $R_1 \cap R_2$ is discussed. It is shown that the basic relations known for polyhedra remain true in this more general situation. This is done, by the way, not only for the absolute homology of R, R_1, R_2 and $R_1 \cap R_2$ but also for the homology of the pairs (R, A), $(R_1, A \cap R_1)$, $(R_2, A \cap R_2)$ and $(R_1 \cap R_2, A \cap R_1 \cap R_2)$ for a closed subspace A of R. It goes without saying that such results can be nowadays obtained as a consequence of the exact homology sequence of a triad (Mayer-Vietoris sequence), but it is necessary to bear in mind that exact sequences as such (generalizing, in fact, the reasoning mentioned) first appeared as late as in 1942.

At the end of Chapter III, the zero-dimensional homologies $\check{H}_0(R)$ are considered and, in particular, the influence of the connectedness properties of R on the group $\check{H}_0(R)$ is studied. Under natural restrictions, the coincidence of zero-dimensional Betti number with the number of the quasi-components of R is established.

Čech's paper was valued at that time not only for its results but also for its close connections with the other directions in homology theory. It was soon discovered that in the category of metric compacta, Čech's homology groups coincide with the groups of L. Vietoris (see Chapter VII, § 6 in [25]). P. S. Alexandroff emphasizes in § 3B of [8] that in his articles [5, 6 and 7] he has built a homology theory of abstract projection spectra giving the homology groups of a compactum in case of the open covering projection spectrum of this compactum. These groups are also equivalent to the Čech's ones. It also turned out that Kurosh's [20], Alexander's [3] and Kolmogorov's [18] homology theories for compact spaces are variants of Čech's theory (see Chapter VII, § 7, in [25]).

A disadvantage of Čech's theory which was soon observed lies in breaking the invariance property under homotopy. This disadvantage is encountered only in the homology of non-compact spaces. The reason is that the structure of the system of finite open coverings of a normal space R coincides, as noted later, with the structure of the analogous system of Čech's compactification βR; hence, the homology groups $\check{H}_n(R)$ of such a space are in fact the $\check{H}_n(\beta R)$; for example, in the particular case when the space is the real line $\mathbb{R}$ (evidently contractible to a point), $\check{H}_1(\mathbb{R})$ is not only distinct from zero but forms a rather voluminous group (C. H. Dowker [12], 1937).

It follows that it is not reasonable to apply the Čech construction outside the category of compact spaces. The fact that the other general constructions of homology mentioned above were adapted from the very beginning only to the compact spaces speaks in favour of this conclusion. Kolmogorov's theory makes an exception as it is defined by means of the "infinite" cycles of locally compact spaces (this is the so-called second kind homology theory); these homologies of locally compact spaces, however, are converted into the usual ones after the transition to one-point compactifications (see, for example, the review [29]).

It is clear that it was rather difficult to notice this disadvantage of Čech's theory when the general homology theories were born. For comparison it is useful to recollect that, simultaneously with the development of Čech's ideas, the singular homology theory was developed, also with considerable difficulties. At the beginning of the 1930's, for example, the singular chain of a space R in this theory was understood as a system consisting of a finite complex K, a simplicial chain of K and a continuous mapping of K into R. But such homology groups cannot be evidently obtained as homologies of a chain complex. The groups of singular chains were first defined by S. Lefschetz [23]. They were generated by singular simplexes, that is, by affine equivalence classes of continuous mappings of oriented simplexes into R. But it was E. Čech who was the first to pay attention to the following obstacle: both the orientations of the one-dimensional singular simplex obtained by folding of the segment coincide. This meant that the Lefschetz's chain complex contained elements of order two, i.e., it was not free! This obstacle was overcome later in S. Eilenberg's paper [15] where the oriented simplexes in the definition of singular chain were replaced by ordered ones.

To amend the Čech definition of the homology for non-compact spaces, it was proposed more than once to use in the initial definition also infinite open coverings as well as the finite coverings. Another approach was less noticeable for a long time: It was proposed to use the direct limit of homologies of all compact subspaces $C \subset R$ for the definition of homologies of a space R. These were Čech homologies $\check{H}_n^c$ with compact supports. This approach proved to be most natural. The fact is that one has compactness in any definition of the homology. For example, we have the finite simplicial chains (and hence also the cycles) of simplicial complexes, the finite singular chains (having compact supports), the compact Steenrod-Sitnikov chains, etc. The second kind homologies of locally compact spaces are converted into the customary ones after transition to the one-point compactifications of such spaces. Thus, all the natural homologies are homologies with compact supports. It is known there are many situations in which Čech's homologies of compacta coincide with the usual homologies of N. Steenrod (for compact groups of the coefficients, for the coefficients in a field, for algebraically compact groups of the coefficients, and so on). In any such case outside the compact category the groups $\check{H}_n^c$ (but not $\check{H}_n$, generally speaking) coincide with the usual Steenrod-Sitnikov groups H_n^c.

Nowadays, it is universally recognized, however, that the main disadvantage of Čech's homology is the absence of the exactness property for homological sequences of pairs of spaces. This disadvantage was discovered in the 40's as a consequence of the inexactness of the inverse limit functor $\varprojlim$. But one ought to bear on mind that E. Čech himself chiefly used the coefficients in the field $\mathbb{Q}$ of the rational numbers, both in his first paper and in the other ones concerning the general homology theory. In the appendix to his first paper E. Čech notes that the same results can be obtained for the homologies with coefficients in other fields, in particular, in Z_p of any prime order p. Moreover, the author notes that the other cyclic groups Z_n also can be used as coefficients but herewith the meaning of the term "Betti number" is lost. Since the Čech homologies of compact spaces coincide with Steenrod's, for such coefficients the groups $\check{H}_n$ for compacta (and hence also the groups $\check{H}_n^c$ for general spaces) do not have the mentioned disadvantage.

E. Čech used his general theory only in cases where the coefficient groups are $\mathbb{Q}$ or Z_p. In the appendix to his first paper and in the later ones concerning general homology theory, E. Čech already pointed to various difficulties arising when attempting to use the integer coefficients, and formulated some additional requirements and conditions needed for attaining final results. But it is known nowadays that the Čech groups $\check{H}_n$ do not posses the exactness property for integer coefficients! Thus E. Čech foresaw the flaws of his theory for general groups of coefficients and did not try to develop it in this too general situation.

As it was already noted above, a general theory of inverse spectra was formed later on the basis of Čech's general homology theory. Various notions and constructions which were used by E. Čech found their natural generalizations. In particular, the importance of the notion of the "essential element" of a spectrum which is an evident generalization of the notions of the "essential cycle" and of the "essential class of homology" became clear. The same is true about the Čech's result which states that the property of an element to be essential in an inverse spectrum of finite-dimensional vector spaces is equivalent to the property to be a coordinate of some limit element of such spectrum. This result proved to be valid for all the spectra satisfying certain conditions of the compactness type (see the article 27 of § 6 in chapter II in [25]). It turned out to be the principal moment in the demonstrations of the exactness property of Čech's theory in the situations when this theory is really exact. It is known now (see [19]) that Čech's homologies $\check{H}_n^c$ possess the exactness property (and hence are equivalent to Steenrod-Sitnikov homologies H_n^c) under just the condition that the coefficient group G is algebraically compact (i.e., G is a direct summand of a compact topological group). Whenever $\check{H}_*^c \neq H_*^c$, it is possible to find a metric compactum, R, a point in R and a suitable dimension n so that there is an element in the group $H_n(R)$ essential in the inverse spectrum $\{H_n^c(U)\}$ corresponding to all the neighborhoods U of that point in R, in spite of the fact that the inverse limit $\varprojlim_U H_n^c(U)$ is zero (see the note after Lemma 8 in [28]).

If Čech's homology groups $\check{H}_n$ or $\check{H}_n^c$ differ from H_n^c, they play only an auxiliary part in the modern topology. In contrast to the groups $\check{H}_n(R)$, defined by nerves of coverings, the groups $H_n(R)$ of a compact space R can be obtained as the homologies of a chain complex coinciding with the inverse limit of the chain complexes of the nerves for a special system $\{\alpha\}$ of coverings with the uniquely defined projections. The seeming simplicity of such a description of the groups H_n is false. During its development, the theory of H_* had to overcome many difficulties (see in this connection Chapter 5 of the review [29]). The natural epimorphism $H_*^c \to \check{H}_*^c$ is always present.

For a compact Hausdorff space we have the exact sequence (see[29])

$$0 \to \varprojlim_{\alpha}{}^{1} H_{n+1}(\alpha) \to H_n(R) \to \check{H}_n(R) \to 0 .$$

It is clear that the principal constructions contained in the work of E. Čech can be applied with the same effect to the cohomology that appeared soon after his paper. Unlike the inverse limit, the direct limit $\varinjlim$ is exact, and this circumstance gave impulse to the further tempestuous development of the cohomology theory and its numerous successful applications.

As long as the applications were limited to the category of compact spaces (the duality laws in manifolds, the relations connected with Pontrjagin's character theory, etc.) there were no difficulties. The difficulties started with attempts to apply the cohomology theory to non-compact spaces. They were analogous to those in homology, namely breaking the homotopical invariance property. For example, the integer cohomology of the real line $\mathbb{R}$ in dimension one has the cardinality of continuum [13]. In [13], C. H. Dowker analyses the conditions that provide to a certain degree for the homotopy property. These conditions unnaturally restrict the classes of admissible spaces, and they have to be modified to specific "uniform" homotopies. At the same time C. H. Dowker shows that the classical theorems of Brushlinsky and Hopf concerning the homotopical classification of mappings into the circle and spheres remain valid for mappings of arbitrary paracompact Hausdorff spaces on the condition that all the open coverings (not only the finite ones) are used in the cohomology definition.

Developing this idea in another paper [14], C. H. Dowker presents a systematic construction of modernized Čech cohomology (including the cohomology of the pair (R, A) with A not necessarily closed). He also shows that the cohomology thus obtained satisfies all the standard requirements. Since then this theory, thanks to numerous widely spread applications, became very important and has remained so up to now. The necessity of taking into account the infinite coverings in the cohomology definition is also confirmed by a later result of V. Bártík. In his article [9] V. Bártík proved that the modernized Čech cohomologies of an arbitrary paracompact Hausdorff space with a coefficient group G in dimension n coincide (as in the case of polyhedra) with the set of homotopical classes of mappings of R into the Eilenberg-MacLane polyhedron $K(G, n)$.

It is clear that the connection between Čech cohomologies defined only by finite open coverings and those using arbitrary ones is described by the homomorphism

$\beta : \check{H}^n(\beta R) \to \check{H}^n(R)$ corresponding to the natural inclusion $R \subset \beta R$. Some natural reasons for which the mapping β is not epimorphic for infinite-dimensional spaces R are pointed out in the review [30] (see § 7 of Chapter 3 there). However, if the coefficient group G is finitely generated, for $n > 1$, this mapping is isomorphic for any finite-dimensional paracompact Hausdorff space R (and this is also valid for the cohomology of pairs (R, A) with closed subsets A), see V. Bártík, [10]. The same result is valid for arbitrary n if the group G is finite. The structure of β for $n = 1$ (under the condition that $\dim R < \infty$) is described by A. Calder in [11]. For infinite groups G it suffices to use only those coverings in the definition of cohomology the power of which does not exceed the power of G (V. Bártík). See also §7 of chapter 3 in the review [30].

In the course of time new methods were developed and, as a result, the sheaf cohomology arose. There are many strong arguments (see, for example, [29] and [30]) confirming that the sheaf cohomology is the "best possible" one. It represents the unique "universal" theory in the sense that it admits a comparison with any other individual cohomology theory. However, Čech's groups $\check{H}^n$ turn out to be isomorphic to the sheaf ones in most of the more natural situations: a) for paracompact Hausdorff spaces; b) for paracompactifying families of supports; c) in any case when the topological space R has sufficiently many arbitrarily small open sets U with $\check{H}^q(U) = 0$ for all $q > 0$ (Cartan's theorem). Such a case arises in algebraic geometry when one investigates the cohomology of algebraic manifolds with coefficients in coherent algebraic sheaves; this is why Čech's cohomology is typically used in this area.

Today there are plenty of approaches to sheaf theory and sheaf cohomology. We must stress, however, that the original construction of the sheaf cohomology belonging to J. Leray represents a variant of Čech's theory: defined at the start as Čech type cohomologies with coefficients in a presheaf F, the cohomologies turn out to depend not so much on the presheaf F as, more essentially, on the sheaf $\mathcal{F}$ generated by F.

The comparatively simple and sufficiently complete modern exposition of the cohomology theory $\check{H}^*$, including the non-constant coefficients, is given in the review [29], § 2 of Chapter 2 and Chapter 4.

References

[1] Alexander J.W., *Combinatorial analysis situs, I.*, Trans. Amer. Math. Soc. **28, 2** (1926), 301 – 329.

[2] Alexander J. W., *On the ring of a compact metric space*, Proc. Nat. Acad. Sci. U.S.A. **21, 8** (1935), 511 – 512.

[3] Alexander J. W., *A theory of connectivity in terms of graitings*, Ann. of Math. **39, 4** (1938), 883 – 912.

[4] ALEXANDROFF P. S., *Untersuchungen über Gestalt und Lage abgeschlossener Mengen beliebiger Dimension*, Ann. of Math. **30, 1** (1928), 101 - 187.

[5] ALEXANDROFF P. S., *On local properties of closed sets*, Ann. of Math. **36, 1** (1935), 1 - 35.

[6] ALEXANDROFF P. S. AND PONTRJAGIN L., *Les varietés à n-dimensions generalisées*, C. R. Acad. Sci. **202** (1936), 1327 - 1329.

[7] ALEXANDROFF P. S., *O sčetno-kratnych otkrytych otobraženijach*, Doklady AN SSSR **4, 7** (1936), 283 - 288.

[8] ALEXANDROFF P. S., *Teoretiko-množestvennaja topologija, Matematika v SSSR za sorok let*, Fizmatgiz, Moskva (1959), 230 - 263.

[9] BÁRTÍK V., *Kogomologii Aleksandrova-Čecha i otobraženija v poliedry Eilenberga - Makleina*, Matem. sbornik **76, 2** (1968), 231 - 238.

[10] BÁRTÍK V., *On the bijectivity of the canonical transformation* $[\beta X, Y] \to [X, Y]$, Quart. J. of Math. **29, 113** (1978), 77 - 91.

[11] CALDER A., *Cohomology of finite covers*, Trans. Amer. Math. Soc. **218** (1976), 349 - 352.

[12] DOWKER C. H., *Hopf's theorem for non-compact spaces*, Proc. Nat. Acad. Sci. U.S.A. **23, 5** (1937), 293 - 294.

[13] DOWKER C. H., *Mapping theorems for non-compact spaces*, Amer. Journal of Math. **69, 2** (1947), 200 - 242.

[14] DOWKER C. H., *Čech cohomology theory and the axioms*, Ann. of Math. **51, 2** (1950), 278 - 292.

[15] EILENBERG S., *Singular homology theory*, Ann. of Math. **45,3** (1944), 407 - 447.

[16] HUREWICZ W., *God knows*, Bull. Amer. Math. Soc. **47** (1942), 562.

[17] KOLMOGOROV A. N., *Über Dualität im Aufbau der Kombinatorischen Topologie*, Mat. Sb. **1, 1** (1936), 97 - 102.

[18] KOLMOGOROV A. N., *Les groupes de Betti des espaces localement bicompactes*, C. R. Acad. Sci. Paris. **202** (1936), 1144 - 1147.

[19] KUZ'MINOV V. I., ŠVEDOV I. A., *Gipergomologii prjamogo spektra kompleksov i gruppy gomologij topologičeskich prostranstv*, Sibirsk. Matem. Žurnal **16, 1** (1975), 62 - 74.

[20] KUROSCH A., *Kombinatorischer Aufbau der bikompakten topologischen Räume*, Compositio Math. **2, 3** (1935), 471 - 476.

[21] LEFSCHETZ S., *The residual set of a complex on a manifold and related questions*, Proc. Nat. Acad. Sci. U.S.A. **13, 8** (1927), 614 - 622.

[22] LEFSCHETZ S., *Topology*, Amer. Math. Soc. Coll. Publ. 1930, 12, New York.

[23] LEFSCHETZ S., *On singular chains and cycles*, Bull. Amer. Math. Soc. **39, 2** (1933), 124 - 129.

[24] LEFSCHETZ S., *On the fixed point formula*, Ann. of Math. **38, 4** (1937), 819 - 822.

[25] LEFSCHETZ S., *Algebraic topology*, 1942.

[26] PONTRJAGIN L., *Über den algebraischen Inhalt topologischer Dualitätssätze*, Math. Ann. **105, 2** (1931), 165 - 205.

[27] PONTRJAGIN L., *Sur les groupes topologiques compacts et le cinquième problème de M. Hilbert*, C. R., Acad. Sci. **198** (1934), 238 - 240.

[28] SKLYARENKO E. G., *Teorija gomologij i aksioma točnosti*, Uspehi matem. nauk **24, 5** (1969), 87 - 140.

[29] SKLYARENKO E. G., *Gomologii i kogomologii obščich prostranstv*, Itogi nauki i techniki, sovremennye problemy matematiki, fundamental'nye napravlenija, VINITI **50** (1989), 129 - 266.

[30] SKLYARENKO E. G., *Obščie teorii gomologij i kogomologij. Sovremennoe sostojanie i tipičnye primenenija*, Itogi nauki i techniki. Algebra, topologija, geometrija. VINITI **27** (1989), 125 - 228.

[31] STEENROD N. E., *Universal homology groups*, American J. of Math. **58, 4** (1936), 661 - 701.

[32] VIETORIS L., *Über den hohern Zusammenhang kompakter Räume und eine Klass von zusammenhangstreuen Abbildungen*, Math. Ann. **97** (1927), 454 - 472.

On the papers by E. Čech a) "Théorie générale des variétés et de leurs théorèmes de dualité", Ann. of Math., 1933, 34,4, 621 – 730; b) "Sur les nombres de Betti locaux", Ann. of Math., 1934, 35,3, 678 – 701; c) "Les théorèmes de dualité en topologie", Čas. Pěst. Matematiky a Fysiky, 1935, 64, 17 – 25; d) "Accessibility and homology", Mat. Sb. 1936, 1,5, 661 – 662; e) "On general manifolds", Proc. Nat. Acad. Sci. U.S.A., 1936, 22,2, 110 – 111.

Having presented his new approach to homology, E. Čech demonstrates in the same article several natural applications of his theory. The applications of the new theory, first of all to the duality relations in manifolds and to the problems of position of closed subsets in euclidean spaces, are considered in other papers of E. Čech on general homology theory.

There was no cohomology in topology at that time. For some time the groups connected with cohomology by means of Pontrjagin's character theory were used instead in the duality laws for manifolds. Taking into account those circumstances, we see that the principal duality laws for piecewise linear manifolds were already well-studied, and that their various generalizations were discussed in literature. That shed some more light on the nature of the duality.

From a general standpoint, the duality is also studied by E. Čech. Similarly, as in the first paper, the papers under consideration deal with the coefficient groups for homology which are either the field $\mathbb{Q}$ of rational numbers or the cyclic group Z_p of the prime order p. First of all, almost in all of his work E. Čech replaced the usual manifolds R by the R *mod* S where S is a closed subset of a metric compactum R with an arbitrary structure at the points of S and with a local structure at the points of $R \setminus S$, defined by certain homological conditions including requirements of homological local connectedness type and a homological structure around the points. Together with the papers of S. Lefschetz [4] and R. L. Wilder [7], those of E. Čech stand behind the theory of generalized manifolds which appeared later.

The compacta studied in a) are called abstract manifolds. They are supposed to be triangulated at the complement of S. Associating with each q-simplex σ in $R \setminus S$ an $(n-q)$-dimensional complex K_σ surrounding σ (where $n = \dim(R \setminus S)$) the author introduces coefficients of incidence for such complexes, considers the $(n-q)$-dimensional chains generated by them, the connections between q-chains and $(n-q)$-chains, studies the Kronecker's index corresponding to q-cycles and $(n-q)$-cycles (including also the relative cycles). In these terms the author analyses certain local conditions imposed in homological terms in some interval of the dimensions $k \leq q \leq n$, defines abstract manifolds (among them the abstract manifolds of a certain degree of "regularity" depending on k) and establishes the duality relations in a suitable range of the dimensions.

Among other techniques, the local Betti numbers $\beta_m(a, R)$ are used. These numbers are defined (see b)) in terms of the ranks of Čech homology groups $\check{H}_m(R, R \setminus U)$ by means of the limit transition over all the neighbourhoods of the point $a \in R \setminus S$. In modern terms the number $\beta_m(a, R)$ is the rank of the homology group $H^c_m(R, R \setminus a)$. The local structure of R at the points of $R \setminus S$ is defined by the requirement of the local connectedness in the highest dimension n, by the condition $\beta_n(a, R) = 0$ and by the condition $\beta_n(a, A) = 0$ for points a belonging to the intersection of any closed subset $A \subset R$ with the closure of its complement $R \setminus A$. Such compactum R is called a generalized manifold *mod* S of order zero. The author defines orientable generalized manifolds of order zero, considers orienting cycles and classes of homology (which are called principal ones), and studies the fundamental properties of such classes. If, moreover, R is homologically locally m-connected and $\beta_m(a, R) = 0$ in the diapason of the dimensions $n - q \leq m < n$, such a compactum R is called the generalized manifold of order q.

The class of the generalized manifolds (of order q) contains the corresponding abstract manifolds (which may happen not to be manifolds in the customary sense). It must be noted that in the case when the coefficient group is a field, homological manifolds in the modern sense are virtually the Čech generalized manifolds of the order $q = n$.

From the modern point of view the principal relation of the duality obtained by E. Čech is in the following: The homology group of the pair $\check{H}_p(R, S)$ coincides with the group of homomorphisms into the coefficient field of the homology group $H^c_{n-p}(R \setminus S)$ (over the "finite" cycles) of the complement $R \setminus S$ and in fact (since there is no torsion) with the cohomology group $H^{n-p}(R \setminus S)$. This relation is known today as the Poincaré-Lefschetz duality (it was obtained first by S. Lefschetz in the case when S was the boundary of a manifold R). The number p in the Čech's result depends on the diapason of the imposed local conditions. For example, if R is a generalized manifold then its order must be larger than or equal to the least of two numbers p and $n - p$.

Under a suitable acyclicity condition the duality laws of J. W. Alexander and L. S. Pontrjagin are generalized to abstract manifolds. All these results confirm the observation (which was done much later) that in the theory of duality the paramount role must belong to the duality of Poincaré, and that all the other duality relations are either particular forms of Poincaré duality or its special consequences (see, for example, § 5 of Chapter 8 in the review [5]).

In 1935, cohomology was introduced into topology ([1], [2]) and its role in the theory of duality was pointed out [3]. In his note e) E. Čech shows that the fundamental duality relation $H_p(R) = H^{n-p}(R)$ remains valid for any absolute oriented abstract manifold R. He notes that the "dual homology group" $H^{n-p}(R)$ (this term was sometimes used for cohomology at the beginning) can be represented as the character group of the customary $(n-p)$-dimensional homology group (with dual coefficients), but that it is much better to define $H^{n-p}(R)$ directly.

The applications of homology in Čech's work are not limited only to the duality laws. For instance, E. Čech investigates in detail the local Betti numbers $\beta_k(a, R)$ and the results are assuredly of great independent interest. In particular, he has studied the local structure of a compactum R at those points in which the zero-dimensional numbers $\beta_0(a, R)$ are equal to 0, 1 or ∞. In the case when the space R is a union $A \cup B$ of closed subspaces A and B, E. Čech has studied the relations between $\beta_k(a, A \cap B)$ and $\beta_{k+1}(a, R)$ for $a \in A \cap B$ under the condition that $\beta_p(a, A) = \beta_p(a, B) = 0$ in the cases $p = k$ and $p = k + 1$. Later such relations were established as corollaries of the exact homological sequences of triads.

Considering Betti numbers of the subspace $U \setminus A$ for all the neighbourhoods U of a point a belonging to a fixed closed subset $A \subset R$, using the limit transition over all such U, E. Čech defines certain new local Betti numbers $\alpha_p(a, R \setminus A)$ which evidently depend on the inclusion $A \subset R$ and characterize the position of A in R. The connection between $\alpha_0(a, R \setminus A)$ and the number of connected components of subsets $U \setminus A$ for arbitrarily small neighbourhoods U is established. The condition that the numbers $\alpha_p(a, \mathbb{R}^n \setminus A)$ vanish for all $p \leq m$ provides the total accessibility of a subset A of the Euclidean space $R = \mathbb{R}^n$ at a point $a \in A$ if $m = n - 1$, and a weaker property of accessibility if $m = n - 2$ (this result is dealt with in the note d)).

In the case when R is a generalized manifold of order zero, under various additional (alternate) assumptions (which are satisfied by usual manifolds) a painstaking analysis leads (in dependence on the assumptions) to one of the inequalities $\alpha_0(a, R \setminus A) \leq \beta_{n-1}(a, A)$ or $\alpha_0(a, R \setminus A) \geq \beta_{n-1}(a, A)$. Analogous relations are obtained for the local numbers $\alpha_p(a, R \setminus A)$ and $\beta_q(a, A)$, where $q = n - p - 1$ in the case when R is a generalized manifold of the order $t = \min(p, q + 1)$. Thus for homological manifolds (in particular, for the usual ones) and any $p \leq n - 1$ and $q = n - p - 1$ we have the equality $\alpha_p(a, R \setminus a) = \beta_q(a, A)$. This fact is a rather thin display of Poincaré type duality. Since $\beta_q(a, A)$ is the rank of q-dimensional homologies of the pair $(A, A \setminus a)$, this result compared with modern ideas has the following explanation: In accordance with a Poincaré-Lefschetz kind duality in (generalized) manifolds the q-dimensional homologies of the pair $(A, A \setminus a)$ coincide with the $(p+1)$-dimensional cohomologies of the complementary pair $((R \setminus a) \cup a, R \setminus a)$ (see, for example, § 5.8 of Chapter 8 in the review [5]), i.e., with the local cohomologies at the point a in the space $(R \setminus A) \cup a$. But the latter coincide with the direct limit of the p-dimensional cohomologies of the subsets $U \setminus A$ (see § 3.2 of Chapter 4 in the review [6]). Now it suffices to take into account the fact that for the fields of coefficients the ranks of the homology and cohomology groups $H_p(U \setminus A)$ and $H^p(U \setminus A)$ (when both of them are finite) coincide (see § 4.2 of Chapter 8 in the review [5]).

Čech's results on duality, on the local homology and on the position of closed subsets have not lost their importance up to now.

References

[1] ALEXANDER J. W., *On the ring of a compact metric space*, Proc. Nat. Acad. Sci. U.S.A. **21, 8** (1935), 511 – 512.

[2] KOLMOGOROV A. N., *Les groupes de Betti des espaces localement bicompactes*, C. R. Acad. Sci. Paris. **202** (1936), 1144 – 1147.

[3] KOLMOGOROV A. N., *Über Dualität im Aufbau der Kombinatorischen Topologie*, Mat. Sb. **1, 1** (1936), 97 – 102.

[4] LEFSCHETZ S., *On generalized manifolds*, Amer. J. of Math. **55, 4** (1933), 469 – 504.

[5] SKLYARENKO E. G., *Gomologii i kogomologii obščich prostranstv*, Itogi nauki i techniki, sovremennye problemy matematiki, fundamental'nye napravlenija, VINITI **50** (1989), 129 – 266.

[6] SKLYARENKO E. G., *Obščie teorii gomologij i kogomologij. Sovremennoe sostojanie i tipičnye primenenija*, Itogi nauki i techniki. Algebra, topologija, geometrija. VINITI **27** (1989), 125 – 228.

[7] WILDER R. L., *Generalized closed manifolds in n-space*, Ann. of Math. **35, 4** (1934), 876 – 903.

On the paper by E. Čech "Les groupes de Betti d'un complexe infini", Fund. Math. 1935, 25, 33 – 44.

This paper deals with the "classic" simplicial homology. The author considers the following problem: To what extent the integer homologies of a simplicial complex K and a concrete group G define the homologies $H_n(K;G)$ of this complex with coefficients in G.

At that time the solution was known only in the particular case when the complex K was finite and G was a cyclic group (J. W. Alexander [1]). At present (in the case of any group G), the answer to the question is the following splitting short exact sequence

$$0 \to H_n(K;Z) \otimes G \to H_n(K;G) \to Tor(H_{n-1}(K;Z), G) \to 0.$$

Now it is the well-known universal coefficient formula contained practically in every modern book on homology theory.

This is what was obtained by E. Čech, although in other symbols and terms. He pointed out a direct summand $B_1^n(G)$ of $H_n(K;G)$ which is defined exclusively by the groups $H_n(K;Z)$ and G. The detailed description of $B_1^n(G)$ is given. It is also shown that the factor-group $H_n(K;G)/B_1^n(G) = B_2^n(G)$ is defined by both the torsion subgroup in $H_{n-1}(K;Z)$ and the coefficient group G only. Some methods of calculation of this group are given. Some examples are also analysed: for instance, when the group $H_n(K;Z)$ or the torsion subgroup of $H_{n-1}(K;Z)$ is finitely generated, when G is a division group, etc.

Although the tensor product was defined in algebra by H. Whitney [2] in 1938, it appeared first in fact in this article of E. Čech. Moreover, as we can see, the torsion product operation Tor for abelian groups was also introduced here, although it was not called so. Another description of the functor Tor for abelian groups, in terms of generators and relations was given in 1954 by S. Eilenberg and S. MacLane. Much later the paramount importance of the functor Tor in homological algebra was pointed out: this functor turned out to be the left derived functor of the tensor product functor $\otimes$.

REFERENCES

[1] ALEXANDER J.W., *Combinatorial analysis situs, I.*, Trans. Amer. Math. Soc. **28, 2** (1926), 301 – 329.

[2] WHITNEY H., *Tensor products of Abelian groups*, Duke Math. J. **4, 3** (1938), 495 – 528.

On the paper by E. Čech "Multiplication on a complex", Annals of Math. 1936, 37,3, 681 – 697.

It is well known that in their communications at the First International Topological Conference (Moscow, September 1935) J. W. Alexander and A. N. Kolmogorov introduced the notion of cohomology and defined a product of cohomologies (see [1] and [2]). This multiplicative operation was denoted later by the symbol $\smile$ and was named the cup-product. To a considerable extent formal, the multiplication theory in the cohomologies appeared as a result of the development in 1930 – 35 of the intersection theory in the homologies of manifolds which corresponded to intuitive geometric ideas and was based on them. The problem arose to find a description of the multiplication fit to be used in concrete tasks concerning, in particular, cohomologies of polyhedra.

E. Čech was the first to find the required solution. In his work E. Čech presents a general scheme of the construction of a set of $\smile$-product operations in simplicial cochains of a simplicial complex K, shows the connection between each of such operations and the co-boundary homomorphism, and studies the behavior of cocycles and coboundaries under such operations. In particular, for any p-cocycle ξ and q-cocycle η the product $\xi \smile \eta$ is a $(p+q)$-cocycle the cohomology class of which does not depend on the choice of ξ and η in their cohomology classes. It is shown that all the considered multiplications of simplicial cochains lead to the same product operation $\smile$ in the cohomologies of the complex K.

Among the large number of multiplication operations in cochains E. Čech distinguishes one (see § 7) which is rated nowadays as "classical" and included virtually in any book on algebraic topology. The distinguished operation is the following: for $(a_0, \ldots, a_p, a_{p+1}, \ldots, a_{p+q})$ to be the vertexes of an oriented $(p+q)$-simplex we have $(\xi \smile \eta)(a_0, \ldots, a_p, a_{p+1}, \ldots, a_{p+q}) = \xi(a_0, \ldots, a_p) \cdot \eta(a_p, \ldots, a_{p+q})$.

In the original definitions it was presupposed that the coefficient domain is a ring. E. Čech was the first to observe, however, that the multiplication can be defined in all situations when we have a pairing operation for elements of abelian groups A and B with the range in a group G (i.e. a bilinear mapping $A \times B \to G$, see § 5). There arises a cup-product of the form

$$\smile : H^p(K; A) \otimes H^q(K; B) \to H^{p+q}(K; G).$$

Taking into account the orientation of the simplexes, E. Čech receives the relation $h_1 \smile h_2 = (-1)^{pq} h_2 \smile h_1$ as an immediate corollary of his construction.

Dualizing his construction of the $\smile$-multiplication E. Čech defines new product operations — the multiplication of $(p+q)$-chains (with coefficients in A) and q cochains (with coefficients in B) with the results being p-chains with coefficients in G). There appear similar connections of such operations with the boundary and co-boundary homomorphisms, with cycles, cocycles, boundaries and coboundaries. Thus E. Čech was first to define the $\frown$-multiplication

$$\frown: H_{p+q}(K; A) \otimes H^q(K; B) \to H_p(K; G).$$

Among the constructions of the $\frown$-products in § 10 he distinguished one which was rated in the sequel as "classical". It has the following description: for a $(p+q)$-chain f and q-cochain ξ the p-chain $f \frown \xi$ contains a simplex $(a_0, \dots, a_p)$ with the coefficient which is the sum of all products of the kind $f(a_0, \dots, a_p, a_{p+1}, \dots, a_{p+q}) \cdot \xi(a_p, a_{p+1}, \dots, a_{p+q})$ (over all $(p+q)$-simplexes containing p-simplex $(a_0, \dots, a_p)$ as face).

Analogous constructions of mutiplications were proposed later by H. Whitney in [3] (where also the symbols "$\smile$" and "$\frown$" appeared). Both the "classical" multiplication formulas for chains and cochains mentioned above have been used in different concrete calculations since.

Let Γ_n be an orienting homology class of an oriented n-dimensional piecewise linear manifold K (possibly an abstract one in the construction of previous articles of the author). For the integer homology and cohomology there arises the homomorphism $H^{n-p}(K) \to H_p(K)$ defined by the $\frown$-multiplication

$$\frown: \Gamma^n \frown H^{n-p}(K) \to H_p(K).$$

It is shown that this mapping is monomorphic for "p-regular" abstract manifolds K and epimorphic for "$(p-1)$-regular" manifolds. Thus this mapping is isomorphic for any customary manifold K. This isomorphism is one of the most known interpretations of Poincaré duality. In terms of this interpretation E. Čech gives a lucid idea of the intersections of cycles in manifolds (some $(p+q-n)$-cycles naturally corresponding in the general disposition to any p-cycles and q-cycles).

The basic constructions which are contained in the work of E. Čech are widely used in topology till now, hence the work itself has not lost its actuality up to the present day. It goes without saying that while studying the work it is necessary to take into account changes of terminology (for example "cohomology" instead of "dual homology", "cocycles" instead of "dual cycles", etc).

References

[1] Alexander J. W., *On the ring of a compact metric space*, Proc. Nat. Acad. Sci. U.S.A. **21, 8** (1935), 511 – 512.

[2] Kolmogoroff A. N., *Homologiering des Komplexes und des lokal-bikompakten Räumes*, Mat. Sb. **1, 5** (1936), 701 – 706.

[3] Whitney H., *On products in a complex*, Ann. of Math. **39, 2** (1938), 397 – 432.

GENERAL HOMOLOGY THEORY IN AN ARBITRARY SPACE

EDUARD ČECH

In the last years several authors[1] have developed a homology theory in a *compact metric* space. This is the way Alexandroff proceeds: By virtue of the Borel theorem one can cover R by a *finite* system

$$U_1, U_2, \ldots, U_k \tag{1}$$

of open sets[2] whose norm (= maximum of diameters of the sets (1)) is smaller than a given positive number. Starting from (1), one can now derive a (abstract) complex N whose vertices $a_1, a_2, \ldots, a_k$ correspond to the sets (1), and the vertices $a_{\nu_0}, a_{\nu_1}, \ldots, a_{\nu_n}$ determine an n-simplex of N if and only if the corresponding sets $U_{\nu_0}, U_{\nu_1}, \ldots, U_{\nu_n}$ have a common point. Then one considers a sequence of coverings of the type (1) whose norms tend to zero, each covering of the sequence being obtained from the preceding one by a subdivision, and one forms the sequence

$$N_1, N_2, N_3, \ldots \tag{2}$$

of the corresponding complexes (Projektionsfolge). Each cycle $C^n_{\nu+1}$ situated in $N_{\nu+1}$ determines a cycle $C^n_\nu = \pi C^n_{\nu+1}$ in N_ν, the *projection* of $C^n_{\nu+1}$. A sequence of cycles contained in the complexes (2) constitutes a cycle in R (Projektionszyklus, Vollzyklus) if there is for each ν in N_ν the homology $C^n_\nu \sim \pi C^n_{\nu+1}$.

However, if the space R is arbitrary, one can also consider finite coverings (1) formed by open sets (= open nets). Here also, each net can be considered as a complex. The only difference consists in the fact that there does not exist any more a sequence of the type (2) formed by "arbitrarily small" nets. Instead of the sequence (2), one considers here the family of *all* the open nets.[3] The main aim of this paper is to show how one can arrive at a homology theory in an arbitrary

[1] P. Alexandroff, Untersuchungen über Gestalt und Lage abgeschlossener Mengen beliebiger Dimension, Annals of Math., (2)30, 1929; pp. 101–187, where the previous papers by the same author are cited.

S. Lefschetz, Topology, Amer. Math. Soc. Coll. Publ., vol. 12, 1930, Chap. 7, §4.

L. Vietoris, Über den höheren Zusammenhang kompakter Räume und eine Klasse von zusammenhangstreuen Abbildunden, Math. Annalen, 97, 127, pp. 454–472.

For the particular case where R is a part of the Euclidean plane see already L. E. J. Brouwer, Beweis der Invarianz der geschlossenen Kurve, Math. Annalen, 72, 1912, pp. 422–425.

[2] Alexandroff considers *closed* sets. But this difference is not essential (see this paper, V, 8).

[3] It is known that Hurewicz has shown the importance of the family of all the open nets in the theory of dimension of separable metric spaces (see W. Hurewicz, Proc. Acad. Amsterdam 30, 1927, p. 425, or K. Menger, Dimensionstheorie, Chap. V).

EDUARD ČECH

topological space. Moreover, the presentation does not assume any knowledge of the previous papers.

The paper is divided into five chapters. In Chap. I, I recall without proofs some properties of modules. In Chap. II, I explain the general theory in a very abstract way. I suppose here ***nothing*** about the nature of the space R. On the other hand, I suppose having given a *fundamental family of nets*, i. e. a family of finite coverings of R subject to the unique condition that to arbitrary two coverings of the family there exists a third one which is a common refinement of the two given ones. Each fundamental family of nets enables to introduce a homology theory of cycles. Besides, I consider, following Lefschetz,[4] cycles mod α, α being an arbitrary given subset of R. In Chap. III, I consider the important case where R is a topological space with the fundamental family being that of open nets. I explain here first the relations between the cycles in R and the cycles in a closed subset of R. Then I show that the theory of cycles of dimension zero coincides with the theory of connectivity in the sense of Lennes-Hausdorff. At the end, I consider nets which are *regular*[5] with respect to a subset of R.

To this end, I prove (III 20) two lemmas concerning hereditarily normal topological spaces, which may be of independent interest. In Chap. IV, I present an application of the general theory. Namely, I generalize a theorem by Mayer and Vietoris[6] about the homologies in the sum $R_1 + R_2$ of two complexes to the case where R_1 and R_2 are two normal topological spaces closed in their sum. The method which I use here is that of Mayer and Vietoris, but in the substantially more general case, which I consider, there are difficulties that can be overcome on the one hand by means of the notion of regular net, and on the other hand by a general theorem (II 21) concerning the existence of cycles. The theorem which I prove contains as a particular case the "generalized Phragmén-Brouwer theorem" by Alexandroff.[7] In Chap. V, I remark first that the theory of cycles with rational coefficients (due to Lefschetz[8]), explained in the preceding chapters, can be transferred to the case (considered by Alexander)[9] of cycles whose coefficients are integers reduced mod m. Then I show that the case where the fundamental family consists of *closed* nets is essentially identical with the case (which I consider) of open nets. At the end I remark that a new homology theory can be obtained by choosing a fixed additive family of subsets in R. Particularly, one can obtain a theory which is in the same relation to that explained in Chap. III as the notion of a *continuum* to that of a *connected* set is.

[4] Topology, Chap. I, n° 14.

[5] Cf. Lefschetz, Topology, p. 91 (normal neighborhood N^L).

[6] W. Mayer, Über abstrakte Topologie, Monatshefte f. Math. u. Phys. 36, 1929, pp. 1–42. L. Vietoris, Über die Homologiegruppen der Vereinigung zweier Komplexe, ibidem, 37, 1930, pp. 159–162.

[7] Untersuchungen etc., p. 178.

[8] Topology, Chap. VII.

[9] J. W. Alexander, Combinatorial analysis situs, I. Transactions Amer. Math. Soc. , 28, 1926, pp. 301–329.

GENERAL HOMOLOGY THEORY IN AN ARBITRARY SPACE

I. Modules.

1. $\mathfrak{R}$ denotes the set of rational numbers.

An arbitrary *non-empty* set M is called a *module* if there are defined two operations: 1° the *sum* $a+b \in M$ of two elements $a, b \in M$; 2° the *product* $ra \in M$ of an element $a \in M$ with a number $r \in \mathfrak{R}$. Further, we suppose that

1° with respect to the addition, M is a commutative group whose zero element is denoted by $0'$;

2° for $a, b \in M$; $r, s \in \mathfrak{R}$, it holds

$$r(a+b) = ra + rb, \qquad (r+s)a = ra + sa$$
$$r(sa) = (rs)a, \qquad 1a = a.$$

Hence, one deduces in particular that $0a = 0'$, $r0' = 0'$, whereas for $r \neq 0$, $a \neq 0'$ one has $ra \neq 0'$. In the next, we shall write simply 0 instead of $0'$.

2. An element $a \in M$ is said to *depend* on $A \subset M$ if $a = 0$ or $a = \sum_1^n r_\nu a_\nu$ for a convenient choice of $n = 1, 2, 3, \ldots, r_\nu \in \mathfrak{R}$, $a_\nu \in A$ $(1 \leqq \nu \leqq n)$.

A subset $A \subset M$ is said to be *independent* if it is not possible to find pairwise dictinct elements $a_1, \ldots, a_n \in M$ $(n = 1, 2, 3, \ldots)$ and numbers $r_1, \ldots, r_n \in \mathfrak{R}$, at least one of which is $\neq 0$, in such a way that $\sum_1^n r_n a_\nu = 0$. This condition is satisfied if $A = 0$ is the empty set. A set $A \subset M$ is a *basis* of the module M if 1° A is independent; 2° every element of M depends on A.

3. Every module possesses at least one basis. The number h of elements of a basis (it is a natural number or an aleph) is the same for all the bases. We shall call h *rank* of the module M. A module M is called *finite* if its rank is finite.

4. A set $M_1 \subset M$, $M_1 \neq 0$ is a module if and only if

1° a, $b \in M_1$ implies $a + b \in M_1$;

2° $a \in M_1$, $r \in \mathfrak{R}$ implies $ra \in M_1$. Then M_1 is called *submodule* of the module M.

5. Let A_1 be a basis of a submodule M_1 of the module M. Then there exists a basis $A \supset A_1$ of the module M.

6. Let h, h_1 be the ranks of a module M and its submodule M_1, respectively. Then $h_1 \leqq h$. Thus, if the module M is finite, then M_1 is also finite. In this case the equality $h_1 = h$ implies $M_1 = M$.

7. A *decreasing* sequence of submodules of a *finite* module M is always *finite* (if the rank of M is h, then the sequence has at most $h+1$ elements).

8. Let $\mathfrak{M}$ be a family (non-empty) of submodules of a *finite* module M, and let P be the product (= common part) of all the modules of the family $\mathfrak{M}$. Then there exist elements $M_1, M_2, \ldots, M_k$ (k is finite) of the family $\mathfrak{M}$ such that $P = M_1 M_2 \ldots M_k$. If $P \neq 0$, then it is a submodule of M.

9. Let M, M' be two modules. Let f be a univalent function defined in the domain M and such that: 1° for $a \in M$, there is $f(a) \in M'$; 2° for $a' \in M'$, there exists an element $a \in M$ such that $a' = f(a)$; 3° for a, $b \in M$, there is $f(a+b) = f(a) + f(b)$; 4° for $a \in M$, $r \in \mathfrak{R}$, there is $f(ra) = rf(a)$. Then the

module M' is called a *homomorphic* image of the module M. If $a \neq b$ implies $f(a) \neq f(b)$, then the modules M and M' are called *isomorphic*.

10. Let h and h' be the ranks of modules M and M', respectively. If M' is a homomorphic image of M, then $h' \leqq h$ (in particular, a homomorphic image of a finite module is a finite module). If M and M' are isomorphic, then $h' = h$. The converse is also true.

11. Let M_1 be a submodule of a module M. It is possible to consider as *equal* two elements $a, b \in M$ such that $a - b \in M_1$. In this way one can obtain from M a new module which we shall denote by $M - M_1$. Obviously $M - M_1$ is a homomorphic image of M.

12. Let M_1, M_2 be two submodules of a module M possessing the following property: For each $a \in M$, there exists one and only one pair a_1, a_2 such that $a_1 \in M_1$, $a_2 \in M_2$, $a = a_1 + a_2$. Then the module M is said to be the *direct sum* of the modules M_1, M_2, and we write $M = M_1 + M_2$.

13. Let M_1 be a submodule of a module M. Then there exists a submodule M_2 of M such that $M = M_1 + M_2$. The module M_2 is isomorphic to $M - M_1$.

14. Let M be a module, and let $L \subset M$. We say that L is a *linear system* if there exists an element $a_0 \in L$ and a submodule M_1 of M such that, for $a \in M$, there is $a \in L$ if and only if $a - a_0 \in M_1$.

15. Let Λ be a family (non-empty) of linear systems contained in a *finite* module M, and let Q be the intersection of all the systems belonging to the family Λ. Then there exist elements $L_1, L_2, \ldots, L_k$ (k finite) of the family Λ such that $Q = L_1 L_2 \ldots L_k$. If $Q \neq 0$, then it is a linear system.

II. Homologies with respect to a fundamental family of nets.

1. Let R be an arbitrary given set. Let Z be a given family satisfying the following two axioms:

1° Every element $\mathfrak{U}$ of Z is a *finite* system $U_1, U_2, \ldots, U_k$ of subsets of R such that $\sum_1^k U_\nu = R$; $U_\nu \neq 0$;

2° Given $\mathfrak{U}_1$, $\mathfrak{U}_2 \in Z$, there exists an element $\mathfrak{V} \in Z$ such that $V \in \mathfrak{V}$ implies $V \subset U_1 U_2$ for a convenient choice of $U_1 \in \mathfrak{U}_1$, $U_2 \in \mathfrak{U}_2$.

Each element $\mathfrak{U}$ of Z will be called a *net*. The family Z will be called the *fundamental family of nets*. The elements U of a net $\mathfrak{U}$ will be called the *vertices* of the net $\mathfrak{U}$, and also the $(0, \mathfrak{U})$-*simplices*. Usually, we shall denote a net by the letters $\mathfrak{U}$, $\mathfrak{V}$, $\mathfrak{W}$. The vertices of $\mathfrak{U}$ will be denoted, e. g., by the letter U (if need be, with an index).

2. For a net $\mathfrak{U}$ and a natural number n, an $(n, \mathfrak{U})$-*simplex* is by definition a symbol of the form

$$(U_0, U_1, \ldots, U_\nu, \ldots, U_n),$$

where the U_ν's (which will be called *vertices of the simplex*) are pairwise distinct vertices of $\mathfrak{U}$ such that the set

$$\prod_0^n U_\nu \tag{1}$$

GENERAL HOMOLOGY THEORY IN AN ARBITRARY SPACE

is not empty. If $(\nu_0, \nu_1, \ldots, \nu_n)$ is a permutation of the indices $(0, 1, \ldots, n)$, we shall set

$$(U_{\nu_0}, U_{\nu_1}, \ldots, U_{\nu_n}) = \pm(U_0, U_1, \ldots U_n) \tag{2}$$

with the upper (lower) sign in the case of an even (odd) permutation. The set (1) will be called the *kernel* of the simplex. We shall denote an $(n, \mathfrak{U})$-simplex by $S^n(\mathfrak{U})$ or by S^n (possibly with a lower index). The kernel of the simplex S^n will be denoted by $J(S^n)$.

3. Let $\mathfrak{U}$ be a net, $n = 0, 1, 2, \ldots$. Let $S_1^n, S_2^n, \ldots, S_{\alpha_n}^n$ be all the $(n, \mathfrak{U})$-simplices (from the two simplices, as in 2(2), we write only one). An $(n, \mathfrak{U})$-chain is by definition a symbol of the form

$$\sum_1^{\alpha_n} r_\nu S_\nu^n,$$

where $r_\nu \in \mathfrak{R}$. We shall denote an $(n, \mathfrak{U})$-chain by $K^n(\mathfrak{U})$ or by K^n (possibly with a lower index). According to the evident conventions, the set of all the $(n, \mathfrak{U})$-chains forms a *finite module*. For almost all values of n, there is $\alpha_n = 0$. Consequently, there exists only one $(n, \mathfrak{U})$-chain $K^n = 0$.

4. The *boundary* of a $(0, \mathfrak{U})$-chain is zero. We use here the notation $F(K^0) = 0$ or $K^0 \to 0$. Now let $n > 0$. The *boundary* of an $(n, \mathfrak{U})$-simplex $S^n = (U_0, U_1, \ldots, U_n)$ is the $(n-1, \mathfrak{U})$-chain

$$F(S^n) = \sum_0^{n-1} (-1)^\nu S_\nu^{n-1},$$

where S_ν^{n-1} is an $(n-1, \mathfrak{U})$-simplex whose symbol one obtains from $(U_0, U_1, \ldots, U_n)$ by omitting the vertex U_ν. The *boundary* of an $(n, \mathfrak{U})$-chain

$$K^n = \sum_1^{\alpha_n} r_\nu S_\nu^n$$

is the $(n-1, \mathfrak{U})$-chain

$$F(K^n) = \sum_1^{\alpha_n} r_\nu F(S_\nu^n)\,.$$

Instead of $K^{n-1} = F(K^n)$, we shall also write $K^n \to K^{n-1}$. With respect to the operation F, a certain submodule of the module of $(n-1, \mathfrak{U})$-chains is a *homomorphic* image of the module of all the $(n, \mathfrak{U})$-chains.

5. From now on, the letter A will denote a given subset of R. We shall say that an $(n, \mathfrak{U})$-chain

$$K^n = \sum_1^{\alpha_n} r_\nu S_\nu^n$$

is *contained in* A (notation: $K^n \subset A$) if for each value of ν, one of the following two cases arises: 1° $r_\nu = 0$; 2° $A \cdot J(S_\nu^n) \neq 0$. This condition is always satisfied

EDUARD ČECH

in the case $A = R$. The $(n,\mathfrak{U})$-chains contained in A form a *module*. Obviously $K^n \subset A$ implies $F(K^n) \subset A$.[10]

6. From now on, the letter α denotes a given subset of A. The notation $K_1^n(\mathfrak{U}) = K_2^n(\mathfrak{U})$ mod α means that $K_1^n(\mathfrak{U}) - K_2^n(\mathfrak{U}) \subset \alpha$. In this case, if $K_1^n(\mathfrak{U}) \subset A$, then also $K_2^n(\mathfrak{U}) \subset A$. If we consider as equal two $(n,\mathfrak{U})$-chains which are equal mod α, then the set of all the $(n,\mathfrak{U})$-chains forms again a module (see I, 11). We shall write

$$K^n \to K^{n-1} \bmod \alpha \tag{1}$$

in order to indicate that $F(K^n) = K^{n-1}$ mod α. In the relation (1), it is permitted to replace any chain by another one equal to it mod α.

7. An $(n,\mathfrak{U})$-chain $K^n \subset A$ will be called an *$(n,\mathfrak{U})$-cycle mod α in A* if $K^n \to 0$ mod α. [In the case $A = R$ we shall simply speak about an $(n,\mathfrak{U})$-cycle mod α.] The $(n,\mathfrak{U})$-cycles mod α in A form a module. We shall denote the $(n,\mathfrak{U})$-cycles mod α by $C^n(\mathfrak{U})$ or by C^n, possibly with a lower index. In the case $\alpha = 0$, we shall speak about *absolute* $(n,\mathfrak{U})$-cycles. Obviously, an absolute $(n,\mathfrak{U})$-cycle is also an $(n,\mathfrak{U})$-cycle mod α for any choice of α.

8. It can be proved without difficulty[11] that $F[S^{n+1}(\mathfrak{U})]$ is an absolute $(n,\mathfrak{U})$-cycle. Thus, an $(n,\mathfrak{U})$-chain C^n is an $(n,\mathfrak{U})$-cycle mod α in A if there exists an $(n+1,\mathfrak{U})$-chain $K^{n+1} \subset A$ such that $K^{n+1} \to C^n$ mod α (hence it follows $C^n \subset A$). Every $(n,\mathfrak{U})$-cycle C^n mod α in A having this property will be called *homologic to zero* mod α in A, and we shall write

$$C^n \sim 0 \bmod \alpha \text{ in } A.$$

[We shall omit "in A" if $A = R$.] The notation

$$C_1^n \sim C_2^n \bmod \alpha \text{ in } A \tag{1}$$

means that C_1^n and C_2^n are $(n,\mathfrak{U})$-cycles mod α in A such that $C_1^n - C_2^n \sim 0$ mod α in A. The $(n,\mathfrak{U})$-cycles homologic to zero mod α in A form obviously a submodule of the module of all $(n,\mathfrak{U})$-cycles mod α in A. Considering as equal two cycles C_1^n, C_2^n related by (1), which we shall do everywhere in what follows, the $(n,\mathfrak{U})$-cycles mod α in A form (see I 11) a finite module. In particular, the relation (1) holds if $C_1^n = C_2^n$ mod α.

9. A net $\mathfrak{V}$ is a *refinement* of a net $\mathfrak{U}$ if each vertex V of $\mathfrak{V}$ is a part of some vertex U of $\mathfrak{U}$. If $\mathfrak{U}_1, \mathfrak{U}_2, \ldots, \mathfrak{U}_k$ are given nets (in finite number), then according to n° 1, axiom 2°, there exists a simultaneous refinement $\mathfrak{V}$ of all the nets $\mathfrak{U}_\nu$. Obviously, a refinement of a refinement of a net $\mathfrak{U}$ is a refinement of the net $\mathfrak{U}$.

10. Let $\mathfrak{V}$ be a refinement of a net $\mathfrak{U}$. Then to each vertex V of $\mathfrak{V}$ we can assign a well determined vertex $\pi V = U \supset V$ of the net $\mathfrak{U}$. The operation π will be called the *projection* of the net $\mathfrak{V}$ into the net $\mathfrak{U}$. We shall write $\pi = \mathrm{Pr.}(\mathfrak{V},\mathfrak{U})$. For the given nets $\mathfrak{U}$, $\mathfrak{V}$, there can exist *several* projections of $\mathfrak{V}$ into $\mathfrak{U}$.

11. Let $\mathfrak{V}$ be a refinement of a net $\mathfrak{U}$, $\pi = \mathrm{Pr.}(\mathfrak{V},\mathfrak{U})$. Let

$$S^n = (V_0, V_1, \ldots, V_n)$$

[10] It is important to remark that the relations $K^n \subset A_1$, $K^n \subset A_2$ do not imply $K^n \subset A_1 A_2$.
[11] See e. g. Lefschetz, Topology, p. 19.

be an $(n, \mathfrak{V})$-simplex. If the vertices

$$\pi V_0, \pi V_1, \ldots, \pi V_n$$

of the net $\mathfrak{U}$ are not pairwise distinct, we set $\pi S^n = 0$. In the opposite case

$$\pi S^n = (\pi V_0, \pi V_1, \ldots, \pi V_n)$$

is an $(n, \mathfrak{U})$-simplex. If

$$K^n = \sum_1^{\alpha_n} r_\nu S_\nu^n$$

is an $(n, \mathfrak{V})$-chain, its projection will be by definition the $(n, \mathfrak{U})$-chain

$$\pi K^n = \sum_1^{\alpha_n} r_\nu \pi S_\nu^n.$$

By virtue of the operation π, a certain submodule of the module of all $(n, \mathfrak{U})$-chains is a *homomorphic* image of the module of all $(n, \mathfrak{V})$-chains. For each $(n, \mathfrak{V})$-simplex S^n, we have the relation

$$\pi F(S^n) = F(\pi S^n),$$

which is evident if $\pi S^n \neq 0$, but which holds also[12] in the case $\pi S^n = 0$. Therefore, $\pi F(K^n) = F(\pi K^n)$ for every $(n, \mathfrak{V})$-chain K^n. If $K^n \subset A$ or $K^n \subset \alpha$, then there is obviously also $\pi K^n \subset A$ or $\pi K^n \subset \alpha$. From all these remarks it follows: if C^n is an $(n, \mathfrak{V})$-cycle mod α in A, then πC^n is an $(n, \mathfrak{U})$-cycle mod α in A. If, moreover, $C^n \sim 0$ mod α in A, then there is also $\pi C^n \sim 0$ mod α in A.

12. Let $\mathfrak{V}$ be a refinement of a net $\mathfrak{U}$, and let $\pi_1 = \text{Pr.}(\mathfrak{V}, \mathfrak{U})$, $\pi_2 = \text{Pr.}(\mathfrak{V}, \mathfrak{U})$. Let us arrange all the vertices of $\mathfrak{V}$ into a well determined finite sequence

$$V_1, V_2, \ldots, V_k,$$

and let us set $\pi_1 V_\nu = U'_\nu$, $\pi_2 V_\nu = U''_\nu$. Now, let

$$S^n = (V_{\nu_0}, V_{\nu_1}, \ldots, V_{\nu_n})$$

be an $(n, \mathfrak{V})$-simplex. We may suppose that $\nu_0 < \nu_1 \cdots < \nu_n$. Let us set for a moment

$$P(S^n) = \sum_{i=0}^{n} (-1)^{i-1} (U''_{\nu_0} U''_{\nu_1} \ldots U''_{\nu_{i-1}} U''_{\nu_i} U'_{\nu_i} U'_{\nu_{i+1}} \ldots U'_{\nu_{n-1}} U'_{\nu_n}),$$

agreeing that each symbol on the right hand side whose vertices are not pairwise distinct means zero. Then $P(S^n)$ is an $(n+1, \mathfrak{U})$-chain. If $K^n = \sum_1^{\alpha_n} r_\nu S_\nu^n$ is an $(n, \mathfrak{V})$-chain, let us set $P(K^n) = \sum_1^{\alpha_n} r_\nu P(S_\nu^n)$. One can easily prove[13] that

$$P(S^n) \to \pi_2 S^n - \pi_1 S^n - P[F(S^n)] \ ,$$

[12] See e. g. Lefschetz, op. c., Chap. II, n° 2.

[13] Cf. Lefschetz, op. c. , Chap. II, n° 8.

EDUARD ČECH

whence for each $(n,\mathfrak{V})$-chain K^n

$$P(K^n) \to \pi_2 K^n - \pi_1 K^n - P[F(K^n)].$$

In particular, let us consider an $(n,\mathfrak{V})$-cycle C^n mod α in A. Then $C^n \subset A$, $F(C^n) \subset \alpha$, whence $P(C^n) \subset A$, $P[F(C^n)] \subset \alpha$, so that the relation (1) gives

$$\pi_2 C^n \sim \pi_1 C^n \text{ mod } \alpha \text{ in } A.$$

We have agreed to consider as equal two $(n,\mathfrak{U})$-cycles mod α in A homologic mod α in A. Therefore, *we can always choose arbitrarily the projection* Pr.$(\mathfrak{V},\mathfrak{U})$.

13. Let $\mathfrak{V}$ be a refinement of a net $\mathfrak{U}$, and let $\mathfrak{W}$ be a refinement of the net $\mathfrak{V}$. Let $\pi = \text{Pr.}(\mathfrak{V},\mathfrak{U})$, $\pi' = \text{Pr.}(\mathfrak{W},\mathfrak{U})$. Let us suppose that an $(n,\mathfrak{U})$-cycle $C^n(\mathfrak{U})$ mod α in A has the property that $\pi' C^n(\mathfrak{W}) \sim C^n(\mathfrak{U})$ mod α in A, for a convenient choice of the $(n,\mathfrak{W})$-cycle $C^n(\mathfrak{W})$ mod α in A. Then $\pi C^n(\mathfrak{V}) \sim C^n(\mathfrak{U})$ mod α in A, for a convenient choice of the $(n,\mathfrak{W})$-cycle $C^n(\mathfrak{V})$ mod α in A.

In fact, let $\pi'' = \text{Pr.}(\mathfrak{W},\mathfrak{V})$. According to 12, we can suppose that $\pi' = \pi\pi''$.[14] Then it is sufficient to set $C^n(\mathfrak{V}) = \pi'' C^n(\mathfrak{W})$.

14. An $(n,\mathfrak{U})$-cycle mod α in A will be called *essential* if, for each refinement $\mathfrak{V}$ of $\mathfrak{U}$, there exists an $(n,\mathfrak{V})$-cycle $C^n(\mathfrak{V})$ mod α in A such that $\pi C^n(\mathfrak{V}) \sim C^n(\mathfrak{U})$, where $\pi = \text{Pr.}(\mathfrak{V},\mathfrak{U})$. If one considers as equal two $(n,\mathfrak{U})$-cycles homologic mod α in A, then the essential $(n,\mathfrak{U})$-cycles mod α in A form a finite module. This important module will be denoted by $M_n(A,\mathfrak{U};\alpha)$. In the case $A = R$ the symbol A will be omitted. Similarly, in the case $\alpha = 0$, the symbol α will be omitted. Thus, e. g., $M_n(\mathfrak{U}) = M_n(R,\mathfrak{U};0)$.

15. A refinement $\mathfrak{V}$ of a net $\mathfrak{U}$ is called *normal* if, for every $n = 0,1,2,\dots$ and for every $(n,\mathfrak{V})$-cycle $C^n(\mathfrak{V})$ mod α in A, the $(n,\mathfrak{U})$-cycle $\pi C^n(\mathfrak{V})$ mod α in A $[\pi = \text{Pr.}(\mathfrak{V},\mathfrak{U})]$ is essential. Thus, the notion of normality depends on A and α. According to 13, every refinement of a normal refinement of a net $\mathfrak{U}$ is a normal refinement of $\mathfrak{U}$.

16. For an arbitrary given net $\mathfrak{U}$, there exists a normal refinement of $\mathfrak{U}$ (for a given choice of A, α).

Proof. The normality condition is trivial for all the sufficiently large values of n (for all the values of n, for which the $(n,\mathfrak{U})$-simplices do not exist). Therefore, it is sufficient (cf. the remark at the end of N° 15) to prove the existence of a refinement satisfying the normality condition for a *given* value of n. For each refinement $\mathfrak{V}$ of the net $\mathfrak{U}$, let $\Gamma(\mathfrak{V})$ be the set of all the $(n,\mathfrak{U})$-cycles $C^n(\mathfrak{U})$ mod α in A for which there exists an $(n,\mathfrak{V})$-cycle $C^n(\mathfrak{V})$ mod α in A such that $C^n(\mathfrak{U}) \sim \pi C^n(\mathfrak{V})$ mod α in A, where $\pi = \text{Pr.}(\mathfrak{V},\mathfrak{U})$. For any choice of $\mathfrak{V}$, $\Gamma(\mathfrak{V})$ is a submodule of the *finite* module $M_n(A,\mathfrak{U};\alpha)$. Moreover, if $\mathfrak{W}$ is a refinement of $\mathfrak{V}$, then, according to 13, there is $\Gamma(\mathfrak{W}) \subset \Gamma(\mathfrak{V})$.

Obviously, the set E of all *essential* $(n,\mathfrak{U})$-cycles mod α in A coincides with the common part of all $\Gamma(\mathfrak{V})$, where $\mathfrak{V}$ runs through all the refinements of $\mathfrak{U}$. According to I 8, there exist refinements $\mathfrak{V}_1,\mathfrak{V}_2,\dots,\mathfrak{V}_k$ (k is finite) of $\mathfrak{U}$ such that

[14] The operation $\pi\pi''$ is obtained by performing first the operation π'' and then π.

GENERAL HOMOLOGY THEORY IN AN ARBITRARY SPACE

$E = \prod_{1}^{k} \Gamma(\mathfrak{V}_\nu)$. Let $\mathfrak{W}$ be a common refinement of all the nets $\mathfrak{V}_\nu$ $(1 \leqq \nu \leqq k)$. Then $\Gamma(\mathfrak{W}) \subset \Gamma(\mathfrak{V}_\nu)$, whence $\Gamma(\mathfrak{W}) \subset E$ (and naturally $\Gamma(\mathfrak{W}) = E$). Consequently, the refinement $\mathfrak{W}$ of $\mathfrak{U}$ has the desired property.

17. More generally, according to the remark at the end of N° 15, one has: if $\mathfrak{U}_1, \ldots, \mathfrak{U}_k$ is a finite number of given nets, then there exists a common normal refinement of all $\mathfrak{U}_\nu$.

18. Let $\mathfrak{V}$ be a refinement of a net $\mathfrak{U}$, $\pi = \text{Pr.}(\mathfrak{V}, \mathfrak{U})$. Let $C^n(\mathfrak{U})$ be an essential $(n, \mathfrak{U})$-cycle mod α in A. Then there exists an *essential* $(n, \mathfrak{V})$-cycle $C^n(\mathfrak{V})$ mod α in A such that $\pi C^n(\mathfrak{V}) \sim C^n(\mathfrak{U})$ mod α in A.

Proof. Let (16) $\mathfrak{W}$ be a *normal* refinement of the net $\mathfrak{V}$, $\pi' = \text{Pr.}(\mathfrak{W}, \mathfrak{V})$. Then $\pi\pi' = \text{Pr.}(\mathfrak{W}, \mathfrak{U})$. Because the cycle $C^n(\mathfrak{U})$ is essential, there exists an $(n, \mathfrak{W})$-cycle $C^n(\mathfrak{W})$ mod α in A such that $C^n(\mathfrak{U}) \sim \pi\pi' C^n(\mathfrak{W})$ mod α in A. Setting $C^n(\mathfrak{V}) = \pi' C^n(\mathfrak{W})$, one has $C^n(\mathfrak{U}) \sim \pi C^n(\mathfrak{V})$ mod α in A. Moreover, the cycle $C^n(\mathfrak{V})$ is essential because $\mathfrak{W}$ is a normal refinement of $\mathfrak{V}$.

19. The theorem which has been proved can be obviously formulated as follows: If $\mathfrak{V}$ is a refinement of $\mathfrak{U}$, then the module $M_n(A, \mathfrak{U}; \alpha)$ is a homomorphic image of the module $M_n(A, \mathfrak{V}; \alpha)$ under the operation $\pi = \text{Pr.}(\mathfrak{V}, \mathfrak{U})$.

20. For *each* net $\mathfrak{U}$, let be given an $(n, \mathfrak{U})$-cycle $C^n(\mathfrak{U})$ mod α in A, and let us suppose that the following condition is satisfied: If $\mathfrak{V}$ is a refinement of $\mathfrak{U}$, $\pi = \text{Pr.}(\mathfrak{V}, \mathfrak{U})$, then $C^n(\mathfrak{U}) \sim \pi C^n(\mathfrak{V})$ mod α in A. The set $\{C^n(\mathfrak{U})\}$ of all the cycles $C^n(\mathfrak{U})$ will be called *(n, R)-cycle mod α in A*. In the case $A = R$, the atribute "in A" will be omitted. In the case $\alpha = 0$, we shall speak about *absolute* (n, R)-cycles. By virtue of the obvious conventions, the set of all the (n, R)-cycles mod α in A constitues a *module*. The *homology* $\{C^n(\mathfrak{U})\} \sim 0$ mod α in A means that $C^n(\mathfrak{U}) \sim 0$ mod α in A for each net $\mathfrak{U}$. The homology $\{C_1^n(\mathfrak{U})\} \sim \{C_2^n(\mathfrak{U})\}$ means that $\{C_1^n(\mathfrak{U})\} - \{C_2^n(\mathfrak{U})\} = \{C_1^n(\mathfrak{U}) - C_2^n(\mathfrak{U})\} \sim 0$ mod α in A. If one considers as equal two homologic cocycles mod α in A, then the set of all the (n, R)-cycles mod α in A constitues again (see I 11) a module. This important module will be denoted by $M_n(A, R; \alpha)$, and its rank (which is a natural number or an aleph) will be denoted by $P_n(A, R; \alpha)$. In the case $A = R$, the letter A will be omitted. Similarly in the case $\alpha = 0$, the letter α will be omitted. The number $P_n(R; \alpha)$ is the *n-th Betti number* of R mod α. The number $P_n(R)$ is the n-th absolute Betti number of R.

21. For *each* net $\mathfrak{U}$ let be given a linear system[15] (see I 14) $L^n(\mathfrak{U})$ of $(n, \mathfrak{U})$-cycles mod α in A, and let us suppose that the following condition is satisfied: If $\mathfrak{V}$ is a refinement of $\mathfrak{U}$, $\pi = \text{Pr.}(\mathfrak{V}, \mathfrak{U})$, then there is $L^n(\mathfrak{V}) \subset L^n(\mathfrak{U})$. Therefore, for each net $\mathfrak{U}$, there exists an (n, R)-cycle $\{C^n(\mathfrak{U})\}$ mod α in A such that $C^n(\mathfrak{U}) \in L^n(\mathfrak{U})$.

The proof of this theorem will be the subject of N°s 22-27.

22. For each net $\mathfrak{U}$, let $L_1^n(\mathfrak{U})$ be the common part of all the sets $\pi L^n(\mathfrak{V})$,[16] where $\mathfrak{V}$ runs through all the refinements of $\mathfrak{U}$, $\pi = \text{Pr.}(\mathfrak{V}, \mathfrak{U})$. Obviously, each $\pi L^n(\mathfrak{V})$ is a linear system of $(n, \mathfrak{U})$-cycles mod α in A, and if $\mathfrak{W}$ is a refinement

[15] We consider two homologic cycles mod α in A as equal, so that the relations $C_1^n(\mathfrak{U}) \in L^n(\mathfrak{U})$, $C_2^n(\mathfrak{U}) \sim C_1^n(\mathfrak{U})$ mod α in A imply $C_2^n(\mathfrak{U}) \in L^n(\mathfrak{U})$.

[16] $\pi L^n(\mathfrak{V})$ is the set of all the $(n, \mathfrak{U})$-cycles $C^n(\mathfrak{U})$ mod α in A such that $C^n(\mathfrak{U}) \sim \pi C^n(\mathfrak{V})$ for a convenient choice of $C^n(\mathfrak{V}) \in L^n(\mathfrak{V})$.

EDUARD ČECH

of $\mathfrak{V}$, $\pi' = \text{Pr.}(\mathfrak{W}, \mathfrak{U})$, then one has $\pi' L^n(\mathfrak{W}) \subset \pi L^n(\mathfrak{V})$ (cf. 13). According to I 15, there exist refinements $\mathfrak{V}_1, \mathfrak{V}_2, \cdots \mathfrak{V}_k$ (k is finite) of $\mathfrak{U}$ such that $L_1^n(\mathfrak{U}) = \prod_1^k \pi_\nu L^n(\mathfrak{V}_\nu)$, where $\pi_\nu = \text{Pr.}(\mathfrak{V}_\nu, \mathfrak{U})$. Let $\mathfrak{V}$ be a common refinement of the nets $\mathfrak{V}_\nu, \pi = \text{Pr.}(\mathfrak{V}, \mathfrak{U})$. Then $\pi L^n(\mathfrak{V}) \subset \pi_\nu L^n(\mathfrak{V}_\nu)$, and, therefore, $\pi L^n(\mathfrak{V}) \subset L_1^n(\mathfrak{U})$, whence $\pi L^n(\mathfrak{V}) = L_1^n(\mathfrak{U})$ by virtue of the very definition of $L_1^n(\mathfrak{U})$.

23. A refinement $\mathfrak{V}$ of a net $\mathfrak{U}$ is called *favourable* if $\pi L^n(\mathfrak{V}) = L_1^n(\mathfrak{U})$. We are going to see that every net $\mathfrak{U}$ has a favourable refinement. Obviously, each refinement of a favourable refinement of a net $\mathfrak{U}$ is a favourable refinement of $\mathfrak{U}$. Thus, if there is given a *finite* set of nets, then there exists a common favourable refinement of all the given nets.

24. Let $\mathfrak{V}$ be a refinement of a net $\mathfrak{U}$, $\pi = \text{Pr.}(\mathfrak{V}, \mathfrak{U})$. Then $\pi L_1^n(\mathfrak{V}) = L_1^n(\mathfrak{U})$.

Proof. Let $\mathfrak{W}$ be a common favourable refinement of the nets $\mathfrak{U}$, $\mathfrak{V}$, and let $\pi' = \text{Pr.}(\mathfrak{W}, \mathfrak{V})$. Then $\pi\pi' = \text{Pr.}(\mathfrak{W}, \mathfrak{U})$. According to the very definition of the favourable refinement, one has $\pi' L^n(\mathfrak{W}) = L_1^n(\mathfrak{V})$, $\pi\pi' L^n(\mathfrak{W}) = L_1^n(\mathfrak{U})$, whence $\pi L_1^n(\mathfrak{V}) = L_1^n(\mathfrak{U})$.

25. Let us arrange the fundamental family Z of nets into a well ordered transfinite sequence

$$\mathfrak{U}_0, \mathfrak{U}_1, \ldots \mathfrak{U}_\omega, \mathfrak{U}_{\omega+1}, \ldots, \mathfrak{U}_\xi, \ldots \qquad (\xi < \gamma).$$

Therefore one can create a transfinite sequence

$$(1) \qquad C_0^n, C_1^n, \ldots, C_\omega^n, C_{\omega+1}^n, \ldots, C_\xi^n \ldots \qquad (\xi < \gamma),$$

where $C_\xi^n \in L_1^n(\mathfrak{U}_\xi)$, so that the following condition P is satisfied: If $\eta_1, \eta_2, \ldots, \eta_k$ is a finite set of ordinal numbers less than γ, then there exists a common refinement $\mathfrak{V}$ of the nets $\mathfrak{U}_{\eta_\nu}$ $(1 \leqq \nu \leqq k)$ and a cycle $C^n(\mathfrak{V}) \in L_1^n(\mathfrak{V})$ such that $\pi_\nu C^n(\mathfrak{V}) \sim C_{\eta_\nu}^n$ mod α in A $[1 \leqq \nu \leqq k;\ \pi_\nu = \text{Pr.}(\mathfrak{V}, \mathfrak{U}_{\eta_\nu})]$.

The proof of this statement will be given in the next n^{os}. Then the assertion of n° 21 will be proved, for $\{C_\xi^n\}$ is an (n, R)-cycle mod α in A. In fact, if $\mathfrak{U}_{\eta_2}$ is a refinement of $\mathfrak{U}_{\eta_1}$, $\pi = \text{Pr.}(\mathfrak{U}_{\eta_2}, \mathfrak{U}_{\eta_1})$, then according to the property P, there exists a common refinement $\mathfrak{V}$ of the nets $\mathfrak{U}_{\eta_1}$, $\mathfrak{U}_{\eta_2}$ and a cycle $C^n(\mathfrak{V}) \in L_1^n(\mathfrak{V})$ such that $\pi_1 C^n(\mathfrak{V}) \sim C_{\eta_1}^n$, $\pi_2 C^n(\mathfrak{V}) \sim C_{\eta_2}^n$ mod α in A, where $\pi_1 = \text{Pr.}(\mathfrak{V}, \mathfrak{U}_{\eta_1})$, $\pi_2 = \text{Pr.}(\mathfrak{V}, \mathfrak{U}\eta_2)$. Thus one can suppose that $\pi_1 = \pi\pi_2$ so that $C_{\eta_1} \sim \pi\pi_2 C^n(\mathfrak{V}) \sim \pi C_{\eta_2}^n$ mod α in A.

26. The transfinite sequence 25 (1) can be constructed by transfinite induction. The cycle $C_0^n \in L_1^n(\mathfrak{U}_0)$ can be chosen arbitrarily. Let us suppose that, for a given ordinal number $\xi < \gamma$, we have already determined all the terms $C_\eta^n \in L_1^n(\mathfrak{U}_\eta)$, $\eta < \xi$ of the sequence 25 (1) in such a way that the following property is satisfied: If $\eta_1, \eta_2, \ldots, \eta_k$ is a finite number of ordinal numbers less than ξ, then there exists a common refinement $\mathfrak{V}$ of the nets $\mathfrak{U}_{\eta_\nu}$ $(1 \leqq \nu \leqq k)$ and a cycle $C^n(\mathfrak{V}) \in L_1^n(\mathfrak{V})$ such that $\pi_\nu C^n(\mathfrak{V}) \sim C_{\eta_\nu}^n$ mod α in A, where $\pi_\nu = \text{Pr.}(\mathfrak{V}, \mathfrak{U}_{\eta_\nu})$. The question is only, whether it is possible to find a cycle $C_\xi^n \in L_1^n(\mathfrak{U}_\xi)$ in such a way that, for $\eta_1, \ldots, \eta_k < \xi$ (k is finite), there exists always a common refinement $\mathfrak{W}$ of the $k+1$ nets $\mathfrak{U}_\xi, \mathfrak{U}_{\eta_1}, \ldots, \mathfrak{U}_{\eta_k}$ and a cycle $C^n(\mathfrak{W}) \in L_1^n(\mathfrak{W})$ such that $\pi'_\nu C^n(\mathfrak{W}) \sim C_{\eta_\nu}^n$, $\pi C^n(\mathfrak{W}) \sim C_\xi^n$ mod α in A, where $\pi'_\nu = \text{Pr.}(\mathfrak{W}, \mathfrak{U}_{\eta_\nu})$, $\pi = \text{Pr.}(\mathfrak{W}, \mathfrak{U}_\xi)$.

GENERAL HOMOLOGY THEORY IN AN ARBITRARY SPACE

27. For *given* $\eta_1, \eta_2, \dots, \eta_k < \xi$, let us denote by $\Lambda(\eta_1, \eta_2, \dots, \eta_k)$ the set of all the cycles $C_\xi^n \in L_1^n(\mathfrak{U}_\xi)$ having the just described property. Then $\Lambda(\eta_1, \eta_2, \dots, \eta_k) \neq 0$. In fact, let $\mathfrak{V}$ be a common refinement of the nets $\mathfrak{U}_{\eta_1}, \mathfrak{U}_{\eta_2}, \dots, \mathfrak{U}_{\eta_k}$ such that for a convenient choice of $C^n(\mathfrak{V}) \in L_1^n(\mathfrak{V})$, there is $\pi_\nu C^n(\mathfrak{V}) \sim C_{\eta_\nu}^n$ mod α in A, where $\pi_\nu = \text{Pr.}(\mathfrak{V}, \mathfrak{U}_{\eta_\nu})$. Let $\mathfrak{W}$ be a common refinement of the nets $\mathfrak{U}_\xi$ and $\mathfrak{V}$, and let $\pi = \text{Pr.}(\mathfrak{W}, \mathfrak{U})$, $\overline{\pi} = \text{Pr.}(\mathfrak{W}, \mathfrak{V})$. Then $\pi'_\nu = \pi_\nu \overline{\pi} = \text{Pr.}(\mathfrak{W}, \mathfrak{U}_{\eta_\nu})$. According to 24, there is $\overline{\pi} L_1^n(\mathfrak{W}) = L_1^n(\mathfrak{V})$. Thus, there exists a cycle $C^n(\mathfrak{W}) \in L_1^n(\mathfrak{W})$ such that $\overline{\pi} C^n(\mathfrak{W}) \sim C^n(\mathfrak{V})$ and $\pi'_n C^n(\mathfrak{W}) \sim C_{\eta_\nu}^n$ mod α in A. Setting $C_\xi^n = \pi C^n(\mathfrak{W}) \in \pi L_1^n(\mathfrak{W}) = L_1^n(\mathfrak{U}_\xi)$, one has $C_\xi^n \in \Lambda(\eta_1, \eta_2, \dots, \eta_k)$. Having thus proved that $\Lambda(\eta_1, \eta_2, \dots, \eta_k) \neq 0$, one can easily verify that $\Lambda(\eta_1, \eta_2, \dots, \eta_k)$ is a *linear system* of $(n, \mathfrak{U}_\xi)$-cycles mod α in A. The common part of an arbitrary finite number of such linear systems

$$\Lambda(\eta_1^{(r)}, \eta_2^{(r)}, \dots, \eta_{k_r}^{(r)}) \qquad (r = 1, 2, \dots, h)$$

is alwyas $\neq 0$, for it contains obviously the linear system

$$\Lambda(\eta_1^{(1)}, \eta_2^{(1)}, \dots, \eta_{k_1}^{(1)}, \eta_1^{(2)}, \dots, \eta_{k_h}^{(h)}).$$

According to I 15, one deduces from this that the common part Λ of *all* $\Lambda(\eta_1, \dots, \eta_k)$ (for all the possible choices of a finite number of ordinal numbers $\eta_1, \dots, \eta_k$, all less than ξ) is also $\neq 0$. Obviously, it suffices to choose arbitrarily $C_\xi^n \in \Lambda$.

28. Let $\mathfrak{U}_0$ be a given net, and let C_0^n be a given *essential* $(n, \mathfrak{U}_0)$-cycle mod α in A. Then there exists an (n, R)-cycle $\{C^n(\mathfrak{U})\}$ mod α in A such that $C^n(\mathfrak{U}_0) = C_0^n$.

Proof. Let $\mathfrak{V}$ be an arbitrary net. Let us denote by $L^n(\mathfrak{V})$ the set of all $(n, \mathfrak{V})$-cycles mod α in A having the following property: There exists a common refinement $\mathfrak{V}_1$ of the two nets $\mathfrak{V}, \mathfrak{U}_0$ and a $(n, \mathfrak{V}_1)$-cycle $C^n(\mathfrak{V}_1)$ mod α in A such that $\pi_1 C^n(\mathfrak{V}_1) \sim C^n(\mathfrak{V})$, $\pi_0 C^n(\mathfrak{V}_1) \sim C_0^n$ mod α in A, where $\pi_1 = \text{Pr.}(\mathfrak{V}_1, \mathfrak{V})$, $\pi_0 = \text{Pr.}(\mathfrak{V}_1, \mathfrak{U}_0)$. Obviously, $L^n(\mathfrak{V}) \neq 0$ because C_0^n is essential. One can easily see that $L^n(\mathfrak{V})$ is a linear system of $(n, \mathfrak{V})$-cycles mod α in A. Furthermore, let us suppose that the net $\mathfrak{V}$ is a refinement of a net $\mathfrak{U}$, and let us choose $C^n(\mathfrak{V}) \in L^n(\mathfrak{V})$. We keep the preceding notations. Let $\pi = \text{Pr.}(\mathfrak{V}, \mathfrak{U})$. Then $\pi'_1 = \pi\pi_1 = \text{Pr.}(\mathfrak{V}_1, \mathfrak{U})$. Setting $C^n(\mathfrak{U}) = \pi C^n(\mathfrak{V})$, one has $\pi'_1 C^n(\mathfrak{V}_1) \sim C^n(\mathfrak{U})$, $\pi_0 C^n(\mathfrak{V}_1) \sim C_0^n$ mod α in A, whence $C^n(\mathfrak{U}) \in L^n(\mathfrak{V})$. Thus, $\pi L^n(\mathfrak{V}) = L^n(\mathfrak{U})$. Then, according to 21, there exists an (n, R)-cycle $\{C^n(\mathfrak{U})\}$ mod α in A such that $C^n(\mathfrak{U}) \in L^n(\mathfrak{U})$ for every net $\mathfrak{U}$. But $C^n(\mathfrak{U}_0) \in L^n(\mathfrak{U}_0)$ implies obviously that $C^n(\mathfrak{U}_0) \sim C_0^n$ mod α in A, so that one can set $C^n(\mathfrak{U}_0) = C_0^n$.

29. The result proved above can be obviously formulated as follows: Let $\mathfrak{U}_0$ be a fixed net. Assigning to each (n, R)-cycle $\{C^n(\mathfrak{U})\}$ mod α in A the $(n, \mathfrak{U}_0)$-cycle $C^n(\mathfrak{U}_0)$ mod α in A, then the finite module $M_n(A, \mathfrak{U}_0; \alpha)$ represents a *homomorphic image* of the module $M_n(A, R; \alpha)$. If the Betti number $P_n(A, R; \alpha)$ is finite (and only in this case), one can obviously choose the net $\mathfrak{U}_0$ in such a way that the modules $M_n(A, R; \alpha)$ and $M_n(A, \mathfrak{U}_0; \alpha)$ are *isomorphic*. Each refinement of such a net has obviously the same property.

30. Let Z_1 be a subset of the fundamental family Z of nets such that, for an arbitrary given net $\mathfrak{U} \in Z$, there exists *in* Z_1 a refinement $\mathfrak{V}$ of $\mathfrak{U}$. The family Z_1

satisfies obviously the axioms 1° and 2° of $n°$ 1. We shall say that Z_1 is a *complete family of nets* (with respect to the fundamental family Z.)

For the moment, let us call $(n, R)^*$-cycle the notion which differs from the (n, R)-cycle only in that respect that the fundamental family Z is replaced by Z_1. From each (n, R)-cycle $\{C^n(\mathfrak{U})\}$ mod α in A, one gets an $(n, R)^*$-cycle mod α in A by neglecting the nets not belonging to the family Z_1. Conversely, let $\{C^n(\mathfrak{V})\}$ be an arbitrarily given $(n, R)^*$-cycle mod α in A. If $\mathfrak{U} \in Z$ is a net not belonging to Z_1, let us choose one of its refinements $\mathfrak{V} \in Z_1$, and let us set $C^n(\mathfrak{U}) = \pi C^n(\mathfrak{V})$, $\pi = \text{Pr.}(\mathfrak{V}, \mathfrak{U})$. It can be easily seen that, in this way, one obtains an (n, R)-cycle $\{C^n(\mathfrak{U})\}$ mod α in A which is well determined by the $(n, R)^*$-cycle $C^n(\mathfrak{V})$ up to a homology mod α in A. *Replacing the fundamental family Z of nets by the complete family Z_1, the module $M_n(A, R; \alpha)$ does not change.*

III. Homologies in the topological spaces.

1. From now on, R denotes a *topological space*, i. e. a set (whose elements are called points) where one has given to certain subsets the name *open sets* (in R). The complementary set $R - A$ of an open set A in R is called *closed* (in R). Moreover, we suppose the following four axioms:

1° The empty set is both open and closed.

2° A set consisting of one point is closed.

3° The sum of an arbitrary family of open sets is open.

4° The product of two open sets is open.

The least closed set containing a given set $A \subset R$ is called *closure* of A (in R), and is denoted by $\overline{A}$. There is $A = \overline{A}$ if and only if A is closed.

If A is a given subset of R, then each set of the form AU, where U is open in R, is called *open* in A. By virtue of this definition, each subset of a topological space is a topological space.

We shall assume that the most elementary properties of topological spaces are known.

2. An (*open*) *net* in R is a system consisting of a finite number of open non-empty sets whose sum coincides with the whole space R. From now on, the fundamental family of nets Z will be composed of all the open nets. The axiom 1° of the Chap. II, $n°$ 1 is evident. But the axiom 2° is also satisfied. It is sufficient to take for $\mathfrak{V}$ the system of sets U_1U_2, where U_1 runs through all the vertices of $\mathfrak{U}_1$, and U_2 runs through those of $\mathfrak{U}_2$. One can easily find that it is possible, in this whole chapter, to replace the fundamental family Z by an arbitrarily chosen complete family (II 30) $Z_1 \subset Z$.

3. Obviously an open set intersects $A \subset R$ if it intersects $\overline{A}$, and vice versa. Hence, one deduces[17] that the homology theory of (n, R)-cycles mod α in A ($\alpha \subset A \subset R$) remains unchanged if one replaces α, A by α_1, A_1, so that $\alpha_1 \subset A_1$, $\overline{\alpha} = \overline{\alpha_1}$, $\overline{A} = \overline{A_1}$. From this, it follows that one can suppose, without the loss of generality, that the sets α and A are *closed* in R.

4. Let $\mathfrak{U}$ be a net in R. Replacing each vertex U of $\mathfrak{U}$ by its intersection $u = AU$ with $A \subset R$, and omitting the empty intersections, one obtains a net $\mathfrak{u} = A \cdot \mathfrak{U}$ in A.

[17] It is necessary to notice the evident fact that the kernel (II 2) of a simplex is an open set.

GENERAL HOMOLOGY THEORY IN AN ARBITRARY SPACE

5. Let A be closed in R, and let $\mathfrak{u}$ be a net in A. Then there exists a net $\mathfrak{U}$ in R such that $\mathfrak{u} = A \cdot \mathfrak{U}$.

It is sufficient to replace each vertex u of $\mathfrak{u}$ by an open set in R such that $u = A \cdot U$ and to add the vertex $U = R - A$.

6. Let A be closed in R, let $\mathfrak{u}$ be a net in A, and let $\mathfrak{U}_1$, $\mathfrak{U}_2$ be nets in R such that $A \cdot \mathfrak{U}_1 = A \cdot \mathfrak{U}_2 = \mathfrak{u}$. Then there exists a common refinement $\mathfrak{V}$ of $\mathfrak{U}_1$ and $\mathfrak{U}_2$ such that $A \cdot \mathfrak{V} = \mathfrak{u}$.

The vertices V of $\mathfrak{V}$ can be obtained as follows: 1° $V = U_1U_2$, where $U_1 \in \mathfrak{U}_1$, $U_2 \in \mathfrak{U}_2$, $AU_1 = AU_2 \neq 0$; 2° $V = (R - A)U_1U_2$, where $U_1 \in \mathfrak{U}_1$, $U_2 \in \mathfrak{U}_2$.

7. Let A be closed in R, let $\mathfrak{U}$ be a net in R, and let $\mathfrak{A} = A\mathfrak{U}$. Let S^n be an $(n, \mathfrak{U})$-simplex in A, i.e. $A \cdot J(S^n) \neq 0$. Replacing each vertex U of S^n by $u = A \cdot Z$, one obtains from S^n an $(n, \mathfrak{A})$-simplex $s^n = A \cdot S^n$, *provided that all the* $n + 1$ *vertices u are distinct*. In the opposite case, one sets $s^n = 0$. More generally, let $K^n = \sum r_\nu S^n_\nu$ be an $(n, \mathfrak{U})$-chain in A. Then $k^n = A \cdot K^n = \sum r_\nu \cdot AS^n_\nu$ is an $(n, \mathfrak{A})$-chain. Obviously, $F(AK^n) = AF(K^n)$. Hence, one deduces: if C^n is an $(n, \mathfrak{U})$-cycle mod α in A, then AC^n is an $(n, \mathfrak{A})$-cycle. If, in addition, $C^n \sim 0$ mod α in A, then also $AC^n \sim 0$ mod α.

8. Conversely, let $\mathfrak{u}$ be a net in $A = \overline{A} \subset R$, and let $c^n(\mathfrak{u})$ be an $(n, \mathfrak{u})$-cycle mod α. According to 5, one can determine a net $\mathfrak{u}$ in R such that $\mathfrak{A} = A \cdot \mathfrak{U}$. Obviously, there exists an $(n, \mathfrak{U})$-cycle $C^n(\mathfrak{U})$ mod α in A such that $c^n(\mathfrak{u}) = A \cdot C^n(\mathfrak{U})$. The cycle $C^n(\mathfrak{U})$ is not completely determined, but one can easily see that $c^n(\mathfrak{u}) = A \cdot C^n_1(\mathfrak{U}) = A \cdot C^n_2(\mathfrak{U})$ implies that $C^n_1(\mathfrak{U}) \sim C^n_2(\mathfrak{U})$ mod α in A. More generally, one can easily find out that $c^n_1(\mathfrak{u}) = A \cdot C^n_1(\mathfrak{U})$, $c^n_2(\mathfrak{u}) = A \cdot C^n_2(\mathfrak{U})$, so that $c^n_1(\mathfrak{u}) \sim c^n_2(\mathfrak{u})$ mod α implies $C^n_1(\mathfrak{U}) \sim C^n_2(\mathfrak{U})$ mod α, and vice versa.

9. Let $\{c^n(\mathfrak{u})\}$ be an (n, A)-cycle mod α $(A = \overline{A})$. For each net $\mathfrak{U}$ in R, let us choose (8) an $(n, \mathfrak{U})$-cycle $C^n(\mathfrak{U})$ mod α in A in such a way that $AC^n = c^n(A\mathfrak{U})$. Each cycle $C^n(\mathfrak{U})$ is (8) determined up to a homology mod α in A. Let $\mathfrak{V}$ be a refinement of $\mathfrak{U}$, $\pi = \mathrm{Pr.}(\mathfrak{V}, \mathfrak{U})$. One can easily see that $\pi C^n(\mathfrak{V})$ is an admissible value for $C^n(\mathfrak{U})$, which gives $C^n(\mathfrak{U}) \sim \pi C^n(\mathfrak{V})$ mod α in A. Thus, $\{C^n(\mathfrak{U})\}$ is an (n, R)-cycle mod α in A. Obviously, the cycle $\{C^n(\mathfrak{U})\}$ is determined by $\{c^n(\mathfrak{u})\}$ up to a homology mod α in A. Let us set $\{c^n(\mathfrak{u})\} = A \cdot \{C^n(\mathfrak{U})\}$.

10. Conversely, let $\{C^n(\mathfrak{U})\}$ be an (n, R)-cycle mod α in $A = \overline{A}$. There exists an (n, A)-cycle $\{c^n(\mathfrak{u})\}$ mod α such that $\{c^n(\mathfrak{u})\} = A \cdot \{C^n(\mathfrak{U})\}$.

Proof. Let $\mathfrak{u}$ be a net in A. Let us choose (5) the net $\mathfrak{U}$ in R in such a way that $\mathfrak{u} = A \cdot \mathfrak{U}$, and let us set $c^n(\mathfrak{u}) = A \cdot C^n(\mathfrak{U})$. It is necessary to prove that the $(n, \mathfrak{u})$-cycle $c^n(\mathfrak{u})$ is well determined up to a homology mod α. Thus, let $\mathfrak{u} = A \cdot \mathfrak{U}_1 = A \cdot \mathfrak{U}_2$. We must prove that $A \cdot C^n(\mathfrak{U}_1) \sim A \cdot C^n(\mathfrak{U}_2)$ mod α. According to 6, we can obviously suppose that $\mathfrak{U}_2$ is a refinement of $\mathfrak{U}_1$, $\pi = \mathrm{Pr.}(\mathfrak{U}_2, \mathfrak{U}_1)$. Using the method of Chap. II $n°$ 12, one finds that $A\pi C^n(\mathfrak{U}_1) \sim AC^n(\mathfrak{U}_2)$ mod α. But $\pi C^n(\mathfrak{U}_2) \sim C^n(\mathfrak{U}_1)$ mod α in A, so that (8) $A\pi C^n(\mathfrak{U}_2) \sim AC^n(\mathfrak{U}_1)$ mod α. Hence, finally, $AC^n(\mathfrak{U}_1) \sim AC^n(\mathfrak{U}_2)$ mod α.

Let $\mathfrak{w}$ be a refinement of $\mathfrak{u}$. One can easily see that it is possible to determine $\mathfrak{U}$, $\mathfrak{V}$ in such a way that $\mathfrak{w} = A \cdot \mathfrak{V}$, $\mathfrak{u} = A \cdot \mathfrak{U}$ and that $\mathfrak{V}$ is a refinement of $\mathfrak{U}$, $\pi = \mathrm{Pr.}(\mathfrak{V}, \mathfrak{U})$. Because $\pi C^n(\mathfrak{V}) \sim C^n(\mathfrak{U})$ mod α in A, it can be easily verified that $\pi' c^n(\mathfrak{w}) \sim c^n(\mathfrak{u})$ mod α, where $\pi' = \mathrm{Pr.}(\mathfrak{w}, \mathfrak{u})$.

Thus, $\{c^n(\mathfrak{u})\}$ is the required (n, A)-cycle mod α.

EDUARD ČECH

11. The above considerations prove that *if A is closed in R*, then there is a bijective correspondence between the modules[18] $M_n(A, R; \alpha)$ and $M_n(A; \alpha)$. In particular, $P_n(A, R; \alpha) = P_n(A, \alpha)$.

12. Let p be a given point of R. For each net $\mathfrak{U}$ in R, we choose an absolute $(0, \mathfrak{U})$-cycle $C^0(\mathfrak{U})$ consisting of a single $(0, \mathfrak{U})$-simplex U such that $p \in U$. Obviously, $\{C^0(\mathfrak{U})\}$ is an absolute $(0, R)$-cycle. We shall denote it by $\{p\}$. This cycle is not completely determined (for a net can have several vertices containing p), but it is certainly determined up to a homology.

13. Let A, B be two subsets of R (possibly one-point sets). For the sake of brevity, we say that a net $\mathfrak{U}$ in R *separates* A from B if for no vertices U_1, U_2 of $\mathfrak{U}$ such that $AU_1 \neq 0$, $BU_2 \neq 0$ there is a homology $U_1 \sim U_2$. Obviously, each refinement $\mathfrak{V}$ of such a net $\mathfrak{U}$ separates also A from B. If the net $\mathfrak{U}$ separates A from B, let U be the sum of all $(0, \mathfrak{U})$-simplices homologic to a $(0, \mathfrak{U})$-simplex meeting A, and let V be the sum of the other $(0, \mathfrak{U})$-simplices. Thus, we have obviously

$$(1) \qquad R = U + V; \quad UV = 0; \quad A \subset U, B \subset V; \qquad U, V \text{ open in } R.$$

Conversely, under the conditions (1), the net consisting of U, V separates A from B.

It can be seen that the space R is *connected* if and only if any two points of its are never separated by any net. Let us recall that each point p of R belongs always to a *maximal connected* subset Γ of R. Γ is called *component* of R.

Let p be a given point of R. Let Q be the set of all the points $q \in R$ which are not separated from p by any net. The set Q is a quasicomponent of R in the sense of Hausdorff.[19] It is known that each quasicomponent of R is a component of R or a sum of components of R. If the number of quasicomponents is *finite*, they coincide with the components.

14. If the points p and q belong to the same quasicomponent of P, then $\{p\} \sim \{q\}$.

Let $\{p\} = \{C_1^0(\mathfrak{U})\}$, $\{q\} = \{C_2^0(\mathfrak{U})\}$. No net $\mathfrak{U}$ can separate p from q. Therefore $C_1^0(\mathfrak{U}) \sim C_2^0(\mathfrak{U})$.

15. Let $\alpha \subset R$. A quasicomponent Q is called *essential* mod α if there exists a net $\mathfrak{U}$ separating Q from α. If the component Q is not essential mod α and if $p \in Q$, then, obviously, no net separates p from α.

If α meets only a *finite* number of quasicomponents of R and $Q \cdot \alpha = 0$, then the quasicomponent Q is essential. In fact, for each quasicomponent Q_ν meeting α, there exists obviously a net $\mathfrak{U}_\nu$ separating Q from Q_ν. A common refinement $\mathfrak{U}$ of the nets $\mathfrak{U}_\nu$ then separates Q from α.

16. Let $p \in R$, $\alpha \subset R$. If the quasicomponent Q containing p is not essential mod α, then $\{p\} \sim 0$ mod α.

Let $\{p\} = \{C^0(\mathfrak{U})\}$. Since no net $\mathfrak{U}$ can separate p from α, then, obviously, $C^0(\mathfrak{U}) \sim 0$ mod α.

17. Let $Q_1, Q_2, \ldots, Q_k$ be distinct quasicomponents essential mod α, and let $p_\nu \in Q_\nu$. Then the $(0, R)$-cycles $\{p_\nu\}$ are independent mod α.

[18] For the notation see II 20.

[19] Grundzüge der Mengenlehre, 1914, pp. 248–249.

Proof. Let $\{p_\nu\} = \{C^0_\nu(\mathfrak{U})\}$. Let $\mathfrak{U}_\nu$ $(1 \leqq \nu \leqq k)$ be a net separating p_ν from α, and let $\mathfrak{V}_{\mu\nu}$ $(1 \leqq \mu < \nu \leqq k)$ be a net separating p_μ from p_ν. Let $\mathfrak{W}$ be a common refinement of all the nets $\mathfrak{U}_\nu$, $\mathfrak{V}_{\mu\nu}$. Then the net $\mathfrak{W}$ separates simultaneously each point p_ν from α and from all the other points p_μ. Hence, one deduces easily that a homology $\sum r_\nu C^0_\nu(\mathfrak{W}) \sim 0$ mod α implies $r_1 = \cdots = r_k = 0$.

18. The preceding considerations lead us easily to the following general theorem: The number of essential quasicomponents mod α is equal to $P_0(R;\alpha)$. Particular case: If α meets only a finite number k of quasicomponents of R, then the total number of quasicomponents of R is equal to $P_0(R;\alpha) + k$. The number of quasicomponents of R is equal to $P_0(R)$, etc.

19. From now on, we shall suppose that the topological space R is *hereditarily normal.*[20] This means that, besides the axioms 1°–4° of n° 1, we have the following axiom:

5° if two sets $A, B \subset R$ are separated by a net $\mathfrak{u}$ in $A + B$,[21] then they are also separated by a net $\mathfrak{U}$ in R.

Each subset A of R represents a hereditarily normal topological space.[22]

20. **Lemma** $h'(h = 1,2,3,\dots)$: Let $u_1, u_2, \dots, u_h$ be open (closed) sets in $A \subset R$. Let $\prod_1^h u_\nu = 0$. Then there exist open sets $U_1, U_2, \dots, U_h$ in R such that $U_\nu \supset u_\nu$ and $\prod_1^h U_\nu = 0$.

Lemma $h''(h = 1,2,3,\dots)$: Let $u_1, u_2, \dots, u_h$ be open (closed) sets in $A \subset R$. Let V be an open set in R such that $V \supset \prod_1^h u_\nu$. Then there exist open sets $U_1, U_2, \dots, U_h$ in R such that $U_\nu \supset u_\nu$, $\prod_1^h U_\nu = V$.

The lemmas being obvious for $h = 1$, it suffices to deduce 1° h'' from h'; 2° $(h+1)'$ from h''.

Let us suppose a t f i r s t the validity of h' as well as the hypothesis of h''. Let us set $A' = A - \prod_1^h u_\nu$, $u'_\nu = A' \cdot u_\nu$. Then the sets u'_ν are open (closed) in A', and there is $\prod_1^h u'_\nu = 0$. According to h', there exist open sets U'_ν in R such that $U'_\nu \supset u'_\nu$, $\prod_1^h U'_\nu = 0$. It can be easily seen that it is sufficient to set $U_\nu = U'_\nu + V$.

S e c o n d l y , let us suppose the validity of h'' as well as the hypothesis of $(h+1)'$. Obviously,

$$\overline{u}_{h+1} \cdot v = u_{h+1} \cdot \overline{v}_{h+1} = 0; \quad (v = \prod_1^h u_\nu).$$

[20] This notion was introduced by Tietze (Math. Annalen, 88, p. 301). The designation is that of Urysohn (Math. Annalen, 94, p. 265).

[21] The hypothesis can obviously be formulated in the following form: $A \cdot B = 0$ and A, B are open in $A + B$. Further formulation is $A\overline{B} + B\overline{A} = 0$.

[22] See Urysohn, l. c., p. 284.

EDUARD ČECH

Because the space R is hereditarily normal, one can prove (see 13 (1)) that there exist open sets V and U_{h+1} in R such that $V \supset \prod_1^h u_\nu$, $U_{h+1} \supset u_{h+1}$, $V \cdot U = 0$. According to h'', there exist open sets $U_1, U_2, \dots, U_h$ in R such that $U_\nu \supset u_\nu$, $\prod_1^h U_\nu = V$. One can see that the sets $U_1, U_2, \dots, U_{h+1}$ have the desired properties.

21. Let $A \subset R$. Let $u_1, u_2, \dots, u_k$ (k is finite) be open (closed) sets in A. Then there exist open sets $V_1, V_2, \dots, V_k$ in R such that: 1° $V_\nu \supset u_\nu$ $(1 \leqq \nu \leqq k)$; 2° there is $V_{\nu_1} \cdot V_{\nu_2} \cdot \ldots \cdot V_{\nu_h} = 0$ for every combination $(\nu_1, \nu_2, \dots, \nu_h)$ of the indices $1, 2, \dots, k$ $(1 \leqq h \leqq k)$ such that $u_{\nu_1} \cdot u_{\nu_2} \cdot \ldots \cdot u_{\nu_h} = 0$.

Proof. Let us suppose that the symbol $\kappa = (\nu_1, \nu_2, \dots, \nu_h)$ runs through all the combinations for which $u_{\nu_1} \cdot u_{\nu_2} \cdot \ldots \cdot u_{\nu_h} = 0$. According to the lemma h', there exist open sets $U_{\nu_1}^{(\kappa)}, U_{\nu_2}^{(\kappa)}, \dots, U_{\nu_h}^{(\kappa)}$ in R such that $U_{\nu_1}^{(\kappa)} \supset u_{\nu_1}; \dots, U_{\nu_h}^{(\kappa)} \supset u_{\nu_h}$ and $U_{\nu_1}^{(\kappa)} \cdot U_{\nu_2}^{(\kappa)} \cdot \ldots \cdot U_{\nu_h}^{(\kappa)} = 0$. For any value of ν $(1 \leqq \nu \leqq k)$, let us set

$$V_\nu = \prod_\kappa U_\nu^{(\kappa)},$$

where the index κ runs through *those* of its values $(\nu_1, \nu_2, \dots, \nu_h)$, which contain the given index ν. It can be easily seen that the sets $V_1, V_2, \dots, V_k$ have the desired properties.

22. A net $\mathfrak{U}$ in R is called *regular* with respect to a set $A \subset R$ if every $(n, \mathfrak{U})$-simplex S^n of $\mathfrak{U}$ has the following property: *either* no vertex of S^n meets A, *or* the kernel of S^n meets A. It can be seen (see 3) that this property depends only on the *closure* $\overline{A}$ of A. Thus, one can suppose A to be closed.

23. Let $A \subset R$, and let $\mathfrak{U}$ be a net in R. Then there exists a refinement $\mathfrak{W}$ of $\mathfrak{U}$, regular with respect to A, and such that $A \cdot \mathfrak{U} = A \cdot \mathfrak{W}$.

Moreover, if $\alpha \subset R$ is a closed set, one can choose $\mathfrak{W}$ in such a way that for each vertex W of $\mathfrak{W}$ it holds either $WA = 0$, or $W\alpha = 0$, or finally $WA\alpha \neq 0$.

Proof. One can suppose that A is closed. Let us denote by $U_1, U_2, \dots, U_k$ all the vertices of $\mathfrak{U}$ meeting A, and let us set $u_\nu = A \cdot U_\nu$. By virtue of the theorem of n° 21, let us associate with these sets $u_1, u_2, \dots, u_k$ open in A open sets $V_1, V_2, \dots, V_k$ in R, and let us set $W_\nu = U_\nu V_\nu$ if $u_\nu \alpha \neq 0$, $W_\nu = U_\nu V_\nu (R - \alpha)$ in the opposite case. To the sets W thus defined, let us add the sets $W = U \cdot (R - A)$, where U runs through *all* the vertices of $\mathfrak{U}$. One can easily see that the sets W form a refinement $\mathfrak{W}$ having the desired property.

IV. Homologies in the sum of two spaces.

1. Let R be a normal topological space. Let R_1 and R_2 be two *closed* subsets of R such that $R = R_1 + R_2$. Let us set $R_3 = R_1 R_2$. Let α be a closed subset of R. Let us set (for $i = 1, 2, 3$) $\alpha_i = R_i \alpha$, so that α_i is a closed subset of R_i.

2. Let us denote by N_3 the family of all (open) nets in R_3 regular (see III 22) with respect to α_3. According to II 23, N_3 is a complete family (see II 30) of open nets in R_3. Hence, it follows easily that the family N' consisting of all open nets

GENERAL HOMOLOGY THEORY IN AN ARBITRARY SPACE

$\mathfrak{U}$ in R such that $R_3 \cdot \mathfrak{U} \in N_3$ is also complete. Let N'' be the family consisting of those nets $\mathfrak{U} \in N'$ which are regular with respect to R_3, and which, in addition, have the following property: If U is a vertex of $\mathfrak{U}$, then $UR_3 = 0$, or $U\alpha = 0$, or finally $U\alpha_3 = 0$. According to III 23, N'' is a complete family of nets in R. From each net $\mathfrak{U} \in N''$, we construct a new net $\mathfrak{V}$ in R in the following way: if U is a vertex of $\mathfrak{U}$ such that $UR_3 = 0$, one replaces it by the two vertices $U_1 = U(R-R_1)$, $U_2 = U(R-R_2)$ [there is $U_1 + U_2 = U(R-R_3) = U$]. Let us denote by N the family consisting of all nets $\mathfrak{V}$ constructed in this way from all the nets $\mathfrak{U} \in N''$.

In the next, *we consider*, instead of the family consisting of all open nets in R, *the family N as the fundamental family of nets in R*. This is allowed (see II 30 and III 2), because N is obviously a complete family of open nets in R.

3. The fundamental family has then the following properties ($\mathfrak{U} \in N$):

1° For each vertex U of $\mathfrak{U}$ there is

$$UR_3 \neq 0, \quad \text{or} \quad UR_1 = 0, \quad \text{or finally} \quad UR_2 = 0.$$

2° If each vertex of an $(n,\mathfrak{U})$-simplex S^n meets R_3, then $R_3 \cdot J(S^n) \neq 0$.

3° If each vertex of an $(n,\mathfrak{U})$-simplex S^n meets α_3, then $\alpha_3 \cdot J(S^n) \neq 0$.

4° For each vertex U of $\mathfrak{U}$ such that $UR_3 \neq 0$, $U\alpha \neq 0$ there is $U\alpha_3 \neq 0$.

4. From the properties 1°, 2° of N° 3, one obtains for $\mathfrak{U} \in N$: 1° Each $(n,\mathfrak{U})$-chain $K^n(\mathfrak{U})$ can be expressed in the form

$$K^n(\mathfrak{U}) = K_1^n(\mathfrak{U}) - K_2^n(\mathfrak{U}) \; ; \quad K_1^n(\mathfrak{U}) \subset R_1 \; ; \quad K_2^n(\mathfrak{U}) \subset R_2 \; .$$

2° If $K^n(\mathfrak{U}) \subset R_1$ and $K^n(\mathfrak{U}) \subset R_2$, then $K^n(\mathfrak{U}) \subset R_3$.

5. Let $C^n(\mathfrak{U})$ be an $(n,\mathfrak{U})$-cycle mod α in R_i ($i = 1,2,3$). Then $C^n(\mathfrak{U})$ is an $(n,\mathfrak{U})$-cycle mod α_i in R_i.

Proof. Let us set $F(C^n(\mathfrak{U})) = \sum r_\nu S_\nu^{n-1}$, where we can suppose that in each term on the right hand side there is $r_\nu \neq 0$. Thus, $S_\nu^{n-1} \subset R_i$, $S_\nu^{n-1} \subset \alpha$. Firstly, let $i = 3$. Each vertex U of each S_ν^{n-1} meets R_3 and α, so that (according to 4° in N° 3) it meets also α_3. Hence, it results (according to 3° in N° 3) that $S_\nu^{n-1} \subset \alpha_3$, from where $F(C^n(\mathfrak{U})) \subset \alpha_3$, $C^n(\mathfrak{U}) \to 0$ mod α_3. Secondly, let $i = 1$ (the case $i = 2$ can be treated in the same way). If, for some value of ν, the simplex S_ν^{n-1} has a vertex U such that $UR_3 = 0$, then, according to 1° of N° 3, there is $UR_1 = 0$ or $UR_2 = 0$. But $S_\nu^{n-1} \subset R_1$ implies $UR_1 \neq 0$. Therefore, $J(S_\nu^{n-1}) \subset U \subset R - R_2 \subset R_1$, whence $J(S_\nu^{n-1}) \cdot \alpha = J(S_\nu^{n-1}) \cdot \alpha_1$. But $J(S_\nu^{n-1}) \cdot \alpha \neq 0$, because $S_\nu^{n-1} \subset \alpha$. Consequently, $J(S_\nu^{n-1}) \cdot \alpha_1 \neq 0$. On the other hand, if every vertex of S_ν^{n-1} meets R_3, the relation $S_\nu^{n-1} \subset \alpha$ implies (according to 4° of N° 3) that every vertex of S_ν^{n-1} meets α_3, and therefore (according to 3° of N° 3) $0 \neq J(S_\nu^{n-1}) \cdot \alpha_3 \subset J(S_\nu^{n-1}) \cdot \alpha_1$. This shows that for each value of the index ν, one has the inclusion $J(S_\nu^{n-1}) \subset \alpha_1$, which means that $C^n(\mathfrak{U}) \to 0$ mod α_1.

6. Let $C^{n+1}(\mathfrak{U})$ be an arbitrarily given $(n+1,\mathfrak{U})$-cycle mod α ($\mathfrak{U} \in N$). According to 4, one can set

$$C^{n+1}(\mathfrak{U}) = K_1^{n+1}(\mathfrak{U}) - K_2^{n+1}(\mathfrak{U}) \tag{1}$$

EDUARD ČECH

with

$$(2) \qquad K_1^{n+1}(\mathfrak{U}) \subset R_1, \quad K_2^{n+1}(\mathfrak{U}) \subset R_2.$$

One has $F[K_1^{n+1}(\mathfrak{U})] = F[K_2^{n+1}(\mathfrak{U})] \bmod \alpha$. Let us denote by $C^n(\mathfrak{U})$ an $(n, \mathfrak{U})$-chain arising from $F(K_1^{n+1}(\mathfrak{U}))$ [and also from $F(K_2^{n+1}(\mathfrak{U}))$] by removing all the $(n, \mathfrak{U})$-simplices contained in α and by adding then an arbitrary $(n, \mathfrak{U})$-chain *contained in* α_3.

Obviously,

$$(3) \qquad C^n(\mathfrak{U}) = F[K_1^{n+1}(\mathfrak{U})] \bmod \alpha_1, \quad C^n(\mathfrak{U}) = F[K_2^{n+1}(\mathfrak{U})] \bmod \alpha_2 .$$

According to (2) and (3), one has $C^n(\mathfrak{U}) \subset R_1$, $C^n(\mathfrak{U}) \subset R_2$, whence (4) $C^n(\mathfrak{U}) \subset R_3$. According to (3), $C^n(\mathfrak{U})$ is an $(n, \mathfrak{U})$-cycle mod α. Hence, (see 5) $C^n(\mathfrak{U})$ is an $(n, \mathfrak{U})$-cycle mod α_3 in R_3.Besides, the relations (2) and (3) give

$$(4) \qquad C^n(\mathfrak{U}) \sim 0 \bmod \alpha_1 \text{ in } R_1 , \quad C^n(\mathfrak{U}) \sim 0 \bmod \alpha_2 \text{ in } R_3.$$

For the sake of brevity, let us write

$$C^n(\mathfrak{U}) = \Phi[C^{n+1}(\mathfrak{U})]$$

in order to indicate that the cycle $C^n(\mathfrak{U})$ was deduced from the cycle $C^{n+1}(\mathfrak{U})$ in the way explained above.

7. The function Φ is not univalent. In general, one can set, instead of 6 (1),

$$C^{n+1}(\mathfrak{U}) = \overline{K}_2^{n+1}(\mathfrak{U}) - \overline{K}_2^{n+1}(\mathfrak{U}),$$

where

$$\overline{K}_i^{n+1}(\mathfrak{U}) = K_i^{n+1}(\mathfrak{U}) + K_3^{n+1}(\mathfrak{U}), \quad (i = 1, 2)$$

with $K_3^{n+1}(\mathfrak{U})$ being a chain such that $K_3^{n+1}(\mathfrak{U}) \subset R_1$, $K_3^{n+1}(\mathfrak{U}) \subset R_2$, and consequently, (4) $K_3^{n+1}(\mathfrak{U}) \subset R_3$. Then, instead of $C^n(\mathfrak{U})$, one has generally

$$\overline{C}^n(\mathfrak{U}) = C^n(\mathfrak{U}) + F[K_3^{n+1}(\mathfrak{U})] \bmod \alpha_3.$$

Consequently: If $C^n(\mathfrak{U}) = \Phi[C^{n+1}(\mathfrak{U})]$, then $\overline{C}^n(\mathfrak{U}) = \Phi[C^{n+1}(\mathfrak{U})]$ if and only if $\overline{C}^n(\mathfrak{U}) \sim C^n(\mathfrak{U}) \bmod \alpha_3$ in R_3.

8. Conversely, let $C^n(\mathfrak{U})$ be an $(n, \mathfrak{U})$-cycle mod α_3 in R_3 such that

$$C^n(\mathfrak{U}) \sim 0 \bmod \alpha_1 \text{ in } R \,; \quad C^n(\mathfrak{U}) \sim 0 \bmod \alpha_2 \text{ in } R_2 .$$

Then there exists, for $i = 1$ and for $i = 2$, an $(n+1, \mathfrak{U})$-chain $K_i^{n+1}(\mathfrak{U}) \subset R_i$ such that

$$F[K_i^{n+1}(\mathfrak{U})] = C^n(\mathfrak{U}) \bmod \alpha_i .$$

Setting

$$C^{n+1}(\mathfrak{U}) = K_1^{n+1}(\mathfrak{U}) - K_2^{n+1}(\mathfrak{U}),$$

GENERAL HOMOLOGY THEORY IN AN ARBITRARY SPACE

one has obviously $C^n(\mathfrak{U}) = \Phi[C^{n+1}(\mathfrak{U})]$.

9. If

$$C^{n+1}(\mathfrak{U}) = \sum_1^k r_\nu C_\nu^{n+1}(\mathfrak{U}), \quad C_\nu^n(\mathfrak{U}) = \Phi[C_\nu^{n+1}(\mathfrak{U})] ,$$

then obviously

$$\sum_1^k r_\nu C_\nu^n(\mathfrak{U}) = \Phi[C^{n+1}(\mathfrak{U})].$$

10. Let $C^{n+1}(\mathfrak{U})$, $C_1^{n+1}(\mathfrak{U})$, $C_2^{n+2}(\mathfrak{U})$ be $(n+1,\mathfrak{U})$-cycles mod α such that

$$(1) \qquad C^{n+1}(\mathfrak{U}) = C_1^{n+1}(\mathfrak{U}) - C_2^{n+1}(\mathfrak{U});\ C_1^{n+1}(\mathfrak{U}) \subset R_1;\ C_2^{n+1}(\mathfrak{U}) \subset R_2 .$$

Then $0 = \Phi[C^{n+1}(\mathfrak{U})]$. Conversely, if $0 = \Phi[C^{n+1}(\mathfrak{U})]$, then there exist two $(n+1,\mathfrak{U})$-cycles mod α $C_i^{n+1}(\mathfrak{U})$ $(i = 1,2)$ such that (1) holds.

11. Let $C^{n+1}(\mathfrak{U})$ be an $(n+1,\mathfrak{U})$-cycle mod α homologic to zero mod α. Then $0 = \Phi[C^{n+1}(\mathfrak{U})]$.

Proof. There exist an $(n+2,\mathfrak{U})$-chain $K^{n+2}(\mathfrak{U})$ and an $(n+1,\mathfrak{U})$-chain $\overline{K}^{n+1}(\mathfrak{U})$ such that

$$C^{n+1}(\mathfrak{U}) = F[K^{n+2}(\mathfrak{U})] + \overline{K}^{n+1}(\mathfrak{U}) ; \quad \overline{K}^{n+1}(\mathfrak{U}) \subset \alpha.$$

According to 4, one can set

$$K^{n+2}(\mathfrak{U}) = K_1^{n+2}(\mathfrak{U}) - K_2^{n+2}(\mathfrak{U}) ; \quad \overline{K}^{n+1}(\mathfrak{U}) = \overline{K}_1^{n+1}(\mathfrak{U}) - \overline{K}_2^{n+1}(\mathfrak{U}) ,$$

where

$$K_i^{n+2}(\mathfrak{U}) \subset R_i ; \qquad \overline{K}_i^{n+1} \subset R_i \quad (i = 1,2) .$$

Obviously, one can arrange that $\overline{K}_i^{n+1}(\mathfrak{U}) \subset \alpha$.

Setting

$$C_i^{n+1}(\mathfrak{U}) = F[K_i^{n+2}(\mathfrak{U})] + \overline{K}_i^{n+1}(\mathfrak{U}),$$

the conditions 10 (1) are satisfied. Besides, $C_i^{n+1}(\mathfrak{U})$ is an $(n+1,\mathfrak{U})$-cycle mod α in R_i.

12. Let $\mathfrak{V} \in N$ be a refinement of a net $\mathfrak{U} \in N$, $\pi = \mathrm{Pr.}(\mathfrak{V},\mathfrak{U})$. Let $C^{n+1}(\mathfrak{U})$ be an $(n+1,\mathfrak{U})$-cycle mod α, and let $C^n(\mathfrak{U}) = \Phi[C^{n+1}(\mathfrak{V})]$. It can be easily verified that

$$\pi C^n(\mathfrak{U}) = \Phi[\pi C^{n+1}(\mathfrak{U})] .$$

13. Now, let $\{C^{n+1}(\mathfrak{U})\}$ be an $(n+1,R)$-cycle mod α. For each net[23] $\mathfrak{U}$, let $C^n(\mathfrak{U}) = \Phi[C^{n+1}(\mathfrak{U})]$. Each $C^n(\mathfrak{U})$ is an (n,R)-cycle mod α_3 in R_3 (6) determined precisely up to a homology mod α_3 in R_3. According to (12), $\{C^n(\mathfrak{U})\}$ is an (n,R)-cycle mod α_3 in R_3. Let us set $\{C^n(\mathfrak{U})\} = \Phi[\{C^{n+1}(\mathfrak{U})\}]$.

According to 6 (4) one has

$$(1) \qquad \{C^n(\mathfrak{U})\} \sim 0 \bmod \alpha_i \text{ in } R_i \quad (i = 1,2) .$$

[23]Let us recall that we consider only the nets $\mathfrak{U} \in N$.

EDUARD ČECH

14. Conversely, let $\{C^n(\mathfrak{U})\}$ be an (n, R)-cycle mod α_3 in R_3 such that the two relations 13 (1) are satisfied. For each net $\mathfrak{U} \in N$, let us denote by $L^{n+1}(\mathfrak{U})$ the set of $(n+1, R)$-cycles mod α in R such that $C^n(\mathfrak{U}) = \Phi[C^{n+1}(\mathfrak{U})]$. By virtue of (8), $L^{n+1}(\mathfrak{U}) \neq 0$. According to 9 and 11, $L^{n+1}(\mathfrak{U})$ is a linear system of $(n+1, R)$-cycles mod α. If $\mathfrak{V} \in N$ is a refinement of $\mathfrak{U} \in N$, $\pi = \mathrm{Pr.}(\mathfrak{V}, \mathfrak{U})$, one has $\pi L^{n+1}(\mathfrak{U}) \subset L^{n+1}(\mathfrak{U})$. By virtue of the theorem formulated in Chap. II, N° 21, it is possible to choose $C^{n+1}(\mathfrak{U}) \in L^{n+1}(\mathfrak{U})$ in such a way that one obtains an $(n+1, R)$-cycle mod α $\{C^{n+1}(\mathfrak{U})\}$ such that

$$\{C^n(\mathfrak{U})\} = \Phi[\{C^{n+1}(\mathfrak{U})\}] .$$

15. Let $\mu_n(R_3; \alpha)$ be the submodule of the module $M_n(R_3, R; \alpha_3)$ consisting of all the (n, R)-cycles mod α_3 in R_3 homologic to zero mod α_1 in R_1, as well as mod α_2 in R_2. By means of the function Φ, the module $\mu_n(R_3; \alpha_3)$ is (see 9, 11, 13 and 14) a homomorphic image of the module $M_{n+1}(R; \alpha)$. More precisely, according to 10, the module $\mu_n(R_3, R; \alpha_3)$ is isomorphic to the module (see I 11) $M_{n+1}(R; \alpha) - M^*_{n+1}(R; \alpha)$, where $M^*_{n+1}(R; \alpha)$ consists of the $(n+1, R)$-cycles mod α of the form $\{C^n_1(\mathfrak{U})\} - \{C^n_2(\mathfrak{U})\}$, $\{C^n_i(\mathfrak{U})\}$ $(i = 1, 2)$ being an $(n+1, R)$-cycle mod α_i in R_i. According to I 12, the module $M_{n+1}(R; \alpha)$ is a direct sum of the module $M^*_{n+1}(R; \alpha)$ and a module isomorphic to $\mu_n(R_3; \alpha_3)$. Consequently,

$$P_{n+1}(R; \alpha) = P^*_{n+1}(R; \alpha) + \pi_n(R_3; \alpha_3) , \tag{1}$$

where $P^*_{n+1}(R; \alpha)$ is the rank of the module $M^*_{n+1}(R; \alpha)$, and $\pi_n(R_3; \alpha_3)$ is that of the module $\mu_n(R_3; \alpha_3)$.

16. Let $\{C^{n+1}_i(\mathfrak{U})\}$ $(i = 1, 2)$ be an $(n+1, R)$-cycle mod α_i in R_i. Let $\{C^{n+1}_1(\mathfrak{U})\} \sim \{C^{n+1}_2(\mathfrak{U})\}$ mod α. Then there exists an $(n+1, R)$-cycle $C^{n+1}_3(\mathfrak{U})$ mod α_3 in R_3 such that $\{C^{n+1}_i(\mathfrak{U})\} \sim \{C^{n+1}_3(\mathfrak{U})\}$ mod α_i in R_i for $i = 1$ and for $i = 2$.

Proof. For each $\mathfrak{U} \in N$ there exists an $(n+2, \mathfrak{U})$-chain $K^{n+2}_1(\mathfrak{U}) - K^{n+2}_2(\mathfrak{U})$ such that

$$C^{n+1}_1(\mathfrak{U}) - C^{n+1}_2(\mathfrak{U}) = F[K^{n+2}_1(\mathfrak{U}) - K^{n+2}_2(\mathfrak{U})] \text{ mod } \alpha.$$

One can suppose (see 4) that $K^{n+2}_i(\mathfrak{U}) \subset R_i$ $(i = 1, 2)$. Hence, it follows easily (cf. 6) that there exists an $(n+1, \mathfrak{U})$-cycle $C^{n+1}_3(\mathfrak{U})$ mod α_3 in R_3 such that for $i = 1, 2$ one has

$$C^{n+1}_3(\mathfrak{U}) = C^{n+1}_i(\mathfrak{U}) - F[K^{n+2}_i(\mathfrak{U})] \text{ mod } \alpha ,$$

i. e.

$$C^{n+1}_3(\mathfrak{U}) \sim C^{n+1}_i(\mathfrak{U}) \text{ mod } \alpha_i \text{ in } R_i .$$

The cycle $C^{n+1}_3(\mathfrak{U})$ is not uniquely determined. Denoting by $L^{n+1}(\mathfrak{U})$ the set of all its values, one can easily see that $L^{n+1}(\mathfrak{U})$ is a linear system of $(n+1, \mathfrak{U})$-cycles mod α_3 in R_3. Moreover, if $\mathfrak{V} \in N$ is a refinement of $\mathfrak{U} \in N$, $\pi = \mathrm{Pr.}(\mathfrak{V}, \mathfrak{U})$, one has manifestly $\pi L^{n+1}(\mathfrak{V}) \subset L^{n+1}(\mathfrak{U})$. Therefore, by virtue of II 21, one can choose $C^{n+1}_3(\mathfrak{U}) \in L^{n+1}(\mathfrak{U})$, for each $\mathfrak{U} \in N$, in such a way that $\{C^{n+1}_3(\mathfrak{U})\}$ represents an $(n+1, R)$-cycle mod α_3 in R_3 having obviously the desired property.

17. Let us denote, for a moment, by $\overline{M}_{n+1}(R_1, R_2)$ the direct sum (I 12) of two modules isomorphic to

$$M_{n+1}(R_1, R; \alpha_1) \quad \text{and} \quad M_{n+1}(R_2, R; \alpha_2),$$

respectively. $\overline{M}_{n+1}(R_1, R_2)$ can be considered as consisting of the pairs $[\{C_1^{n+1}(\mathfrak{U})\}, \{C_2^{n+1}(\mathfrak{U})\}]$, where (for $i = 1,2$) $\{C_i^{n+1}(\mathfrak{U})\}$ is an $(n+1, R)$-cycle mod α_i in R_i, and where two pairs $[\{C_1^{n+1}(\mathfrak{U})\}, \{C_2^{n+1}(\mathfrak{U})\}]$ and $[\{\overline{C}_1^{n+1}(\mathfrak{U})\}, \{\overline{C}_2^{n+1}(\mathfrak{U})\}]$ are equal if and only if $\{\overline{C}_i^{n+1}(\mathfrak{U})\} \sim \{C_i^{n+1}(\mathfrak{U})\}$ mod α_i in R_i for $i = 1$ and for $i = 2$. Obviously, the module $M_{n+1}^*(R; \alpha)$ is a homomorphic image of $\overline{M}_{n+1}(R; \alpha)$, so that (see I 13) $\overline{M}_{n+1}(R_1, R_2)$ is the direct sum of a module isomorphic to $\overline{M}_{n+1}^*(R; \alpha)$ and of the module $M_{n+1}^0(R_1, R_2)$ consisting of the pairs $[\{C_1^{n+1}(\mathfrak{U})\}, \{C_2^{n+1}(\mathfrak{U})\}]$ whose image in $M_{n+1}^*(R; \alpha)$ is zero. Therefore, (see III 11)

$$(1) \qquad P_{n+1}(R_1; \alpha_1) + P_{n+1}(R_2; \alpha_2) = P_{n+1}^*(R; \alpha) + P_{n+1}^0(R_1, R_2),$$

where $P_{n+1}^0(R_1, R_2)$ denotes the rank of the module $M_{n+1}^0(R_1, R_2)$. But, according to 16, the module $M_{n+1}(R_1, R_2)$ consists of the pairs $[\{C_3^{n+1}(\mathfrak{U})\}, \{C_3^{n+1}(\mathfrak{U})\}]$, where $C_3^{n+1}(\mathfrak{U})$ is an $(n+1, R)$-cycle mod α_3 in R_3, and where the pair just described is considered to be equal to the couple $[\{\overline{C}_3^{n+1}(\mathfrak{U}), \{\overline{C}_3^{n+1}(\mathfrak{U})\}]$ if and only if

$$\{\overline{C}_3^{n+1}(\mathfrak{U})\} - \{C_3^{n+1}(\mathfrak{U})\} \sim 0 \text{ mod } \alpha_i \text{ in } R_i$$

for $i = 1$ and for $i = 2$. Therefore, the module $M_{n+1}^0(R_1, R_2)$ is a homomorphic image of the module $M_{n+1}(R_3, R; \alpha_3)$, so that the cycles of the module $M_{n+1}(R_3, R;$ $\alpha_3)$ whose image in the module $M_{n+1}^0(R_1, R_2)$ is zero form the module $\mu_{n+1}(R_3; \alpha_3)$ (the notation is that of N° 15). Consequently, (see III 11)

$$(2) \qquad P_{n+1}(R_3; \alpha_3) = P_{n+1}^0(R_1, R_2) + \pi_{n+1}(R_3; \alpha_3) \; .$$

18. From 15 (1) and from 17 (1), (2), one obtains the final result:

$$\begin{aligned} P_{n+1}(R_1; \alpha_1) + P_{n+1}(R_2; \alpha_2) + \pi_n(R_3; \alpha_3) + \pi_{n+1}(R_3; \alpha_3) = \\ = P_{n+1}(R; \alpha) + P_{n+1}(R_3; \alpha_3), \end{aligned}$$

where $\pi_n(R_3; \alpha_3)$ is the rank of the module consisting of all the (n, R)-cycles mod α_3 in R_3 homologic to zero mod α_1 in R_1, as well as mod α_2 in R_2, with two such cycles being considered as equal if they are homologic mod α_3 in R_3.

V. Supplements.

1. In what preceded, the coefficients of all the cycles were taken in the set $\mathfrak{R}$ of *rational numbers*. It is not possible to replace $\mathfrak{R}$ by the set of *integers*, for then the theorems I 7 and I 8 are no more valid, and, consequently, the important theorem II 16 ceases to be true. But one can choose as a module a fixed *arithmetic module* $m = 2, 3, 4, \cdots$, and replace $\mathfrak{R}$ by the set $\mathfrak{R}_m$ consisting of all the integers with *two*

EDUARD ČECH

integers congruent mod m considered as equal. If $m = m_1 \cdot m_2$, the two factors m_1 and m_2 being *prime to each other*, one can easily see that the module consisting of the cycles mod m of any of the various modules cosidered in the preceding Chapters is a *direct sum* of the module consisting of the cycles mod m_1, and that consisting of the cycles mod m_2. [This follows immediately from the well known fact that the ring $\mathfrak{R}_m$ is a direct sum of the two rings $\mathfrak{R}_{m_1}$, $\mathfrak{R}_{m_2}$]. Therefore we can restrict ourselves to the case where $m = p^k$, p being a prime number and $k = 1, 2, 3, \ldots$.In the case $k = 1$ ($m = p$) the theory mod m is *formally* identical with that presented in the preceding Chapters. The reason for this is simply that the set $\mathfrak{R}_p$ forms a *field* equally with the set $\mathfrak{R}$. In the case $m = p^k$, $k > 1$, there are differences consisting mainly in the fact that a module is no more characterized by a *unique* cardinal number (its rank). Besides the Betti numbers, it is necessary to consider the torsion coefficients. Nevertheless, the essential content of the theory remains the same.

2. In Chap. II, we took as a fundamental family of nets in a topological space R the family consisting of all the *open* nets. One can also base the theory on the *closed* nets (these are, naturally, the nets whose vertices are closed subsets of R). We are now going to prove (in the n°s 3–8) that if R is a *hereditarily normal topological space, then the two procedures are absolutely equivalent.*

3. Let R be a hereditarily normal topological space, and let α be a given closed subset of R. A closed net $\mathfrak{U}_f$ and an open net $\mathfrak{U}_g$ in R are called *isologic* if there exists a bijective correspondence between the vertices $u_1, u_2, \ldots, u_k$ of $\mathfrak{U}_f$ and the vertices $U_1, U_2, \ldots, U_k$ of $\mathfrak{U}_g$ having the following properties:

1° $\mu_\nu \subset U_\nu$ $(1 \leqq \nu \leqq k)$;

2° $U_{\nu_1} \cdot U_{\nu_2} \cdot \ldots \cdot U_{\nu_h} \neq 0$ implies $u_{\nu_1} \cdot u_{\nu_2} \cdot \ldots \cdot u_{\nu_h} \neq 0$;

3° $\alpha \cdot U_{\nu_1} \cdot U_{\nu_2} \cdot \ldots \cdot U_{\nu_h} \neq 0$.

Therefore, one has obviously a bijective correspondence between the $(n, \mathfrak{U}_f)$-cycles mod α and the $(n, \mathfrak{U}_g)$-cycles mod α, and this correspondence preserve the homologies mod α. We shall say that one *transports* a cycle of $\mathfrak{U}_f$ to a cycle of $\mathfrak{U}_g$, and vice versa.

4. Let $\mathfrak{U}_f$ be a closed net in R. Then there exists an open net $\mathfrak{U}_g$ isologic to $\mathfrak{U}_f$.

Proof. Let us denote by $u_1, u_2, \ldots, u_k$ the vertices of $\mathfrak{U}_f$. Applying to the closed subsets $\alpha, u_1, u_2, \ldots, u_k$ of R the theorem of Chap. III, n° 21 ($A = R$), one obtains open subsets $V, U_1, U_2, \ldots, U_k$ of R. It can be easily seen that $U_1, U_2, \ldots, U_k$ form an open net $\mathfrak{U}_g$ isologic to $\mathfrak{U}_f$.

5. Let $\mathfrak{U}_g$ be an open net in R. Then there exists a closed net $\mathfrak{U}_f$ isologic to $\mathfrak{U}_g$.

Proof. Let us denote by $U_1, U_2, \ldots, U_k$ the vertices of U_g. According to a lemma by Menger,[24] there exist open sets $V_1, V_2, \ldots, V_k$ such that 1° $\overline{V}_\nu \subset U_\nu$; 2° $\sum_1^k V_\nu = R$. For each combination $(\nu_1, \nu_2, \ldots, \nu_h)$ of the indices $1, 2, \ldots, k$ such that the set $U_{\nu_1} \cdot U_{\nu_2} \cdot \ldots \cdot U_{\nu_h}$ is not empty, let us choose a point $p_{\nu_1, \nu_2, \ldots, \nu_h} \in U_{\nu_1} \cdot U_{\nu_2} \cdot \ldots \cdot U_{\nu_h}$. We choose this point in α always when it is possible. For $\nu = 1, 2, \ldots, k$, let us

[24]K. Menger, Dimensions theorie, "Bemerkung" , pp. 156–160. Menger supposes that the space R is separable. But it can be easily seen that the proof is valid in every completely normal space (even, more generally, in every normal space).

GENERAL HOMOLOGY THEORY IN AN ARBITRARY SPACE

denote by I_ν the finite set consisting of the points $p_{\nu_1,\nu_2,\dots,\nu_h}$ such that one of the indices $\nu_1,\nu_2,\dots,\nu_h$ coincides with ν. Let us set $u_\nu = \overline{V}_\nu + I_\nu$. One can easily see that $u_1,u_2,\dots,u_k$ are the vertices of a closed net $\mathfrak{U}_f$ isologic to $\mathfrak{U}_g$.

6. Let $\{C^n(\mathfrak{U}_g)\}$ be a given $(n,R)_g$-cycle[25] mod α. Let us associate with each closed net $\mathfrak{U}_f$ an open net $\mathfrak{U}_g = I(\mathfrak{U}_f)$ isologic to $\mathfrak{U}_f$, and let us denote by $C^n(\mathfrak{U}_f)$ the $(n,\mathfrak{U}_f)$-cycle mod α obtained by transporting $C^n(\mathfrak{U}_g)$ of $\mathfrak{U}_g$ to $\mathfrak{U}_f$. Let $\mathfrak{V}_f$ be a refinement of $\mathfrak{U}_f$, and let $\mathfrak{U}_g = I(\mathfrak{U}_f)$, $\mathfrak{U}_g = I(\mathfrak{V}_f)$. Let us denote by $u_1,u_2,\dots,u_h$ $(v_1,v_2,\dots,v_k)$ the vertices of $\mathfrak{U}_f$ $(\mathfrak{V}_f)$, and by $U_1,U_2,\dots,U_h$ $(V_1,V_2,\dots,V_k)$ the corresponding vertices of $\mathfrak{U}_g$ $(\mathfrak{V}_g)$. Let $\pi = \mathrm{Pr.}(\mathfrak{V}_f,\mathfrak{U}_f)$; $\pi v_\nu = u_{\pi(\nu)}$. Let us replace each vertex V_ν of $\mathfrak{V}_g$ by the open set $V \cdot U_{\pi(\nu)}$. In this way, the net $\mathfrak{V}_g$ transforms to a new open net $\mathfrak{W}_g$ which is obviously a common refinement of the two nets $\mathfrak{U}_g$ and $\mathfrak{V}_g$. It can also be easily seen that $\mathfrak{W}_g$ is isologic to $\mathfrak{V}_f$. Let us set

$$\pi'[V_\nu \cdot U_{\pi(\nu)}] = V_\nu\,, \quad \pi''[V_\nu \cdot U_{\pi(\nu)}] = U_{\pi(\nu)}\,,$$

so that $\pi' = \mathrm{Pr.}(\mathfrak{W}_g,\mathfrak{V}_g)\,,\ \pi'' = \mathrm{Pr.}(\mathfrak{W}_g,\mathfrak{U}_g)$. Let us set

$$C'^n(\mathfrak{V}_g) = \pi' C^n(\mathfrak{W}_g)\,, \qquad C''^n(\mathfrak{U}_g) = \pi'' C^n(\mathfrak{W}_g)\,,$$

so that

$$C^n(\mathfrak{V}_g) \sim C'^n(\mathfrak{V}_g) \bmod \alpha\,, \quad C^n(\mathfrak{U}_g) \sim C''^n(\mathfrak{U}_g) \bmod \alpha\,.$$

Hence, it follows

$$C^n(\mathfrak{V}_f) \sim C'^n(\mathfrak{V}_f) \bmod \alpha\,, \quad C^n(\mathfrak{U}_f) \sim C''^n(\mathfrak{U}_f) \bmod \alpha\,, \tag{1}$$

where the cycles $C'^n(\mathfrak{V}_f)$ and $C''^n(\mathfrak{U}_f)$ are obtained by transporting $C'^n(\mathfrak{V}_g)$ of $\mathfrak{V}_g$ to $\mathfrak{V}_f$, and $C''^n(\mathfrak{U}_g)$ of $\mathfrak{U}_g$ to $\mathfrak{U}_f$. Now, it is obvious that $C''^n(\mathfrak{U}_f) = \pi C'^n(\mathfrak{U}_f)$, so that the homologies (1) imply

$$C^n(\mathfrak{U}_f) \sim \pi C^n(\mathfrak{V}_f) \bmod \alpha\,. \tag{2}$$

Therefore, $\{C^n(\mathfrak{U}_f)\}$ is an $(n,R)_f$-cycle mod α. This cycle is not absolutely determined, for the operation $\mathfrak{U}_g = I(\mathfrak{U}_f)$ is not absolutely determined. If one sets $\mathfrak{U}_f = \mathfrak{V}_f$ in the previous consideration, the relation (2) (where π is the identity) shows that the cycle $\{C^n(\mathfrak{U}_f)\}$ is well determined up to a homology mod α.

7. Let $\{C^n(\mathfrak{U}_f)\}$ be a given $(n,R)_f$-cycle mod α. Let us associate with every open net $\mathfrak{U}_g$ a closed net $\mathfrak{U}_f = I(\mathfrak{U}_g)$ isologic to $\mathfrak{U}_g$, and let us denote by $C^n(\mathfrak{U}_g)$ the $(n,\mathfrak{U}_g)$-cycle mod α obtained by transporting $C^n(\mathfrak{U}_f)$ of $\mathfrak{U}_f$ to $\mathfrak{U}_g$. Let $\mathfrak{V}_g$ be a refinement of $\mathfrak{U}_g$, and let $\mathfrak{U}_f = I(\mathfrak{U}_g)$, $\mathfrak{V}_f = I(\mathfrak{V}_g)$. Let us choose a common refinement $\mathfrak{W}_f$ of the two nets $\mathfrak{U}_f$ and $\mathfrak{V}_f$, and let $\pi' = \mathrm{Pr.}(\mathfrak{W}_f,\mathfrak{V}_f)$, $\pi'' = \mathrm{Pr.}(\mathfrak{W}_f,\mathfrak{U}_f)$. Let $\mathfrak{W}_g^*$ be an open net isologic to $\mathfrak{W}_f$. Let W^* be an arbitrary vertex of $\mathfrak{W}_g^*$, and w the corresponding vertex of $\mathfrak{W}_f$. Let V be the vertex of $\mathfrak{V}_g$ corresponding to the vertex $\pi'(\mathfrak{U})$ of $\mathfrak{V}_f$. Let us replace W^* by the set $W = W^* \cdot V$. Proceeding in this way with all the vertices W^* of $\mathfrak{W}_g$, this net transforms to a new open net $\mathfrak{W}_g$,

[25] The index g (f) means always that one takes the open (closed) nets as a fundamental family of nets.

EDUARD ČECH

which is obviously a refinement of $\mathfrak{W}_g$. One can easily see that $\mathfrak{W}_g$ is isologic to W_f.

Let us denote by $C'^n(\mathfrak{V}_g)$ $[C''^n(\mathfrak{U}_g)]$ the cycle obtained by transporting $\pi' C^n(\mathfrak{W}_f)$ $[\pi'' C^n(\mathfrak{W}_f)]$ of $\mathfrak{V}_f$ to $\mathfrak{V}_g$ [of $\mathfrak{U}_f$ to $\mathfrak{U}_g$]. One can easily see (cf. 6 (1)) that

$$(1) \qquad C^n(\mathfrak{V}_g) \sim C'^n(\mathfrak{V}_g) \bmod \alpha\,, \quad C^n(\mathfrak{U}_g) \sim C''^n(\mathfrak{U}_g) \bmod \alpha\,.$$

Let $\pi = \text{Pr.}(\mathfrak{W}_g, \mathfrak{U}_g)$, $\overline{\pi} = \text{Pr.}(\mathfrak{W}_g, \mathfrak{V}_g)$. Then $\pi\overline{\pi} = \text{Pr.}(\mathfrak{W}_g, \mathfrak{U}_g)$. Let us denote by $C_0^n(\mathfrak{W}_g)$ the cycle obtained by transporting $C^n(\mathfrak{W}_f)$ of $\mathfrak{W}_f$ to $\mathfrak{W}_g$, and let us set $C^n(\mathfrak{V}_g) = \overline{\pi} C_0^n(\mathfrak{W}_g)$, $C_0^n(\mathfrak{U}_g) = \pi\overline{\pi} C_0^n(\mathfrak{W}_g)$, so that

$$(2) \qquad C_0^n(\mathfrak{U}_g) = \pi C_0^n(\mathfrak{V}_g)\,.$$

The reasoning of Chap. II, n° 12 is obviously applicable for showing that

$$(3) \qquad C'^n(\mathfrak{V}_g) \sim C_0^n(\mathfrak{V}_g) \bmod \alpha, \qquad C''^n(\mathfrak{U}_g) \sim C_0^n(\mathfrak{U}_g) \bmod \alpha\,.$$

From the relations (1), (2), (3) one can deduce

$$(4) \qquad C^n(\mathfrak{U}_g) \sim \pi C^n(\mathfrak{V}_g) \bmod \alpha\,.$$

Therefore, $\{C^n(\mathfrak{U}_g)\}$ is an $(n, R)_g$-cycle mod α. Setting $\mathfrak{U}_g = \mathfrak{V}_g$ in the previous reasoning, the relation (4) shows that this cycle is well determined up to a homology mod α.

8. The operations considered in the two preceding n^{os} being obviously inverse to each other, one can see that there exists a bijective correspondence between the two modules $M_n(R;\alpha)_g$ and $M_n(R;\alpha)_f$. It is easy to see that this correspondence is an *isomorphism*. Consequently, $P_n(R;\alpha)_g = P_n(R;\alpha)_f$. The two fundamental families consisting of open and closed nets, respectively, are therefore equivalent. The open nets seem to be more convenient in the applications; cf. the theorem III, 11, which played an important role in Chap. IV.

9. Let us return to the general homology theory presented in Chap. II. Let us suppose that we have chosen a certain family κ of subsets of R with the property that $A_1 \in \kappa$, $A_2 \in \kappa$ imply $A_1 + A_2 \in \kappa$. By an $(n, R)_\kappa$-cycle mod α (with α being a given subset of R) we want to understand an (n, R)-cycle $\{C^n(\mathfrak{U})\}$ mod α such that there exists a set $A \in \kappa$ satisfying $C^n(\mathfrak{U}) \subset A$ (in the sense of Chap. II, n° 5) for every net $\mathfrak{U}$. An $(n, R)_\kappa$-cycle $\{C^n(\mathfrak{U})\}$ mod α will be considered homologic to zero mod α only if there exists a set $B \in \kappa$ such that, for each net $\mathfrak{U}$, there is $K^{n+1}(\mathfrak{U}) \to C^n(\mathfrak{U})$ mod α, *the* $(n+1, \mathfrak{U})$*-chain being contained in* B. By virtue of the additive property of the family κ, the $(n, R)_\kappa$-cycles mod α constitute a *module* $M_n(R;\alpha)_\kappa$, when considering as equal two cycles homologic in the sense which has just been described. The rank $P_n(R;\alpha)_\kappa$ of this module is the n-th *Betti number of the type* κ *of* R. If $R \in \kappa$, the new theory does not differ from the previous one.

10. There is one important particular case of the definitions presented above. Let R be a *metrizable* space, where we take the open nets as the fundamental family. Let us set $\alpha = 0$. Let κ be the family of *compact* subsets of R, i. e. of subsets $A \subset R$ such that from each sequence $x_n \in A$ one can choose another one y_n possessing a point limit $y \in A$. According to a known theorem of Hausdorff, the

quasicomponents of a metrizable compact set are themselves compact and coincide with the components. Therefore, according to III, 14 and 17, there is $\{p\} \sim \{q\}$ in the sense of the theory of the type κ if and only if there exists a *continuum* (= compact connected set) containing p and q. Therefore, the number $P_0(R)_\kappa$ is the number of constituents of R, i. e. of maximal semicontinua, where the name semicontinuum designates a set $A \subset R$ in which every pair of points belongs to a continuum $C \subset A$.

BETTI GROUPS OF AN INFINITE COMPLEX

EDUARD ČECH

1. Let K be an infinite complex. For $n = 0, 1, 2, \dots$, let us denote by σ_i^n $(i = 1, 2, 3, \dots)$ the n-simplices of K. These simplices are oriented in an arbitrary, but fixed way.

Let $\mathfrak{G}$ be a given abelian group. The composition law in $\mathfrak{G}$ (and in all the other groups considered in this paper, which are all abelian) will be considered as an *addition*. Let us call an $(n, \mathfrak{G})$-*chain* each linear form

$$C^n = \sum_i g_i \sigma_i^n , \quad g_i \in \mathfrak{G} \tag{1}$$

having only a finite number of the coefficients g_i different from zero.[1]

Setting

$$\sum_i g_i \sigma_i^n + \sum_i g_i' \sigma_i^n = \sum_i (g_i + g_i') \sigma_i^n ,$$

the $(n, \mathfrak{G})$-chains constitute an abelian group, which will be denoted by $C^n(\mathfrak{G})$.

The letter $\mathfrak{C}$ denotes the additive abelian group of the integers. Let us set $C^n(\mathfrak{C}) = C^n$. If $C^n = \sum_i a_i \sigma_i^n$ is an $(n, \mathfrak{C})$-chain and $g \in \mathfrak{G}$, then $\sum_i a_i g \cdot \sigma_i^n$ is an $(n, \mathfrak{G})$-chain denoted by gC^n.

The *boundary* FC^0 of a $(0, \mathfrak{G})$-chain is equal to zero. For $n > 0$, the *boundary* $F\sigma_i^n$ of an n-simplex σ_i^n is an $(n-1, \mathfrak{C})$-chain

$$F\sigma_i^n = \sum_k \eta_{ik}^n \sigma_k^{n-1} .$$

It is not necessary to recall here the precise form of the integers η_{ik}^n. We shall only make use of the well known fact that for $n \geqq 2$ one has

$$\sum_k \eta_{ik}^n \eta_{kh}^{n-1} = 0 \qquad (i, h = 1, 2, 3, \dots) . \tag{2}$$

For $n > 0$, the *boundary* FC^n of the $(n, \mathfrak{G})$-chain (1) is the $(n-1, \mathfrak{G})$-chain

[1] In general, every sum considered in this paper has only a finite number of terms different from zero.

$$FC^n = \sum_i g_i \cdot F\sigma_i^n \,.$$

By virtue of (2), one has for $n \geqq 2$ and for every $(n, \mathfrak{G})$-chain C^n

$$FFC^n = 0 \,, \tag{3}$$

which is obvious for $n = 1$.

An $(n, \mathfrak{G})$-chain C^n is called an $(n, \mathfrak{G})$-*cycle* if $FC^n = 0$. The $(n, \mathfrak{G})$-cycles constitute a subgroup of $\boldsymbol{C}^n(\mathfrak{G})$, which will be denoted by $\boldsymbol{\Gamma}^n(\mathfrak{G})$. By virtue of (3), for each $C^{n+1} \in \boldsymbol{C}^{n+1}(\mathfrak{G})$, FC^{n+1} is an $(n, \mathfrak{G})$-cycle. The $(n, \mathfrak{G})$-cycles having the form FC^{n+1} are called *homologic to zero* (~ 0). They constitute a subgroup of the group $\boldsymbol{\Gamma}^n(\mathfrak{G})$, which will be denoted by $\boldsymbol{H}^n(\mathfrak{G})$. The quotient group (Faktorgruppe)

$$\boldsymbol{\Gamma}^n(\mathfrak{G})/\boldsymbol{H}^n(\mathfrak{G})$$

will be denoted by $\boldsymbol{B}^n(\mathfrak{G})$. This is the *n-th Betti group with respect to the coefficient domain* $\mathfrak{G}$. Let us set

$$\boldsymbol{\Gamma}^n(\mathfrak{C}) = \boldsymbol{\Gamma}^n \,, \quad \boldsymbol{H}^n(\mathfrak{C}) = \boldsymbol{H}^n \,, \quad \boldsymbol{B}^n(\mathfrak{C}) = \boldsymbol{B}^n \,.$$

$\boldsymbol{B}^n$ is the *n-th ordinary Betti group*. The elements B^n of $\boldsymbol{B}^n$ whose order is finite, i. e. the elements for which there exists $c \in \mathfrak{C}$ such that $c > 0$, $cB^n = 0$, constitute a subgroup of the group $\boldsymbol{B}^n$ called the *n-th torsion group*. It will be denoted by $\boldsymbol{T}^n$. It is known that $\boldsymbol{T}^0 = 0$. Let us set also $\boldsymbol{T}^{-1} = 0$. If the dimension m of the complex K is finite, there is $\boldsymbol{B}^n = 0$ for $n > m$ and $\boldsymbol{T}^n = 0$ for $n \geqq m$.

If the abelian groups $\mathfrak{G}_1$ and $\mathfrak{G}_2$ are isomorphic, then the groups $\boldsymbol{B}^n(\mathfrak{G}_1)$ and $\boldsymbol{B}^n(\mathfrak{G}_2)$ are also isomorphic. In particular, if $\mathfrak{G}$ is a cyclic group of infinite order, then the group $\boldsymbol{B}^n(\mathfrak{G})$ is isomorphic to $\boldsymbol{B}^n$. If the abelian group $\mathfrak{G}$ is a direct sum of two subgroups $\mathfrak{G}_1$ and $\mathfrak{G}_2$, then the group $\boldsymbol{B}^n(\mathfrak{G})$ is isomorphic to the direct sum of the two groups $\boldsymbol{B}^n(\mathfrak{G}_1)$ and $\boldsymbol{B}^n(\mathfrak{G}_2)$.

2. Alexander has considered[2] the case where (1) *the complex K is finite*, (2) *$\mathfrak{G}$ is a cyclic group of finite order.* He has shown that, under these hypotheses, the *structure*[3] *of the group $\boldsymbol{B}^n(\mathfrak{G})$ is completely determined by the structure of the three groups $\mathfrak{G}$, $\boldsymbol{B}^n$, and $\boldsymbol{T}^{n-1}$*. It is easy to eliminate the hypothesis (2) from the Alexander's considerations. But the hypothesis (1) is used quite essentially, when making use of the reduction of a matrix with integer coefficients to a canonical form. Nevertheless, I am going to show that the result by Alexander is completely general, and that both the conditions (1) and (2) are superfluous.

Let us begin with a definition. If Γ_i^n are $(n, \mathfrak{C})$-cycles (in finite number), and if $g_i \in \mathfrak{G}$, then $\sum_i g_i \Gamma_i^n$ is an $(n, \mathfrak{G})$-cycle. Let us call *pure* each cycle of the form

$$\sum_i g_i \Gamma_i^n, \qquad g_i \in \mathfrak{G}, \quad \Gamma_i^n \in \boldsymbol{\Gamma}^n.$$

[2] *Combinatorial Analysis Situs*, Trans. Amer. Math. 28, 301–329 (1926). Cf. also A. W. Tucker, *Modular homology characters*, Proc. Nat. Acad. Sc. 18, 467–471 (1932).

[3] Two groups have the same structure if they are isomorphic.

BETTI GROUPS OF AN INFINITE COMPLEX

The pure $(n, \mathfrak{G})$-cycles constitute an abelian group, which will be denoted by $\Gamma_1^n(\mathfrak{G})$. Obviously,

$$\Gamma^n(\mathfrak{G}) \supset \Gamma_1^n(\mathfrak{G}) \supset H^n(\mathfrak{G}).$$

The quotient group

$$\Gamma_1^n(\mathfrak{G})/H^n(\mathfrak{G})$$

will be denoted by $B_1^n(\mathfrak{G})$, and will be called *n-th pure Betti group with respect to the coefficient domain* $\mathfrak{G}$. Having this, I show in this paper the following three theorems:

Theorem I. *There exists a subgroup* $B_2^n(\mathfrak{G})$ *of the group* $B^n(\mathfrak{G})$ – let us call it *n-th complementary Betti group with respect to the coefficient domain* $\mathfrak{G}$ – *such that the group* $B^n(\mathfrak{G})$ *is a direct sum of the two subgroups* $B_1^n(\mathfrak{G})$ *and* $B_2^n(\mathfrak{G})$. The group $B_2^n(\mathfrak{G})$ is not uniquely determined, but its structure is determined without ambiguity, $B_2^n(\mathfrak{G})$ being isomorphic to the quotient group

$$B^n(\mathfrak{G})/B_1^n(\mathfrak{G}).$$

Conversely, the structure of $B^n(\mathfrak{G})$ is completely determined by the structure of the two groups $B_1^n(\mathfrak{G})$ and $B_2^n(\mathfrak{G})$.

Theorem II. *The structure of the group* $B_1^n(\mathfrak{G})$ *is completely determined by the structure of the two groups* $\mathfrak{G}$ *and* B^n. *More precisely: Let us suppose that the group* B^n *is given by the generating elements* α_i *(*$i = 1, 2, 3, \ldots$*, possibly ad inf.) and by the defining relations* $\sum_i a_{ik}\alpha_i = 0$ *(*$k = 1, 2, 3, \ldots$*, possibly ad inf.)*[4], *where* $a_{ik} \in \mathfrak{C}$, *and, for each* k, *there is only a finite number of values of* i *such that* $a_{ik} \neq 0$. *One obtains a group* X *isomorphic to the group* $B_1^n(\mathfrak{G})$ *in the following way. Let us attach to each* i *a symbol* x_i. *The elements of* X *are the symbols* $\sum_i g_i x_i$ *having only a finite number of coefficients* $g_i \in \mathfrak{G}$ *different from zero. The addition in* X *is defined by*

$$\sum_i g_i x_i + \sum_i g_i' x_i = \sum_i (g_i + g_i') x_i \ ,$$

and the defining relations are

$$\sum_i a_{ik} g \cdot x_i = 0 \quad \textit{for} \quad g \in \mathfrak{G} \quad (k = 1, 2, 3 \ldots) \ .$$

[4] This means that the elements of B^n have the form $\sum_i c_i\alpha_i$, where there is only a finite number of coefficients $c_i \in \mathfrak{G}$ which are $\neq 0$, that $\sum_i c_i\alpha_i + \sum_i c_i'\alpha_i = \sum_i (c_i + c_i')\alpha_i$, and, finally, that $\sum_i c_i\alpha_i = 0$ if and only if there exist a finite number of integers b_k such that $c_i = \sum_k b_k a_{ik}$ for $i = 1, 2, 3, \ldots$ In the case, where there is no defining relation, one says that the generating elements α_i are *linearly independent*.

EDUARD ČECH

Theorem III. *The structure of the group $B_2^n(\mathfrak{G})$ is completely determined by the structure of the two groups $\mathfrak{G}$ and T^{n-1}. More precisely: Let us suppose that the group T^{n-1} is given by the generating elements β_i ($i = 1, 2, 3, \ldots$, possibly ad inf.), and by the defining relations $\sum_i b_{ik}\beta_i = 0$ ($k = 1, 2, 3 \ldots$, possibly ad inf.), where $b_{ik} \in \mathfrak{C}$, and, for each k, there is only a finite number of values of i such that $b_{ik} \neq 0$. One obtains a group Y isomorphic to the group $B_2^n(\mathfrak{G})$ in the following way. Let us attach to each k a symbol y_k. The elements of Y are the symbols $\sum_k g_k y_k$ having only a finite number of coefficients $g_k \in \mathfrak{G}$ different from zero. The coefficients are not arbitrary, but related by the equations*

$$\sum_k b_{ik} g_k = 0 \quad (i = 1, 2, 3, \ldots) .$$

The addition in Y is defined by

$$\sum_k g_k y_k + \sum_k g'_k y_k = \sum_k (g_k + g'_k) y_k ,$$

and the defining relations are

$$\sum_k c_k g \cdot y_k = 0 \quad \text{for each} \quad g \in \mathfrak{G} ,$$

where c_k are integers (from which only a finite number is $\neq 0$) such that

$$\sum_k b_{ik} c_k = 0 \quad (i = 1, 2, 3, \ldots) .$$

3. The proofs are based on the

Lemma. *Let us suppose that the abelian group $\mathfrak{G}$ possesses a finite or countable number of linearly independent generators. Let $\mathfrak{H}$ be a subgroup of $\mathfrak{G}$. Then $\mathfrak{H}$ possesses a finite or countable system of linearly independent generators.*

Proof. Let g_i ($i = 1, 2, 3, \ldots$, possibly ad inf.) be given linearly independent generators of the group $\mathfrak{G}$. Let us set

$$h_1 = \sum_{\nu=1}^{\lambda_1} a_{1\nu} g_\nu \quad (a_{1\nu} \in \mathfrak{C}) ,$$

choosing λ_1 and $a_{1\nu}$ in such a way that (1) $h_1 \in \mathfrak{H}$, (2) λ_1 = minimum, (3) $a_{1\nu_1} > 0$, $a_{1\nu_1}$ = minimum. Let us suppose that one has already determined the indices $\lambda_1 < \lambda_2 < \cdots < \lambda_i$ as well as the elements $h_1, h_2, \ldots, h_i$ of $\mathfrak{H}$. Let us set then

$$h_{i+1} = \sum_{\nu=1}^{\lambda_{i+1}} a_{i+1,\nu} g_\nu \quad (a_{i+1,\nu} \in \mathfrak{C}) ,$$

choosing λ_{i+1} and $a_{i+1,\nu}$ in such a way that (1) $h_{i+1} \in \mathfrak{H}$, (2) $\lambda_{i+1} > \lambda_i$, λ_{i+1} = minimum, (3) $a_{i+1,\lambda_i+1} > 0$, a_{i+1,λ_i+1} = minimum. Proceeding in this way, one

BETTI GROUPS OF AN INFINITE COMPLEX

obtains a finite or countable sequence $h_1, h_2, h_3, \dots$. Let us set $V(0) = 0$. If $h \in \mathfrak{H}$ and $h \neq 0$, let $h = \sum b_\nu g_\nu$. The largest index ν such that $b_\nu > 0$ has the form $\nu = \lambda_i$. Let us set $V(h) = i$.

The h_i's are linearly independent. In fact, if $h = \sum_{\mu=1}^{i} c_\mu h_\mu$, $c_\mu \in \mathfrak{C}$ and $c_i \neq 0$, one has $h = \sum_{\nu=1}^{\lambda_i} b_\nu g_\nu$ and $b_\nu \in \mathfrak{C}$, $b_{\lambda_i} = c_i a_{i\lambda_i} \neq 0$, whence $h \neq 0$.

Each element h of the group $\mathfrak{H}$ has the form $\sum_{\mu=1}^{i} c_\mu h_\mu$, $c_\mu \in \mathfrak{C}$. This is obvious if $V(h) = 0$. Let $V(h) = i > 0$, and let us suppose that the assertion is true for each $h' \in \mathfrak{H}$ such that $V(h') < i$. One has $h = \sum_{\nu=1}^{\lambda_i} b_\nu g_\nu$. Let us determine the integers q and r in such a way that $b_{\lambda_i} = q a_{i,\lambda_i} + r$, $0 \leqq r < a_{i,\lambda_i}$. One has $h - qh_i = \sum_{\nu=1}^{\lambda_i} b'_\nu g_\nu$, $b'_\nu \in \mathfrak{C}$, $0 \leqq b'_{\nu_i} = r < a_{i,\lambda_i}$, whence, according to the choice of a_{i,λ_i}, $b'_{\nu_i} = 0$. Therefore, $V(h - qh_i) < i$, whence $h - qh_i = \sum_\mu c_\mu h_\mu$, $h = qh_i + \sum_\mu c_\mu h_\mu$.

4. *Proof of the theorem I.* For $n = 0$, one has obviously $\boldsymbol{B}_1^0(\mathfrak{G}) = \boldsymbol{B}^0(\mathfrak{G})$, $\boldsymbol{B}_2^0(\mathfrak{G}) = 0$. Let $n > 0$. The $(n-1, \mathfrak{C})$-chains $1 \cdot \sigma_i^{n-1}$ constituting obviously a finite or countable system of linearly independent generating elements of the group $\boldsymbol{C}^{n-1}$, it results from the lemma that the subgroup $\boldsymbol{H}^{n-1}$ of the group $\boldsymbol{C}^{n-1}$ possesses a finite or countable system of linearly independent generating elements h_k ($k = 1, 2, 3, \dots$, possibly ad inf.). Since $h_k \in \boldsymbol{H}^{n-1}$, there exist $(n, \mathfrak{C})$-chains D_k^n such that $FD_k^n = h_k$. The chains D_k^n constitute a system of generators of a subgroup – let us denote it by $\boldsymbol{D}^n$ – of the group $\boldsymbol{C}^n$.

Let $C^n \in \boldsymbol{C}^n$. Thus, $FC^n \in \boldsymbol{H}^{n-1}$. Therefore, there exist numbers $a_k \in \mathfrak{C}$ (in finite number) such that $FC^n = \sum_k a_k h_k$, so that $\Gamma^n = C^n - \sum_k a_k D_k^n \in \boldsymbol{\Gamma}^n$. On the other hand, let $\Gamma^n \in \boldsymbol{\Gamma}^n$, $D^n \in \boldsymbol{D}^n$, $\Gamma^n + D^n = 0$. Hence, it results that $FD^n = F(\Gamma^n + D^n) = 0$. But one has $D^n = \sum_k a_k D_k^n$, whence $\sum_k a_k h_k = FD^n = 0$. Therefore, $a_k = 0$, and consequently $D^n = 0$ and $\Gamma^n = 0$. Therefore, the group $\boldsymbol{C}^n$ is a direct sum of $\boldsymbol{\Gamma}^n$ and $\boldsymbol{D}^n$. Hence, one deduces easily that the group $\boldsymbol{\Gamma}^n(\mathfrak{G})$ is a direct sum of the group $\boldsymbol{\Gamma}_1^n(\mathfrak{G})$ and the group $\boldsymbol{\Gamma}_2^n(\mathfrak{G})$, this last one being defined as the set of all $(n, \mathfrak{G})$-cycles of the form $\sum_k g_k D_k^n$, $g_k \in \mathfrak{G}$. Hence it follows easily that the group $\boldsymbol{B}^n(\mathfrak{G})$ is a direct sum of the group $\boldsymbol{B}_1^n(\mathfrak{G})$ and a group $\boldsymbol{B}_2^n(\mathfrak{G})$ isomorphic to the group $\boldsymbol{\Gamma}_2^n(\mathfrak{G})$.

5. *Proof of the theorem II.* Let us suppose that the group is given by the generators α_i ($i = 1, 2, 3, \dots$, possibly ad inf.) and by the defining relations $\sum_i a_{ik}\alpha_i = 0$ ($k = 1, 2, 3, \dots$, possibly ad inf.). In other words: (1) there exist $(n, \mathfrak{C})$-cycles Γ_i^n such that for each $(n, \mathfrak{C})$-cycle Γ^n there is a homology of the form $\Gamma^n - \sum_i b_i \Gamma_i^n \sim 0$, (2) one has $\sum_i b_i \Gamma_i^n \sim 0$ if and only if there exist integers c_k (in finite number) such that $b_i = \sum_k a_{ik} c_k$. Hence, it follows that the elements of the group $\boldsymbol{\Gamma}_1^n(\mathfrak{G})$ have

the form $\sum_i g_i \Gamma_i^n$, $g_i \in \mathfrak{G}$, and that one has $\sum_i g_i \Gamma_i^n \sim 0$ if and only if there exist $g'_k \in \mathfrak{G}$ (in finite number) such that $g_i = \sum_k a_{ik} g'_k$. But this implies the validity of the theorem II.

6. *Proof of the theorem III.* For $n = 0$, one has $B_2^0(\mathfrak{G}) = 0$, $T^{-1} = 0$. Let $n > 0$. Let us suppose that the group T^{n-1} is given by the generators β_i ($i = 1, 2, 3, \ldots$, possibly ad inf.) and by the defining relations $\sum_i b_{ik}\beta_i = 0$ ($k = 1, 2, 3, \ldots$, possibly ad inf.). The theorem III deduces from the generators β_i and the relations $\sum_i b_{ik}\beta_i = 0$ a group Y, and asserts that the two groups $B_2^n(\mathfrak{G})$ and Y are isomorphic. First, we are going to prove that the structure of the group Y is determined without ambiguity by the structure of the groups $\mathfrak{G}$ and T^{n-1}.

First, let us suppose that one has chosen, for the generators β_i, the system of *all* the elements of the group T^{n-1}. Concerning the defining relations $\sum_i b_{ik}\beta_i = 0$, let us suppose they have been chosen in an arbitrary (but fixed) way. If one adds to the relations $\sum_i b_{ik}\beta_i = 0$ *all* the other relations existing among the β_i's, the group Y must be replaced by a new group Y'. The elements of Y were of the form $\sum_k g_k y_k$ with $g_k \in \mathfrak{G}$ being such that $\sum_k b_{ik} g_k = 0$ for $i = 1, 2, 3, \ldots$. One had also the defining relations $\sum_k c_k g \cdot y_k = 0$ ($g \in \mathfrak{G}$), where the c_k's were integers such that $\sum_k b_{ik} c_k = 0$ for $i = 1, 2, 3, \ldots$. The $\sum_i b_{ik}\beta_i = 0$'s constitute a system of defining relations for the generators β_i of the group T^{n-1}, the other relations existing among the β_i's being of the form $\sum_{ik} v_{hk} b_{ik} \beta_i = 0$, $v_{hk} \in \mathfrak{C}$ (for a given value of h, the v_{hk}'s $\neq 0$ are in finite number). The elements of Y' have the form $\sum_k g_k y_k + \sum_h g'_h y'_h = 0$, $g_k \in \mathfrak{G}$ and $g'_h \in \mathfrak{G}$ being such that $\sum_k b_{ik}(g_k + \sum_h v_{hk} g'_h) = 0$. One has also the defining relations $\sum c_h g \cdot y_k + \sum c'_h g \cdot y'_h = 0$ ($g \in \mathfrak{G}$), where the integers c_k and c'_k are such that $\sum_k b_{ik}(c_k + \sum_h v_{hk} c'_h) = 0$ for $i = 1, 2, 3, \ldots$. Setting

$$\varphi\Big(\sum_k g_k y_k + \sum_h g'_h y'_h\Big) = \sum_k \Big(g_k + \sum_h v_{hk} g'_h\Big) y_k,$$

one finds easily that φ is a bijective and isomorphic correspondence between the two groups Y' and Y.

Let us suppose now that the generators β_i and the defining relations $\sum_i b_{ik}\beta_i = 0$ of the group T^{n-1} are chosen arbitrarily. To finish the proof of the fact that the structure of the group Y depends only on the structure of $\mathfrak{G}$ and T^{n-1}, it is sufficient to prove that if one adds to the generators β_i all the other elements $\beta'_j = \sum_i u_{ij}\beta_i$ of the group T^{n-1}, joining simultaneously to $\sum_i b_{ik}\beta_i = 0$ the new relations $\beta'_j - \sum_i u_{ij}\beta_i = 0$, the corresponding group Y' is identical with Y. The elements of the group Y' have the form $\sum_k g_k y_k + \sum_j g'_j y'_j$, $g_k \in \mathfrak{G}$ and $g'_j \in \mathfrak{G}$ being such that $\sum_k b_{ik} g_k - \sum_j u_{ij} g'_j = 0$, $g'_j = 0$. One has also the defining relations

BETTI GROUPS OF AN INFINITE COMPLEX

$\sum_k c_k g \cdot y_k + \sum_j c'_j g \cdot y'_j = 0$ ($g \in \mathfrak{G}$), the integers c_k and c'_j being such that $\sum_k b_{ik} c_k - \sum_j u_{ij} c'_j = 0$, $c'_j = 0$. Therefore, the elements of Y' are simply $\sum_k g_k y_k$, $g_k \in \mathfrak{G}$ being such that $\sum_k b_{ik} g_k = 0$, and the defining relations are $\sum_k c_k g \cdot y_k = 0$ ($g \in \mathfrak{G}$), the integers c_k being such that $\sum_k b_{ik} c_k = 0$. Consequently, the groups Y' and Y are identical.

Now, it is sufficient to prove the validity of the theorem III by choosing, in a convenient way, the generators and the defining relations of the group T^{n-1}.

Since the $(n-1, \mathfrak{C})$-chains $1 \cdot \sigma_i^{n-1}$ constitute a finite or countable system of linearly independent generators of the group T^{n-1}, it results from the lemma that there exists a finite or infinite sequence of linearly independent generators C_i^{n-1} of the group $L^{n-1} \subset C^{n-1}$ consisting of all the $(n-1, \mathfrak{C})$-cycles a certain multiple of which is homologic to zero. The index k running through the same values as the index i $(1, 2, 3, \ldots,$ possibly ad inf.), there exist, for each k, integers b_{ik} such that (1) $b_{ik} = 0$ for $i > k$, (2) $b_{kk} > 0$, (3) $\sum_i b_{ik} C_i^{n-1} \sim 0$. Let us choose the integers b_{ik} in such a way that the value of b_{kk} is minimal, and let us set $h_k = \sum_i b_{ik} C_i^{n-1}$, so that $h_k \in H^{n-1}$. But the h_k's have been deduced from the C_i^{n-1}'s precisely in the same way as the h_i's from g_i's in the proof of the lemma. It is sufficient to replace there the two groups $\mathfrak{G}$ and $\mathfrak{H}$ by L^{n-1} and H^{n-1}, respectively.[5] Hence, it results that the h_k's constitute a system of linearly independent generators of the group H^{n-1}.

Since

$$T^{n-1} = L^{n-1}/H^{n-1},$$

to the C_i^{n-1}'s there correspond generators β_i of the group T^{n-1}, the defining relations being $\sum_i b_{ik} \beta_i = 0$. The elements of the group Y have the form $\sum_k g_k y_k$, $g_k \in \mathfrak{G}$ being such that $\sum_k b_{ik} g_k = 0$ for each i. One has $\sum_k g_k y_k = 0$ only if all the g_k's are $= 0$. In fact, the defining relations of the group Y are $\sum_k c_k g \cdot y_k = 0$ ($g \in \mathfrak{G}$), the integers c_k being such that (1) $c_k = 0$ starting from a certain value of k, (2) $\sum_k b_{ik} c_k = 0$ for each i. Since $b_{ik} = 0$ for $i > k$, and $b_{kk} > 0$, the conditions (1) and (2) imply that $c_k = 0$ for each k.

On the other hand, according to the proof of the theorem I, the group $B_2^n(\mathfrak{G})$ is isomorphic to the group $\Gamma_2^n(\mathfrak{G})$ consisting of all the $(n, \mathfrak{G})$-cycles of the form $\sum_k g_k D_k^n$, $D_k^n \in C$ being chosen in such a way that $F D_k^n = h_k$. The h_k's being linearly independent, the D_k^n's are also linearly independent. It results easily that one can have $\sum_k g_k D_k^n = 0$ only if $g_1 = g_2 = \cdots = 0$. But one has $F \sum_k g_k D_k^n = \sum_k g_k h_k = \sum_{ik} b_{ik} g_k C_i^{n-1}$. The C_i^{n-1}'s being linearly independent, it results that $\sum_k g_k D_k^n$ is an $(n, \mathfrak{G})$-cycle if and only if $\sum_k b_{ik} g_k = 0$ for each i. Consequently the

[5] In the present case one has obviously $\lambda_i = i$ in the proof quoted.

two groups $\boldsymbol{\Gamma}_2^n$ and $\boldsymbol{Y}$ are isomorphic. The two groups $\boldsymbol{B}_2^n(\mathfrak{G})$ and $\boldsymbol{\Gamma}_2^n(\mathfrak{G})$ are also isomorphic, and the theorem is proved.

7. Remarks. I. Let us suppose that the group $\boldsymbol{B}^n$ possesses a finite system of generators (which is the case particularly if the complex K is finite). For $\mu \in \mathfrak{C}$, $\mu > 0$, let us denote by $\mathfrak{G}(\mu)$ the subgroup of $\mathfrak{G}$ consisting of all the elements of the form μg ($g \in \mathfrak{G}$). It is known that the group $\boldsymbol{B}^n$ possesses a finite number of generators α_i ($1 \leqq i \leqq m$) such that the defining relations have the form $\mu_i \alpha_i = 0$ ($1 \leqq i \leqq m$), where $\mu_i \in \mathfrak{C}$, $\mu_i \geqq 0$. One deduces easily from the theorem II that the group $\boldsymbol{B}_1^n(\mathfrak{G})$ is a direct sum of m groups $\boldsymbol{X}_i$ ($1 \leqq i \leqq m$), where (1) if $\mu_i = 0$, then the group $\boldsymbol{X}_i$ is isomorphic to $\mathfrak{G}$, (2) if $\mu_i > 0$, then the group $\boldsymbol{X}_i$ is isomorphic to $\mathfrak{G}/\mathfrak{G}(\mu_i)$.

II. Let us suppose that the group $\boldsymbol{T}^{n-1}$ possesses a *finite* number of generators (which is the case particularly if the complex K is finite). For $\mu \in \mathfrak{C}$, $\mu > 0$, let us denote by $\mathfrak{G}[\mu]$ the subgroup of $\mathfrak{G}$ consisting of all the elements $g \in \mathfrak{G}$ such that $\mu g = 0$. It is known that the group $\boldsymbol{T}^{n-1}$ possesses a finite number of generators β_i ($1 \leqq i \leqq m$) such that the defining relations are of the form $\mu_i \beta_i = 0$ ($1 \leqq i \leqq m$), where $\mu_i \in \mathfrak{C}$, $\mu_i > 0$. One can easily deduce from the theorem III that the group $\boldsymbol{B}_2^n(\mathfrak{G})$ is a direct sum of m groups $\boldsymbol{Y}_i$ ($1 \leqq i \leqq m$), the group $\boldsymbol{Y}_i$ being isomorphic to $\mathfrak{G}[\mu_i]$.

III. *Let us suppose that the group $\mathfrak{G}$ has the following property: $g \in \mathfrak{G}$ and $\mu \in \mathfrak{C}$, $\mu > 0$ being chosen arbitrarily, then there exists always $g' \in \mathfrak{G}$ such that $g = \mu g'$. Then the structure of the group $\boldsymbol{B}_1^n(\mathfrak{G})$ is completely determined by the structure of the two groups $\mathfrak{G}$ and $\boldsymbol{B}^n/\boldsymbol{T}^n$. More precisely: $\mathfrak{G}$ having the above mentioned property, then the group $\boldsymbol{B}_1^n(\mathfrak{G})$ is isomorphic to the group X^* which is obtained from the group $\boldsymbol{B}^n/\boldsymbol{T}^n$ in the same way as the group $\boldsymbol{X}$ has been obtained from the group $\boldsymbol{B}^n$ in the theorem II.*

The most important example of a group $\mathfrak{G}$ having the above mentioned property is the additive group of the real numbers reduced mod 1, playing an important part in the recent works of Pontrjagin.

Proof. It can be easily seen that one can choose a system of generators and defining relations of the group $\boldsymbol{B}^n$ in the following way: (1) the generators are α_i ($i = 1, 2, 3, \ldots$, possibly ad inf.) and α'_j ($j = 1, 2, 3, \ldots$, possibly ad inf.), (2) the defining relations are $\sum_i a_{ik}\alpha_i + \sum_j a'_{jk}\alpha'_j = 0$ ($k = 1, 2, 3, \ldots$, possibly ad inf.) and $\sum_j u_{jh}\alpha'_j = 0$ (the index k running through the same values as the index g), (3) the α'_i's constitute a system of generators of the group $\boldsymbol{T}^n$, and the $\sum_j u_{jh}\alpha'_j = 0$'s constitute the defining relations of the group $\boldsymbol{T}^n$, (4) $u_{jh} = 0$ for $j > h$, $u_{hh} > 0$. Obviously, there exists a system of generators α_i^* of the group $\boldsymbol{B}^n/\boldsymbol{T}^n$ such that the defining relations are $\sum_i a_{ik}\alpha_i^* = 0$. The elements of the group $\boldsymbol{X}$ are $\sum_i g_i x_i + \sum_j g'_j x'_j$ ($g_i \in \mathfrak{G}$, $g'_j \in \mathfrak{G}$), and the defining relations are $\sum_i a_{ik} g \cdot x_i + \sum_j a'_{jk} g \cdot x'_j = 0$, $\sum_j u_{jh} g \cdot x'_j = 0$ ($g \in \mathfrak{G}$). The elements of the group X^* are $\sum_i g_i x'_i$ ($g_i \in \mathfrak{G}$), and the defining relations are $\sum_i a_{ik} g \cdot x_i^* = 0$ ($g \in \mathfrak{G}$).

BETTI GROUPS OF AN INFINITE COMPLEX

Let us consider the elements X' of $\boldsymbol{X}$ having the particular form $X' = \sum_j g'_j x'_j$ ($g'_j \in \mathfrak{G}$). Let us set $f(X') = 0$ if all the g'_j's are equal to zero. In the opposite case, let $f(X')$ be equal to the largest index j such that $g'_j \neq 0$. We are going to prove that $X' = 0$. This is obvious for $f(X') = 0$. Let us suppose it is true for $f(X') < m$, and let us consider an element $X' = \sum_j g'_j x'_j$ such that $f(X') = m$. Since $u_{mm} > 0$, there exists an element $g \in \mathfrak{G}$ such that $g'_m = u_{mm} g$. But one has $\sum_j u_{jm} g \cdot x'_j = 0$, whence $X' = X'' = \sum_j g'_j x'_j - \sum_j u_{jm} g \cdot x'_j$. Because $u_{mm} g = g'_m$, $u_{jm} = 0$ for $j > m$, $f(X') = m$, one has $f(X'') < m$, whence $X' = X'' = 0$.

Now, let us set

$$\varphi(\sum_i g_i x_i + \sum_j g'_j x'_j) = \sum g_i x^*_i .$$

One can easily verify that φ is a bijective and isomorphic correspondence between the two groups $\boldsymbol{X}$ and $\boldsymbol{X}^*$. But the group $\boldsymbol{X}$ is isomorphic to the group $\boldsymbol{B}^n_1(\mathfrak{G})$ by virtue of the theorem II.

IV. *The abelian group $\mathfrak{G}$ being arbitrary, let us denote by $\mathfrak{H}$ the subgroup consisting of all the elements of $\mathfrak{G}$ with finite order. Then the group $\boldsymbol{B}^n_2(\mathfrak{G})$ is isomorphic to the group $\boldsymbol{B}^n_2(\mathfrak{H})$.*

In particular, if all the elements of $\mathfrak{G}$ have infinite order (e. g., if $\mathfrak{G}$ is a subgroup of the additive group of all complex numbers), then $\boldsymbol{B}^n_2(\mathfrak{G}) = 0$.

Proof. One can suppose (see the proof of the theorem III) that the group $\boldsymbol{T}^{n-1}$ is given by the generators β_i ($i = 1, 2, 3, \ldots$, possibly ad inf.) such that the defining relations have the form $\sum_i b_{ik}\beta_i = 0$ (the index k running through the same values as the index i), where $b_{kk} > 0$, $b_{ik} = 0$ for $i > k$. We have seen that the group $\boldsymbol{B}^n_2(\mathfrak{G})$ is isomorphic to the group $\boldsymbol{Y}$ consisting of all linear forms $\sum_k g_k y_k$ whose coefficients $g_k \in \mathfrak{G}$ are such that

$$\sum_k b_{ik} g_{ik} = 0 . \tag{4}$$

One has $\sum_k g_k y_k = 0$ only if $g_k = 0$ for each k. One must only prove that $g_k \in \mathfrak{H}$, i. e. that the order of each g_k is finite. But this results easily by recurrence from the equations (4), when taking into account the conditions $b_{kk} > 0$, $b_{ik} = 0$ for $i > k$.

V. *Let $m \in \mathfrak{C}$, $m > 0$. Let us suppose that the group $\mathfrak{G}$ is such that $mg = 0$ for each $g \in \mathfrak{G}$. The structure of the group $\boldsymbol{B}^n(\mathfrak{G})$ is completely determined by the structure of the three groups $\mathfrak{G}$, $\boldsymbol{B}^n(\mathfrak{C}_m)$, and $\boldsymbol{B}^{n-1}(\mathfrak{C}_m)$, where $\mathfrak{C}_m$ denotes the additive group of the integers reduced mod m.* In fact, if the group $\mathfrak{G}$ has the above property, one can easily see that the theorems I, II, and III remain true (with the proofs being essentially the same) when replacing $\mathfrak{C}$ by $\mathfrak{C}_m$.[6]

[6] One must replace $\mathfrak{C}$ by $\mathfrak{C}_m$ also in the definition of the group $\boldsymbol{B}^n_1(\mathfrak{G})$.

ANNALS OF MATHEMATICS
Vol. 37, No. 3, July, 1936

MULTIPLICATIONS ON A COMPLEX

BY EDUARD ČECH

(Received February 20, 1936)

In their communications at the First International Topological Conference (Moscow, September 1935), J. W. Alexander and A. Kolmogoroff introduced the notion of a dual cycle[1] and defined a product of a dual p-cycle and a dual q-cycle, this product being a dual $(p + q)$-cycle. A different multiplication of the same sort is considered in this paper. It may be shown that the Alexander-Kolmogoroff product, augmented by the dual boundary of a suitable $(p + q - 1)$-chain, is equal to the $\binom{p+q}{p}$th multiple of the product here introduced.[2] Moreover, I consider also a product of an ordinary n-cycle and a dual p-cycle $(n \geqq p)$, this product being an ordinary $(n - p)$-cycle. There is a simple algebraic relationship between the two kinds of multiplication, which I shall explain elsewhere. As an application of the general theory, I give a new approach to the duality and intersection theory of a combinatorial manifold, given in a simplicial subdivision. The theory works exclusively in the given subdivision.

This is a preliminary paper of a purely combinatorial nature. In a later paper, I shall apply the same methods to general topological spaces, and in particular to the very general "manifolds" defined in my recent note in the *Proceedings of the National Academy of Sciences* (U. S. A.).

Many of the results of this paper were found independently by H. Whitney, but his methods of proof seem not much related to mine.

1. Let there be given a complex K. We shall designate by σ_i^p $(p = 0, 1, 2, \cdots)$ the (oriented) p-simplices of K and by $\eta_{i\,j}^{p}$ $(= 0, 1, -1)$ the incidence coefficient of σ_i^{p+1} and σ_j^p.

The word *group* will always designate an additively written abelian group. If $\mathfrak{A}$ is a group, then a $(p, \mathfrak{A})$-chain is a symbol of the form $a_i\sigma_i^p$, $a_i \in \mathfrak{A}$, where, as always in this paper, one has to sum over every subscript appearing twice.

The *boundary* FA^p of a $(p, \mathfrak{A})$-chain $A^p = a_i\sigma_i^p$ is zero if $p = 0$, and it is the $(p - 1, \mathfrak{A})$-chain

$$FA^p = \eta_{i\,j}^{p-1} a_i \sigma_j^{p-1}$$

[1] As a matter of fact the *Topology* of S. Lefschetz (1930), contains an essentially equivalent notion (pp. 282–286).

[2] In his paper "On the Connectivity Ring of an Abstract Space" in this number of the *Annals of Mathematics*, pp. 698–708, J. W. Alexander has modified his definition, and it is now in agreement with the one here presented.

if $p > 0$. If $FA^p = 0$, we say that A^p is an *ordinary* $(p, \mathfrak{A})$*-cycle*. The $(p, \mathfrak{A})$-chain FA^{p+1} is an ordinary $(p, \mathfrak{A})$-cycle for every $(p + 1, \mathfrak{A})$-chain A^{p+1}. Two ordinary $(p, \mathfrak{A})$-cycles A_1^p and A_2^p are said to be of the same homology class, or to be homologous to each other (in symbols $A_1^p \sim A_2^p$) if there exists a $(p + 1, \mathfrak{A})$-chain A_0^{p+1} such that

$$A_1^p - A_2^p = FA_0^{p+1}.$$

The *dual boundary* F^*A^p of a $(p, \mathfrak{A})$-chain $A^p = a_i\sigma_i^p$ is the $(p + 1, \mathfrak{A})$-chain

$$F^*A^p = \eta_{ji}^p a_i \sigma_j^{p+1}.$$

If $F^*A^p = 0$, we say that A^p is a *dual* $(p, \mathfrak{A})$*-cycle*. The $(p + 1, \mathfrak{A})$-chain F^*A^p is a dual $(p + 1, \mathfrak{A})$-cycle for every $(p, \mathfrak{A})$-chain A^p. Two dual $(p, \mathfrak{A})$-cycles A_1^p and A_2^p are said to be of the same homology class, or to be homologous to each other (in symbols $A_1^p \sim A_2^p$) (1) in the case $p = 0$ only if they are identical, (2) in the case $p > 0$ if there exists a $(p - 1, \mathfrak{A})$-chain A_0^{p-1} such that

$$A_1^p - A_2^p = F^*A_0^{p-1}.$$

2. Let $\mathfrak{B}$ be a given group. Let B^q be a given dual $(q, \mathfrak{B})$-cycle. By an *auxiliary construction* we mean an operation attaching to every simplex σ_i^p $(p = 0, 1, 2, \cdots)$ a $(p + q, \mathfrak{B})$-chain $B^{p+q}(\sigma_i^p)$ such that the following three conditions are satisfied. *First*, if the coefficient of a $(p + q)$-simplex τ^{p+q} in $B^{p+q}(\sigma_i^p)$ is different from zero, then σ_i^p must be a face of τ^{p+q}. *Second*, we must have

$$B^q = \sum_i B^q(\sigma_i^0). \tag{2.1}$$

Third, we must have for every simplex σ_i^p $(p = 0, 1, 2, \cdots)$

$$F^*B^{p+q}(\sigma_i^p) = \sum_j \eta_{ji}^p B^{p+q+1}(\sigma_j^{p+1}). \tag{2.2}$$

3. We shall prove that *the auxiliary construction is always possible*. Let there be given a fixed ordering of the vertices of K. Let σ^p be a given p-simplex, written as

$$\sigma^p = (v_0, v_1, \cdots, v_p)$$

corresponding to the given ordering of vertices (i.e. v_0 precedes v_1 etc.). We shall define the $(p + q, \mathfrak{B})$-chain $B^{p+q}(\sigma^p)$ as follows. The only $(p + q)$-simplices appearing in $B^{p+q}(\sigma^p)$ will have, corresponding to the given ordering of vertices, the form

$$(v_0, v_1, \cdots, v_p, \cdots, v_{p+q}), \tag{3.1}$$

i.e. the first $p + 1$ vertices will be those of σ^p. The coefficient of any such simplex (3.1) in $B^{p+q}(\sigma^p)$ will be equal to the coefficient of the q-simplex $(v_p, \cdots, v_{p+q})$ in B^q.

The first two properties of the auxiliary construction being evident, we have only to prove (2.2) for

$$\sigma_i^p = \sigma^p = (v_0, v_1, \cdots, v_p).$$

The only $(p + q + 1)$-simplices τ^{p+q+1} appearing on either side of (2.2) must all have σ^p as their common face and, moreover, corresponding to the given ordering of vertices, the vertex v_p must be either the $(p + 1)^{\text{th}}$ or the $(p + 2)^{\text{th}}$ vertex of τ^{p+q+1}. We have to prove that any such τ^{p+q+1} has equal coefficients on both sides of (2.2). This being quite evident in the case where v_p is the $(p + 2)^{\text{th}}$ vertex of τ^{p+q+1}, we only have to examine the case when, in the given order of vertices, we have

$$\tau^{p+q+1} = (v_0, v_1, \cdots, v_p, \cdots, v_{p+q+1}).$$

Let b_{p+i} $(0 \leqq i \leqq q + 1)$ be the coefficient, in the $(q, \mathfrak{B})$-chain B^q, of the oriented q-simplex obtained from $(v_p, \cdots, v_{p+q+1})$ by omitting the vertex v_{p+i}. The coefficients of τ^{p+q+1} in both sides of (2.2) are respectively equal to

$$(-1)^{p+1}b_{p+i} \text{ and to } (-1)^{p+1}b_p.$$

But since B^q is a dual $(q + 1, \mathfrak{B})$-cycle, the coefficient of the $(q + 1)$-simplex $(v_p, \cdots, v_{p+q+1})$ in F^*B^q must vanish, i.e.

$$(-1)^i b_{p+i} = 0 \quad \text{or} \quad (-1)^{p+i}b_{p+i} = (-1)^{p+1}b_p.$$

4. Let us suppose that the dual $(q, \mathfrak{B})$-cycle B^q is identically zero. $B^{p+q}(\sigma_i^p)$ being the elements of an auxiliary construction chosen in any manner corresponding to $B^q = 0$, we shall prove that we may attach to every p-simplex $\sigma_i^p (p = 1, 2, 3, \cdots)$ a $(p + q - 1, \mathfrak{B})$-chain $C^{p+q-1}(\sigma_i^p)$ such that the following three conditions are satisfied. *First*, if the coefficient of a $(p + q - 1)$-simplex τ^{p+q-1} in $C^{p+q-1}(\sigma_i^p)$ is different from zero, then σ_i^p must be a face of τ^{p+q-1}. *Second*, we have for every 0-simplex σ_i^0

$$B^q(\sigma_i^0) = \eta_{ji}^0 C^q(\sigma_j^1). \tag{4.1}$$

Third, we have for every p-simplex σ_i^p, where $p = 1, 2, 3, \cdots,$

$$B^{p+q}(\sigma_i^p) = \eta_{ji}^p C^{p+q}(\sigma_j^{p+1}) + F^*C^{p+q-1}(\sigma_i^p). \tag{4.2}$$

We begin by the construction of $C^q(\sigma_j^1)$. Let τ^q be any q-simplex and let $b_i(\tau^q)$ be its coefficient in $B^q(\sigma_i^0)$. If σ_i^0 is not a vertex of τ^q, we have $b_i(\tau^q) = 0$. Moreover, since $B^q = 0$, it follows from (2.1) that $\sum_i b_i(\tau^q) = 0$. Therefore, $b_i(\tau^q) \cdot \sigma_i^0$ is an ordinary $(0, \mathfrak{B})$-cycle of the q-simplex τ^q having zero as the sum of its coefficients. It is well known that such a $(0, \mathfrak{B})$-cycle is equal to the boundary of a $(1, \mathfrak{B})$-chain of the q-simplex τ^q. Therefore there exists, for every 1-simplex σ_j^1, an element $c_j(\tau^q)$ of the group $\mathfrak{B}$ such that (1) $c_j(\tau^q) = 0$ if σ_j^1 is not a face of τ^q, (2) $b_i(\tau^q) = \eta_{ji}^0 c_j(\tau^q)$ for every σ_i^0. Let us put

$$C^q(\sigma_j^1) = \sum c_j(\tau^q)\tau^q,$$

the summation running over all q-simplices τ^q. Then σ_j^1 is a face of every q-simplex appearing in $C^q(\sigma_j^1)$ and the relations (4.1) hold true.

If we put $C^{q-1}(\sigma_i^0) = 0$, the relation (4.2) corresponding to $p = 0$ reduces to (4.1). Therefore, we may suppose our construction carried through up to the relations (4.2), where p is given, and we have to construct $(p + q + 1)$-chains $C^{p+q+1}(\sigma_k^{p+2})$ satisfying the analogous relations

$$(4.3) \qquad B^{p+q+1}(\sigma_j^{p+1}) = \eta_{kj}^{p+1} C^{p+q+1}(\sigma_k^{p+2}) + F^*C^{p+q}(\sigma_j^{p+1}).$$

Since $F^*C^{p+q-1}(\sigma_i^p)$ is a dual $(p + q, \mathfrak{B})$-cycle, it follows from (4.2) that

$$F^*B^{p+q}(\sigma_i^p) = \eta_{ji}^p F^*C^{p+q}(\sigma_j^{p+1}).$$

Comparing with (2.2) we get

$$(4.4) \qquad \eta_{ji}^p B^{p+q+1}(\sigma_j^{p+1}) - F^*C^{p+q}(\sigma_j^{p+1}) = 0.$$

Now let τ^{p+q+1} be any $(p + q + 1)$-simplex and let $b_j(\tau^{p+q+1})$ be its coefficient in the $(p + q + 1, \mathfrak{B})$-chain

$$B^{p+q+1}(\sigma_j^{p+1}) - F^*C^{p+q}(\sigma_j^{p+1}).$$

If σ_j^{p+1} is not a face of τ^{p+q+1}, we have $b_j(\tau^{p+q+1}) = 0$. Moreover, it follows from (4.4) that $\eta_{ji}^p b_j(\tau^{p+q+1}) = 0$. Therefore, $b_j(\tau^{p+q+1})\sigma_j^{p+1}$ is an ordinary $(p + 1, \mathfrak{B})$-cycle of the $(p + q + 1)$-simplex τ^{p+q+1}. It is well known that such a $(p + 1, \mathfrak{B})$-cycle is equal to the boundary of a $(p + 2, \mathfrak{B})$-chain of the simplex τ^{p+q+1}. Therefore there exists, for every $(p + 2)$-simplex σ_k^{p+2}, an element $c_k(\tau^{p+q+1})$ of the group $\mathfrak{B}$ such that (1) $c_k(\tau^{p+q+1}) = 0$ if σ_k^{p+2} is not a face of τ^{p+q+1}, (2) $b_j(\tau^{p+q+1}) = \eta_{kj}^{p+1} c_k(\tau^{p+q+1})$ for every σ_j^{p+1}. Let us put

$$C^{p+q+1}(\sigma_k^{p+2}) = \sum c_k(\tau^{p+q+1}) \cdot \tau^{p+q+1}$$

the summation running over all $(p + q + 1)$-simplices τ^{p+q+1}. Then σ_k^{p+2} is a face of every $(p + q + 1)$-simplex appearing in $C^{p+q+1}(\sigma_k^{p+2})$ and the relations (4.3) hold true.

5. Let there be given three groups $\mathfrak{A}$, $\mathfrak{B}$ and $\mathfrak{C}$. Let there be given a law attaching to every couple a, b, where $a \in \mathfrak{A}$ and $b \in \mathfrak{B}$, an element $c \in \mathfrak{C}$, called the *product* of a and b and designated by ab or $a \cdot b$. Furthermore, let us suppose the validity of the distributive laws

$$(a_1 + a_2)b = a_1b + a_2b, \qquad a(b_1 + b_2) = ab_1 + ab_2.$$

In such circumstances, we put $\mathfrak{C} = (\mathfrak{A}, \mathfrak{B})$ and say that there is given an $(\mathfrak{A}, \mathfrak{B})$-*multiplication.*

Any $(\mathfrak{A}, \mathfrak{B})$-multiplication defines an "*inverse*" $(\mathfrak{B}, \mathfrak{A})$-multiplication (with the same group $\mathfrak{C}$), if we define the new product ba to be equal to the original product ab.

6. Let there be given an $(\mathfrak{A}, \mathfrak{B})$-multiplication. Let

$$A^p = a_i \sigma_i^p$$

be a dual $(p, \mathfrak{A})$-cycle. Let B^q be a dual $(q, \mathfrak{B})$-cycle. We shall define their product A^pB^q as a dual $[p + q, (\mathfrak{A}, \mathfrak{B})]$-cycle which, however, will be affected with a slight indetermination. To this end, we start with B^q and choose an auxiliary construction (sect. 2), which is always possible by sect. 3. Then we put

$$A^pB^q = a_i B^{p+q}(\sigma_i^p).$$

From (2.2) we have

$$F^*a_i B^{p+q}(\sigma_i^p) = a_i F^*B^{p+q}(\sigma_i^p) = \eta_{ji}^p a_i B^{p+q+1}(\sigma_j^{p+1})$$

which is equal to zero, since $\eta_{ji}^p a_i = 0$, A^p being a dual p-cycle. Therefore, $F^*(A^pB^q) = 0$, i.e., the product A^pB^q is indeed a dual $(p + q)$-cycle.

It can easily be seen that the product A^pB^q is *not* uniquely determined, depending really on the choice of the auxiliary construction. But the *homology class* of the product A^pB^q *is* determined without ambiguity, i.e. any two values $(A^pB^q)_1$ and $(A^pB^q)_2$ are connected by the homology

$$(A^pB^q)_1 \sim (A^pB^q)_2.$$

This fact is an easy consequence of the following statement: If $B^q = 0$, then $A^pB^q \sim 0$ for any choice of the auxiliary construction. We proceed to the proof of that statement. If $B^q = 0$, we saw in sect. 4 that there exist chains $C^{p+q-1}(\sigma_i^p)$ such that (4.1) and (4.2) hold true. If $p = 0$, it follows from (4.1) that

$$A^0B^q = a_i B^q(\sigma_i^0) = \eta_{ji}^0 a_i C^q(\sigma_i^0) = 0,$$

because $\eta_{ji}^0 a_i = 0$. If $p > 0$, it follows from (4.2) that

$$F^*a_i C^{p+q-1}(\sigma_i^p) = a_i F^*C^{p+q-1}(\sigma_i^p) = a_i B^{p+q}(\sigma_i^p) - \eta_{ji}^p a_i C^{p+q}(\sigma_j^{p+1})$$
$$= a_i B^{p+q}(\sigma_i^p) = A^pB^q$$

because $\eta_{ji}^p a_i = 0$. Therefore $A^pB^q = F^*a_i C^{p+q-1}(\sigma_i^p) \sim 0$.

The homology class of the product A^pB^q is uniquely determined by the homology classes of A^p and B^q. This is an easy consequence of the following statement: If either $A^p \sim 0$ or $B^q \sim 0$, then $A^pB^q \sim 0$. Let us first suppose that $A^p \sim 0$. If $p = 0$, then $A^p = 0$, which implies $A^pB^q = 0$. If $p > 0$, then there exists a $(p - 1, \mathfrak{A})$-chain $\alpha_j \sigma_j^{p-1}$ such that $A^p = a_i \sigma_i^p = F^*(\alpha_j \sigma_j^{p-1})$, i.e. $a_i = \eta_{ij}^{p-1} \alpha_j$. According to (2.2), we have

$$F^*\alpha_j B^{p+q-1}(\sigma_j^{p-1}) = \alpha_j F^*B^{p+q-1}(\sigma_j^{p-1}) = \eta_{ij}^{p-1} \alpha_j B^{p+q}(\sigma_i^p) = a_i B^{p+q}(\sigma_i^p) = A^pB^q$$

so that $A^pB^q = 0$. Now we suppose that $B^q \sim 0$. If $q = 0$, then $B^q = 0$, which we know to imply $A^pB^q \sim 0$. If $q > 0$, then there exists a $(q - 1, \mathfrak{B})$-chain H^{q-1} such that $B^q = F^*H^{q-1}$. One finds easily $(q - 1, \mathfrak{B})$-chains $H^{q-1}(\sigma_i^0)$ such that (1) σ_i^0 is a vertex of every $(q - 1)$-simplex appearing in $H^{q-1}(\sigma_i^0)$,

(2) $H^{q-1} = \sum_i H^{q-1}(\sigma_i^0)$. If we put $B^q(\sigma_i^0) = F^*H^{q-1}(\sigma_i^0)$ and $B^{p+q}(\sigma_i^p) = 0$ for $p > 0$, we evidently have an auxiliary construction in the sense of sect. 2. With this choice of auxiliary construction, we have $A^pB^q = 0$ if $p > 0$, and

$$A^0B^q = F^*a_iH^{q-1}(\sigma_i^0) \sim 0 \text{ if } p = 0.$$

7. Let there be given an ordering of the vertices of the complex K. Then we can use the particular auxiliary construction described in sect. 3, which leads to the following simple definition of the product A^pB^q. Given a $(p + q)$-simplex σ^{p+q}, we write it as

$$\sigma^{p+q} = (v_0, v_1, \cdots, v_p, \cdots, v_{p+q})$$

according to the given ordering of the vertices. Let a be the coefficient of the p-simplex $(v_0, v_1, \cdots, v_p)$ in the $(p, \mathfrak{A})$-chain A^p; let b be the coefficient of the q-simplex $(v_p, \cdots, v_{p+q})$ in the $(q, \mathfrak{B})$-chain B^q. Then ab is the coefficient of σ^{p+q} in the product A^pB^q.

This definition leads to a simple proof of the *commutative law*:

$$B^qA^p \sim (-1)^{pq}A^pB^q. \tag{7.1}$$

Here we suppose that, $\mathfrak{A}$ and $\mathfrak{B}$ being two groups, A^p is a dual $(p, \mathfrak{A})$-cycle and B^q is a dual $(q, \mathfrak{B})$-cycle. Furthermore, an $(\mathfrak{A}, \mathfrak{B})$-multiplication is given, and hence an inverse $(\mathfrak{B}, \mathfrak{A})$-multiplication also (sect. 5). The products A^pB^q and B^qA^p are formed according to the first and second of these multiplications, respectively. To prove (7.1), we fix the value of A^pB^q according to a given ordering of the vertices, and fix B^qA^p according to the *inverse* ordering of the vertices. Let a $(p + q)$-simplex

$$(v_0, v_1, \cdots, v_p, \cdots, v_{p+q})$$

be written in the original ordering of the vertices. Since

$$(v_p, \cdots, v_0) = (-1)^{\frac{1}{2}p(p+1)}(v_0, \cdots, v_p),$$

$$(v_{p+q}, \cdots, v_p) = (-1)^{\frac{1}{2}q(q+1)}(v_p, \cdots, v_{p+q}),$$

$$(v_{p+q}, \cdots, v_p, \cdots, v_0) = (-1)^{\frac{1}{2}(p+q)(p+q+1)}(v_0, \cdots, v_p, \cdots, v_{p+q}),$$

$$\tfrac{1}{2}(p + q)(p + q + 1) = \tfrac{1}{2}p(p + 1) + \tfrac{1}{2}q(q + 1) + pq,$$

it is readily seen that, with our particular choice of the auxiliary construction, we have $B^qA^p = (-1)^{pq}A^pB^q$. It seems difficult to prove the commutative law (7.1) directly from the general definition given in sect. 6.

The *distributive laws*

$$(A_1^p + A_2^p)B^q \sim A_1^pB^q + A_2^pB^q, \tag{7.2}$$

$$A^p(B_1^q + B_2^q) \sim A^pB_1^q + A^pB_2^q \tag{7.3}$$

are immediate consequences of either of the two definitions of the product.

Now suppose that three groups $\mathfrak{A}_1$, $\mathfrak{A}_2$ and $\mathfrak{A}_3$ are given. Let there be given an $(\mathfrak{A}_1, \mathfrak{A}_2)$-multiplication and an $(\mathfrak{A}_2, \mathfrak{A}_3)$-multiplication. Further, putting

$$(\mathfrak{A}_1, \mathfrak{A}_2) = \mathfrak{A}_{12}, \qquad (\mathfrak{A}_2, \mathfrak{A}_3) = \mathfrak{A}_{23},$$

let us suppose that there is given an $(\mathfrak{A}_{12}, \mathfrak{A}_3)$-multiplication and an $(\mathfrak{A}_1, \mathfrak{A}_{23})$-multiplication. Suppose, finally, that the associative law

$$a_1a_2 \cdot a_3 = a_1 \cdot a_2a_3$$

holds true for $a_1 \in \mathfrak{A}_1$, $a_2 \in \mathfrak{A}_2$, $a_3 \in \mathfrak{A}_3$. Then we have, if $A_i^{p_i}$ $(i = 1, 2, 3)$ is a dual $(p_i, \mathfrak{A}_i)$-cycle, the *associative law*

$$A_1^{p_1}A_2^{p_2} \cdot A_2^{p_3} \sim A_1^{p_1} \cdot A_2^{p_2}A_3^{p_3}. \tag{7.4}$$

The proof based on a given ordering of the vertices is quite trivial. A proof based directly on our general definition of the product is not difficult, however.

It would be interesting to prove, using only definitions based on the ordering of the vertices, that the homology class of the product A^pB^q is independent of the choice of the ordering.[3]

8. Let there be given an $(\mathfrak{A}, \mathfrak{B})$-multiplication. If $A^p = a_i\sigma_i^p$ is a $(p, \mathfrak{A})$-chain and if $B^p = b_i\sigma_i^p$ is a $(p, \mathfrak{B})$-chain, let us put

$$\varphi(A^p, B^p) = a_ib_i \in (\mathfrak{A}, \mathfrak{B}).$$

If A^{p+1} is a $(p + 1, \mathfrak{A})$-chain and if B^p is a $(p, \mathfrak{B})$-chain, it is readily seen that

$$\varphi(FA^{p+1}, B^p) = \varphi(A^{p+1}, F^*B^p); \tag{8.1}$$

similarly we have

$$\varphi(A^p, FB^{p+1}) = \varphi(F^*A^p, B^{p+1}) \tag{8.2}$$

for any $(p, \mathfrak{A})$-chain A^p and any $(p + 1, \mathfrak{B})$-chain B^{p+1}.

9 Let there be given an $(\mathfrak{A}, \mathfrak{B})$-multiplication. Let A^{p+q} be an *ordinary* $(p + q, \mathfrak{A})$-cycle. Let B^q be a *dual* $(q, \mathfrak{B})$-cycle. We shall define a product $A^{p+q}B^q$ (not quite uniquely determined), which will be an *ordinary* $[p, (\mathfrak{A}, \mathfrak{B})]$-cycle. We choose an auxiliary construction $B^{p+q}(\sigma_i^p)$ associated with B^q (sect. 2), and we put

$$A^{p+q}B^q = c_i\sigma_i^p,$$

where (see sect. 8)

$$c_i = (-1)^{pq}\varphi[A^{p+q}, B^{p+q}(\sigma_i^p)].$$

That $A^{p+q}B^q$ is an ordinary $[p, (\mathfrak{A}, \mathfrak{B})]$-cycle, is trivial if $p = 0$. If $p > 0$,

[3] Such a proof has now been given by J. W. Alexander; see his paper cited above.

it follows from (2.2) and (8.1) that, for any $(p - 1)$-simplex σ_j^{p-1},

$$(-1)^{pq}\eta_{ij}^{p-1}c_i = \eta_{ij}^{p-1}\varphi[A^{p+q}, B^{p+q}(\sigma_i^p)] = \varphi[A^{p+q}, \eta_{ij}^{p-1}B^{p+q}(\sigma_i^p)]$$
$$= \varphi[A^{p+q}, F^*B^{p+q-1}(\sigma_j^{p-1})] = \varphi[FA^{p+q}, B^{p+q-1}(\sigma_j^{p-1})] = \varphi[0, B^{p+q-1}(\sigma_j^{p-1})] = 0,$$

i.e. $F(A^{p+q}B^q) = 0$.

Suppose that $B^q = 0$. If $p = 0$, it follows from (4.1) that

$$\varphi[A^q, B^q(\sigma_i^0)] = \eta_{ji}^0\varphi[A^q, C^q(\sigma_j^1)],$$

so that

$$A^qB^q = F(\gamma_j\sigma_j^1), \qquad \gamma_j = \varphi[A^q, C^q(\sigma_j^1)],$$

i.e. $A^qB^q \sim 0$. If $p > 0$, it follows from (4.2) that

$$\varphi[A^{p+q}, B^{p+q}(\sigma_i^p)] = \eta_{ji}^p\varphi[A^{p+q}, C^{p+q-1}(\sigma_j^{p+1})] + \varphi[A^{p+q}, F^*C^{p+q-1}(\sigma_i^p)].$$

But the last summand is zero, from (8.1), since $FA^{p+q} = 0$. Therefore

$$A^{p+q}B^q = F(\gamma_j\sigma_j^{p+1}), \qquad \gamma_j = (-1)^{pq}\varphi[A^{p+q}, C^{p+q-1}(\sigma_j^{p+1})],$$

i.e. again $A^{p+q}B^q \sim 0$.

It follows readily from the preceding proof that, in any case, the homology class of the $[p, (\mathfrak{A}, \mathfrak{B})]$-cycle $A^{p+q}B^q$ is independent of the choice of the auxiliary construction. As a matter of fact, this homology class is uniquely determined by the homology classes of the ordinary $(p + q, \mathfrak{A})$-cycle A^{p+q} and the dual $(q, \mathfrak{B})$-cycle B^q. It is sufficient to prove that $A^{p+q}B^q \sim 0$, if either $A^{p+q} \sim 0$ or $B^q \sim 0$. If $A^{p+q} \sim 0$, there exists a $(p + q + 1, \mathfrak{A})$-chain H^{p+q+1} such that $A^{p+q} = FH^{p+q+1}$. It follows easily from (2.2) and (8.1) that

$$A^{p+q}B^q = F(\gamma_j\sigma_j^{p+1}) \sim 0, \gamma_j = (-1)^{pq}\varphi[H^{p+q+1}, B^{p+q+1}(\sigma_j^{p+1})].$$

If $B^q \sim 0$ and $q = 0$, we have $B^q = 0$, which we know to imply $A^{p+q}B^q \sim 0$. If $B^q \sim 0$ and $q > 0$, we choose the auxiliary construction as at the end of sect. 6: $B^q(\sigma_i^0) = F^*H^{q-1}(\sigma_i^0)$ and $B^{p+q}(\sigma_i^p) = 0$ for $p > 0$. If $p > 0$, we have then $A^{p+q}B^q = 0$. If $p = 0$, we have again $A^qB^q = 0$ from (8.1), since $FA^q = 0$.

If A^p is a *dual* $(p, \mathfrak{A})$-cycle and if B^{p+q} is an *ordinary* $[(p + q), \mathfrak{B}]$-cycle, we put

$$A^pB^{p+q} = c_i\sigma_i^q,$$

where

$$c_i = \varphi[A^{p+q}(\sigma_i^p), B^{p+q}],$$

the $(p + q, \mathfrak{A})$-chains $A^{p+q}(\sigma_i^q)(q = 0, 1, 2, \cdots)$ being the elements of an auxiliary construction associated with A^p. Again, the product is an ordinary $[q, (\mathfrak{A}, \mathfrak{B})]$-cycle and only its homology class is uniquely determined, this class being indeed given by the mere knowledge of the homology classes of the factors. If A^{p+q} is an ordinary $(p + q, \mathfrak{A})$-cycle and if B^q is a dual $(q, \mathfrak{B})$-cycle, we have evidently

(9.1) $$A^{p+q}B^q \sim (-1)^{pq}B^qA^{p+q},$$

where the left-hand member is defined according to the given $(\mathfrak{A}, \mathfrak{B})$-multiplication and the right-hand member according to the inverse $(\mathfrak{B}, \mathfrak{A})$-multiplication.

10. Let there be given an ordering of the vertices of the complex K. The particular auxiliary construction described in sect. 3 leads to following simple definition of the product A^pB^{p+q} of a dual $(p, \mathfrak{A})$-cycle A^p and an ordinary $(p + q, \mathfrak{B})$-cycle B^{p+q}. Given a q-simplex σ^q, we write it as

$$\sigma^q = (v_0, v_1, \cdots, v_q)$$

according to the given ordering of the vertices, and consider all the $(p + q)$-simplices

$$\sigma_k^{p+q} = (v_0, v_1, \cdots, v_q, \cdots, v_{p+q})$$

having σ^q as their common face and such that, in the given ordering, v_q precedes any vertex of σ_k^{p+q} which is not a vertex of σ^q. For every such σ_k^{p+q} put

$$\sigma_k^p = (v_q, \cdots, v_{p+q}).$$

Let a_k be the coefficient of σ_k^p in A^p; let b_k be the coefficient of σ_k^{p+q} in B^{p+q}. Then the coefficient of σ^q in A^pB^{p+q} is equal to

$$\sum_k a_kb_k.$$

Now let us consider the product $A^{p+q}B^q$ of an ordinary $(p + q, \mathfrak{A})$-cycle A^{p+q} and a dual $(q, \mathfrak{B})$-cycle B^q. This time we use the auxiliary construction based on the *inverse* ordering of the vertices, but we describe the result in terms of the original ordering. Given a p-simplex σ^p, we write it as

$$\sigma^p = (v_q, \cdots, v_{p+q})$$

according to the given ordering of the vertices, and consider all the $(p + q)$-simplices

$$\sigma_k^{p+q} = (v_0, v_1, \cdots, v_q, \cdots, v_{p+q})$$

having σ^p as their common face and such that, in the given ordering, v_q follows any vertex of σ_k^{p+q} which is not a vertex of σ^p. For every such σ_k^{p+q}, put

$$\sigma_k^q = (v_0, \cdots, v_q).$$

Let a_k be the coefficient of σ_k^{p+q} in A^{p+q}; let b_k be the coefficient of σ_k^q in B^q. Then the coefficient of σ^p in $A^{p+q}B^q$ is equal to

$$\sum_k a_kb_k.$$

These definitions, in connection with that given at the beginning of sect. 7

(for the product of two dual cycles) lead to a simple proof of the *associative laws*:

(10.1) $$A_1^{p_1+p_2+p_3} B_2^{p_2} \cdot B_3^{p_3} \sim A_1^{p_1+p_2+p_3} \cdot B_2^{p_2} B_3^{p_3},$$

(10.2) $$B_1^{p_1} A_2^{p_1+p_2+p_3} \cdot B_3^{p_3} \sim B_1^{p_1} \cdot A_2^{p_1+p_2+p_3} B_3^{p_3},$$

(10.3) $$B_1^{p_1} B_2^{p_2} \cdot A_3^{p_1+p_2+p_3} \sim B_1^{p_1} \cdot B_2^{p_2} A_3^{p_1+p_2+p_3}.$$

Here we suppose given three groups $\mathfrak{A}_1$, $\mathfrak{A}_2$, $\mathfrak{A}_3$, an $(\mathfrak{A}_1, \mathfrak{A}_2)$-multiplication, an $(\mathfrak{A}_2, \mathfrak{A}_3)$-multiplication, an $(\mathfrak{A}_{12}, \mathfrak{A}_3)$-multiplication with $\mathfrak{A}_{12} = (\mathfrak{A}_1, \mathfrak{A}_2)$ and an $(\mathfrak{A}_1, \mathfrak{A}_{23})$-multiplication with $\mathfrak{A}_{23} = (\mathfrak{A}_2, \mathfrak{A}_3)$. It is supposed that $a_1a_2 \cdot a_3 = a_1 \cdot a_2a_3$ for $a_i \in \mathfrak{A}_i$ $(i = 1, 2, 3)$. $A_i^{p_1+p_2+p_3}$ $(i = 1, 2, 3)$ is an ordinary $(p_1 + p_2 + p_3, \mathfrak{A}_i)$-cycle and $B_i^{p_i}$ $(i = 1, 2, 3)$ is a dual $(p_i, \mathfrak{A}_i)$-cycle. Of course, any of the three formulas (10.1), (10.2) and (10.3) implies the others using (7.1) and (9.1). We omit writing explicitly the trivial *distributive laws.*

11. In the remaining part of this paper the coefficients of all chains are taken from the additive group of all integer numbers. Moreover, we suppose that $K = M_n$ is an orientable simple n-circuit, i.e. that the following four conditions are satisfied. *First,* each simplex of M_n is either an n-simplex or a face of an n-simplex. *Second,* each $(n - 1)$-simplex of M_n is a common face of precisely two n-simplices of M_n. *Third,* any two n-simplices of M_n may be connected by a sequence of n-simplices of M_n such that any two consecutive n-simplices of the sequence have a common $(n - 1)$-face. *Fourth,* the n-simplices σ_i^n of M_n can be given such orientations that their sum $\Gamma^n = \sum_i \sigma_i^n$ is an ordinary n-cycle. (We always suppose the orientation of the n-simplices chosen in this manner.)

If σ_i^p is any p-simplex of M_n, we denote by Lk. $[\sigma_i^p]$ its *link*, i.e. the subcomplex of M_n composed of all the simplices τ of M_n having no common vertex with σ_i^p but having the property that there exists a simplex of M_n having both τ and σ_i^p among its faces.

If $0 \leqq p \leqq n$, we say that M_n is *p-regular* if the following two conditions are satisfied. *First* (requiring nothing if $p = n$ or $p = n - 1$), the link Lk. $[\sigma_i^p]$ on any p-simplex of M_n is an orientable simple $(n - p - 1)$-circuit. Second (requiring nothing if $p = 0$), for each k such that $0 \leqq k \leqq p - 1$, any dual $(n - p - 1)$-cycle of any Lk. $[\sigma_i^k]$ is homologous to zero in Lk. $[\sigma_i^k]$. It is easily seen that the orientable combinatorial n-manifolds are identical with orientable simple n-circuits, which are p-regular for any $0 \leqq p \leqq n$.

12. For $0 \leqq p \leqq n$, we denote by $\mathfrak{B}_p$ the group of all the homology classes of ordinary p-cycles of M_n and by $\bar{\mathfrak{B}}_p$ the group of all the homology classes of dual p-cycles of M_n.

Given any dual $(n - p)$-cycle B^{n-p} of M_n $(0 \leqq p \leqq n)$, we put

$$\psi_p(B^{n-p}) = \Gamma^n \cdot B^{n-p},$$

where $\Gamma^n = \sum_i \sigma_i^n$. Evidently, ψ_p is a homomorphic mapping of the group $\bar{\mathfrak{B}}_{n-p}$ on a subgroup $\psi_p(\bar{\mathfrak{B}}_{n-p})$ of the group $\mathfrak{B}_p$.

13. *If M_n is p-regular, then the mapping ψ_p is* $1 - 1$, so that the group $\bar{\mathfrak{B}}_{n-p}$ is isomorphic with a subgroup [i.e. $\psi_p(\bar{\mathfrak{B}}_{n-p})$] of the group $\mathfrak{B}_p$.

It is sufficient to prove that $\Gamma^n B^{n-p} \sim 0$ implies $B^{n-p} \sim 0$.

Let $B^{n-p+k}(\sigma_i^k)$ be the elements of a given auxiliary construction associated with the dual $(n - p)$-cycle B^{n-p}. Since $\Gamma^n \cdot B^{n-p} \sim 0$, there exists a $(p + 1)$-chain $c_j\sigma_j^{p+1}$ such that $\Gamma^n \cdot B^{n-p} = (-1)^{p(n-p)}F(c_j\sigma_j^{p+1})$, i.e.

$$\varphi[\Gamma^n, B^n(\sigma_i^p)] = \eta_{ji}^p c_j .$$

For any σ_j^{p+1}, let us choose an n-simplex τ^n such that σ_j^{p+1} is a face of τ^n, and put $H^n(\sigma_j^{p+1}) = c_j\tau^n$. Since $\Gamma^n = \sum_i \sigma_i^n$, we have $\varphi[\Gamma_j^n H^n(\sigma_j^{p+1})] = c_j$ and, therefore,

$$\varphi[\Gamma^n, B_0^n(\sigma_i^p)] = 0 , \tag{13.1}$$

where

$$B_0^n(\sigma_i^p) = B^n(\sigma_i^p) - \eta_{ji}^p H^n(\sigma_j^{p+1}) . \tag{13.2}$$

Evidently σ_i^p is a face of each n-simplex appearing in the n-chain $B_0^n(\sigma_i^p)$. Therefore there exists in the link Lk. $[\sigma_i^p]$ an $(n - p - 1)$-chain $C^{n-p-1}(\sigma_i^p)$ such that the n-chain $B_0^n(\sigma_i^p)$ can be obtained from the $(n - p - 1)$-chain C^{n-p-1} by replacing each $(n - p - 1)$-simplex

$$(v_{p+1}, \cdots , v_n)$$

by the n-simplex

$$(v_0, \cdots , v_p, v_{p+1}, \cdots , v_n)$$

where

$$(v_0, \cdots , v_p) = \sigma_i^p . \tag{13.3}$$

Since the complex Lk. $[\sigma_i^p]$ contains no $(n - p)$-simplex, the $(n - p - 1)$-chain $C^{n-p-1}(\sigma_i^p)$ of the complex Lk. $[\sigma_i^p]$ must be a dual $(n - p - 1)$-cycle. Moreover, the equation (13.1) signifies that the sum of the coefficients of $C^{n-p-1}(\sigma_i^p)$ is equal to zero. Since M_n is p-regular, Lk. $[\sigma_i^p]$ is an orientable simple $(n - p - 1)$-circuit, which implies readily the existence of an $(n - p - 2)$-chain $D^{n-p-2}(\sigma_i^p)$ in the complex Lk. $[\sigma_i^p]$ such that

$$F^*D^{n-p-2}(\sigma_i^p) = (-1)^{p+1}C^{n-p-1}(\sigma_i^p) . \tag{13.4}$$

Let $H^{n-1}(\sigma_i^p)$ signify the $(n - 1)$-chain which arises from the $(n - p - 2)$-chain $D^{n-p-2}(\sigma_i^p)$ by replacing each $(n - p - 2)$-simplex

$$(v_{p+1}, \cdots , v_{n-1})$$

by the $(n-1)$-simplex

$$(v_0, \cdots, v_p, v_{p+1}, \cdots, v_{n-1}),$$

supposing the validity of (13.3). Then (13.4) implies that

$$F^*H^{n-1}(\sigma_i^p) = B_0^n(\sigma_i^p). \tag{13.5}$$

Moreover, σ_i^p is a face of every $(n-1)$-simplex appearing in the $(n-1)$-chain $H^{n-1}(\sigma_i^p)$.

Now, let us put

$$B_{p-1}^n(\sigma_i^p) = 0,$$

$$B_{p-1}^{n-1}(\sigma_j^{p-1}) = B^{n-1}(\sigma_j^{p-1}) - \eta_{ij}^{p-1} H^{n-1}(\sigma_i^p)$$

and

$$B_{p-1}^{n-p+k}(\sigma_i^k) = B^{n-p+k}(\sigma_i^k) \quad \text{for} \quad p-1 \neq k \neq p.$$

From (13.2) and (13.5) it is easily seen that the chains $B_{p-1}^{n-p+k}(\sigma_i^k)$ form an auxiliary construction associated with B^{n-p}.

Now let us suppose that (as we have found to be possible in the case $r = p-1$) we have found chains $B_r^{n-p+k}(\sigma_i^k)(1 \leqq r \leqq p-1)$ forming an auxiliary construction associated with B^{n-p} and such that $B_r^{n-p+r+1}(\sigma_i^{r+1}) = 0$. By the definition of an auxiliary construction, we have

$$F^*B_r^{n-p+r}(\sigma_i^r) = 0 \tag{13.6}$$

for each σ_i^r. Since σ_i^r is a face of each $(n-p+r)$-simplex appearing in $B_r^{n-p+r}(\sigma_i^r)$, there exists in the link Lk. $[\sigma_i^r]$ an $(n-p-1)$-chain $C^{n-p-1}(\sigma_i^r)$ such that the $(n-p+r)$-chain $B_r^{n-p+r}(\sigma_i^r)$ can be obtained from the $(n-p-1)$-chain $C^{n-p-1}(\sigma_i^r)$ by replacing each $(n-p-1)$-simplex

$$(v_{r+1}, \cdots, v_{n-p+r})$$

by the $(n-p+r)$-simplex

$$(v_0, \cdots, v_r, v_{r+1}, \cdots, v_{n-p+r}),$$

where

$$(v_0, \cdots, v_r) = \sigma_i^r. \tag{13.7}$$

Now the equation (13.6) signifies that $C^{n-p-1}(\sigma_i^r)$ is a dual $(n-p-1)$-cycle of the complex Lk. $[\sigma_i^r]$. Since M_n is p-regular, it follows that there exists an $(n-p-2)$-chain $D^{n-p-2}(\sigma_i^r)$ of the complex Lk. $[\sigma_i^r]$ such that

$$F^*D^{n-p-2}(\sigma_i^r) = (-1)^{r+1}C^{n-p-1}(\sigma_i^r). \tag{13.8}$$

Let $H^{n-p+r-1}(\sigma_i^r)$ denote the $(n-p+r-1)$-chain which arises from the $(n-p-2)$-chain $D^{n-p-2}(\sigma_i^r)$ by replacing each $(n-p-2)$-simplex

$$(v_{r+1}, \cdots, v_{n-p+r-1})$$

by the $(n-p+r-1)$-simplex

$$(v_0, \cdots, v_r, v_{r+1}, \cdots, v_{n-p+r-1}) .$$

supposing the validity of (13.7). Then (13.8) implies that

$$F^*H^{n-p+r-1}(\sigma_i^r) = B_r^{n-p+r}(\sigma_i^r) . \tag{13.9}$$

Now, let us put

$$\begin{aligned} B_{r-1}^{n-p+r}(\sigma_i^r) &= 0 , \\ B_{r-1}^{n-p+r-1}(\sigma_j^{r-1}) &= B_r^{n-p+r-1}(\sigma_j^{r-1}) - \eta_{i\,j}^{r\,r-1} H^{n-p+r-1}(\sigma_i^r) \end{aligned} \tag{13.10}$$

and

$$B_{r-1}^{n-p+k}(\sigma_i^k) = B_r^{n-p+k}(\sigma_i^k) \quad \text{for} \quad r-1 \neq k \neq r .$$

It follows readily from (13.9) that the chains $B_{r-1}^{n-p+k}(\sigma_i^k)$ form an auxiliary construction associated with B^{n-p} and such that (13.10) holds true.

Applying the preceding argument successively for $r = p-1, p-2, \cdots, 2, 1$, we obtain an auxiliary construction $B_0^{n-p+k}(\sigma_i^k)$ associated with B^{n-p} and such that $B_0^{n-p+1}(\sigma_i^1) = 0$. Applying the same argument again in the case $r = 0$, we have (13.9), written now as

$$F^*H^{n-p-1}(\sigma_i^0) = B_0^{n-p}(\sigma_i^0) .$$

But since $B_0^{n-p}(\sigma_i^0)$ are elements of an auxiliary construction associated with B^{n-p}, we have $B^{n-p} = \sum_i B_0^{n-p}(\sigma_i^0) = F^* \sum_i H^{n-p-1}(\sigma_i^0)$, whence $B^{n-p} \sim 0$.

14. *If M_n is $(p-1)$-regular,*[4] *then the group $\psi_p(\bar{\mathfrak{B}}_{n-p})$ is the whole group $\mathfrak{B}_p$,* so that the group $\mathfrak{B}_p$ is a homomorphic image of the group $\bar{\mathfrak{B}}_{n-p}$. Comparing this with the result of the preceding section we see that, *if M_n is both $(p-1)$-regular and p-regular, the groups $\mathfrak{B}_p$ and $\bar{\mathfrak{B}}_{n-p}$ are isomorphic.*

Let $C^p = c_i\sigma_i^p$ be an ordinary p-cycle of M_n, so that $\eta_{i\,j}^{p\,p-1}c_i = 0$. We shall find a dual $(n-p)$-cycle B^{n-p} and an auxiliary construction $B^{n-p+k}(\sigma_i^k)$ associated with it such that $\Gamma^n \cdot B^{n-p} = C^p$, i.e.

$$\varphi[\Gamma^n, B^n(\sigma_i^p)] = c_i . \tag{14.1}$$

The construction of n-chains $B^n(\sigma_i^p)$ satisfying (14.1) is quite evident; it is sufficient to choose for each σ_i^p an n-simplex τ^n having σ_i^p among its faces and to put $B^n(\sigma_i^p) = c_i\tau^n$. Since $\eta_{i\,j}^{p\,p-1}c_i = 0$, we have for each σ_j^{p-1}

$$\varphi[\Gamma^n, \eta_{i\,j}^{p\,p-1} B^n(\sigma_i^p)] = 0 . \tag{14.2}$$

Since σ_j^{p-1} is a face of every n-simplex appearing in $\eta_{i\,j}^{p\,p-1}B^n(\sigma_i^p)$ and since the $(p-1)$-regularity of M_n implies that the link Lk. $[\sigma_j^{p-1}]$ is an orientable simple $(n-p)$-circuit, we can start with (14.2) and repeat the same argument which, in the preceding section and starting with (13.1), led us to (13.5). We thus

[4] Any M_n is supposed to be (-1)-regular.

obtain, for every σ_j^{p-1}, an $(n-1)$-chain $B^{n-1}(\sigma_j^{p-1})$ such that σ_j^{p-1} is a face of each simplex appearing in $B^{n-1}(\sigma_j^{p-1})$ and such that

$$F^*B^{n-1}(\sigma_j^{p-1}) = \eta_{ij}^{p-1}B^n(\sigma_i^p) .$$

More generally, let us suppose that, for a given $r(1 \leqq r \leqq p-1)$, we have succeeded in attaching to every σ_i^k $(r \leqq k \leqq p)$ an $(n-p+k)$-chain $B^{n-p+k}(\sigma_i^k)$ having the two following properties. *First*, σ_i^k is a face of each $(n-p+k)$-simplex appearing in $B^{n-p+k}(\sigma_i^k)$. *Second*, we have for $r \leqq k \leqq p-1$

$$F^*B^{n-p+k}(\sigma_i^k) = \eta_{ji}^k B^{n-p+k+1}(\sigma_j^{k+1}) . \tag{14.3}$$

It follows that

$$F^*\eta_{ij}^{r-1}B^{n-p+r}(\sigma_i^r) = 0 . \tag{14.4}$$

Since σ_j^{r-1} is a face of every $(n-p+r)$-simplex appearing in $\eta_{ij}^{r-1}B^{n-p+r}(\sigma_i^r)$ and since the $(p-1)$-regularity of M_n implies that every dual $(n-p-1)$-cycle of the complex Lk. $[\sigma_j^{r-1}]$ is homologous to zero in Lk. $[\sigma_j^{r-1}]$, we can start with (14.4) and repeat the same argument which, in the preceding section and starting with (13.6), led us to (13.9). We obtain thus, for every σ_j^{r-1}, an $(n-p+r-1)$-chain $B^{n-p+r-1}(\sigma_j^{r-1})$ such that σ_j^{r-1} is a face of each simplex appearing in $B^{n-p+r-1}(\sigma_j^{r-1})$ and such that (14.3) holds true for $k = r-1$.

Starting with the chains $B^n(\sigma_i^p)$ and $B^{n-1}(\sigma_i^{p-1})$ already found, and applying the preceding argument successively for $r = p-1, p-2, \cdots, 2, 1$, we find chains $B^{n-p+k}(\sigma_i^k)$ $(0 \leqq k \leqq p)$ such that σ_i^k is a face of each simplex appearing in $B^{n-p+k}(\sigma_i^k)$ and such that (14.3) holds true for $0 \leqq k \leqq p-1$. In particular, for $k = 0$, (14.3) says that

$$F^*B^{n-p}(\sigma_i^0) = \eta_{ji}^0 B^{n-p+1}(\sigma_j^1) .$$

Since $\sum_i \eta_{ji}^0 = 0$ for every σ_j^1, we have $F^* \sum_i B^{n-p}(\sigma_i^0) = 0$, i.e.

$$B^{n-p} = \sum_i B^{n-p}(\sigma_i^0)$$

is a dual $(n-p)$-cycle. Of course our chains $B^{n-p+k}(\sigma_i^k)$ form an auxiliary construction associated with B^{n-p} and we have $\Gamma^n \cdot B^{n-p} = C^p$.

15. Let $0 \leqq p \leqq n$, $0 \leqq q \leqq n$. Suppose that M_n is r-regular both for $r = p$ and for $r = q$. Let C^p be an ordinary p-cycle belonging to the family $\psi_p(\mathfrak{B}_{n-p})$; let D^q be an ordinary q-cycle belonging to the family $\psi_q(\mathfrak{B}_{n-q})$; if M_n is r-regular also for $r = p-1$ and $r = q-1$, we know (sect. 14) that the cycles C^p and D^q are unrestricted.

We shall define the *intersection* of C^p and D^q and we shall designate it by $C^p \times D^q$. In the case $p + q < n$ we simply put

$$C^p \times D^q = 0 .$$

In the case $p + q \geqq n$, we shall define $C^p \times D^q$ as an ordinary $(p+q-n)$-cycle, but only its homology class will be uniquely determined.

Since C^p belongs to $\psi_p(\bar{\mathfrak{B}}_{n-p})$, there exists a dual $(n-p)$-cycle A^{n-p} such that

$$\Gamma^n A^{n-p} \sim C^p. \tag{15.1}$$

Since D^q belongs to $\psi_q(\bar{\mathfrak{B}}_{n-q})$, there exists a dual $(n-q)$-cycle B^{n-q} such that

$$\Gamma^n B^{n-q} \sim D^q. \tag{15.2}$$

We know (see sect. 13) that the homology classes of A^{n-p} and B^{n-q} are uniquely defined.

This being done, we put

$$C^p \times D^q \sim \Gamma^n \cdot A^{n-p} B^{n-q}. \tag{15.3}$$

It follows from (10.1) and (15.1) that

$$C^p \times D^q \sim C^p B^{n-q}. \tag{15.4}$$

The *distributive laws*

$$\begin{aligned} (C_1^p + C_2^p) \times D^q &\sim (C_1^p \times D^q) + (C_2^p \times D^q), \\ C^p \times (D_1^q + D_2^q) &\sim (C^p \times D_1^q) + (C^p \times D_2^q) \end{aligned} \tag{15.5}$$

are evident. The *commutative law*

$$D^q \times C^p \sim (-1)^{(n-p)(n-q)} C^p \times D^q \tag{15.6}$$

follows from (7.1) and (15.3). If M_n is also s-regular and if E^s is an ordinary s-cycle belonging to the family $\psi_s(\bar{\mathfrak{B}}_{n-s})$, we see from (7.4), (10.1) and (15.3) the validity of the *associative law*

$$(C^p \times D^q) \times E^s \sim C^p \times (D^q \times E^s). \tag{15.7}$$

16. Let M_n be an orientable combinatorial n-manifold and let M_n' be its barycentrical subdivision. It is well known that M_n' is also an orientable combinatorial n-manifold. We shall show that, on the manifold M_n', our definition of intersection of ordinary cycles is equivalent to the classical definition.

Let $\sigma_i^p (0 \leqq p \leqq n)$ denote the simplices of M_n. We choose the orientation of the n-simplices σ_i^n in such manner that $\gamma^n = \sum_i \sigma_i^n$ is an ordinary n-cycle on M_n; we choose arbitrarily the orientation of the p-simplices $\sigma_i^p (1 \leqq p \leqq n-1)$ and, as usual, we denote by η_{ij}^p the incidence coefficient of σ_i^{p+1} and $\sigma_j^p (0 \leqq p \leqq n-1)$.

Now let us recall the definition of the complex M_n'. The vertices of M_n' are identical with the simplices $\sigma_i^p (0 \leqq p \leqq n)$ of M_n. The vertices $\sigma_{i_0}^{p_0}$, $\sigma_{i_1}^{p_1}, \cdots, \sigma_{i_r}^{p_r}$ of M_n', where $p_0 \leqq p_1 \leqq \cdots \leqq p_r$, form an r-simplex of M_n' if and only if (1) $p_0 < p_1 < \cdots < p_r$, (2) $\sigma_{i_s}^{p_s}$ is a face of $\sigma_{i_{s+1}}^{p_{s+1}}$ for $0 \leqq s \leqq r-1$.

Put

$$\Gamma^n = \sum \eta^0_{i_1 i_0} \eta^1_{i_2 i_1} \cdots \eta^{n-1}_{i_n i_{n-1}} (\sigma^0_{i_0}, \sigma^1_{i_1}, \cdots, \sigma^n_{i_n})$$

the summation running over all the n-simplices of M'_n. It is well known that Γ^n is an ordinary n-cycle of M'_n (usually called the barycentrical subdivision of γ^n).

The classical intersection of two ordinary cycles on M'_n is obtained by choosing each factor in a particular way in its homology class, which we must describe in detail.

Let $H^p = a_i \sigma^p_i$ be an ordinary p-cycle of M_n. Put

$$C^p = \sum \eta^0_{i_1 i_0} \eta^1_{i_2 i_1} \cdots \eta^{p-1}_{i_p i_{p-1}} a_{i_p} (\sigma^0_{i_0}, \sigma^1_{i_1}, \cdots, \sigma^p_{i_p}),$$

the summation running over all the p-simplices of M'_n having the indicated form $(\sigma^0_{i_0}, \sigma^1_{i_1}, \cdots, \sigma^p_{i_p})$. Let $K^{n-q} = b_i \sigma^{n-q}_i$ be a dual $(n-q)$-cycle of M_n. Put

$$D^q = \sum \eta^{n-q}_{i_{n-q+1} i_{n-q}} \cdots \eta^{n-1}_{i_n i_{n-1}} b_{i_{n-q}} (\sigma^{n-q}_{i_{n-q}}, \sigma^{n-q+1}_{i_{n-q+1}}, \cdots, \sigma^n_{i_n}),$$

the summation running over all the q-simplices of M'_n having the indicated form $(\sigma^{n-q}_{i_{n-q}}, \sigma^{n-q+1}_{i_{n-q+1}}, \cdots, \sigma^n_{i_n})$.

In the classical theory of combinatorial manifolds it is shown that C^p is an ordinary p-cycle on M'_n, that D^q is an ordinary q-cycle on M'_n, and that we may choose the ordinary p-cycle H^p on M_n and the dual $(n-q)$-cycle K^{n-q} on M_n in such a manner that C^p and D^q are homologous to arbitrarily given ordinary p-cycle and q-cycle on M'_n. The classical intersection of C^p and D^q is zero if $p + q < n$; in the case $p + q \geqq n$, it is equal to

$$C^p \times D^q = \sum \eta^{n-q}_{i_{n-q+1} i_{n-q}} \cdots \eta^{p-1}_{i_p i_{p-1}} a_{i_p} b_{i_{n-q}} (\sigma^{n-q}_{i_{n-q}}, \cdots, \sigma^p_{i_p}), \tag{16.1}$$

the summation running over all the $(p + q - n)$-simplices of M'_n having the indicated form $(\sigma^{n-q}_{i_{n-q}}, \cdots, \sigma^p_{i_p})$.

The case $p + q < n$ being trivial, we have to show that, if $p + q \geqq n$, (16.1) holds true according to our definition of intersection.

We now choose an ordering ω of the vertices of M_n and define an $(n-q)$-chain B^{n-q} on M'_n as follows. Let

$$\tau^{n-q} = (\sigma^{h_0}_{i_0}, \sigma^{h_1}_{i_1}, \cdots, \sigma^{h_{n-q}}_{i_{n-q}})$$

be an $(n-q)$-simplex of M'_n. Let v_λ be the first vertex of the h_λ-simplex $\sigma^{h_\lambda}_{i_\lambda}$. $(0 \leqq \lambda \leqq n - q)$, relatively to the ordering ω. If the v_λ's $(0 \leqq \lambda \leqq n - q)$ are not all different from each other, then the coefficient of τ^{n-q} in B^{n-q} will be zero. In the other case,

$$(v_0, v_1, \cdots, v_{n-q}) \tag{16.2}$$

is an $(n-q)$-simplex of M_n and the coefficient of τ^{n-q} in B^{n-q} will be equal to the coefficient of (16.2) in K^{n-q}. It is not difficult to verify that B^{n-q} is a dual cycle on M'_n.

Now we order the set of all the vertices of M_n' in such a manner that σ_i^h precedes σ_j^k, whenever $h < k$; this can be done in many ways. We form the product $\Gamma^n B^{n-q}$ in the manner explained in sect. 10, using our ordering of the vertices of M_n'. We easily verify that

$$\Gamma^n B^{n-q} = D^q,$$

so that

$$C^p \times D^q \sim C^p B^{n-q}$$

from (15.2) and (15.3). Now if we form the product $C^p B^{n-q}$ again in the manner explained in sect. 10, using the same ordering of the vertices of M_n', we easily verify that (16.1) holds true.

THE INSTITUTE FOR ADVANCED STUDY.

ON GENERALIZED MANIFOLDS.

By S. LEFSCHETZ.

The object of the present paper is to extend to a larger class of spaces certain results recently obtained for topological manifolds.† The extension consists in replacing the requirement that every point possess a combinatorial cell for neighborhood by certain weaker conditions on the chains through the point. Roughly speaking they amount to demanding that locally any p-chain be deformable (in a certain very general sense) into one which does not meet any assigned q-space ($=q$ dimensional space), where $p+q<n$, the dimension of the manifold. This extension is made in Part III of the present paper. In Part I we take up again, partly as a preparation to the second Part, the homology theory of metric spaces from the standpoint initiated in our Colloquium Lectures *Topology*, Ch. VII. The notation and terminology are as in our book.‡

§ 1. THE APPROXIMATING COMPLEXES OF A METRIC SPACE.

1. The homology properties of a compact metric space are intimately related to the homology properties of certain subchains of an infinite complex, the fundamental complex of the space (*Topology*, Ch. VII), or to certain sequences of chains of approximating complexes (Alexandroff). We shall first show how these may be selected in a certain convenient way for the sequel.

Let for the present $\mathcal{R}$ be a compact metric n-space and let U, V, W, denote generically its open sets, and $F(U)$, $F(V)$, $F(W)$, their boundaries.

We shall repeatedly consider various aggregates of subsets, $\Sigma=\{A^\alpha\}$, of $\mathcal{R}$. The mesh of Σ is max diam A^α. If the set of A's covers $\mathcal{R}$ we call Σ a *covering*, an *ϵ-covering* if its mesh $<\epsilon$. Of particular importance are the finite coverings by open sets ($=$ f. c. o. s.).

Each set A^α of the aggregate Σ may be considered as an abstract point,

† S. Lefschetz and W. W. Flexner, *Proceedings of the National Academy*, Vol. 16 (1930), pp. 530-533; W. W. Flexner, *Annals of Mathematics*, Ser. 2, Vol. 32 (1931), pp. 393-406, 539-548.

‡ A very extensive paper by Čech on the same general topic was presented simultaneously with the present one to the *Annals of Mathematics* where his paper is now appearing. While there are many contacts between the two, they differ essentially in method and scope. Čech deals indeed with a much more general type of space, but the restriction to locally compact metric spaces which we have imposed here, has enabled us to proceed much more quickly to the point.

and we may then introduce for each intersection $A^{a_0}\cdots A^{a_p}\neq 0$ an abstract p-simplex $\sigma_p = A^{a_0}\cdots A^{a_p}$. It will be convenient to designate the intersection also by σ_p : $\sigma_p = 0$ signifies then that the sets $A^{a_0},\cdots,A^{a_p}$ do not intersect.

The aggregate $\{\sigma\}$ has the property that with each σ every face of σ also belongs to the set. Hence $\{\sigma\}$ is a closed simplicial (abstract) complex Φ, the *skeleton* of Σ. If another aggregate $\Sigma' = \{A'^{a}\}$ has for skeleton Φ' a complex whose structure is that of a subcomplex of Φ, we shall briefly say that its skeleton is a subcomplex of Φ. The *dimension* of Φ is the highest integer ν such that there is at least one aggregate of $\nu+1$ intersecting A's. ν is also called the *order* of Σ. Clearly of course Φ is finite when and only when Σ is finite.

Suppose in particular that $\Sigma = \{U^{a}\}$ is an ϵ-f. c. o. s. It is called *irreducible* (Alexandroff) when there is no ϵ-f. c. o. s. whose skeleton is a proper subcomplex of Φ. If Σ is reducible there is an ϵ-f. c. o. s. Σ^1 whose skeleton is a proper subcomplex Φ^1 of Φ. If Σ^1 is in turn reducible there is an ϵ-f. c. o. s. Σ^2 whose skeleton is a proper subcomplex Φ^2 of Φ, etc. Since Φ has only a finite number of subcomplexes the process must stop after a finite number of steps. Therefore there exists an irreducible ϵ-f. c. o. s. whatever ϵ. If the order of the initial covering is the least possible for an ϵ-f. c. o. s. it will also be the order of the ultimate irreducible covering.

We recall that as $\epsilon\to 0$ the least order ν tends to an upper limit n or else $\to\infty$. In the first case dim $\mathfrak{R} = n$, in the second case dim $\mathfrak{R}=\infty$.

Let $\Sigma = \{U^{a}\}$ be a f. c. o. s. whose skeleton Φ is the same as for $\{\bar{U}^{a}\}$. Then there exists a constant η, the *characteristic constant* of Σ, such that: (a) if a set A on $\mathfrak{R}$ whose diameter $<\eta$ meets a certain number of U's, these U's have a non-vacuous intersection; (b) any point x of $\mathfrak{R}$ is on at least one U such that $d(x,\mathfrak{R}-U)>\eta$. As a consequence of (b) if diam $A<\eta$ then some $U\supset A$.

2. Taking $n =$ dim $\mathfrak{R}$ finite, let ϵ be so small that the least order of an ϵ-f. c. o. s. is n, and let Σ be an irreducible ϵ-f. c. o. s. There exists another ϵ-f. c. o. s. of order n, $\Sigma' = \{V^{a}\}$ consisting of as many sets as Σ and such that for every a we have $V^{a}\Subset U^{a}$.† Clearly Σ' is an ϵ-f. c. o. s. whose skeleton is Φ or a subcomplex of Φ, and since Σ is irreducible it can only be Φ. Therefore *the order of* Σ *is* n. In other words an irreducible f. c. o. s. whose mesh is sufficiently small is of order n. Observe incidentally that $\{\bar{V}^{a}\}$ has the same skeleton as $\{V^{a}\}$.

† Menger, *Dimensionstheorie*, p. 160. We shall use his "strong inclusion" symbol $\Subset$ ($A\Subset B$ means that $\bar{A}\subset B$).

Consider now a sequence $\{\Sigma^i\}$, where $\Sigma^i = \{U^{i\alpha}\}$ is an irreducible ϵ_i-f. c. o. s. such that: (a) $\epsilon_1 = \epsilon$; (b) if η_i is the characteristic constant of Σ^i we have $\epsilon_{i+1} < \frac{1}{2}\eta_i$ and $< \frac{1}{2}\epsilon_i$; (c) $\{\bar{U}^{i\alpha}\}$ has the same skeleton as Σ^i. As a consequence Σ^i is of order n and for every $U^{i+1,\beta}$ there is a $U^{i\alpha} \supset U^{i+1,\beta}$. Let Φ^i be the skeleton of Σ^i; choose for each $U^{i+1,\beta}$ a definite $U^{i\alpha} \supset U^{i+1,\beta}$ and define a transformation t_i of the vertices of Φ^{i+1} into vertices of Φ^i whereby the vertex $U^{i+1,\beta}$ goes into the vertex $U^{i\alpha}$. Let $\sigma_p = U^{i+1,\beta_0} \cdots U^{i+1,\beta_p}$ be a simplex of Φ^{i+1}. As a consequence, if $U^{i\alpha_h} = t_i U^{i+1,\beta_h}$ then $U^{i\alpha_h} \supset U^{i+1,\beta_h}$ and hence $\sigma'_q = U^{i\alpha_0} \cdots U^{i\alpha_p}$ is a simplex of Φ^i. (It may happen that several of the vertices $U^{i\alpha_h}$ coincide, in which case $q < p$). Thus if certain vertices U^{i+1} belong to a σ_p of Φ^{i+1} the transformed vertices $t_i U^{i+1}$ are vertices of a σ_q $(q \leqq p)$ of Φ^i. Consequently t_i may be extended to a simplicial transformation τ_i of Φ^{i+1} into Φ^i or into a subcomplex of Φ^i. We call τ_i a *projection* of Φ^{i+1} onto Φ^i, and more generally $\tau_i \tau_{i+1} \cdots \tau_{i+j-1} \Phi^{i+j}$ a projection of Φ^{i+j} onto Φ^i. The latter is also a simplicial transformation of Φ^{i+j} into Φ^i or into a subcomplex of Φ^i.

3. I say that in fact $\tau_i \Phi^{i+1} = \Phi^i$, that is every simplex of Φ^i is the transform of a simplex of Φ^{i+1}, or, in other words, Φ^i is completely covered by $\tau_i \Phi^{i+1}$. For let us suppose that $\tau_i \Phi^{i+1} = \Psi$, a proper subcomplex of Φ^i. There exists then a simplex $\sigma_p = U^{i\alpha_0} \cdots U^{i\alpha_p} \subset \Phi^i - \Psi$. Denote generically by V^α the sum of all the sets $U^{i+1,\beta}$ which make up $\tau_i^{-1} U^{i\alpha}$; clearly $V^\alpha \subset U^{i\alpha}$. Since every U^{i+1} corresponds to one (and only one) V, $\Sigma = \{V^\alpha\}$ is an ϵ_i-f. c. o. s. and it has a subcomplex Φ' of Φ as its skeleton. I say that σ_p is not a cell of Φ'. For otherwise we would have $V^{\alpha_0} \cdots V^{\alpha_p} \neq 0$, and hence there would exist a $U^{i+1,\beta_0} \cdots U^{i+1,\beta_p} \neq 0$, where U^{i+1,β_h} is a constituent of V^{α_h}. Since $\tau_i U^{i+1,\beta_h} = U^{i\alpha_h}$, we would then have in $\sigma'_p = U^{i+1,\beta_0} \cdots U^{i+1,\beta_p}$ a simplex of Φ^{i+1} such that $\tau_i \sigma'_p = \sigma_p$ and hence $\sigma_p \subset \Psi = \tau_i \cdot \Phi^{i+1}$, contrary to assumption.

Under the circumstances then $\sigma_p \not\subset \Phi'$. It follows that Φ' is a proper subcomplex of Φ^i and also the skeleton of an ϵ_i-f. c. o. s. But this is ruled out since Σ^i is irreducible. Hence σ_p cannot exist, and $\Psi = \tau_i \Phi^{i+1} = \Phi^i$.

Definitions. A sequence $\{B^i\}$ of elements, (sets, complexes, etc.) such that $B^i \subset \Phi^i$ and $\tau_i B^{i+1} = B^i$ is called a *projection-sequence* (of sets, of complexes, etc.).

Given any (non-singular) chain C_p we shall designate by $| C_p |$ the complex made up of the cells of the chain. A sequence $\{C_p{}^i\}$ will be called a projection-sequence of chains or cycles whenever

$$\tau_i C_p{}^{i+1} = C_p{}^i.$$

4. Let $U^{i\delta} \supset x$. There exists a set $U^{i-1,\gamma} \supset U^{i\delta}$ such that $\tau_{i-1}U^{i\delta} = U^{i-1,\gamma}$; a set $U^{i-2,\beta}$ similarly related to $U^{i-1,\gamma}$, etc., clear up to a certain set $U^{1\alpha}$. Let k_i be for each i the class of all sets $U^{1\alpha}$ thus obtained. The classes k_i are all finite, $\neq 0$ and $k_i \supset k_{i+1}$. Therefore from a certain i on $k_i = k_{i+1} = \cdots$. Consequently there exists an infinite sequence $\{U^{i\alpha_i}\}$ such that $U^{i\alpha_i} \supset U^{i+1,\alpha_{i+1}}$, $\tau_i U^{i+1,\alpha_{i+1}} = U^{i\alpha_i}$, $\Pi U^{i\alpha_i} = x$. Let $U^{i\alpha_0}, \cdots, U^{i\alpha_p}$ be all the sets of Σ^i occurring in any such sequence corresponding to the same point x and let $V^i = U^{i\alpha_0} \cdots U^{i\alpha_p} \neq 0$, so that $\sigma_p{}^i = U^{i\alpha_0} \cdots U^{i\alpha_p}$ is a simplex of Φ^i. Since every $U^{i\alpha}$ here occurring is the τ_i transform of a similar $U^{i+1,\beta}$ we have $\tau_i\sigma^{i+1} = \sigma^i$, hence $\{\sigma^i\}$ is a projection-sequence of simplexes. Moreover clearly $V^i \supset V^{i+1}, \Pi V^i = x$.

Conversely if $\{\sigma^{*i}\}$ is a projection-sequence of simplexes, and if V^{*i} is the intersection of the sets U^i associated with σ^{*i}, then $V^{*i} \supset V^{*i+1} \supset x$, hence $\Pi V^{*i} = x$. Clearly also the sets U^i associated with σ^{*i} are among those associated with σ^i, hence σ^{*i} is σ^i or a face of σ^i. We call $\{\sigma^{*i}\}$ and $\{\sigma^i\}$ respectively *projection-sequence* and *maximal projection-sequence* for the point x.

5. Owing to the choice of $\{\Sigma^i\}$ we may use $\{\Phi^i\}$ to map the space $\mathfrak{R}$ topologically on an Euclidean S_r, $r \geqq 2n+1$.† Choosing $r \geqq 2n+2$ we may even carry out the mapping so as to be able to construct the joining cell of any simplex of Φ^{i+1} with its transform under τ_i (deformation cell corresponding to τ_i), and from there, as the sum of all these cells, the $(n+1)$-complex K or *fundamental complex* of $\mathfrak{R}$, (*Topology*, p. 327) which will be an infinite complex on S_r. The part of K obtained on removing $\Phi^i, \Phi^{i-1}, \cdots$ and the cells joining them will be denoted by N^i and the finite complex $K - N^i$ by K^i.

In practice we shall find it more convenient to have a representation of $\mathfrak{R}$ and K on the Hilbert parallelatope

$$\mathcal{H}: \quad 0 \leqq x_i \leqq 1/i, \qquad (i = 1, 2, \cdots + \infty).$$

This image is to be constructed as follows. As proved by Urysohn $\mathfrak{R}$ has a topological image $\mathfrak{R}'$ on $\mathcal{H}$. Consider now the following homeomorphism of $\mathcal{H}$:

$$T: \quad x'_i = \tfrac{1}{2}(x_i + 1/i)$$

which transforms it into the subset

$$\mathcal{H}': \quad 1/2i \leqq x_i \leqq 1/i.$$

Then $T\mathfrak{R}' = \mathfrak{R}''$ is a topological image of $\mathfrak{R}$ which possesses no point for which any x_i is zero. We identify henceforth $\mathfrak{R}$ with $\mathfrak{R}''$.

† See our paper in the *Annals of Mathematics*, Vol. 32 (1931), p. 528.

Let us denote by S^i the subset of $\mathfrak{H}$ consisting of all points for which $x_k = 0$ when $k > (2n + n_i)$, where n_i increases so rapidly that we may carry out the construction of K, given in *Topology*, p. 325, in such manner that $\Phi^i \subset S^i - S^{i-1}$. As a consequence $K \cdot \mathfrak{R} = 0$. Now, with closures referring to $\mathfrak{H}$, the only limit-points of $\bar{K}$ not on K are on $\mathfrak{R}$, hence $\mathfrak{R} = \bar{K} - K$. It is in order to fulfill this condition that the complex K has been constructed in the above special manner.

6. It is convenient to join each point of Φ^{i+1} to its transform by τ_i by a segment in $\mathfrak{H}$. The sum of these segments coincides with K. An infinite arc consisting of a sequence of projecting segments for $\tau_1, \tau_2, \cdots$ plus their co-terminal end-points, will be called a *projecting line*. The projecting lines all start at Φ^1, which we designate henceforth by Φ, and continue indefinitely throughout K.

If $\{B^i\}$ is a projection sequence of sets or complexes, the set $\mathfrak{B}$ obtained by adding to the sequence the projecting segments of the points of the B's is called a *projection-set*. If the B's are complexes, the projecting segments of a definite p-cell of B^{i+1} make up a $(p+1)$-cell; these are the joining cells of B^{i+1} and $\tau_i B^{i+1}$ (No. 5). The sum of the closures of all these cells is a *projection-complex* $\mathcal{K}$. If B^i is a subcomplex of Φ^i for every i, $\mathcal{K}$ is a subcomplex of K.

We are primarily interested in the relation between various subcomplexes of K and certain associated sets of $\mathfrak{R}$. Properly speaking instead of a subcomplex of K we might well take any subset of K, but actually the subcomplexes will suffice for our purpose.

With any subcomplex L of K we may associate the closed subset $F = \bar{L} \cdot \mathfrak{R}$, and we observe immediately that this set F depends solely upon the "infinite" part of L, i. e. it is unchanged when a finite complex is added to or removed from L. In the sense of *Topology*, Ch. VII, $\mathfrak{R}$ is associated with the total ideal element of K, and F with a certain closed ideal element of the complex.

Suppose that we construct a new fundamental K' for $\mathfrak{R}$, that we suppose as before on $\mathfrak{H}$, and such that $K' \cdot \mathfrak{R} = 0$. Applying to K the deformation theorem of *Topology*, p. 328 † (proved for chains but applicable to complexes), we can reduce L to a subcomplex L' of K' by a deformation that $\rightarrow 0$ for any particular cell of K as that cell $\rightarrow \mathfrak{R}$. Therefore $F = \bar{L}' \cdot \mathfrak{R} = \bar{L} \cdot \mathfrak{R}$, i. e. the set F is in a large measure independent of the complex K.

† In the proof *loc. cit.*, A_j should be mapped on $\tau_{i-1}\lambda_i A_j$. Owing to the condition $\epsilon_i < \frac{1}{2}\eta_{i-1}$ which we have imposed, C_p^{i-1} will still be mapped as before on a subchain of Φ^i.

7. We shall now reverse the situation: starting with any particular closed set F we shall associate with it a certain projection-complex L, such that $F = \bar{L} \cdot \mathfrak{R}$ and $\dim F = p = \dim L - 1$, which is the maximum value possible for p.

According to Menger (*Dimensionstheorie*, p. 158) there is a f. c. o. s. of order $\leqq p$ of F (not of $\mathfrak{R}$), $\Sigma'^i = \{V^{ia}\}$, such that there is one and only one V^{ia} on any U^{ia} that meets F. When i exceeds a certain value the skeleton Φ'^i of Σ'^i is a p-complex. Associate with each V^{ia} the vertex U^{ia} of the set of same name. Now when a certain aggregate of sets V^i intersect, the same holds as regards the corresponding sets U^i. Hence Φ'^i will thus become a subcomplex of Φ^i. Now take all the subcomplexes Ψ'^1 of Φ^1 which are the projections of a Φ'^i. Since their number is infinite and the number of subcomplexes of Φ^1 is finite, at least one, Ψ^1 is the projection of an infinity of complexes Φ'^i. Consider the subcomplexes Ψ'^2 of Ψ^2 such that $\tau_1\Psi'^2 = \Psi^1$. There is an infinity of complexes Φ'^i, $i \geqq 2$, projected onto Ψ^1 and their projections on Φ^2 are each a Ψ'^2. Therefore at least one of the latter, Ψ^2, is the projection of an infinity of complexes Φ'^i, etc. By this obvious process we obtain an infinite projection-sequence $\{\Psi^i\}$, where Ψ^i is a subcomplex of Φ^i which is the projection of a Φ'^j, and $\dim \Psi^i \leqq \dim \Phi'^j \leqq p$. Since Φ'^j is the skeleton of an ϵ_j-f. c. o. s. of F, the latter may be $6\epsilon_j$-deformed into F.† Moreover, referring to the representation in $\mathfrak{H}$, Φ'^j can be ξ_i-deformed into Ψ^i, ($\xi_i \to 0$ with $1/i$). Hence F can be ζ_i-deformed into Ψ^i, ($\zeta_i \to 0$ with $1/i$). Therefore Ψ^i is the skeleton of a θ_i-f. c. o. s. of $F(\theta_i \to 0$ with $1/i)$ (Alexandroff, *loc. cit.*, p. 18). As a consequence if we put in the joining cells of the Ψ's, we obtain a fundamental complex L for F. We have $\dim L = p + 1$, for it is $\geqq p + 1$ since $\dim F = p$, and $\leqq p + 1$, since $\dim \Psi^i \leqq p$.

Since we have but little information regarding the meshes or the characteristic constants of the coverings of F whose skeleta are the Ψ's, it is not easy to show that the deformation theorem applies to L. Therefore for the homology theory another similar $(p + 1)$-complex L^* is more suitable. It is constructed as follows: take the skeleton of the aggregate $\{U^{ia} \cdot F\}$ (i fixed) and remove from it all cells of dimension $> p$. What is left is a subcomplex Ω^i of Φ^i, and we have immediately, owing to the mode of constructing the Φ's, $\tau_i\Omega^{i+1} \subset \Omega^i$. The complex L^* consists of all the Ω's plus their joining cells. It is clearly a $(p + 1)$-subcomplex of K, which we shall call the *generalized fundamental complex* of the set F. The proof of the deformation theorem is directly applicable to L^* for all cycles or complexes of dimension $\leqq p$. Since

† P. Alexandroff, *Annals of Mathematics*, Vol. 30 (1928-29), p. 13.

$\dim F = p$, F possesses a fundamental $(p+1)$-complex L' to which the deformation theorem is applicable. For example L' may be built up out of a subset of the Ψ's. It follows (see No. 9), that the q-cycles, $q > p$, of F are all $\equiv 0$ and hence they need not concern us further.

8. *The chains and cycles of K.* The only chains of K with which we shall be concerned are its subchains, no others being considered. Whatever C_p we have: $C_p = C'_p + C_p''$, where C'_p is the part of C_p on K^i and C_p'' the rest. It is convenient to write: $C'_p = K^i \cdot C_p$, $C_p'' = N^i \cdot C_p$. The part of $F(C'_p)$ which is on Φ^i will be designated by $\Phi^i \cdot C_p$ and called the *trace* of C_p on Φ^i.

Let us suppose that we have on K an aggregate of chains $\{C_q{}^i\}$, $q = 0, 1, \cdots, p$; $i = 1, 2, \cdots$, such that: (a) $C_p = \Sigma C_p{}^i$ is a true chain of K, i. e. includes no cell of K taken with an infinite coefficient; (b) for every $C_q{}^i$, we have

$$C_q{}^i \to \Sigma \eta^q{}_{ij} C^j{}_{q-1}. \tag{8.1}$$

The aggregate $\{C_q{}^i\}$ is called an *elementary decomposition* of C_p. An example is of course the decomposition of C_p into its cells. For later purposes a more general decomposition is introduced here.

Two decompositions $\{C_q{}^i\}$, $\{C'_q{}^i\}$ of two chains C_p, C'_p are said to have the same structure if they correspond to one another chain for chain (for every $C_q{}^i$ one and only one chain $C'_q{}^i$ and conversely) and if the corresponding incidence numbers $\eta^q{}_{ij}$ are the same. That is to say if the sets are labelled in such manner that $C_q{}^i$ and $C'_q{}^i$ are the associated chains in the correspondence then they have the same incidence matrices $\| \eta^q{}_{ij} \|$.

Suppose now that we have two decompositions $\{C_q{}^i\}$, $\{C'_q{}^i\}$ of C_p, C'_p whose structure is the same, and let there exist for every $C_q{}^i$ a $(q+1)$-chain $\mathfrak{D} C_q{}^i$, called a *deformation-chain*, such that

$$\mathfrak{D}\, C_q{}^i \to C'_q{}^i - C_q{}^i - \Sigma \eta^q{}_{ij} \mathfrak{D}\, C^j{}_{q-1}. \tag{8.2}$$

If we agree to write

$$\mathfrak{D}\, \Sigma\, a_i C_q{}^i = \Sigma\, a_i \mathfrak{D}\, C_q{}^i \tag{8.3}$$

then (8.2) assumes the form

$$\mathfrak{D}\, C_q{}^i \to C'_q{}^i - C_q{}^i - \mathfrak{D} F(C_q{}^i). \tag{8.4}$$

Under the circumstances the passage from C_p to C_p' is called a *deformation* of C_p into C'_p, and $\mathfrak{D}\,(C_p)$ is called the *deformation-chain* of C_p.

A deformation of a subcomplex L of K into another L' could be defined substantially along similar lines. We would merely replace the chains $C_q{}^i$ by the cells E_q^i of L, and in (8.2), (8.3), (8.4), the C's would be cells and the $\mathfrak{D}\,C$'s would continue to be chains but otherwise the rest would be as before. Then $\Sigma\,|\,\mathfrak{D}E\,| + L + L'$ would be called the *deformation-complex* $\mathfrak{D}\,L$ of L.

All this is entirely in line with the treatment of deformations in *Topology*, p. 78, except that there we had only cells and obtained (8.2) from direct geometric considerations, essentially by considering the deformation as a "singular" translation, whereas (8.2) serves directly to define the deformation. This departure is justified on the ground that (8.2) is the central property of a deformation as regards the applications to any homology theory.

For purposes of reference, if we agree to neglect everywhere chains on L, or else if we only consider integral chains mod m or both we have associated deformations and deformation-chains mod L, mod m, mod (L, m), as the case may be.

If in a given deformation $\mathfrak{D}$ every deformation chain is of diameter $< \epsilon$ we have a so-called *ϵ-deformation.*

By analogy with ordinary deformations we shall say that $\mathfrak{D}$ leaves a chain C_q^i *invariant* or *does not displace the chain,* whenever the chain $\mathfrak{D}\,C_q^i = 0$.

9. *The chains and cycles of the space* $\mathfrak{R}$. Taking substantially the point of view of *Topology*, Ch. VII, § 4, we consider a $(p+1)$-chain C_{p+1} of K as defining a p-chain c_p of $\mathfrak{R}$, a cycle mod Φ, Γ_{p+1} of K as defining an absolute p-cycle γ_p of $\mathfrak{R}$. In particular if

$$K \supset C_{p+2} \rightarrow \Gamma_{p+1} \quad \text{mod}\ \Phi,$$

and if C_{p+2} determines c_{p+1} of $\mathfrak{R}$ we write

$$c_{p+1} \rightarrow \gamma_p, \qquad \gamma_p = F(c_{p+1}),$$

and say "γ_p bounds c_{p+1}". A special case is where Γ_{p+1} is finite, for it is then ≈ 0 mod Φ, since it can be deformed along the projecting lines onto Φ. We say that γ_p is homologous to zero: $\gamma_p \approx 0$, whenever it is a finite or infinite sum of bounding cycles. The extension to cycles mod A, A closed, is in the usual manner: Γ_{p+1} is then a cycle mod L, where L is any subcomplex of K such that $\bar{L}\cdot\mathfrak{R} = A$.

The p-th homology group $\mathfrak{H}_p$ (absolute or mod A) is the quotient group $\mathfrak{G}_p \div \mathfrak{G}'_p$ of the Abelian group $\mathfrak{G}_p$ of the *p-cycles* (written additively) by the group $\mathfrak{G}'_p$ of the cycles ≈ 0. The bases and homology characters are defined as usual.

For $p = n$ the bounding relations between the cycles are reduced to the identical linear relations between them. In terms of the n-cycles it is possible to define the generalized absolute orientable n-circuit (*Topology*, p. 76): it is a compact metric n-space $\mathcal{R}$ such that $R_n(\mathcal{R}) = 1$, and $R_n(A) = 0$ for any *proper* closed subset A of $\mathcal{R}$. As a consequence the circuit has a base for the n-cycles consisting of a single γ_n, i. e. every n-cycle of $\mathcal{R}$ is of the form $t\gamma_n$. In place of γ_n we might as well take $-\gamma_n$ and either one of the pairs $(\mathcal{R}, \gamma_n)$, $(\mathcal{R}, -\gamma_n)$ is called an *oriented* circuit, the passage from one to the other being described as a reversal of orientation. The non-orientable circuit is obtained by taking the cycles mod 2, and similarly for the circuits mod m. Analogous notions hold for the circuit mod A, A closed, the circuit conditions being $R_n(\mathcal{R}, A) = 1$, $R_n(B, A) = 0$, where B is now any proper closed subset of $\mathcal{R}$ which $\supset A$.

10. We have taken the chains and cycles of $\mathcal{R}$ as represented by actual chains or cycles mod Φ of K. Their characteristic part corresponds however to the infinite portion of the representative C_{p+1} or Γ_{p+1}. As a matter of fact the difference is not great: we may always suppress, say $\Phi^1, \cdots, \Phi^k$, with all the cells joining them, and consider Φ^k as the new Φ, thus converting any Γ with a finite boundary into a cycle mod Φ. Another way of looking at the matter is as follows: under our conventions for chains the suppression of any finite part of C_{p+1} is not to affect c_p. As for a γ_p it is then to be represented by a C_{p+1} with finite boundary C_p. But if we slide the points of C_p along the projecting lines down onto Φ, and add the deformation-chain, which is finite, to C_{p+1}, we have a cycle mod Φ, Γ_{p+1}, which also represents γ_p.

The set $\mathcal{R} \cdot |\overline{C_{p+1}}|$, where as before the closure refers to $\mathfrak{H}$, is a closed subset of $\mathcal{R}$ associated with c_p, that we shall denote by $|c_p|$. This set depends solely on c_p, and not on the particular fundamental complex K chosen (No. 6).†

By the points of c_p we shall always mean the points of $|c_p|$. In particular a set A is said to intersect c_p whenever it intersects $|c_p|$, to be $\subset c_p$ or to $\supset c_p$ whenever $A \subset |c_p|$ or $\supset |c_p|$ as the case may be.

Let A be a closed set. By a p-cycle mod A we shall mean a c_p such that $F(c_p) \subset A$. The cycle is said to *bound* mod A whenever there exists a c_{p+1} such that $F(c_{p+1}) - c_p \subset A$. Finally it is ≈ 0 mod A whenever the cycle is a finite or infinite sum of cycles which bound mod A.

We may also consider the absolute cycles of $\mathcal{R} - A$. Such a cycle is ≈ 0 on $\mathcal{R} - A$ whenever it is ≈ 0 on some closed subset of $\mathcal{R} - A$.

† The p-chains such that $\dim |c_p| \leqq p$ form a topological subclass of the class of all p-chains. These special chains played an important part in the initial version of the present paper. We found it simpler since then, to eliminate them entirely, and to replace them everywhere merely by the projection-chains which are introduced in No. 13. As the properties needed in Part II were only those of projection-chains, the only important modifications required were in Nos. 11, 18, 19 (June, 1933).

11. A deformation of a C_{p+1} into C'_{p+1} on K may serve to define two kinds of deformations $\mathfrak{D}$ of the associated chains c_p, c'_p on $\mathfrak{R}$. The deformation $\mathfrak{D}$ is of the *first kind* whenever the chains of the associated elementary decompositions $\{C_q{}^i\}$, $\{C'_q{}^i\}$, are all finite; it is of the *second kind* when some or all are infinite.

Consider for the present a $\mathfrak{D}$ of the first kind. If the deformation-chain of $C_q{}^i \rightarrow 0$ with $1/i$, we consider the two chains c_p, c'_p as identical. If U is any open set $\supset c_p$, and if L is any subcomplex of K such that $\overline{L} \cdot \mathfrak{R} = \overline{U}$, then for i above a certain value $\mathfrak{D}\, C_q{}^i \subset L$, and hence C_{p+1} has at most a finite subchain on $K - L$.

As an application if C_{p+1} is deformed over $\mathfrak{H}$, according to the deformation theorem of *Topology*, p. 328, into a new chain C'_{p+1} of K, then the chain c'_p defined by C_{p+1} is identical with c_p. For the deformation over $\mathfrak{H}$ gives rise to a certain deformation-chain $\mathfrak{D}\, C_{p+1}$ with a suitable elementary decomposition. If we now reduce $\mathfrak{D}\, C_{p+1}$ to K by the deformation theorem, choosing, as we may, the chains of the decompositions which it demands (the analogues of the chains $C_p{}^i$ of the proof *loc. cit.*) exact sums of chains of the decomposition of C_{p+1}, the sole effect of the deformation on C_{p+1}, C'_{p+1} may be to subdivide them, and this has no influence on c_p, c'_p. As a consequence we have on K a deformation-chain for a deformation of C_{p+1} into C'_{p+1} which is of the first kind. Hence $c_p \equiv c'_p$.

Suppose in particular that we have a closed set A with $L^*{}_A$ as its generalized fundamental complex (No. 7) and let γ_p be a cycle mod A. If Γ_{p+1} is the representative chain of γ_p, $F(\Gamma_{p+1})$ represents the absolute cycle $F(\gamma_p)$ of A. This absolute cycle has a representative image Γ'_p which is a cycle of $L^*{}_A$ mod Φ (No. 7) and by the above

$$K \supset D_{p+1} \rightarrow \Gamma'_p - F(\Gamma_{p+1})\,;$$
$$\Gamma'_{p+1} = \Gamma_{p+1} + D_{p+1} \rightarrow \Gamma'_p; \quad \lceil D_{p+1} \rceil \cdot \mathfrak{R} \subset A.$$

Hence if Γ'_{p+1} represents γ'_p of $\mathfrak{R}$ we have $\gamma'_p - \gamma_p \subset A$ so that γ'_p represents the same cycle mod A as γ_p. Therefore we may represent a cycle mod A by a chain C_{p+1} whose boundary is on the generalized fundamental complex $L^*{}_A$ of the set A. This result will be useful later.

The only deformations occurring in the sequel are of the second kind, and the elementary decompositions and deformation-chains on K will always be in finite number. This will be understood throughout. They determine elementary decompositions $\{c_q{}^i\}$, $\{c'_q{}^i\}$, and deformation-chains $\mathfrak{D}\, c_q{}^i$ for the deformation of c_p into c'_p, and the rest is as in No. 8. In particular

$$\mathfrak{D}\, c_p \rightarrow c'_p - c_p - \mathfrak{D}\, F(c_p)\,; \tag{11.1}$$

$$\mathfrak{D}\gamma_p \approx \gamma'_p - \gamma_p \approx 0 \text{ on } \mathfrak{R}. \tag{11.2}$$

12. With notations as in No. 11, let γ_p be a cycle mod A whose representative Γ_{p+1} has its boundary on $L^*_A + \Phi$. The N S C in order that $\gamma_p \approx 0 \bmod A$, is that for every i

$$\Gamma_{p+1} \approx 0 \bmod (N^i + \Phi + L^*). \tag{12.1}$$

Whether the cycle is ≈ 0 or not when (12.1) holds for any particular i it holds also for the lower values of i. Therefore there is an h, called the *index* of γ_p, such that (12.1) holds for $i \leqq h - 1$ but not for $i \geqq h$. It implies that there exists an infinite cycle $\Gamma'_{p+1} \subset N^{h-1}$ such that

$$\Gamma_{p+1} \approx \Gamma'_{p+1} \bmod \Phi, \tag{12.2}$$

while no such cycle exists for any N^i, $i \geqq h$. In terms of the traces we have at once

$$\Phi^i \cdot \Gamma_{p+1} \approx \Phi^i \cdot \Gamma'_{p+1} \text{ on } \Phi^i, \quad (i \geqq h), \tag{12.3}$$

$$\Phi^i \cdot \Gamma_{p+1} \approx 0 \text{ on } \Phi^i, \quad (i < h). \tag{12.4}$$

Conversely suppose that (12.4) holds for $i < h$ but not for any higher i. We have then

$$\Phi^i \supset D_{p+1} \to \Phi^i \cdot \Gamma_{p+1}, \tag{12.5}$$

$$D'_{p+1} = N^i \cdot \Gamma_{p+1} - D_{p+1} \to 0. \tag{12.6}$$

Since the cycle D'_{p+1} is finite it is $\approx 0 \bmod \Phi$ on K, for it can be projected onto Φ. It follows that (12.2) holds with $\Gamma'_{p+1} = \Gamma_{p+1} - D_{p+1} \subset N^{i-1}$ and hence the index $\geqq h$. On the other hand the index $\leqq h$, since otherwise (12.4) would hold for some $i \geqq h$. Therefore *the index h of Γ_{p+1} is the highest value of $i+1$ for which* (12.4) *holds.*

13. We may consider τ_i as a deformation of Φ^{i+1} into Φ^i over K. The cell joining E_p of Φ^{i+1} with $\tau_i E_p$, suitably oriented, is the deformation-chain of E_p (*Topology*, p. 78), and the deformation-chain of any subchain C_p^{i+1} of Φ^{i+1} is then obtained as *loc. cit.* by the condition that it is a linear chain-function. If we designate this function by $\mathfrak{D}$ we have

$$\mathfrak{D} C_p^{i+1} \to C_p^i - C_p^{i+1} - \mathfrak{D} F(C_p^{i+1}), \tag{13.1}$$

$$C_p^i = \tau_i C_p^{i+1}. \tag{13.2}$$

If k^{i+1} is any subcomplex of Φ^{i+1} the sum of the closed deformation-cells of its cells (deformation-chains of the cells) under τ_i is a complex $\mathfrak{D} k^{i+1}$, the *deformation-complex* of k^{i+1}. If we have an infinite sequence of complexes $\{k^{i+1}\}$, where k^{i+1} is a subcomplex of Φ^{i+1} and $k^i = \tau_i k^{i+1}$ for every i, the sum $k = \Sigma \mathfrak{D} k^{i+1}$ is a *projection-complex.*

Let now $\{C_p^{i+1}\}$ be an infinite sequence of chains where C_p^{i+1} is a subchain of Φ^{i+1} and $C_p^i = \tau_i C_p^{i+1}$ for every i. We have then an associated chain

$$C_{p+1} = \Sigma \mathfrak{D} C_p^i \tag{13.3}$$

defining a chain c_p of $\mathcal{R}$, called a *projection-chain.* If the chains C_p^i for i above a certain value h are cycles Γ_p^i, C_{p+1} defines a γ_p called a *projection-cycle* of $\mathcal{R}$. The chains C_p^j, $j \leqq h$, can be replaced by the projections of Γ_p^{h+1} without modifying γ_p, so that when we have a γ_p we may assume that all the chains C_p^i are cycles.

Let $\{C_p^i\}$ define as above a projection-chain c_p, with C_{p+1} as the associated chain of K. Then $|C_p^i|$ is not necessarily the projection of the complexes $|C_p^{i+j}|$, but their difference is made up of cells of less than p dimensions. It follows that there exists a projection-complex k such, that for each i, $k^i - |C_p^i|$ consists of cells of dimension $< p$, while k^i is the projection of some C_p^{i+j} on Φ^i. The difference $k - C_{p+1}$ will consist of cells of dimension $< p+1$.

Let A, L^*_A be as before and let C_{p+1} be any chain of K with C_p^i as its traces. We may introduce as above the finite chains $\mathfrak{D} C_p^i$ and also the infinite chain

$$C'_{p+1} = C_{p+1} - \Sigma \mathfrak{D} C_p^i.$$

Let C_{p+1} define c_p of $\mathcal{R}$. Whenever $C'_{p+1} \subset L^*_A$ we shall call C_{p+1} a *projection-chain mod* L^*_A, and c_p a *projection-chain mod* A, a *projection-cycle mod* A when $F(c_p) \subset A$. When $A = 0$ we have $L^*_A = 0$, $C'_{p+1} = 0$ and c_p becomes an ordinary projection-chain.

14. *Certain properties of chain-moduli.* By a *modulus* of p-chains of a complex K we understand a system of rational chains of K forming an abelian group with respect to addition. If $\mathfrak{M}, \mathfrak{N}$ are two such moduli, and if $\mathfrak{N} \subset \mathfrak{M}$ then, as usual, $C_p \equiv 0 \bmod \mathfrak{N}$ or $C_p \equiv C'_p \bmod \mathfrak{N}$, mean that $C_p \subset \mathfrak{N}$ or $C_p - C'_p \subset \mathfrak{N}$. If $\mathfrak{M}'$ is a submodulus of $\mathfrak{M}$, by $\mathfrak{M} \equiv \mathfrak{M}' \bmod \mathfrak{N}$, we shall mean that every element of $\mathfrak{M}$ is congruent to an element of $\mathfrak{M}'$ mod $\mathfrak{N}$ and conversely. If we have a modulus $\mathfrak{M}$ on Φ^j the projections of its chains on Φ^i, $i < j$, constitute a modulus $\mathfrak{M}'$ called the *projection* of $\mathfrak{M}$ on Φ^i. Regarding these moduli of the Φ's and their projections we shall prove the following important

THEOREM I. *Let there be given for every h two moduli of p-chains of Φ^h,*

$\mathfrak{M}^h$ and $\mathfrak{N}^h \subset \mathfrak{M}^h$, such that the projection of any $\mathfrak{M}^j$, $j \geqq h$, on Φ^h, is congruent to $\mathfrak{M}^h$ mod $\mathfrak{N}^h$. Then corresponding to every C_p of $\mathfrak{M}^h$, there is a projection-sequence $\{C_p{}^i\}$, $C_p{}^i \subset \mathfrak{M}^i$, such that $C_p \equiv C_p{}^h$ mod $\mathfrak{N}^h$.

If $C_p \subset \mathfrak{N}^h$ we may take a vacuous sequence as the corresponding $\{C_p{}^i\}$. Therefore we may assume that $C_p \not\subset \mathfrak{N}^h$. Under the circumstances C_p is a proper p-chain and so is any chain $C'_p \equiv C_p$ mod $\mathfrak{N}^h$.

Consider then all the chains of $\mathfrak{M}^h$, $D_p \equiv C_p$ mod $\mathfrak{N}^h$, which are projections of chains of some $\mathfrak{M}^j$, $j \geqq h$. The number of projections being infinite and the number of subcomplexes $| D_p |$ of Φ^h finite, at least one of these subcomplexes must carry an infinity of chains D_p. Let $\mathcal{K}$ be such a $| D_p |$ with the least number possible, s, of p-cells and let $E_p{}^1, \cdots, E_p{}^s$ be its p-cells, so that

$$D_p = \Sigma\, t_a E_p{}^a.$$

The chain D_p is the projection of an element say of $\mathfrak{M}^j$. Suppose that there is another similar chain

$$D'_p = \Sigma\, t'_a E_p{}^a$$

which is the projection of an element of $\mathfrak{M}^k$, $k \geqq j$. Then D'_p is likewise the projection of an element of $\mathfrak{M}^j$ and

$$D_p'' = \Sigma (t_a - t'_a) E_p{}^a = \Sigma\, t_a'' \cdot E_p{}^a$$

is the projection of an element of $\mathfrak{M}^j$ which is in $\mathfrak{N}^h$. Therefore, if, no matter how high we take j, there are in $\mathfrak{M}^j$ two elements whose projections D_p, D'_p are different, and both $\equiv C_p$ mod $\mathfrak{N}^h$, there exists always in $\mathfrak{M}^j$ an element whose projection D_p'' is a chain of $\mathcal{K}$ and in $\mathfrak{N}^h$.

Conceivably some, but not all the t'''s vanish for j high enough. There will be one, however, say $t_1'' \neq 0$ for an infinity of j's, hence for every $j > h$, and $D_p - t_1 D_p''/t_1''$ will be a subchain of $\mathcal{K} - E'$ which is $\equiv C_p$ mod $\mathfrak{N}^h$. We have thus a complex whose number of p-cells $< s$, and which carries an infinity of chains such as D_p. As this contradicts the assumption regarding s, it follows that for j above a certain value $D_p'' \equiv 0$, $D_p \equiv D'_p$. Therefore there is a unique chain of $\mathcal{K}$ which is the projection of chains of $\mathfrak{M}^j$, j above a certain value, and $\equiv C_p$ mod $\mathfrak{N}^h$.

Let us now write $D_p{}^h$ for D_p and consider the chains of $\mathfrak{M}^{h+1}$ whose projection on Φ^h is $D_p{}^h$. Their number being clearly infinite, we may again choose one, $D_p{}^{h+1}$, consider the least number possible of its p-cells and show that if it is not unique $D_p{}^h$ can be replaced by a similar chain with a smaller s, etc. We thus obtain a sequence $\{D_p{}^i\}$ $i = h, h+1, \cdots$. The projection sequence

$\{C_p^i\}$ such that $C_p^i = D_p^i$ for $i \geqq h$; $C_p^{h-i} =$ the projection of D_p^h on Φ^{h-i}, has all the properties required by Theorem I.

15. *Remarks.* I. In the proof the fact that the chains are taken with rational coefficients enters in an essential manner when we multiply chains by the number s_a/r_a. *Clearly any ring of coefficients forming a field (i. e. with unique division) would be admissible,* for instance the ring of integers mod p, p a prime. But we could not have integers mod m, m not a prime.

II. Let us call a projection-sequence $\{C_p^i\}$, $C_p^i \subset \mathfrak{M}^i$, *irreducible* whenever for any other similar $\{C'^i_p\}$, such that C'^i_p is a subchain of $| C_p^i |$, necessarily $C'^i_p = tC_p^i$. In that case of course t is independent of i. If we examine our construction we see that the sequence $\{C_p^i\}$ of our theorem has been chosen irreducible. For the irreducibility condition is imposed when $i \geqq h$, and follows, by projection, when $i < h$.

16. If we consider again the elements of $\mathfrak{M}^h$ where h is now fixed, I say that *we can construct for $\mathfrak{M}^h$ a finite base mod $\mathfrak{N}^h$, C_p^{ha}, $\alpha = 1, 2, \cdots, r$, whose elements are members of irreducible projection-sequences $\{C_p^{ia}\}$.*

Let E_p^β denote this time all the p-cells of Φ^h. By the procedure of *Topology*, p. 302 (method of the "first-cell,") and with a suitable numbering of the cells, Theorem II authorizes us to assume that, except for irreducibility, we already have the required base such that in addition

$$(16.1) \qquad C_p^{h\alpha} = E_p^\alpha + \Sigma\, t_{\alpha\gamma} E_p^{r+\gamma}.$$

Consider now the subcomplex Ψ^i of Φ^i consisting of all the cells of Φ^i projected onto $| C_p^{h\alpha} |$ and apply the theorem to $\{\Psi^i\}$ taking as modulus $\mathfrak{M}^{*i}$ the aggregate of the elements of $\mathfrak{M}^i$ that are subchains of Ψ^i. Since the elements of $\mathfrak{M}^{*h}$ all are, mod $\mathfrak{N}^h$, linear combinations of the chains $C_p^{h\alpha}$, and contain only E_p^α among the first r p-cells, they are all, mod $\mathfrak{N}^h$, multiples of $C_p^{h\alpha}$, and those in $\mathfrak{M}^h - \mathfrak{N}^h$ must contain E_p^α. Now by Th. I taken together with No. 15 Remark II, we can find precisely an irreducible $\{C_p^{i\alpha}\}$ such that $C_p^{h\alpha}$ is of the form (16.1), and in particular congruent mod $\mathfrak{N}^h$ to the chain $C_p^{h\alpha}$ in (16.1). Therefore $\{C_p^{i\alpha}\}$, $\alpha = 1, 2, \cdots, r$, behaves as required.

17. Consider the $(p+1)$-subchains C_{p+1} of K, such that $C_{p+1} \cdot \Phi^h \subset \mathfrak{M}^h$ for every h. Thus if $\{C_p^i\}$ are the projection-sequences that we have just considered the closures of the joining cells of the chains C_p^i form a chain such as C_{p+1}. The finite or infinite linear combinations of these chains which are chains form a modulus $\mathfrak{M}$ and the similar chains corresponding to the moduli $\mathfrak{N}^h$ form a submodulus $\mathfrak{N}$ of $\mathfrak{M}$. We shall say that any particular projection-

chain C_{p+1} of $\mathfrak{M}$ is *irreducible* if it contains no similar subchain (member of $\mathfrak{M}$) which is not a multiple of C_{p+1}. The sequences of No. 16 and the corresponding irreducible chains shall be designated by $\{C_p^{ha i}\}$, C_{p+1}^{ha}, so that $C_p^{ha i} = C_{p+1}^{ha} \cdot \Phi^i$.

THEOREM II. *$\mathfrak{M}$ possesses a base mod $\mathfrak{N}$, which is in general infinite, and whose elements are irreducible projection-chains.*

Given any particular sequence $\{C_p^i\}$, $C_p^i \subset \mathfrak{M}^i$, there exists an h, its *index* such that $C_p^i \subset \mathfrak{N}^i$ for every $i < h$ but not for $i = h$. Given $C_{p+1} \subset \mathfrak{M}$, we shall call *index* of C_{p+1} the index of the sequence $\{C_{p+1} \cdot \Phi^i\}$. The $(p+1)$-chains whose index $\geqq h$ form a submodulus $\mathfrak{M}_h$ of $\mathfrak{M}$, and we have $\mathfrak{M}_{h+1} \subset \mathfrak{M}_h \subset \mathfrak{M}$. If $C_{p+1} \subset \mathfrak{M}$ we have

$$C_{p+1} \cdot \Phi^h = \sum t_{ha} C_p^{ha i} \mod \mathfrak{N}^h,$$
$$C'_{p+1} = C_{p+1} - \sum t_{ha} C^{ha}{}_{p+1} \subset \mathfrak{M}_{h+1}.$$

We can treat similarly C'_{p+1}, etc. Ultimately we thus obtain a chain

$$D_{p+1} = \sum_{i \geqq h} t_{ia} C^{ia}{}_{p+1}$$

such that the index of $C_{p+1} - D_{p+1}$ exceeds any positive number. Therefore $C_{p+1} - D_{p+1} \subset \mathfrak{N}$. It is also clear that no element of $\{\{C^{ha}{}_{p+1}\}\}$ can be expressed in terms of those of same or higher index. Therefore $\{\{C^{ha}{}_{p+1}\}\}$ is a base whose elements are irreducible projection-chains.

18. THEOREM III. *There exists a base for the p-cycles mod A, A closed, whose elements are irreducible projection-cycles mod A.*

Let $L^*{}_A$ $(= L^*)$ be the generalized fundamental complex of A. Take for $\mathfrak{M}^h$ the set of all p-cycles of Φ^h mod $\Phi^h \cdot L^*$ such that if Δ_p is any one of them, every Φ^i, $i \geqq h$, contains a chain Δ'_p whose projection on $\Phi^h \approx \Delta_p$ mod $\Phi^h \cdot L^*$ on Φ^h. If Γ_{p+1} is any cycle of K mod $(L^* + \Phi)$ then $\Gamma_{p+1} \cdot \Phi^h \subset \mathfrak{M}^h$ for every h. For take $i > h$, and let $\Gamma_{p+1} \cdot \Phi^i$ be projected onto $\Delta^*{}_p$ of Φ^h. We have

$$\Gamma_{p+1}(\bar{N}^h - N^i) \to \Gamma_{p+1} \cdot \Phi^h - \Gamma_{p+1} \cdot \Phi^i \mod L^*.$$

Moreover if D_{p+1} is the deformation-chain corresponding to the projection

$$D_{p+1} \to \Delta^*{}_p - \Gamma_{p+1} \cdot \Phi^i \mod L^*.$$

Therefore

$$\bar{N}^h - N^i \supset C_{p+1} \to \Gamma_{p+1} \cdot \Phi^h - \Delta^*{}_p \mod L^*.$$

The chain C_{p+1} is finite and by sliding it along the projecting lines we can reduce it to a chain on Φ^h without modifying its boundary mod L^*. Therefore

$$\Gamma_{p+1}\cdot\Phi^h - \Delta^*_p \approx 0 \bmod \Phi^h\cdot L^* \text{ on } \Phi^h;\ \Gamma_{p+1}\cdot\Phi^h \subset \mathfrak{M}^h.$$

The modulus $\mathfrak{M}^h$ also contains all the bounding chains mod $\Phi^h\cdot L^*$ of Φ^h. For E_{p+1} of Φ^h is the projection of some E'_{p+1} of Φ^i; hence $F(E_{p+1}) \subset \mathfrak{M}^h$, and likewise $F(C_{p+1})$ mod $\Phi^h\cdot L^*$ is in $\mathfrak{M}^h$. These bounding cycles form a submodulus $\mathfrak{N}^h$ of $\mathfrak{M}^h$ and the moduli $\mathfrak{M}^h, \mathfrak{N}^h$ are related as in Theorem I. By what we have shown the corresponding moduli $\mathfrak{M}, \mathfrak{N}$ of $(p+1)$-subchains of K are respectively those of the cycles of K mod $(L^* + \Phi)$ and of the bounding cycles of K mod $(L^* + \Phi)$. The required theorem is then a direct consequence of Theorem II.

THEOREM IV. *Any chain c_p is homologous on itself to a projection-cycle mod its own boundary.*

Let $|c_p| = B$, $|F(c_p)| = A$. By Theorem III there is a base $c_p^1, c_p^2, \cdots$, for the p-cycles of B mod A whose elements are irreducible projection-chains which are projection-cycles mod A. Moreover, referring to No. 17, the index $h(\alpha)$ of c_p^α increases indefinitely with α. We have then

$$c_p \approx c'_p = \Sigma\, t_\alpha \cdot c_p^\alpha \bmod A \text{ on } B. \tag{18.1}$$

Let C^α_{p+1} be the projection-chain of K which represents c_p. Among the chains C^α_{p+1} only a finite number have a trace $\neq 0$ on any Φ^i, and each of these chains satisfies the condition for projection-chains. Hence any linear combination of them, and in particular the representative of c'_p, satisfies the same condition. Therefore c'_p is a projection-chain, and (18.1) proves our theorem.

THEOREM V. *If a projection-cycle mod A is ≈ 0 mod A it bounds a projection-chain mod A.*

Let Γ_{p+1} be a projection-cycle mod $(L^* + \Phi)$ representing the projection-cycle mod A, γ_p. Take for $\mathfrak{M}^h$ the modulus of all $(p+1)$-chains on Φ^h whose boundary mod $L^*\cdot\Phi^h$ is a multiple of $\Gamma_{p+1}\cdot\Phi^h$ and for $\mathfrak{N}^h$ the $(p+1)$-cycles of Φ^h mod $L^*\cdot\Phi^h$. Let C_{p+1} be the projection of any element of $\mathfrak{M}^i - \mathfrak{N}^i$ on Φ^h and let $C'_{p+1} \subset \mathfrak{M}^h - \mathfrak{N}^h$. Then

$$tC_{p+1} - C'_{p+1} \to 0 \bmod L^*\cdot\Phi^h,\ t \neq 0.$$

Hence the $(p+1)$-chain at the left is in $\mathfrak{N}^h$. Hence $\mathfrak{M}^h, \mathfrak{N}^h$ are related in the proper way, and by Theorem I there is a projection-chain C_{p+2} such that

$$t_h(C_{p+2}\cdot\Phi^h) \to \Gamma_{p+1}\cdot\Phi^h \bmod L^*\cdot\Phi^h.$$

Since $\{F(C_{p+2}\cdot\Phi^h)\}$ and $\{\Gamma_{p+1}\cdot\Phi^h\}$ are both projection-sequences (up to a chain of $L^* \cdot \Phi^h$) t_h has a value t independent of h, and $tC_{p+2} \to \Gamma_{p+1} \bmod (L^* + \Phi)$. Therefore tC_{p+2} represents a projection c_{p+1} of $\mathfrak{R}$ which $\to \gamma_p \bmod A$.

19. *Connectedness and circuits.* Let again A, L^*_A be a closed set and its generalized fundamental complex, and let x, y be two points of $\mathfrak{R} - A$. If $\{\sigma^i\}$ is a projection-sequence for x, any vertex A^i of σ^i is the projection on Φ^i of a vertex A^{i+1} of σ^{i+1}. Therefore x has a projection-sequence $\{A^i\}$ consisting of vertices of the Φ's, and there is a similar sequence $\{B^i\}$ for y.

Now the N S C in order that $\mathfrak{R}$ be not disconnected by A is that for every i above a certain value there exist a sequence $U^{i\beta_1}, \cdots, U^{i\beta_r}$ of the sets of the covering Σ^i, in which any two consecutive sets intersect and $U^{i\beta_1} = A^i$, $U^{i\beta_r} = B^i$. This condition is equivalent to $A^i \approx B^i$ on Φ^i for i sufficiently high, and hence for every i. Since $\{A^i\}$, $\{B^i\}$ are projection-sequences they are the traces of projection-cycles Γ_1, Γ'_1 homologous on $K - L^*_A \bmod \Phi$ and representing respectively x, y. By means of Theorem IV and V we find immediately that the above condition is equivalent to $x \approx y$ on $\mathfrak{R} - A$. *Therefore the N S C in order that $\mathfrak{R} - A$ be connected is that $R_0(\mathfrak{R} - A) = 1$.*

Another N S C f,or the connectedness of $\mathfrak{R} - A$ is that *the open complexes $(K - L^*_A) \cdot \Phi^i = \Psi^i$* be all connected. For if they are connected we always have $A^i \approx B^i$ on Ψ^i whatever x, y and hence $x \approx y$ on $\mathfrak{R} - A$ so that $\mathfrak{R} - A$ is connected. Conversely if $\mathfrak{R} - A$ is connected we always have $A^i \approx B^i$ on Ψ^i. But any two vertices A^i, B^i of Ψ^i belong to two projection-sequences $\{A^i\}$, $\{B^i\}$; hence any two vertices of Ψ^i are homologous on Ψ^i; therefore $R_0(\Psi^i) = 1$, and Ψ^i is connected.

20. THEOREM VI. *An n-circuit $\mathfrak{R} - A$ is connected and n-dimensional at all points.*

If $\mathfrak{R} - A$ is disconnected every Ψ^i, for i above a certain value, is the sum of two open complexes without common cells. As a consequence, by suppressing a suitable finite part of K, we shall have a new K such that $K - L^*_A = K' + K''$, where K' and K'' are open complexes without common cells. Let γ_n be the fundamental n-cycle mod A of the circuit and let Γ_{n+1} be its representative cycle mod $(L^*_A + \Phi)$. The chains

$$\Gamma'_{n+1} = K' \cdot \Gamma_{n+1}, \qquad \Gamma''_{n+1} = K'' \cdot \Gamma_{n+1}$$

are similar cycles which represent cycles mod A, γ'_n and γ''_n, such that

$$\gamma_n = \gamma'_n + \gamma''_n, \qquad |\gamma'_n| \cdot |\gamma''_n| \subset A.$$

Any point x of γ'_n has a neighborhood $U \subset \mathfrak{R} - A - |\gamma''_n|$. Hence $B = \mathfrak{R} - U$ is a closed set $\supset A$ and also γ''_n, so that $R_n(B; A) \neq 0$, which contradicts one of the circuit conditions. Therefore $\mathfrak{R} - A$ is connected.

Regarding the dimension of $\mathfrak{R} - A$, let $x \subset \mathfrak{R} - A$, $\dim_x \mathfrak{R} = p < n$. We can find an open set $U \supset x$ such that $U \subset \mathfrak{R} - A$, $\dim F(U) \leqq p - 1$. It follows that if we suppress a suitable finite part of K, the new $K - L^*_A$ shall be disconnected into two subcomplexes K', K'' by a projection-complex L which is a fundamental complex for $F(U)$ and whose dimension is therefore $\leqq p$ (No. 7). Since $p < n$ neither K' nor K'' will have $(n+1)$-cells with n-faces on L. Hence $K' \cdot \Gamma_{n+1}$ and $K'' \cdot \Gamma_{n+1}$ are separately cycles mod $(L^*_A + \Phi)$ determining cycles mod A, γ'_n and γ''_n whose intersection $\subset A$, and we have the same contradiction as before.

21. *Extension to locally compact spaces.* Practically all our results may be extended to a locally compact separable space $\mathfrak{R}$. It is known that such a space is metric and that it can be mapped topologically on a compact space $\mathfrak{R}^*$ with a point x^* removed. That is to say $\mathfrak{R}$ can be identified with $\mathfrak{R}^* - x^*$. Moreover, topologically speaking $\mathfrak{R}^*$ is unique.† If U^* is a neighborhood of x^* on $\mathfrak{R}^*$, $U = U^* - x^*$ is an open set of $\mathfrak{R}$ and $F(U^*) = F(U)$. There exists then another such set $V^* \subset U^*$, and if $V = V^* - x^*$, we have also on $\mathfrak{R}$: $V \subset U$. Since $\mathfrak{R}$ is n-dimensional we can find an open set W of $\mathfrak{R}$ such that $V \subset W \subset U$, $\dim F(W) < n$. Therefore if $W^* = W + x^*$, we have $\dim F(W^*) < n$, $x^* \subset W^* \subset U^*$. In other words given any neighborhood U^* of x^* there is another $W^* \subset U^*$ whose boundary is of dimension $< n$. This shows that $\dim_{x^*} \mathfrak{R}^* \leqq n$. Any point $x \neq x^*$, has relatively to $\mathfrak{R}^*$ a neighborhood which is also a neighborhood relatively to $\mathfrak{R}$ and hence $\dim_x \mathfrak{R}^* = \dim_x \mathfrak{R}$, $\dim \mathfrak{R}^* = \dim \mathfrak{R} = n$.

Let K^* be a fundamental complex of $\mathfrak{R}^*$, and let $\{\sigma^i\}$ be a fundamental sequence for x^* and L^* the sum of the sequence and its joining cells. Then $K = K^* - L^*$ is an open complex which we may consider as a fundamental complex for $\mathfrak{R}$. We now have two types of cycles to consider for $\mathfrak{R}$: (a) the finite p-cycles; they correspond to the $(p+1)$-cycles of $K = K^* - L^*$ mod Φ; (b) the infinite p-cycles of $\mathfrak{R}$; they are represented by the $(p+1)$-cycles of K^* mod $(L^* + \Phi)$. Both types have essentially the same properties as the cycles previously considered. It is also a simple matter to show that they are topological elements of $\mathfrak{R}$ itself, independent of the mode of turning $\mathfrak{R}$ into a compact space $\mathfrak{R}^*$. Let us state in passing that $\mathfrak{R}$ will be called an *open n-circuit* whenever it behaves in the same manner regarding the infinite n-cycles as previously regarding the finite (ordinary) n-cycles.

It is to be observed that the space $\mathfrak{R}$, or rather the set $\mathfrak{R}$, may actually

† Urysohn and Alexandroff, *Mémoire sur les espaces topologiques compacts*, Amsterdam Academy, Verhandeligen, Deel XIV, No. 1, 1929.

be given in the form $\mathcal{R}^* - A$, where $\mathcal{R}$ is compact, metric, but not necessarily n-dimensional, and A is a closed subset of $\mathcal{R}^*$ that may consist of more than one point. L^* is then merely a fundamental complex for A, but otherwise the rest is as before. We can pass to the case where A is a single point by applying to $\mathcal{R}^*$ a continuous single-valued transformation, homeomorphic over $\mathcal{R}^* - A$, and reducing A to a single point. Concurrently we replace the subcomplex $L^* \cdot \Phi^i$ by a single point and L^* by a single projection line. It is clear that this does not affect the cycles which we have introduced above nor their homologies.

PART II. The Generalized Manifold.

22. *Definition.* A generalized n-manifold M_n or M is a locally compact separable n-space with the following properties whose topological character is obvious:

I. M is the sum of a countable aggregate of disjoined n-circuits.

II. The Betti-number $R_n(M, M - x) = 1$ for every $x \epsilon M$.

III. M is locally connected.

IV. Given any closed q-set F on M_n and any open set U there is an open set $V \subset U$, such that every chain c_p, $p < n - q$, on V, whose boundary does not meet F, is deformable over U, without moving its boundary, into a chain c'_p which does not meet F.

If the circuits in I are absolute we call M an *absolute* manifold, otherwise an *open* manifold. If the circuits are all *orientable* M is orientable, otherwise it is *non-orientable.*

Interpretation of the manifold conditions. Condition I requires no comment. Regarding II we may consider, with van Kampen, as p-cycle of a point x a c_p whose boundary $\not\ni x$, i. e. a cycle mod $M - x$ in the sense of *Topology*, the homologies being of the type $\approx M - x$: $c_p \approx 0 \bmod M - x$ means that there is a c_{p+1} such that $F(c_{p+1}) - c_p \not\ni x$. The corresponding Betti-number is the number designated by $R_p(M, M - x)$. Since the fundamental γ_n of M is itself a cycle mod $M - x$ whatever $x \epsilon M$, II signifies that if c_n is any chain whose boundary $\not\ni x$, a certain $c_n - t\gamma_n \not\ni x$.

The local connectedness in III is the so-called local zero-connectedness of *Topology*, p. 90. It means explicitly that for every U there is a $V \subset U$ such that any two points of V are on a connected subset of U.

When we have an absolute M conditions III and IV are equivalent to:

III'. There exists, for every $\epsilon > 0$, a number $\delta(\epsilon) < \epsilon$, such that any two points not farther apart than δ, are on a connected set of diameter $< \epsilon$.

IV′. For every F and ϵ there exists a number $\eta(\epsilon) < \epsilon$, such that any c_p of diameter $< \eta$ whose boundary does not meet F, where $\dim F = q < n - p$, is ϵ-deformable without displacing its boundary, into a chain which does not meet F.

For $0 < p < n$, $F = 0$, this becomes a weak type of local q-connectedness with the p-cell and $(p-1)$-sphere of *Topology* replaced by a p-chain and $(p-1)$-cycle.

Conditions II, III, IV are purely local and serve to characterize the homogeneity properties of M. Condition I on the contrary refers to the whole manifold and serves also to separate the different types.

23. *The Kronecker-index. Definition.* Taking, merely for convenience, an absolute M_n, let c_p, c_{n-p} be two chains on M_n which do not intersect one another's boundaries:

$$|c_p| \cdot |F(c_{n-p})| + |F(c_p)| \cdot |c_{n-p}| = 0. \tag{23.1}$$

Their Kronecker-index is to be a number $(c_p \cdot c_{n-p})$ such that:

(a) $(c_p \cdot c_{n-p}) = 0$ when the chains do not meet.

(b) The index when defined is a bilinear function of the two chains.

(c) If the boundaries of c_p, c_{n-p+1} do not intersect then

$$(c_p \cdot F(c_{n-p+1})) = (-1)^p (F(c_p) \cdot c_{n-p+1}). \tag{23.2}$$

(d) If $x \epsilon M$ and γ_n is the fundamental n-cycle of M then

$$(x \cdot \gamma_n) = 1. \tag{23.3}$$

We shall show that there exists a unique index which is a topological invariant of the two chains and which satisfies conditions (a), $\cdots$, (d), and has the following additional properties:

(e) If c_{p+1} and $F(c_{n-p})$ do not meet

$$(F(c_{p+1}) \cdot c_{n-p}) = 0, \tag{23.4}$$

and similarly with p and $n-p$ interchanged.

(f) The chains being as in the definition:

$$(c_p \cdot c_{n-p}) = (-1)^{p(n+1)} \cdot (c_{n-p} \cdot c_p). \tag{23.5}$$

The existence proof as well as properties (e), (f), will be established by induction.

In the theory of the index for combinatorial manifolds (*Topology*, Ch. IV), (a), (b), (d) enter more or less in the definition, while (c) is proved explicitly. It is in fact essentially formula (20) *loc. cit.*, which plays an all important part there and is a direct consequence of the fundamental boundary relation (18) for intersections of chains. On the contrary here the same relation serves directly to define recurrently the index without passing through intersections of dimension > 0. This is substantially in accord with the definitions suggested *loc. cit.*, p. 216. See also H. A. Newman's recent paper *Cambridge Philosophical Transactions*, Vol. 27 (1931), pp. 491-501.

24. The Kronecker-index $(c_0 \cdot \gamma_n)$, where c_0 is a projection-chain and γ_n the fundamental n-cycle, is readily treated. In the first place if C_0 is a finite subchain of a complex,

$$C_0 = \Sigma\, t_i E_0^i, \tag{24.1}$$

we define its Kronecker-index as in *Topology*, p. 169, by

$$(C_0) = \Sigma\, t_i, \tag{24.2}$$

and we recall that $C_0 \approx 0$ implies that $(C_0) = 0$. If $\mathcal{K}$ is an orientable and oriented combinatorial manifold we have $(C_0) = (C_0 \cdot \mathcal{K})$.

Let now c_0 be a projection-chain of M (projection-zero-cycle), with Γ_1 as its representative projection-cycle mod Φ on K. Since $\tau_i \Gamma_1 \cdot \Phi^{i+1} = \Gamma_1 \cdot \Phi^i$, we have $(\Gamma_1 \cdot \Phi^{i+1}) = (\Gamma_1 \cdot \Phi^i)$. This index is therefore independent of i and its value is by definition (c_0).

We shall show below, that c_0 can be ϵ-deformed whatever ϵ, into a chain consisting of a finite number of points $x_0, \cdots, x_r$, so that

$$c_0 \approx \Sigma\, s_j x_j. \tag{24.3}$$

Since M is connected, if x is any point of M we have $x \approx x_j$, and hence

$$c_0 \approx x\, \Sigma\, s_0 = sx. \tag{24.4}$$

As we have seen (No. 19) x has a projection-sequence $\{A^i\}$ made up of vertices of the Φ's. Owing to (23.4) we have

$$\Gamma_1 \cdot \Phi^i \approx sA^i \text{ on } \Phi^i; \tag{24.5}$$

hence $(\Gamma_1 \cdot \Phi^i) = s(A^i) = s = (c_0)$. Since s is clearly a topological function of c_0 and does not depend in any sense on the fundamental complex K the same holds for (c_0).

If we now define $(c_0 \cdot \gamma_n)$ by the relation

$$(c_0 \cdot \gamma_n) = s(x \cdot \gamma_n), \tag{24.6}$$

we have by the above and (23.3)

(24.7) $$(c_0 \cdot \gamma_n) = (c_0).$$

This disposes of $(c_0 \cdot \gamma_n)$ and shows in particular that it is a topological invariant of c_0.

25. THEOREM VII. (*Deformation theorem*). *Given a projection-chain c_p, a closed q-set F and any ϵ, there exists an ϵ-deformation of c_p into a chain c'_p with an elementary ϵ-decomposition into projection-chains $\{c'_r{}^i\}$, whose zero-chains are isolated points and whose r-chains, $r < n - q$, do not meet F.*

Let c_p be represented by C_{p+1} of K and let E_q^{ha} be the cells of Φ^h. We decompose C_{p+1} in a sum of r chains:

(25.1) $$C_{p+1} = C^1{}_{p+1} + \cdots + C^r{}_{p+1},$$

where r is the number of cells of Φ^h, and where $C^a{}_{p+1}$ is a subchain of the complex consisting of all the cells of K on the set of all projecting lines that meet E^{ha}. A similar decomposition may then be applied to the chains of $F(C^a{}_{p+1})$, etc., until finally we have an elementary decomposition $\{C^a{}_{q+1}\}$ of C_{p+1} characterized by the property that $C^a{}_{q+1}$ is on the set of cells of K that are on the projecting lines through the points of E^{ha}. The chains $C^a{}_{q+1}$ like C_p itself, are projection-chains. There results an elementary decomposition $\{c_q{}^a\}$ of c_p into projection-chains associated with each Φ^h.

Let us now observe that the points of M in whose projection-sequences $\{\sigma^i\}$ the term σ^k is E^{ka} or a face of it, are the points of sets $U^{k\beta}$ with a common point. Their sum is an open set V^{ka} whose diameter $< 2\epsilon_k$, where $\epsilon_k =$ mesh Σ^k. Let η_k be the characteristic-constant of the f. c. o. s. $\{V^{ka}\}$ and take h so high that $\epsilon_h < \frac{1}{2}\delta(\eta_k)$, where δ is the same as in No. 22, III′. As a consequence any two points x, y on a set $V^{h\gamma}$ will be on a connected subset of some $\bar{V}^{ka} \supset V^{h\gamma}$. By No. 19, $x \approx y$ on $\bar{V}^{ka}$. It follows that if $\{\sigma^i\}$, $\{\sigma'^i\}$ are representative sequences for x, y then for $i > h$, σ^i and σ'^i can be joined on Φ^i by a polygonal arc λ (sum of vertices and one-cells of Φ^i) whose projections on Φ^k is on E^{ka}. Hence any two vertices of the subcomplex of Φ^i projected onto E^{ha} can be joined in the above manner by a polygonal arc on Φ^i. For both belong to a pair of simplexes such as σ^i, σ'^i.

26. Henceforth h, k are to be kept fixed. Since M is an n-circuit it is n-dimensional at all points (Theorem VI). Since $q < n$ there exists then on every open set a point not on F. Choose such a point x_a on V^{ha}, and let $\{A^{ai}\}$ be a projection-sequence of vertices for the point x_a.

Consider one of the zero-chains $c_0{}^a$, of the decomposition of c_p. It is defined by means of a certain projection-chain $C_1{}^a$ of K so that $\{C_1{}^a \cdot \Phi^i\}$ is a projection-sequence of zero-chains. From the above follows immediately that when i exceeds a certain value we can find a one-chain D_1^{ai} of Φ^i whose projection is on E^{ka} and such that in addition

$$(26.1) \qquad D_1^{ai} \rightarrow (C_1^a \cdot \Phi^i)A^{ai} - C_1^a \cdot \Phi^i = (c_0^a)A^{ai} - C_1^a \cdot \Phi^i.$$

By Theorem V we may choose for D_1^{ai} a projection-sequence which determines a projection-chain D_2^a of K, and finally a projection-chain d_1^a of $\mathfrak{R}$. As a consequence of (26.1) we have then

$$(26.2) \qquad d_1^a \rightarrow (c_0^a)x_a - c_0^a.$$

We have thus displaced the zero-chains of the elementary decomposition of c_p into points $\not\subset F$. The displacement and the diameters of the chains of the decomposition may be made as small as we please. From this point on and taking account of Theorem V and condition IV of No. 22 for an M_n, the proof of the required deformation theorems proceeds as in *Topology*, p. 93. The only modifications are that singular cells and thei rsingular boundary spheres are replaced by the elementary chains c_q^i and their boundaries.

27. As a first application let $x \subset c_p - F(c_p) \neq 0$, where c_p is otherwise arbitrary. The chain is homologous on itself mod its boundary to a projection-chain c'_p so that $x \not\subset F(c'_p)$. By Theorem VII c_p can be ϵ-deformed whatever ϵ into a projection-chain $c_p'' \not\supset x$. This implies that $c_p \approx 0$ on $M - x$, and also that for every $\xi > 0$ there is an $\eta(\xi)$ such that if x is a point of c_p farther than ξ from $F(c_p)$, then $c_p \approx c'_p$, where c'_p is at a distance $\geqq \eta$ from x.

Referring to No. 22 we have by what precedes,

$$(27.1) \qquad R_p(M - x) = \delta_{np},$$

where δ_{np} is the Kronecker delta ($= 1$ for $p = n$, $= 0$ for $p \neq n$).

Let us now observe that if we apply the construction of Nos. 24, 25 to a γ_p whose diameter is sufficiently small, we may choose all the chains c'^i_q coincident with a single point x. As a consequence if $p > 0$, $c'_p = 0$, and by (11.1) the deformation chain

$$(27.2) \qquad \mathfrak{D}\gamma_p \rightarrow -\gamma_p,$$

Therefore for every open set U there is another $V \subset U$ such that every γ_p, $0 < p < n$, on V is ≈ 0 on U.

The preceding statement is valid for any manifold. For an absolute manifold owing to compactness we have: *for every θ there is a $\tau(\theta)$ such that every γ_p, $p < n$, whose diameter $< \tau$ bounds a chain of diameter $< \theta$.* This is merely another formulation of the weak local p-connectedness property mentioned in No. 22.

28. Let γ_p be one of the irreducible projection-cycles of the base constructed in No. 18 and whose index $i > h$. When we apply the deformation of the preceding numbers with $F = 0$, we find that the deformed cycle $\gamma'_p = 0$, for its chains correspond element for element to the chains of a degenerate simplicial p-cycle. Therefore (27.2) will hold here also and hence $\gamma_p \approx 0$ on M. In particular the base alluded to can only contain a finite number of cycles $\not\approx 0$, namely those whose indices do not exceed a certain value. This proves the important

THEOREM VIII. *The Betti-numbers of an absolute M_n are all finite.*

29. *Determination of the Kronecker-index.* We propose to give a recurrent determination of $(c_p \cdot c_{n-p})$ for two chains which do not intersect one another's boundaries. Taking first $p > 0$ we shall reduce the case in question to the same for $p - 1$ and ultimately to a $(c_0 \cdot c_n)$ where c_0 consists of a finite number of isolated points. This last index shall be treated directly by reduction to the case considered in No. 23. At the same time we shall show that the index has all the properties expected. We assume then first that this holds already for $p - 1$, extend it to p, then take up the case $p = 0$ at the end.

Our first move is to replace c_p, c_{n-p} by projection-chains, homologous respectively to c_p, c_{n-p} on $|c_p|$, $|c_{n-p}|$ mod their boundaries (Theorem IV). To simplify matters we continue to denote the new chains by c_p, c_{n-p}. We merely recall that after the reductions the new sets $|c_p|$, $|c_{n-p}|$, $|F(c_p)|$, $|F(c_{n-p})|$ are subsets of the old. As a consequence in what follows, $|c_p|$ for example, may designate indifferently the new or the old set $|c_p|$.

Let now ξ be the least of the two positive numbers $d(c_p, F(c_{n-p}))$, $d(F(c_p), c_{n-p})$. Since every point x of $|c_p|$ is at least as far as ξ from $F(c_{n-p})$, x has a neighborhood V such that $c_{n-p} \approx 0 \bmod M - V$. Since $|c_p|$ is self-compact it can be covered with a finite number of neighborhoods $V^1, \cdots, V^r$, such that $c_{n-p} \approx 0 \bmod M - V^j$. That is to say there exists a projection-chain

$$(29.1) \qquad c^j_{n-p+1} \rightarrow c_{n-p} \bmod M - V^j.$$

The sets V^j form a f. c. o. s. for $|c_p|$ and there is an analogue ζ of the characteristic constant for that covering: every point x of $|c_p|$ will be on some V^j such that $d(x, M - V^j) > \zeta$.

We shall now choose a certain fixed $\epsilon > 0$, and the determination of the index will depend upon that ϵ. We shall endeavor to show that the index remains the same for all ϵ's sufficiently small, and so we shall not hesitate to take this ϵ arbitrarily small. In particular we shall require that $\epsilon > \frac{1}{4}\xi$, $\frac{1}{4}\zeta$ or $\tau(\frac{1}{2}\zeta)$ where τ is the same function as in No. 27.

By Theorem VII we may ϵ-deform c_p into a projection-chain c'_p with an elementary ϵ-decomposition $\{c'_q{}^i\}$ whose elements (all projection-chains also) of dimension $q < p$ do not meet c_{n-p} and with

(29.2) $$c'_p = \Sigma\, c'_p{}^i.$$

The chains $c'_p{}^i$ are in finite number and their boundaries do not meet c_{n-p}. We shall set by definition $(c_p \cdot c_{n-p}) = (c'_p \cdot c_{n-p})$, and hence, if (23b) is to hold,

(29.3) $$(c_p \cdot c_{n-p}) = (c'_p \cdot c_{n-p}) = \Sigma(c'_p{}^i \cdot c_{n-p}),$$

where in the sum we preserve only the useful terms, namely those corresponding to chains $c'_p{}^i$ which meet c_{n-p}. The problem is to determine the indices in that sum.

Since the points of $c'_p{}^i$ are not farther than $\epsilon < \frac{1}{4}\xi$ from c_p, and since its diameter $< \frac{1}{4}\zeta$, $c'_p{}^i$ is on at least one set V^j and farther than $\frac{1}{2}\zeta$ from the corresponding $M - V^j$. Choose one of the sets V^j of this nature and relabel that V and its c_{n-p+1} entering in (29.1), respectively V'^i, c'^i_{n-p+1}. The situation being as described we have $F(c'^i_{n-p+1}) = c_{n-p} + c'_{n-p}$, where c'_{n-p} does not meet $c'_p{}^i$. Hence, if the basic laws (23a c) for the index are to hold, we must have

(29.4) $$(c'_p{}^i \cdot c_{n-p}) = (-1)^p\,(F(c'_p{}^i) \cdot c'_{n-p+1}),$$

and hence by (23 b)

(29.5) $$(c_p \cdot c_{n-p}) = (-1)^p\,\Sigma\,(F(c'_p{}^i) \cdot c'^i_{n-p+1}).$$

The right hand side is known under the hypothesis of the induction, hence (29.5) determines a value for $(c_p \cdot c_{n-p})$. It remains to be shown that the index thus obtained behaves as expected.

30. We first observe that the index is a linear function of c_p. That is to say if the preceding method has yielded $(c_p \cdot c_{n-p})$ and $(c'_p \cdot c_{n-p})$ then it also yields

(20.1) $$(tc_p + t'c'_p \cdot c_{n-p}) = t(c_p \cdot c_{n-p}) + t'(c'_p \cdot c_{n-p}).$$

We notice also that if c_p and c_{n-p} do not meet and if we take ϵ less than half their distance apart, the value computed for their index is zero, and hence accords with (23 a) independently of the variable elements entering in its determination.

Let us replace V'^i by any other V, say V''^i, behaving in the same manner relatively to $c'_p{}^i$ and let c''^i_{n-p+1} be the corresponding c_{n-p+1}. To show that substituting V''^i for V'^i has not modified the index we must prove that if we substitute c''^i_{n-p+1} for c'^i_{n-p+1} in (29. 5) the index remains the same. Since (23 b) holds for $p-1$ this merely requires that we prove

$$(30.2) \qquad \Sigma\,(F(c'^i_p)\cdot c'^i_{n-p+1}-c''^i_{n-p+1})=0.$$

Let $W=V'^i\cdot V''^i$. This open set $\supset c'^i_p$ and $d(c'^i_p, M-W)>\frac{1}{2}\zeta$. Moreover both chains $c'^i_{n-p+1}, c''^i_{n-p+1}\rightarrow c_{n-p} \bmod M-W$. Hence

$$(30.3) \qquad c'^i_{n-p+1}-c''^i_{n-p+1}\rightarrow 0 \bmod M-W.$$

On the other hand since diam $c'^i_p<\tau(\frac{1}{2}\zeta)$ and since, by hypothesis, (24 e) holds for $p-1$ in place of p, we have

$$(30.4) \qquad (F(c'^i_p)\cdot c'^i_{n-p+1}-c''^i_{n-p+1})=0,$$

from which the required relation (30. 2) follows. If $V'^i=V''^i$ but c'^i_{n-p+1} is replaced by c''^i_{n-p+1} the same reasoning holds. Therefore a modification in the chains c_{n-p+1} likewise leaves the index unaltered.

31. Let us now show that (24 e) holds: if

$$(31.1) \qquad M-F(c_{n-p})\supset c_{p+1}\rightarrow c_p$$

then we have

$$(31.2) \qquad (c_p\cdot c_{n-p})=0.$$

Take a $U\supset c_{p+1}$ and $\Subset M-F(c_{n-p})$, then apply Theorem V with $\bar{U}$ as the basic space. As a consequence we find that we may assume that c_{p+1} is a projection-chain. By Theorem VII c_{p+1} is ϵ-deformable into c'_{p+1} with an ϵ-decomposition $\{c'^i_q\}$ whose chains of dimension $<p$ do not meet c_{n-p}. It is to be observed that the construction of the deformed chains is such that the chains c'^i_r depend solely on those of dimension $<r$. Hence c_p is thus deformed into any c'_p serving to calculate its index in accordance with No. 29. We shall have as the new p- and $(p+1)$-chains

$$(31.3) \qquad c'_{p+1}=\Sigma\, c'^i_{p+1}, \qquad c'_p=\Sigma\, F(c'^i_{p+1}),$$

and therefore

$$(31.4) \qquad (c_p\cdot c_{n-p})=\Sigma\,(F(c'^i_{p+1})\cdot c_{n-p}).$$

If ϵ is small enough the chains c'^j_{p-1} on $F(c'^i_{p+1})$ will meet a single chain c_{n-p+1} that we may call as before c'^i_{n-p+1}. By (29. 5) and No. 30

$$(31.5) \qquad (F(c'^i_{p+1})\cdot c_{n-p})=(-1)^p\,(F(F(c'^i_{p+1}))\cdot c'^i_{n-p+1})\equiv 0,$$

since $F(F)\equiv 0$, and from this follows (31. 2).

As an application suppose that we have obtained, always by means of K, two different ϵ-deformations $\mathfrak{D}', \mathfrak{D}''$ of c_p into c'_p and c''_p serving to calculate the index $(c_p\cdot c_{n-p})$. We have

$$(31.6)\quad \mathfrak{D}'c_p \to c'_p - c_p - \mathfrak{D}'F(c_p), \qquad \mathfrak{D}''c_p \to c''_p - c_p - \mathfrak{D}''F(c_p),$$

$$(31.7)\qquad \mathfrak{D}'c_p - \mathfrak{D}''c_p \to (c'_p - c''_p) - \cdots$$

where, under the limitations upon ϵ, the chain omitted does not meet c_{n-p}. Hence

$$(31.8)\qquad (c'_p - c''_p \cdot c_{n-p}) = 0,$$

whatever the procedure chosen to compute the index. Take as the deformation the process which consists merely in replacing c'_p by the decomposition associated with the deformation $\mathfrak{D}'$, and similarly for c''_p and $\mathfrak{D}''$. As a consequence the index (31.8) becomes merely the difference of the values of the index $(c_p \cdot c_{n-p})$ as computed by means of the two deformations. Therefore these two values are the same. In other words $(c_p \cdot c_{n-p})$ *is independent of the ϵ-deformations used in computing it.*

32. We have already shown that our index possesses properties (23 a) and part of (23 b e). We still have to show that when $p > 0$ properties (23 b c e f) hold.

Since we have established the linearity of the index in c_p, (23 b) will be established if we show that the index is also linear in c_{n-p}. Consider two projection-chains c_{n-p}, c'_{n-p} and let them not meet $F(c_p)$, (c_p a projection-chain), nor let c_p meet their boundaries. Let the ϵ-deformation of c_p into c'_p be so carried out that the q-chains c'^{i}_q, $q < p$, of the decomposition of c'_p meet neither c_{n-p} nor c'_{n-p}. Then it follows at once from the definition of the index by (29.5) that

$$(32.1)\qquad (c_p \cdot tc_{n-p} + t'c'_{n-p}) = (c'_p \cdot tc_{n-p} + t'c'_{n-p})$$
$$= t(c'_p \cdot c_{n-p}) + t'(c'_p \cdot c'_{n-p}) = t(c_p \cdot c_{n-p}) + t'(c_p \cdot c'_{n-p}),$$

which proves the required linearity and hence also that property (b) holds completely.

Consider now property (c): if c_p, c_{n-p+1} have non-intersecting boundaries then (23.2) holds. Here we may take in (29.5) every $c'^{i}_{n-p+1} = c_{n-p+1}$, which yields

$$(32.2)\qquad (c_p \cdot F(c_{n-p+1})) = (-1)^p \Sigma\, (F(c'^{i}_p) \cdot c_{n-p+1}).$$

In the summation in (29.5) only certain chains c'^{i}_p whose sum is c'^{i}_p were preserved, namely those which met c_{n-p}. As we have just shown if c'^{i}_p does not meet c_{n-p} the corresponding contribution of its boundary to the sum in (32.1) is zero; hence the summation may now be extended to all the chains c'^{i}_p. By the linearity of the index for $p-1$, $n-p+1$ we have:

$$(32.3)\qquad \Sigma(F(c'^{i}_p) \cdot c_{n-p+1}) = (F(\Sigma\, c'^{i}_p) \cdot c_{n-p+1})$$
$$= (F(c'_p) \cdot c_{n-p+1}) = (F(c_p) \cdot c_{n-p+1}).$$

This relation together with (32.2) yields (23.2) and proves that property (c) holds.

From (a) and (c) follows that if $F(c_p)$ and c_{n-p+1} do not meet then

$$(c_p \cdot F(c_{n-p+1})) = 0. \tag{32.4}$$

This is the analogue of (23.4) with p and $n-p$ interchanged, and together with the result of No. 31 it embodies the proof of property (e).

We postpone the proof of property (f) till later.

33. We shall now consider the case $p = 0$, i.e. the index $(c_0 \cdot c_n)$, where as before $c_0 \subset M - F(c_n)$. As in No. 24, we have here for an ϵ sufficiently small the analogue of (24.3):

$$c_0 \approx \Sigma s_j x_j \text{ on } M - F(c_n). \tag{33.1}$$

Now for x_j we have by No. 22, condition II,

$$c_n \approx t_j \gamma_n \bmod M - x_j, \tag{33.2}$$

and we shall set

$$(c_0 \cdot c_n) = \Sigma s_j t_j (x_j \cdot \gamma_n) = \Sigma s_j t_j. \tag{33.3}$$

Properties (a), (b) are at once verified for this index and we only have to prove (e), (f). Here also (f) shall be treated later.

The proof of (e) consists of two parts:

(a) if $M - F(c_n) \supset c_1 \to c_0$ then $(c_0 \cdot c_n) = 0$. As in No. 31 we may assume that diam $c_1 < \epsilon$ assigned. Now for c_n and any x there is a neighborhood $V \supset x$ such that $c_n - \lambda\gamma_n \approx 0$ on $M - V$. Since M is compact it may be covered with a finite number of such sets V : $V^1, \cdots, V^r$ with $\lambda = \lambda_i$ on V^i. Let us take $\epsilon < \frac{1}{2}\eta$, where η is the characteristic constant of this f. c. o. s., and let the deformations be $< \frac{1}{2}\eta$. We shall take $c_1 \subset V^h$ and farther than $\frac{1}{2}\eta$ from $M - V^h$. Hence if we calculate the index of $c_0 = F(c_1)$ by our method, the corresponding points x_j are all on V^h, and the associated constants t_j all equal to λ_h. Finally since $c_0 \approx 0$, $(c_0) = 0$. Therefore

$$(c_0 \cdot c_n) = (F(c_1) \cdot c_n) = \lambda_h \Sigma s_j = \lambda_h (c_0) = 0. \tag{33.4}$$

(b) if $c_n \approx 0$ then $(c_0 \cdot c_n) = 0$. This is evident for $c_n \approx 0$ implies that the representative projection-cycle Γ_{n+1} of c_n is $\equiv 0$ and hence $c_n \equiv 0$. Consequently in the homologies $c_n \approx t\gamma_n \bmod M - x$, we always have $t = 0$, so that $(c_0 \cdot c_n) = 0$.

As in No. 31 the first case considered proves here also that $(c_0 \cdot c_n)$ has a value independent of the particular mode of determining it.

34. Let us return to $(c_p \cdot c_{n-p})$. We have obtained its value by an induction on p in which there appear certain intermediary chains c_{p-i}, c_{n-p+i}, so that we have:

$$(c_p \cdot c_{n-p}) = (-1)^p (c_{p-1} \cdot c_{n-p+1}),$$

$$\cdot \quad \cdot \quad \cdot \quad \cdot \quad \cdot \quad \cdot \quad \cdot \quad \cdot \quad \cdot$$

$$(c_{p-i} \cdot c_{n-p+i}) = (-1)^{p-i} (c_{p-i-1} \cdot c_{n-p+i+1}), \tag{34.1}$$

$$\cdot \quad \cdot \quad \cdot \quad \cdot \quad \cdot \quad \cdot \quad \cdot \quad \cdot \quad \cdot$$

$$(c_1 \cdot c_{n-1}) = - (c_0 \cdot c_n);$$

and hence, in the last analysis,

$$(c_p \cdot c_{n-p}) = (-1)^{p(p+1)/2} (c_0 \cdot c_n) = (-1)^{p(p+1)/2} \cdot \lambda \cdot (x \cdot \gamma_n), \tag{34.2}$$

the value of λ being the number given by (33. 3). Observe that if $p > 0$, the various chains of dimension $p-1, \cdots$, zero, introduced in this determination are chains of elementary decompositions in which the zero-chains consist of isolated points taken with finite multiplicities. Therefore in particular c_0 is of this nature. It follows that the numbers s_j, t_j of No. 33, that serve to compute λ are all finite and so is λ. An immediate consequence is the fact that $(c_p \cdot c_{n-p})$ *is independent of the fundamental complex* K, *and hence the Kronecker-index is a topological invariant.* For if we have any index whatever with the properties (a), $\cdots$, (e) of No. 23, it will satisfy the relations (34. 1) and (34. 2). Since λ depends solely on certain homologies but not on K our assertion follows.

Now the above has been obtained as a consequence of an induction on p. By means of (23 c), explicitly proved for our index, we may carry through a similar induction on $n-p$. This leads to a formula analogous to (34. 2)

$$(c_p \cdot c_{n-p}) = (-1)^{[n(n+1)-p(p+1)]/2} \cdot \mu (\gamma_n \cdot x). \tag{34.3}$$

If we apply the process just stated to $(c_{n-p} \cdot c_p)$ we find that the geometric operations carried out for its determination are the same as those used in determining $(c_p \cdot c_{n-p})$ by our initial procedure (induction on p), and that as a consequence the corresponding μ is λ, both being equal to a certain expression $\Sigma\, s_j t_j$ appearing in (33. 3). For each j the number s_j is the multiplicity of a certain point as constituent of c_0 and t_j the coefficient t in a certain homology (33. 2). Therefore

$$(c_{n-p} \cdot c_p) = (-1)^{p(2n-p+1)/2} \lambda \cdot (\gamma_n \cdot x), \tag{34.4}$$

$$(c_{n-p} \cdot c_p) = (-1)^{p(n+1)} (c_p \cdot c_{n-p}) [(\gamma_n \cdot x)/(x \cdot \gamma_n)]. \tag{34.5}$$

In particular for $p = n$, $c_0 = x$, $c_p = \gamma_n$:

(34.6) $$(x \cdot \gamma_n)^2 = (\gamma_n \cdot x)^2,$$
and hence finally
(34.7) $$(\gamma_n \cdot x) = \pm (x \cdot \gamma_n).$$

We shall prove later that the proper sign to be chosen here is +. Assuming this for the present we have from (23 d), $(\gamma_n \cdot x) = 1$ and hence finally

(34.8) $$(c_{n-p} \cdot c_p) = (-1)^{p(n+1)} (c_p \cdot c_{n-p})$$

which is (23 f). Thus except for a certain choice of sign we have finally established that the Kronecker-index has all the properties required.

35. *Duality properties of the absolute M_n.* In order to obtain the extension of Poincaré's duality relation for the Betti-numbers, all that is now needed is a converse of property (c) of No. 23; if a cycle $\gamma_p \not\sim 0$ there is a γ_{n-p} such that $(\gamma_p \cdot \gamma_{n-p}) \neq 0$. A slightly more general result will now be proved.

Consider first the sequence of the images of the skeleta Φ^i which we still call Φ^i, on an S_{2n+1}, whereby one may map topologically a compact metric n-space, here our absolute M_n, on the space S_{2n+1} † and let us modify the construction as follows: We take an S_{2n+2} referred to coördinates $x_1, \cdots, x_{2n+2}$ and assume that our S_{2n+1} is the one given by $x_1 = 0$, so that M is now mapped onto that space. We then project Φ^i onto the space $x_1 = 1/i$, and replace Φ^i by its projection which we henceforth call Φ^i. The joining cells being inserted as before, if their (linear) spaces happen to have intersections of too high dimension, we may slightly displace the vertices of the Φ's in their $(2n+1)$-spaces so as to remove this untoward circumstance. We now have M and K immersed in a certain S_{2n+2}. We may in fact immerse S_{2n+2} in any S_r, $r \geqq 2n+2$ and together with it also both M and K. We shall choose r such that $r-n$ is even.

Let us surround each Φ^i by a closed polyhedral neighborhood $\mathcal{K}^i$ in S_r, take a subdivision $\mathcal{K}'^i$ of $\mathcal{K}^i$ having a subdivision Ψ^i of Φ^i as a subcomplex and such that the $\mathcal{K}'^i$-neighborhood of Ψ^i is normal. By reference to *Topology*, p. 91, it will be seen that both conditions may be fulfilled. Moreover we may assume the $\mathcal{K}$'s taken initially mutually exclusive, so that the closed $\mathcal{K}'$-neighborhoods introduced are all mutually exclusive. To simplify matters we designate henceforth these closed neighborhoods themselves by $\mathcal{K}^i$. Besides being polyhedral these neighborhoods have the following property (*loc. cit.*): if $\mathcal{B}^i = F(\mathcal{K}^i)$ then through every point P of $\mathcal{K}^i - \mathcal{B}^i - \Psi^i$ there passes a unique (open) segment resting on $\mathcal{B}^i$ and Ψ^i and varying continuously with P. We call these segments the *projecting segments* on $\mathcal{K}^i$.

† See *Annals of Mathematics*, vol. 32 (1931), p. 527.

36. Let us assign to each cell E_q^i of Φ^i one of its vertices A^{iq}, and let Ψ'^i be the first derived of Ψ^i. A unique simplicial transformation θ of Ψ'^i into Φ^i is determined by specifying that the vertices of Ψ'^i on E_q^i are all to be transformed by θ into A^{iq}. We shall designate by *projection* of Ψ'^i onto Φ^{i-1}, $\Phi^{i-2}, \cdots$, the simplicial transformation $\tau_{i-1}\theta, \tau_{i-2}\tau_{i-1}\theta, \cdots$. It is to be observed that θ need not be a fixed simplicial transformation, but merely any simplicial transformation of its type.

Let us specify for each i a definite first derived $\mathcal{K}'^i$. It determines a Ψ^i, and also a dual $\mathcal{K}^{*i}$ of $\mathcal{K}^i$. If Γ_{n+1} represents the fundamental cycle γ_n of M, then we have a definite n-cycle $\Gamma_n^i = \Gamma_{n+1} \cdot \Psi^i$ for each i. It is the subdivision induced by Ψ^i on the trace $\Gamma_{n+1} \cdot \Phi^i$ of Γ_{n+1}.

Now referring to *Topology*, Ch. IV, if $C^*{}_q^i$ is any subchain of $\mathcal{K}^{*i}$ there is a uniquely defined intersection-chain $C_s^i = \Gamma_n^i \cdot C^*{}_q^i$, $s = q + n - r$, and we have (*Topology*, p. 169, formula 18):

$$(36.1) \qquad \Gamma_n^i \cdot C^*{}_q^i \to \Gamma_n^i \cdot F(C^*{}_q^i).$$

The intersection and its boundary are both subchains of Ψ'^i and hence they have a unique projection on any Φ^j, $j \leqq i$. Moreover, since a projection is a simplicial transformation, the boundary of the projection is the projection of the boundary.

37. In the argument to follow, in addition to the customary associated chains c_q, C_{q+1}, it will be convenient to make a clear distinction between intersections of chains and traces. We shall therefore designate the former as usual by the "dot" product, and the trace of C_{q+1} of K on Φ^i by $\mathcal{C}_q^i$.

LEMMA. *Let C^α_{p+1}, C^α_{n-p+1} $(\alpha = 1, 2, \cdots, s)$ be projection-chains of K representing chains c_p^α, c^α_{n-p} which do not intersect one another's boundaries so that $(c_p^\alpha \cdot c^\alpha_{n-p})$ is well defined. Suppose that the traces $\mathcal{C}_p^{\alpha i}$ have the following property: whatever h there is a $k_\alpha > h$ such that $\mathcal{C}_p^{\alpha i}$ is the projection of an intersection-chain $\Gamma_n^{k_\alpha} \cdot C^{*\alpha k_\alpha}_{r-n+p}$ where*

$$(37.1) \qquad \sum_\alpha (C^{*\alpha k_\alpha}_{r-n+p} \cdot \mathcal{C}^{\alpha k_\alpha}_{n-p}) = \lambda$$

is independent of h. Then,

$$(37.2) \qquad \sum_\alpha (c_p^\alpha \cdot c^\alpha_{n-p}) = \lambda.$$

Let Λ_p designate the Lemma as stated. We shall reduce Λ_p to Λ_{p-1}, hence to Λ_0, then prove Λ_0.

Dropping α for the present, designate $C^\alpha_{p+1}, \cdots$ by $C_{p+1} \cdots$, and let V^j, c^j_{n-p+1} correspond as in No. 29 to c_{n-p}. We first decompose c_p into elements $\{c_q^a\}$ which are projection-chains whose diameters $< \epsilon$. We then ϵ-deform

c_{n-p} and the chains c'_{n-p+1} simultaneously so as not to impair their relation to one another and to the V's, and also so that they meet only the elements c_p^a of the decomposition of c_p. It is readily seen that all these conditions can be fulfilled with ϵ as small as we please. We choose it $< \frac{1}{2}\zeta$, where ζ is the same as in No. 29. As a consequence we now have a pair c_p, c_{n-p} whose intersection consists of a finite number of disjoined closed sets F^a, one on each c_p^a. Since $\operatorname{diam} F^a < \frac{1}{2}\zeta$, F^a will be covered by a certain set V, which we may call V^a, such that $d(F^a, M - V^a) > \frac{1}{2}\zeta$. Since the F's do not intersect we can find for each F^a an open set W^a such that $F^a \subset W^a \subset V^a$, $\bar{W}^a \cdot \bar{W}^b = 0$ for $a \neq b$. Introduce the closed set $G = M - \Sigma W^a$ and let L be a fundamental projection-complex for G (No. 7) so that $G = \bar{L} \cdot M$. We now remove from C_{p+1} all the p-cells on the Φ's which are on L and also all their joining cells, and call C'_{p+1} the chain left, C''_{p+1} the chain removed and c'_p, c''_p the corresponding chains of M whose sum is c_p. We have

$$c_p = c'_p + c''_p, \quad F(c_p) \subset M - \Sigma V^a \subset M - \Sigma W^a \subset M - c'_p.$$

Hence $| F(c'_p) | \subset | c''_p |$, and by construction the two chains c'_p, c''_p have only boundary points in common. Therefore

$$(37.3) \qquad | c'_p | \cdot | c''_p | = F(c'_p).$$

On the other hand if c'^a_p designates the part of c'_p on $\bar{W}^a$, we have

$$(37.4) \qquad c'_p = \Sigma c'^a_p, \quad | c'^a_p | \cdot | c'^b_p | = 0 \text{ for } a \neq b.$$

Therefore also

$$(37.5) \qquad | c'^a_p | \cdot | c''_p | = F(c'^a_p).$$

As a consequence of (37.4) (second relation), for i sufficiently large $| \mathscr{C}'^{ai}_p |$ and $| \mathscr{C}'^{bi} |$, $a \neq b$, will have Φ^i-neighborhoods without common cells, for otherwise we would have $d(| c'^a_p | \cdot | c'^b_p |) = 0$. We also know that by construction $\mathscr{C}'^{ai}_p$ and $\mathscr{C}''^i_p$ have no common p-cells. Combining with the construction of C'_{p+1}, C''_{p+1}, we have

$$(37.6) \qquad | F(C'^a_{p+1}) | \cdot | C''_{p+1} | = | F(C'^a_{p+1}) | \subset L.$$

38. Until further notice we shall impose upon the simplicial transformation θ of No. 36 the following additional restriction: whenever E_q^i is a cell of $\mathscr{C}''^i_p$ with vertices on L, we choose one of these vertices as the A^{iq} for that cell, that is as the vertex of Φ^i into which θ is to transform all the vertices of Ψ'^i that are on E_q^i.

Consider now C^{*k}_{r-n+p} and let C'^{*k}_{r-n+p} be the chain left on removing from it the cells which do not meet $\mathscr{C}'^k_p$. Due to the mode of separation of the chains $\mathcal{B}'^{ak}_p$, for k sufficiently high C'^{*k}_{r-n+p} will be a sum of disjoined chains C'^{*ak}_{r-n+p} consisting respectively of the cells which meet the chain $\mathcal{B}'^{ak}_p$. Therefore C'^{*ak}_{r-n+p} meets $\mathcal{B}'^{ak}_p$ but not $\mathcal{B}'^{bk}_p$ for $b \neq a$.

Now observe that as regards the cells of $\Gamma_n^k \cdot C^{*k}_{r-n+p}$ that are on the chains $\mathcal{B}'$ the transformation θ preserves the same properties as in the Lemma. However it now takes the p-cells on a $\mathcal{B}''$ and transforms them like cells on an $F(\mathcal{B}')$, i. e. into cells of dimension $< p$ so that the projections of their boundaries do not affect the projections of the chains $F(\Gamma_n^k \cdot C'^{*ak}_{r-n+p})$. It follows that as regards the effect on the intersections $\Gamma_n^k \cdot C'^{*ak}_{r-n+p}$ its performance is as before and that this chain is now projected into $\mathcal{B}'^{ai}_p$. It follows also that $F(\Gamma_n^k \cdot C'^{*ak}_{r-n+p})$ is projected at the same time into $F(\mathcal{B}'^{ai}_p)$. All this holds of course for k large enough, which is all that we need.

It follows from what precedes that we may replace the initial chain c_p by a set of chains c'^a_p whose boundaries behave in a manner similar to that imposed upon c_p by the Lemma.

39. Since $c'^a_p \subset V^a$ we have from our discussion of the index

$$(39.1) \qquad (c_p \cdot c_{n-p}) = \Sigma\, (c'^a_p \cdot F(c^a_{n-p+1})).$$

Similarly since $r - n$ is even by *Topology*, p. 169, formula (20),

$$(39.2) \qquad (C^{*k}_{r-n+p} \cdot \mathcal{B}^k_{n-p}) = \Sigma\, (C'^{*ak}_{r-n+p} \cdot \mathcal{B}^{ak}_{d-u})$$
$$= (-1)^p\, \Sigma\, (F(C'^{*ak}_p) \cdot \mathcal{B}^{ak}_{n-p+1}).$$

Comparing these relations and bringing back the index α, we find

$$(39.3) \qquad \Sigma\, (F(C'^{a\alpha k_\alpha}_{r-n+p}) \cdot \mathcal{B}^{a\alpha k_\alpha}_{n-p+1}) = (-1)^p \cdot \lambda,$$

and the proof of the Lemma is reduced to showing that

$$(39.4) \qquad \Sigma\, (F(c'^{a\alpha}_p) \cdot c^{a\alpha}_{n-p+1}) = (-1)^p \cdot \lambda,$$

the relations between corresponding chains being as for Λ_{p-1}. That is to say we have reduced Λ_p to Λ_{p-1}, and hence to Λ_0.

40. We take up Λ_0, and we shall in fact prove the somewhat more stringent result that Λ_0 holds with all the numbers k_α equal, i. e. with a single chain c_0^α. We may go as far as No. 39 in the same manner as previously. Referring to No. 37 we have to prove that when the diameters of the sets c'^a_0 are small enough, if there is a k arbitrarily high such that $\mathcal{B}_0^{ci}$ is the projection of $\Gamma_n^k \cdot C^{*ak}_{r-n}$, then

$$(40.1) \qquad \Sigma\, I_a(C^{*ak}_{r-n} \cdot \Gamma_n^k) = \Sigma\, I_a(c_0^a \cdot \gamma_n).$$

This will follow if we can show that

$$(40.2)\qquad (C^{*ak}_{r-n}\cdot\Gamma_n^{k}) = (c_0^{a}\cdot\gamma_n).$$

In the first place we have (No. 24)

$$(40.3)\qquad (c_0^{a}) = (c_0^{a}\cdot\gamma_n) = (\mathcal{B}_0^{ai})$$

for i large enough. Also since $\mathcal{B}_0^{ai}$ is the projection of $\Gamma_n^{k}\cdot C^{*ak}_{r-n}$, we have

$$(40.4)\qquad (\mathcal{B}_0^{ai}) = (\Gamma_n^{k}\cdot C^{*ak}_{r-n}) = (C^{*ak}_{r-n}\cdot\Gamma_n^{k}),$$

from which (40.2) and hence (40.1) follow. This proves Λ_0 and hence also the Lemma.

41. An important application of the Lemma is the proof, still lacking, of formula (34.7), and hence of property (23 f), for the index. For take first $p = 0$, and $C^{a}_{p+1} = C_1 =$ a chain made up of a single projecting line whose traces $\mathcal{B}_0^{i}$ are vertices A^{i} of the complexes Φ^{i} such that for i above a certain value A^{i} is the vertex of an E_n of Φ^{i}. Then taking $k_a = i$, $C^{*a}_{r-n} = E^{*}_{r-n}$ the cell of $\mathcal{K}^{i}$ dual to E_n, $\mathcal{B}^{aka}_{r-n} = \Gamma_n^{i}$ and orientations as in *Topology*, Ch. IV, the condition of the Lemma is fulfilled with a single $c_0 = x$ and a single $c_n = \gamma_n$. Therefore

$$(41.1)\qquad (x\cdot V_n) = (E^{*}_{r-n}\cdot E_n) = +1.$$

Choose now $p = n$ and $C^{a}_{p+1} = C_{n+1} = \Gamma_{n+1}$ the projection-chain defining γ_n, $k_a = i$, $C^{*aka}_{r-n+p} = C^{*}{}_r{}^{i}$, the sum of the cells of $\mathcal{K}^{*i}$ oriented like the spaces S_r, $\mathcal{B}^{aka}_{n-p} =$ the same vertex A^{i} as previously. This time the conditions of the Lemma are again fulfilled and we find

$$(41.2)\qquad (\gamma_n\cdot x) = (C^{*}{}_r{}^{i}\cdot A^{i}) = +1 = (x\cdot\gamma_n),$$

which is the result that we required.

42. From the Lemma to the duality formula for the Betti-numbers is but a step. Let γ^{a}_{n-p}, $a = 1, 2, \cdots, R_{n-p}$, be a base for the $(n-p)$-cycles consisting of irreducible projection-cycles (Theorem II). Let Γ^{a}_{n-p+1}, $\mathcal{G}^{ah}_{n-p}$ be the representative cycle mod Φ, and trace on Φ^{h}, associated with γ^{a}_{n-p}. Then for h above a certain value the cycles $\mathcal{G}^{ah}_{n-p}$ are independent on Φ^{h}, hence also independent on $\mathcal{K}^{h}-\mathcal{B}^{h}$. For if say

$$\mathcal{K}^{h}-\mathcal{B}^{h}\supset C_{n-p+1}\to\Sigma\, t_a\,\mathcal{G}^{ah}_{n-p},$$

we could slide down C_{n-p+1} along the projecting segments on $\mathcal{K}^{h}-\mathcal{B}^{h}-\Phi^{h}$ onto Φ^{h}, and obtain a chain on Φ^{h}

$$C'_{n-p+1}\to\Sigma\, t_a\,\mathcal{G}^{ah}_{n-p}$$

so that the cycles $\mathcal{G}^{ah}_{n-p}$ would not be independent on Φ^{h}.

As a consequence of the independence of these cycles on $\mathcal{K}^h - \mathcal{B}^h$, $\mathcal{K}^h$ contains a cycle mod $\mathcal{B}^h$, $\mathcal{G}^{\alpha h}_{r-n+p}$ whose cells intersecting Φ^h consist of cells of the dual $\mathcal{K}^{*h}$, and such that (*Topology*, pp. 140, 174).

$$(42.1) \qquad (\mathcal{G}^{\alpha h}_{r-n+p} \cdot \mathcal{G}^{\beta h}_{n-p}) = \delta_{\alpha\beta}.$$

Consider now the projection $\Gamma_p{}^i$ of all cycles $t\Gamma_n{}^h \cdot \mathcal{G}^{\alpha h}_{r-n+p}$ (α fixed) on a definite Φ^i. As far as the intersections with Φ^h go the chain $\mathcal{G}^{\alpha h}_{r-n+p}$ is a $C^{*h}{}_{r-n+p}$. With $\Gamma_p{}^i$ we associate the numbers $(t\delta_{\alpha\beta})$ and if $\Gamma'_p{}^i$ corresponds to t' and the numbers $(t'\delta_{\alpha\beta})$, we associate with $s\Gamma_p{}^i + s'\Gamma'_p{}^i$ the numbers $((st + s't')\delta_{\alpha\beta})$. In this manner if $\mathfrak{M}^i$ is the modulus generated by the cycles $\Gamma_p{}^i$, there corresponds to each member of $\mathfrak{M}^i$ a definite set $(t\delta_{\alpha\beta})$. Clearly members corresponding to $t = 0$ give rise to a submodulus $\mathfrak{N}^i$ of $\mathfrak{M}^i$. Also by construction the moduli $\mathfrak{M}^i, \mathfrak{N}^i$ are in the very relationship demanded by Theorem I. Therefore there exists a projection-sequence $\{\Gamma_p{}^{\alpha i}\}$ such that the cycle $\Gamma_p{}^{\alpha i}$ is a member of $\mathfrak{M}^i$ corresponding to $t = 1$. This sequence gives rise to a projection-cycle mod Φ, $\Gamma^\alpha{}_{p+1}$, which defines a normal cycle $\gamma_p{}^\alpha$. Owing to (42.1) and to the mode of defining the moduli $\mathfrak{M}^i$, we have by the Lemma

$$(42.2) \qquad (\gamma_p{}^\alpha \cdot \gamma^\beta{}_{n-p}) = \delta_{\alpha\beta}.$$

Hence (No. 23 property *e*) the cycles $\gamma_p{}^\alpha$ are independent and therefore $R_p \geqq R_{n-p}$. Similarly $R_p \leqq R_{n-p}$ and therefore we have proved Poincaré's duality relation for an absolute n-manifold:

$$(42.3) \qquad R_p(M) = R_{n-p}(M).$$

43. *Extension to open manifolds.* Take first an open M_n and let U be an open subset of M whose closure $\bar{U}$ is self-compact. Then if $V \Subset U$, $\bar{V}$ is likewise self-compact. As the manifold conditions hold over U we may apply Theorem VII with the following slight restrictions: $c_p \subset V$, $\epsilon < d(V, M - U)$. From this we conclude, as in No. 28, that there are at most finite numbers: (a) of absolute p-cycles of U independent mod $M - V$; (b) of p-cycles of U mod $M - U$, independent mod $M - V$. We can then show as in the preceding number that the two numbers are equal.

The sequences of open sets $\{U^i\}$ such that $U^{i+1} \Subset U^i$, $\Pi U^i = 0$, may serve to define the different types of ideal elements as we have done in *Topology*, Ch. VII. In the terminology there used let Λ be the total ideal element, and let $\mathcal{L}^1, \mathcal{L}^2$ designate complementary closed and open ideal elements. Let also L be any closed subset of M which $\supset \Lambda$ and let L^1 be any closed subset of L with $\mathcal{L}^1$ for ideal element. Then if $L^2 = L - L^1$, $\mathcal{L}^2$ will be the ideal element of L^2. By means of properties (a), (b), (c), and by unimportant adaptations of the treatment in *Topology*, Ch. VII, § 3, we prove:

FUNDAMENTAL DUALITY THEOREM. *Let* Γ_p, G_{n-p} *be associated cycles of the dual types* $M - L^1 \bmod L^2$ *and* $M - L^2 \bmod L^1$ *in any ring of rational coefficients forming a field. There exists two associated dual bases* $\{\Gamma_p^a\}$, $\{G^\beta_{n-p}\}$ *made up of true normal cycles whose indices satisfy the relations*

$$(\Gamma_p^a \cdot G^\beta_{n-p}) = \delta_{a\beta}. \tag{43.1}$$

Whatever Γ_p, G_{n-p} *we have*

$$\Gamma_p = \Sigma\, (\Gamma_p \cdot G^a_{n-p}) \cdot \Gamma_p^a, \tag{43.2}$$

$$G_{n-p} = \Sigma\, (\Gamma_p^a \cdot G_{n-p}) \cdot G^a_{n-p}, \tag{43.3}$$

and the Betti-numbers satisfy the duality relations

$$R_p(M - L^1, L^2) = R_{n-p}(M - L^2, L^1). \tag{43.4}$$

In particular: (a) *when* $L^1 = \Lambda$, $L^2 = 0$ we have

$$R_p(M - \Lambda) = R_{n-p}(M, \Lambda), \tag{43.5}$$

where the Betti-numbers refer at the left to the finite cycles and at the right to the infinite cycles; (b) *when* $\Lambda = 0$, *i. e. when* M *is absolute, the bases are finite and the duality relation reduces to that of Poincaré*

$$R_p(M) = R_{n-p}(M). \tag{43.6}$$

These results hold also when M *consists of a countable aggregate of circuits.*

The last part of the statement, regarding an M consisting of a countable aggregate of circuits, is an immediate consequence of the following: when $M = \Sigma M^i$, where M^i is a connected n-manifold and $M^i \cdot M^j = 0$, then the p-th homology group of M is the direct sum of those of the manifolds M^i (the groups are assumed written additively).

Annals of Mathematics
Vol. 51, No. 2, March, 1950

ČECH COHOMOLOGY THEORY AND THE AXIOMS

By C. H. Dowker

(Received December 14, 1948)

Čech cohomology theory based on infinite coverings by open sets has proved useful in studying the homotopy classes of maps of general spaces.[1] The purpose of the present paper is to show that this Čech cohomology theory satisfies the Eilenberg-Steenrod axioms.[2] (As is known, the Čech cohomology theory based on finite coverings fails in general[3] to satisfy axiom 4, the homotopy axiom.) The axioms are stated in terms of the relative cohomology groups of a pair (X, A) where A is any subset of a topological space X. However the relative Čech groups are usually defined only relative to closed subsets. In part C we extend the definition of relative Čech cohomology groups to the case of non-closed subsets. Part E contains the proof that the groups so defined satisfy the axioms for arbitrary pairs (X, A).

A. Simplicial complexes

Let G be a fixed discrete coefficient group. We consider pairs (K, L) consisting of an abstract (in general infinite) simplicial complex K and a (closed) subcomplex L. The relative cohomology groups $H^q(K, L) = H^q(K, L, G)$, $q \geqq 0$, are defined using infinite cochains. The coboundary operator maps q-cocycles of L into relative $(q + 1)$-cocycles of (K, L) and this cocycle homomorphism leads to a homomorphism $\delta: H^q(L) \rightarrow H^{q+1}(K, L)$ of the cohomology groups. Let f be a simplicial map of (K, L) into (K_1, L_1), this means that f maps K_1 into K so that L_1 is mapped into L. The simplicial map f induces an inverse homomorphism $f^*: H^q(K_1, L_1) \rightarrow H^q(K, L)$ of the cohomology groups.

The following seven theorems are assumed to be known and are given without proof. Collectively they state that the cohomology theory of abstract simplicial complexes satisfies the axioms.

A1. If $f: (K, L) \rightarrow (K, L)$ is the identity map, then $f^*: H^q(K, L) \rightarrow H^q(K, L)$ is the identity homomorphism.

A2. If $g: (K, L) \rightarrow (K_1, L_1)$ and $f: (K_1, L_1) \rightarrow (K_2, L_2)$, then $(fg)^* = g^*f^*$.

A3. If $f: (K, L) \rightarrow (K_1, L_1)$ then $f^*\delta = \delta(f \mid L)^*: H^q(L_1) \rightarrow H^{q+1}(K, L)$.

A4. Let f and g be simplicial maps of (K, L) into (K_1, L_1) such that, for each simplex s of K, fs and gs are faces of a common simplex s_1 of K_1 and such that, if s is in L, s_1 is in L_1. Then $f^* = g^*$.

A5. The following cohomology sequence is exact:

$$0 \rightarrow H^0(K, L) \rightarrow \cdots \rightarrow H^q(K, L) \xrightarrow{j^*} H^q(K) \xrightarrow{i^*} H^q(L) \xrightarrow{\delta} H^{q+1}(K, L) \rightarrow \cdots.$$

[1] See [1].
[2] See [3], [4] and [6].
[3] Eilenberg and Steenrod [4] show that it satisfies the axioms for compact pairs (X, A).

A6. Let V be an open[4] set of K with $V \subset L$ and let i be the inclusion map of $(K - V, L - V)$ into (K, L). Then i^* is an isomorphism of $H^q(K, L)$ onto $H^q(K - V, L - V)$.

A7. If K consists of a single vertex, $H^q(K) = 0$ for $q > 0$.

We also use the following stronger form of A4.

A8. Let f and g be simplicial maps of (K, L) into (K_1, L_1) and let f_q and g_q be the induced homomorphisms of the group $C_q(K)$ of the integral chains of K into $C_q(K_1)$. If for each q there exists a homomorphism D_{q+1}: $C_q(K) \to C_{q+1}(K_1)$ which maps $C_q(L)$ into $C_{q+1}(L_1)$, such that $\partial D_{q+1} = g_q - f_q - D_q\partial$, then $f^* = g^*$.

PROOF. It is clear that D_{q+1} induces a homomorphism of the group $C_q(K, L)$ of relative chains of (K, L) into $C_{q+1}(K_1, L_1)$. This in turn[5] induces a homomorphism D_q^*: $C^{q+1}(K_1, L_1) \to C^q(K, L)$ of the cochain groups satisfying the condition $\delta D_{q-1}^* = g^q - f^q - D_q^*\delta$. From this it follows that if ϕ^q is a q-cocycle of (K_1, L_1), $g^q\phi^q$ and $f^q\phi^q$ are cohomologous in $Z^q(K, L)$. Hence $g^* = f^*$: $H^q(K_1, L_1) \to H^q(K, L)$.

B. Direct limit groups

By a *directed set* we shall mean a set of elements $\rho, \sigma, \cdots$, with an order relation $<$, such that a) if $\rho < \sigma$ and $\sigma < \tau$ then $\rho < \tau$, b) for any ρ and σ there is a τ such that $\rho < \tau$ and $\sigma < \tau$, and c)[6] for any ρ, $\rho < \rho$.

Let $\{G_\sigma\}$ be a set of abelian groups indexed by a directed set Ω. For each pair of indices ρ, σ for which $\rho < \sigma$, let $\pi_{\sigma\rho}$: $G_\rho \to G_\sigma$ be a homomorphism of G_ρ into G_σ such that $\pi_{\rho\rho}$ is the identity and, if $\rho < \sigma < \tau$, $\pi_{\tau\rho} = \pi_{\tau\sigma}\pi_{\sigma\rho}$. Then the set $S = \{G_\sigma, \pi_{\sigma\rho}, \Omega\}$ of groups, homomorphisms and indices is called a direct spectrum. Two elements, $\gamma_\rho \in G_\rho$ and $\gamma_\sigma \in G_\sigma$ are called equivalent in S, $\gamma_\rho \equiv \gamma_\sigma$, if there exists $\tau > \rho$ and $> \sigma$ such that $\pi_{\tau\rho}\gamma_\rho = \pi_{\tau\sigma}\gamma_\sigma$. In this way the elements of the groups of S form equivalence classes called bundles. The sum $\gamma + \gamma'$ of two bundles is defined as follows: Let $\gamma_\rho \in \gamma$ and $\gamma'_\sigma \in \gamma'$, and choose $\tau > \rho$ and $> \sigma$. Let $\gamma + \gamma'$ be the bundle containing the element $\pi_{\rho\tau}\gamma_\rho + \pi_{\tau\sigma}\gamma'_\sigma$ of G_τ. This addition of bundles can be shown to be unique and the bundles with this addition form an abelian group G called the direct limit group of the spectrum; $G = \varinjlim S$.

Let $S_1 = \{G_\sigma, \pi_{\sigma\rho}, \Omega_1\}$ and $S_2 = \{G_\alpha, \pi_{\beta\alpha}, \Omega_2\}$ be direct spectra. For some pairs σ, α with $\sigma \in \Omega_1$ and $\alpha \in \Omega_2$ let there be given one or more homomorphisms $f_{\alpha\sigma}$: $G_\sigma \to G_\alpha$ and let $f = \{f_{\alpha\sigma}\}$ be the set of these homomorphisms. We write $\sigma \prec \alpha$ whenever there is some homomorphism $f_{\alpha\sigma} \in f$ with these indices. The set f of homomorphisms is called a *map* of S_1 into S_2 provided that

1) for every σ there is some α such that $\sigma \prec \alpha$,
2) if $f_{\alpha\sigma} \in f$ and if $\rho < \sigma \prec \alpha < \beta$, then $\pi_{\beta\alpha} f_{\alpha\sigma} \pi_{\sigma\rho} \in f$, and
3) if $f^1_{\alpha\sigma}$ and $f^2_{\alpha\sigma}$ are in f and if $\gamma_\sigma \in G_\sigma$ there is some $\beta > \alpha$ such that $\pi_{\beta\alpha} f^1_{\alpha\sigma}\gamma_\sigma = \pi_{\beta\alpha} f^2_{\alpha\sigma}\gamma_\sigma$.

---

[4] A set V of simplexes of a complex K is called open if the star of each simplex of V is contained in V.

[5] See [2, p. 413].

[6] Condition c) is not included in the usual definition of directed set.

Let f be any map of S_1 into S_2 and let γ_ρ and γ_σ be equivalent in S_1. Then there is a $\tau > \rho$ and $> \sigma$ such that $\pi_{\tau\rho}\gamma_\rho = \pi_{\tau\sigma}\gamma_\sigma$; let this element of G_τ be called γ_τ. Let $f_{\alpha\rho}$, $f_{\beta\sigma}$ and $f_{\gamma\tau}$ be elements of f. Choose $\delta > \alpha$ and $> \gamma$. Then $\pi_{\delta\alpha}f_{\alpha\rho}$ and $\pi_{\delta\gamma}f_{\gamma\tau}\pi_{\tau\rho}$ are in f, and hence by 3) there is an $\varepsilon > \delta$ such that $\pi_{\varepsilon\delta}\pi_{\delta\alpha}f_{\alpha\rho}\gamma_\rho = \pi_{\varepsilon\delta}\pi_{\delta\gamma}f_{\gamma\tau}\pi_{\tau\rho}\gamma_\rho$, i.e., such that $\pi_{\varepsilon\alpha}f_{\alpha\rho}\gamma_\rho = \pi_{\varepsilon\gamma}f_{\gamma\tau}\gamma_\tau$. Therefore $f_{\alpha\rho}\gamma_\rho \equiv f_{\gamma\tau}\gamma_\tau$. Similarly $f_{\beta\sigma}\gamma_\sigma \equiv f_{\gamma\tau}\gamma_\tau$. Therefore $f_{\alpha\rho}\gamma_\rho \equiv f_{\beta\sigma}\gamma_\sigma$. Thus the elements of a bundle γ in S_1 map into elements of a bundle in S_2 which we call $f^+\gamma$.

B1. *If f is a map of S_1 into S_2, f^+ is a homomorphism of $\varinjlim S_1$ into $\varinjlim S_2$.*

PROOF. Let γ and γ' be bundles in S_1, i.e., elements of $\varinjlim S_1$. Choose σ with some $\gamma_\sigma \in \gamma$ and $\gamma'_\sigma \in \gamma'$. Then $\gamma_\sigma + \gamma'_\sigma \in \gamma + \gamma'$. Let $f_{\alpha\sigma} \in f$; then, since $f_{\alpha\sigma}$ is a homomorphism, $f_{\alpha\sigma}(\gamma_\sigma + \gamma'_\sigma) = f_{\alpha\sigma}\gamma_\sigma + f_{\alpha\sigma}\gamma'_\sigma$. Therefore $f^+(\gamma + \gamma') = f^+\gamma + f^+\gamma'$. Thus f^+ is a homomorphism.

B2. *Let g be a map of S_1 into S_2 and f a map of S_2 into S_3. Let fg be the set of homomorphisms $f_{\mu\alpha}g_{\alpha\sigma}$ with $f_{\mu\alpha} \in f$ and $g_{\alpha\sigma} \in g$. Then fg is a map of S_1 into S_3 and $(fg)^+ = f^+g^+$.*

PROOF. 1) For every $\sigma \in \Omega_1$ there is some $\alpha \in \Omega_2$ such that some $g_{\alpha\sigma} \in g$ and for this α there is some $f_{\mu\alpha} \in f$. Then $f_{\mu\alpha}g_{\alpha\sigma} \in fg$ and hence $\sigma < \mu$.

2) Let $f_{\mu\alpha}g_{\alpha\sigma} \in fg$ and let $\rho < \sigma < \mu < \nu$. Then $\pi_{\nu\mu}f_{\mu\alpha} \in f$ and $g_{\alpha\sigma}\pi_{\sigma\rho} \in g$. Therefore $\pi_{\nu\mu}f_{\mu\alpha}g_{\alpha\sigma}\pi_{\sigma\rho} \in fg$.

3) Let $f_{\mu\alpha}g_{\alpha\sigma}$ and $f_{\mu\beta}g_{\beta\sigma}$ be in fg and let $\gamma_\sigma \in G_\sigma$. Let γ be the bundle containing γ_σ. Then $f_{\mu\alpha}g_{\alpha\sigma}\gamma_\sigma \in f^+g^+\gamma$ and $f_{\mu\beta}g_{\beta\sigma}\gamma_\sigma \in f^+g^+\gamma$. Therefore $f_{\mu\alpha}g_{\alpha\sigma}\gamma_\sigma \equiv f_{\mu\beta}g_{\beta\sigma}\gamma_\sigma$, i.e., there is a $\nu > \mu$ such that $\pi_{\nu\mu}f_{\mu\alpha}g_{\alpha\sigma}\gamma_\sigma = \pi_{\nu\mu}f_{\mu\beta}g_{\beta\sigma}\gamma_\sigma$. Thus fg is a map of S_1 into S_3. Also, since $f_{\mu\alpha}g_{\alpha\sigma}\gamma_\sigma \in (fg)^+\gamma$ and also $f_{\mu\alpha}g_{\alpha\sigma}\gamma_\sigma \in f^+g^+\gamma$, therefore $(fg)^+ = f^+g^+$.

Two maps f and g of S_1 into S_2 are called equivalent if for each element γ_σ of each group G_σ of S_1 there exist $\alpha \in \Omega_2$, $f_{\alpha\sigma} \in f$ and $g_{\alpha\sigma} \in g$ such that $f_{\alpha\sigma}\gamma_\sigma = g_{\alpha\sigma}\gamma_\sigma$.

B3. *Two maps f and g of S_1 into S_2 are equivalent if and only if $f^+ = g^+$.*

PROOF. Let f and g be equivalent and let $\gamma \in \varinjlim S_1$. Let $\gamma_\sigma \in \gamma$. Then, for some $f_{\alpha\sigma}$ and $g_{\alpha\sigma}$, $f_{\alpha\sigma}\gamma_\sigma = g_{\alpha\sigma}\gamma_\sigma$. But $f_{\alpha\sigma}\gamma_\sigma \in f^+\gamma$ and $g_{\alpha\sigma}\gamma_\sigma \in g^+\gamma$. Therefore $f^+\gamma = g^+\gamma$. Conversely let $f^+ = g^+$ and let $\gamma_\sigma \in G_\sigma \in S_1$. Then $f_{\beta\sigma}\gamma_\sigma \equiv g_{\delta\sigma}\gamma_\sigma$. Choose $\alpha > \beta$ and $> \delta$ so that $\pi_{\alpha\beta}f_{\beta\sigma}\gamma_\sigma = \pi_{\alpha\delta}g_{\delta\sigma}\gamma_\sigma$. But $\pi_{\alpha\beta}f_{\beta\sigma} \in f$ and $\pi_{\alpha\delta}g_{\delta\sigma} \in g$. Therefore f and g are equivalent.

If f and g are maps of S_1 into S_2 and if every element of f is an element of g we say that the map f is contained in the map g; $f \subset g$.

B4. *If $f \subset g$ then $f^+ = g^+$.*

PROOF. Let $\gamma_\sigma \in G_\sigma \in S_1$ and choose α so $f_{\alpha\sigma}$ exists. Then $f_{\alpha\sigma} \in f$, $f_{\alpha\sigma} \in g$ and $f_{\alpha\sigma}\gamma_\sigma = f_{\alpha\sigma}\gamma_\sigma$. Thus f and g are equivalent. Therefore, $f^+ = g^+$.

A map f of S_1 into S_2 will be called a *simple map* if for each pair $\sigma < \alpha$ the homomorphism $f_{\alpha\sigma}$ is unique. If g is a simple map of S_1 into S_2 and f is a simple map of S_2 into S_3, fg is not necessarily a simple map.

B5. *Let S_1 and S_2 be direct spectra and let an order relation $\sigma < \alpha$ be defined between elements σ of Ω_1 and α of Ω_2. Whenever $\sigma < \alpha$ let $f_{\alpha\sigma}: G_\sigma \to G_\alpha$ be a uniquely defined homomorphism. Then the set $f = \{f_{\alpha\sigma}\}$ of these homomorphisms is a simple map of S_1 into S_2 if and only if*

1) *for every σ there is an α such that $\sigma < \alpha$, and*

2) *if $\rho < \sigma < \alpha < \beta$, then $\rho < \beta$ and $f_{\beta\rho} = \pi_{\beta\alpha}f_{\alpha\sigma}\pi_{\sigma\rho}$.*

PROOF. The condition 3) of the definition of a map becomes trivial since $f^1_{\alpha\sigma} = f^2_{\alpha\sigma}$.

B6. *Let f be a simple map of S_1 into S_2 and for every $\rho \in \Omega_1$ and $\alpha \in \Omega_2$ let there exist $\sigma > \rho$ and $\beta > \alpha$ such that $f_{\beta\sigma}$ is an isomorphism of G_σ onto G_β. Then f^+ is an isomorphism of $\varinjlim S_1$ onto $\varinjlim S_2$.*

PROOF. 1) The homomorphism f^+ is onto. Let $\gamma'_\alpha \in \gamma' \in \varinjlim S_2$. Choose an isomorphism $f_{\beta\sigma}$ with $\beta > \alpha$. Let $\gamma'_\beta = \pi_{\beta\alpha}\gamma'_\alpha$, then $\gamma'_\beta \in \gamma'$. Then, if γ is the bundle containing $f^{-1}_{\beta\sigma}\gamma'_\beta$, $f^+\gamma = \gamma'$.

2) The kernel of f^+ is zero. Let $\gamma_\rho \in \gamma \in \varinjlim S_1$ and let $f^+\gamma = 0$. Then $f_{\delta\rho}\gamma_\rho = 0$. Hence, for some $\alpha > \delta$, $\pi_{\alpha\delta}f_{\delta\rho}\gamma_\rho = 0$. Hence $f_{\alpha\rho}\gamma_\rho = 0$. Choose an isomorphism $f_{\beta\sigma}$ with $\beta > \alpha$ and $\sigma > \tau$. Then $f_{\beta\sigma}\pi_{\sigma\rho}\gamma_\rho = f_{\beta\rho}\gamma_\rho = \pi_{\beta\alpha}f_{\alpha\rho}\gamma_\rho = 0$. Therefore $\pi_{\sigma\rho}\gamma_\rho = 0$. But $\pi_{\sigma\rho}\gamma_\rho \in \gamma$, therefore $\gamma = 0$.

Let $S_2 = \{G_\sigma, \pi_{\sigma\rho}, \Omega_2\}$ be a direct spectrum and let Ω_1 be a directed subset of Ω_2. Let S_1 consist of those G_σ such that $\sigma \in \Omega_1$, those $\pi_{\sigma\rho}$ for which both σ and ρ are in Ω_1, and the index set Ω_1. Then S_1 is a direct spectrum and is called a subspectrum of S_2. Let $\rho \prec \sigma$ mean that $\rho \in \Omega_1$, $\sigma \in \Omega_2$ and $\rho < \sigma$, and when $\rho \prec \sigma$ let $f_{\sigma\rho} = \pi_{\sigma\rho}$. Then the set $f = \{f_{\sigma\rho}\}$ is a simple map of S_1 into S_2. For, checking B5, we see that 1) if $\sigma \in \Omega_1$, $\sigma \prec \sigma$, and 2) if $\rho < \sigma \prec \alpha < \beta$, then $\rho < \beta$ and hence $\rho \prec \beta$ and, moreover, $f_{\beta\rho} = \pi_{\beta\rho} = \pi_{\beta\alpha}\pi_{\alpha\sigma}\pi_{\sigma\rho} = \pi_{\beta\alpha}f_{\alpha\sigma}\pi_{\sigma\rho}$. This simple map of S_1 into S_2 is called the *inclusion map*. If Ω_1 is cofinal in Ω_2, S_1 is called a *cofinal subspectrum* of S_2. In particular, S_2 is a cofinal subspectrum of itself and the inclusion map of S_2 in S_2 is called the *identity map*.

B7. *The inclusion map f of a cofinal subspectrum induces an isomorphism f^+ of the limit groups. The identity map of a direct spectrum into itself induces the identity isomorphism of the limit group.*

PROOF. The first part of the theorem follows from B6. The second part follows from the fact that, if $\gamma_\sigma \in \gamma$, $f_{\sigma\sigma}\gamma_\sigma = \pi_{\sigma\sigma}\gamma_\sigma \in \gamma$; thus each bundle maps on itself.

The following theorem is useful in studying exact sequences.

B8. *Let f be a simple map of S_1 into S_2 and let g be a simple map of S_2 into S_3. For every $\sigma \in \Omega_1$, $\alpha \in \Omega_2$ and $\mu \in \Omega_3$ let there exist $\tau > \sigma$, $\beta > \alpha$ and $\nu > \mu$ such that $f_{\beta\tau}$ and $g_{\nu\beta}$ exist and such that the kernel of $g_{\nu\beta}$ is the image of $f_{\beta\tau}$. Then the kernel of g^+ is the image of f^+.*

PROOF. 1) The kernel contains the image. Let $\gamma_\sigma \in \gamma \in \varinjlim S_1$. Choose $\tau > \sigma$, $\beta \in \Omega_2$ and $\nu \in \Omega_3$ so that the kernel of $g_{\nu\beta}$ is the image of $f_{\beta\tau}$. Then $g_{\nu\beta}f_{\beta\tau}\pi_{\tau\sigma}\gamma_\sigma = 0$. Thus $g_{\nu\beta}f_{\beta\sigma}\gamma_\sigma = 0$. Hence $g^+f^+\gamma = 0$.

2) The image contains the kernel. Let $\gamma'_\alpha \in \gamma' \in \varinjlim S_2$ and let $g^+\gamma' = 0$. Then there exists $\mu \in \Omega_3$ such that $g_{\mu\alpha}\gamma'_\alpha = 0$. Choose $\tau \in \Omega_1$, $\beta > \alpha$ and $\nu > \mu$ so that the kernel of $g_{\nu\beta}$ is the image of $f_{\beta\tau}$. Then $g_{\nu\beta}\pi_{\beta\alpha}\gamma'_\alpha = g_{\nu\alpha}\gamma'_\alpha = \pi_{\nu\mu}g_{\mu\alpha}\gamma'_\alpha = 0$. Therefore there is some $\gamma_\tau \in G_\tau$ such that $f_{\beta\tau}\gamma_\tau = \pi_{\beta\alpha}\gamma'_\alpha \in \gamma'$. Therefore, if γ is the bundle containing γ_τ, $f^+\gamma = \gamma'$.

NOTATION: In the rest of this paper we shall use the same symbol for a map f of one direct spectrum into another and for the induced homomorphism f^+ of the limit groups.

C. Čech cohomology groups

We consider pairs (X, A) where X is a topological[7] space and A is a subset of X. A covering $(\mathfrak{U}, \mathfrak{B})$ of a pair (X, A) is a set $\mathfrak{U}$ of open sets of X whose union is X and a subset $\mathfrak{B}$ of $\mathfrak{U}$ whose union contains A. A covering $(\mathfrak{U}_1, \mathfrak{B}_1)$ of (X, A) is called a refinement of $(\mathfrak{U}, \mathfrak{B})$, (symbolically: $(\mathfrak{U}, \mathfrak{B}) < (\mathfrak{U}_1, \mathfrak{B}_1)$), if every open set of $\mathfrak{U}_1$ is contained in an open set of $\mathfrak{U}$ and every open set of $\mathfrak{B}_1$ is contained in an open set of $\mathfrak{B}$. In particular $(\mathfrak{U}, \mathfrak{B}) < (\mathfrak{U}, \mathfrak{B})$. If $(\mathfrak{U}_1, \mathfrak{B}_1)$ is a refinement of $(\mathfrak{U}, \mathfrak{B})$ and $(\mathfrak{U}_2, \mathfrak{B}_2)$ is a refinement of $(\mathfrak{U}_1, \mathfrak{B}_1)$, then $(\mathfrak{U}_2, \mathfrak{B}_2)$ is a refinement of $(\mathfrak{U}, \mathfrak{B})$. If $(\mathfrak{U}, \mathfrak{B})$ and $(\mathfrak{U}_1, \mathfrak{B}_1)$ are any two coverings of (X, A) they have a common refinement $(\mathfrak{U}_2, \mathfrak{B}_2)$; e.g., let $\mathfrak{U}_2$ consist of all intersections $U \frown U_1$ of a set of $\mathfrak{U}$ with a set of $\mathfrak{U}_1$ and let $U \frown U_1 \in \mathfrak{B}_2$ if both $U \in \mathfrak{B}$ and $U_1 \in \mathfrak{B}_1$. Thus the coverings $(\mathfrak{U}, \mathfrak{B})$ of (X, A) form a directed set Ω.

We shall frequently use a single letter, say σ, to represent a covering $(\mathfrak{U}, \mathfrak{B})$ of (X, A). in such cases we let $\sigma_1 = \mathfrak{U}$, $\sigma_2 = \mathfrak{B}$, so that $\sigma = (\sigma_1, \sigma_2) = (\mathfrak{U}, \mathfrak{B})$.

With a covering $\sigma = (\mathfrak{U}, \mathfrak{B})$ of (X, A) we associate a simplicial complex K_σ and a subcomplex L_σ of K_σ. The vertices of K_σ are the non-empty open sets[8] of $\mathfrak{U}$; any finite collection of open sets of $\mathfrak{U}$ whose intersection is not empty is a simplex of K_σ; any finite collection of open sets of $\mathfrak{B}$ whose intersection meets A is a simplex of L_σ. The pair (K_σ, L_σ) is called the nerve of the covering $\sigma = (\mathfrak{U}, \mathfrak{B})$ and is designated $N(\mathfrak{U}, \mathfrak{B})$ or $N(\sigma)$ or N_σ. The q-dimensional cohomology group $H^q(K_\sigma, L_\sigma)$ of the pair (K_σ, L_σ) will sometimes be called $H^q(N_\sigma)$ or H^q_σ.

Let σ and τ be coverings of (X, A) such that $\sigma < \tau$, i.e., such that τ is a refinement of σ. For each $U \in \tau_1$, let $\pi_{\tau\sigma} U \in \sigma_1$ be chosen so that $U \subset \pi_{\tau\sigma} U$ and so that, if $U \in \tau_2$, $\pi_{\tau\sigma} U \in \sigma_2$. Let $s = U_0 \cdots U_q$ be any simplex of K and let $p \in U_0 \frown \cdots \frown U_q$. Then $p \in (\pi_{\tau\sigma} U_0) \frown \cdots \frown (\pi_{\tau\sigma} U_q)$ and therefore the (not necessarily distinct) open sets $\pi_{\tau\sigma} U_0, \cdots, \pi_{\tau\sigma} U_q$ are vertices of a simplex $\pi_{\tau\sigma} s$ of K_σ. If s is a simplex of L_τ, each U_i is in τ_2 and hence $\pi_{\tau\sigma} U_i \in \sigma_2$. Also p can be chosen in $U_0 \frown \cdots \frown U_q \frown A \subset (\pi_{\tau\sigma} U_0) \frown \cdots \frown (\pi_{\tau\sigma} U_q) \frown A$; hence $\pi_{\tau\sigma} s$ is a simplex of L_σ. Thus $\pi_{\tau\sigma}$ is a simplicial map,[9] called a projection map, of (K_τ, L_τ) into (K_σ, L_σ). The projection map $\pi_{\tau\sigma}$ is not necessarily unique; let $\pi'_{\tau\sigma}$ be the result of a second choice. Then $p \in (\pi_{\tau\sigma} U_0) \frown \cdots \frown (\pi_{\tau\sigma} U_q) \frown (\pi'_{\tau\sigma} U_0) \frown \cdots \frown (\pi'_{\tau\sigma} U_q)$. Thus the vertices of $\pi_{\tau\sigma} s$ and $\pi'_{\tau\sigma} s$ are vertices of a simplex s_1 of K_σ and thus $\pi_{\tau\sigma} s$ and $\pi'_{\tau\sigma} s$ are faces of s_1. Moreover, if s is in L_τ, s_1 is in L_σ. Hence, by A4, $\pi^*_{\tau\sigma} = \pi'^*_{\tau\sigma}$. Thus the projection homomorphism $\pi^*_{\tau\sigma}: H^q_\sigma \to H^q_\tau$ is independent of the choice of $\pi_{\tau\sigma}$. In particular, if $\pi_{\sigma\sigma}$ is any projection of N_σ into itself, $\pi^*_{\sigma\sigma}$ is the identity isomorphism if H^q_σ onto itself. If $\rho < \sigma < \tau$ and if $\pi_{\sigma\rho}: N_\sigma \to N_\rho$ and $\pi_{\tau\sigma}: N_\tau \to N_\sigma$ are projections, then $\pi_{\sigma\rho}\pi_{\tau\sigma}$ is a projection of N_τ into N_ρ. Hence, if $\pi_{\tau\rho}$ is any projection of N_τ into N_ρ, $\pi^*_{\tau\rho} = (\pi_{\sigma\rho}\pi_{\tau\sigma})^*$. Hence, by A2, $\pi^*_{\tau\rho} = \pi^*_{\tau\sigma}\pi^*_{\sigma\rho}$.

Thus the directed set Ω of coverings σ of (X, A), together with the cohomol-

[7] By topological space we mean a T-space in the sense of Alexandroff-Hopf.

[8] The empty set may be an element of a covering.

[9] We use the same symbol $\pi_{\tau\sigma}$ for a map of the covering τ into the covering σ and for a map of the nerve N_τ into the nerve N_σ.

ogy groups H^q_σ and the projection homomorphisms $\pi^*_{\sigma\rho}$ form a direct spectrum $S^q(X, A) = \{H^q_\sigma, \pi^*_{\sigma\rho}, \Omega\}$. The direct limit group $H^q(X, A)$ of this spectrum is called the q-dimensional Čech cohomology group of the pair (X, A). (If A is the empty set we may write $H^q(X)$ in place of $H^q(X, 0)$.)

It follows from B7 that if we replace Ω by any cofinal subfamily of coverings, and thus replace $S^q(X, A)$ by a cofinal subspectrum, we obtain a limit group isomorphic with $H^q(X, A)$. For example, the coverings $(\mathfrak{U}, \mathfrak{B})$ such that, for $U \in \mathfrak{B}$, $U \frown A \neq 0$ form a cofinal subset of Ω. If A is closed, the coverings $(\mathfrak{U}, \mathfrak{B})$ in which $\mathfrak{B}$ consists of all open sets $U \in \mathfrak{U}$ with $U \frown A \neq 0$ form a cofinal subset of Ω. In particular, if $A = 0$ the coverings $(\mathfrak{U}, 0)$ form a cofinal subset. If X and A are compact the finite coverings form a cofinal subset of Ω.

D. The homomorphisms f^* and δ

For each continuous map f of (X, A) into (Y, B) we define a simple map f^* of $S^q(Y, B)$ into $S^q(X, A)$. If σ is a covering of (Y, B) and α is a covering of (X, A), $\sigma < \alpha$ will mean that for each $U \in \alpha_1$ there is some $V \in \sigma_1$ with $fU \subset V$ and for each $U \in \alpha_2$ there is some $V \in \sigma_2$ with $fU \subset V$. If $\sigma < \alpha$ and if $U \in \alpha_1$ let $f_{\alpha\sigma}U \in \sigma_1$ be chosen so that $fU \subset f_{\alpha\sigma}U$ and so that, if $U \in \alpha_2$, $f_{\alpha\sigma}U \in \sigma_2$. Let $s = U_0 \cdots U_q$ be any simplex of K_α and let $p \in U_0 \frown \cdots \frown U_q$. Then $f(p) \in (fU_0) \frown \cdots \frown (fU_q) \subset (f_{\alpha\sigma}U_0) \frown \cdots \frown (f_{\alpha\sigma}U_q)$. Thus $f_{\alpha\sigma}\, s$ is a simplex of K_σ. If s is a simplex of L_α, each U_i is in α_2 and hence $f_{\alpha\sigma}U_i \in \sigma_2$, and, if p is chosen in A, $f(p) \in fA \subset B$; hence $f_{\alpha\sigma}s$ is a simplex of L_σ. Therefore $f_{\alpha\sigma}$: $(K_\alpha, L_\alpha) \to (K_\sigma, L_\sigma)$ is a simplicial map of N_α into N_σ. If $f'_{\alpha\sigma}$ is any second choice of $f_{\alpha\sigma}$, then for the same point p, $f(p) \in (f'_{\alpha\sigma}U_0) \frown \cdots \frown (f'_{\alpha\sigma}U_q)$. Hence $f_{\alpha\sigma}s$ and $f'_{\alpha\sigma}s$ are faces of a simplex s_1 of K_σ. If s is in L_α, s_1 is in L_σ. Hence, by A4, $f^{*1}_{\alpha\sigma} = f'^{*}_{\alpha\sigma}$. Thus the homomorphism $f^*_{\alpha\sigma}: H^q_\sigma \to H^q_\alpha$ is independent of the choice of $f_{\alpha\sigma}$. Let $f^* = \{f^*_{\alpha\sigma}\}$ be the set of these homomorphisms.

To show that $f^* = \{f^*_{\alpha\sigma}\}$ is a simple map we must verify conditons 1) and 2) of B5.

1) Let σ be any covering of (Y, B). Let $f^{-1}\sigma_1$ be the set of all open sets $f^{-1}U$ of X with $U \in \sigma_1$ and let $f^{-1}\sigma_2$ be the set of all $V \in f^{-1}\sigma_1$ such that $V = f^{-1}U$ for some $U \in \sigma_2$. Then $(f^{-1}\sigma_1, f^{-1}\sigma_2)$ is a covering $f^{-1}\sigma$ of (X, A). From $ff^{-1}U \subset U$ it follows that $\sigma < f^{-1}\sigma$.

2) Let $\rho < \sigma < \alpha < \beta$. Let $\pi_{\sigma\rho}, f_{\alpha\sigma}, \pi_{\beta\alpha}$ be chosen. Then, if $U \in \beta_1$, $U \subset \pi_{\beta\alpha}U$ and hence $fU \subset f\pi_{\beta\alpha}U \subset f_{\alpha\sigma}\pi_{\beta\alpha}U \subset \pi_{\sigma\rho}f_{\alpha\sigma}\pi_{\beta\alpha}U$. If $U \in \beta_2$, then $\pi_{\beta\alpha}U \in \alpha_2$, $f_{\alpha\sigma}\pi_{\beta\alpha}U \in \sigma_2$ and $\pi_{\sigma\rho}f_{\alpha\sigma}\pi_{\beta\alpha}U \in \rho_2$. Thus $\rho < \beta$ and $\pi_{\sigma\rho}f_{\alpha\sigma}\pi_{\beta\alpha}$ is a possible choice for $f_{\rho\beta}$. Hence $f^*_{\rho\beta} = \pi^*_{\beta\alpha}f^*_{\alpha\sigma}\pi^*_{\sigma\rho}$.

Thus f^* is a simple map of $S^q(Y, B)$ into $S^q(X, A)$. We also write f^* for the induced homomorphism of the limit groups, $f^*: H^q(Y, B) \to H^q(X, A)$.

We now further analyse the homomorphisms $f^*_{\alpha\sigma}$. We define a pair[10] $N'_\sigma = (K'_\sigma, L'_\sigma)$ of complexes associated with a covering σ of (Y, B) and a map $f: (X, A) \to (Y, B)$. A finite collection of open sets $U_0, \cdots, U_q$ of σ_1 is a simplex

[10] N'_σ may be regarded as the nerve of an indexed covering of (X, A).

s of K'_σ if $(f^{-1}U_0) \frown \cdots \frown (f^{-1}U_q) \neq 0$, and s is in L'_σ if each U_i is in σ_2 and $(f^{-1}U_0) \frown \cdots \frown (f^{-1}U_q) \frown A \neq 0$. Clearly $(K'_\sigma, L'_\sigma) \subset (K_\sigma, L_\sigma)$; Let $\phi_\sigma: N'_\sigma \to N_\sigma$ be the inclusion map. For any covering α of (X, A) with $\sigma < \alpha$ and any $f_{\alpha\sigma}: N_\alpha \to N_\sigma$, the image $f_{\alpha\sigma}K_\alpha$ is contained in K'_σ and $f_{\alpha\sigma}L_\alpha$ is contained in L'_σ. Therefore the map $f_{\alpha\sigma}$ can be factored into $\phi_\sigma\psi_{\alpha\sigma}$ where $\psi_{\alpha\sigma}$ is a simplicial map of N_α into N'_σ and ϕ_σ is the inclusion map of N'_σ into N_σ.

Let $\gamma = f^{-1}\sigma$. For each vertex U of K'_σ, let $\chi_{\sigma\gamma}U = f^{-1}U$. If $s = U_0 \cdots U_q$ is a simplex of K'_σ, $(\chi_{\sigma\gamma}U_0) \frown \cdots \frown (\chi_{\sigma\gamma}U_q) = (f^{-1}U_0) \frown \cdots \frown (f^{-1}U_q) \neq 0$; hence $\chi_{\sigma\gamma}s$ is a simplex of K_γ. If s is in L'_σ, $(f^{-1}U_0) \frown \cdots \frown (f^{-1}U_q) \frown A \neq 0$, and hence $\chi_{\sigma\gamma}\, s$ is in L_γ. Thus $\chi_{\sigma\gamma}: (K'_\sigma, L'_\sigma) \to (K_\gamma, L_\gamma)$ is a uniquely determined simplicial map of N'_σ into N_γ. If $\sigma < \alpha$ and if V is a vertex of $K\alpha$, then $fV \subset f_{\alpha\sigma}V = \psi_{\alpha\sigma}V$ and $V \subset f^{-1}fV \subset f^{-1}\psi_{\alpha\sigma}V = \chi_{\sigma\gamma}\psi_{\alpha\sigma}V$. If V is a vertex of L_α, $\psi_{\alpha\sigma}V$ is a vertex of L'_σ and $\chi_{\sigma\gamma}\psi_{\alpha\sigma}V$ is a vertex of L_γ. Thus $\chi_{\sigma\gamma}\psi_{\alpha\sigma}$ is a possible choice for $\pi_{\alpha\gamma}: N_\alpha \to N_\gamma$. Hence $\pi^*_{\alpha\gamma} = \psi^*_{\alpha\sigma}\chi^*_{\sigma\gamma}$.

Since $\sigma < \gamma$ we can define $\psi_{\gamma\sigma}: N_\gamma \to N'_\sigma$. In fact, if V is a vertex of K_γ, then $V = f^{-1}U$ for some $U \in \sigma_1$ and, if V is a vertex of L_γ, U can be chosen in σ_2. Then, since $fV = ff^{-1}U \subset U$, we can choose $\psi_{\gamma\sigma}V = U$. Then, since $\chi_{\sigma\gamma}U = f^{-1}U$, $\chi_{\sigma\gamma}\psi_{\gamma\sigma}V = V$. Therefore, for this choice of $\psi_{\gamma\sigma}$, the simplicial map $\chi_{\sigma\gamma}\psi_{\gamma\sigma}: N_\gamma \to N_\gamma$ is the identity. Therefore $\psi^*_{\alpha\sigma}\chi^*_{\sigma\gamma}$ is the identity. If $s = U_0 \cdots U_q$ is a simplex of K'_σ, let $p \in (f^{-1}U_0) \frown \cdots \frown (f^{-1}U_q) = (\chi_{\sigma\gamma}U_0) \frown \cdots \frown (\chi_{\sigma\gamma}U_q)$. Then $f(p) \in (f\chi_{\sigma\gamma}U_0) \frown \cdots \frown (f\chi_{\sigma\gamma}U_q) \subset (\psi_{\gamma\sigma}\chi_{\sigma\gamma}U_0) \frown \cdots \frown (\psi_{\gamma\sigma}\chi_{\sigma\gamma}U_q)$. Also $f(p) \in U_0 \frown \cdots \frown U_q$; thus s and $\psi_{\gamma\sigma}\chi_{\sigma\gamma}s$ are faces of a common simplex of K'_σ. If s is in L'_σ, then $\psi_{\gamma\sigma}\chi_{\sigma\gamma}s$ is in L'_σ and p can be chosen in A; hence s and $\psi_{\gamma\sigma}\chi_{\sigma\gamma}s$ are faces of a common simplex of L'_σ. Hence, by A4, $\psi_{\gamma\sigma}\chi_{\sigma\gamma}$ and the identity map of N'_σ in itself induce the same homomorphism of $H^q(N'_\sigma)$ into itself. Therefore $\chi^*_{\sigma\gamma}\psi^*_{\gamma\sigma}$ is the identity. Therefore, since $\psi^*_{\gamma\sigma}\chi^*_{\sigma\gamma}$ is the identity, $\chi^*_{\sigma\gamma}$ is an isomorphism of $H^q(N_\gamma)$ onto $H^q(N'_\sigma)$ and $\psi^*_{\gamma\sigma}$ is the inverse isomorphism.

We have seen that $\pi^*_{\alpha\gamma} = \psi^*_{\alpha\sigma}\chi^*_{\sigma\gamma}$. Therefore $\pi^*_{\alpha\gamma}\psi^*_{\gamma\sigma} = \psi^*_{\alpha\sigma}\chi^*_{\sigma\gamma}\psi^*_{\gamma\sigma} = \psi^*_{\alpha\sigma}$. Since $f_{\alpha\sigma} = \phi_\sigma\psi_{\alpha\sigma}$, $f^*_{\alpha\sigma} = \psi^*_{\alpha\sigma}\phi^*_\sigma$. Therefore

$$f^*_{\alpha\sigma} = \pi^*_{\alpha\gamma}\psi^*_{\gamma\sigma}\phi^*_\sigma$$

where $\pi^*_{\alpha\gamma}$ is a projection homomorphism, $\psi^*_{\gamma\sigma}$ is an isomorphism and ϕ^*_σ is a inclusion homomorphism.

For each pair (X, A) we define a simple map δ of $S^q(A, 0)$ into $S^{q+1}(X, A)$. If σ is a covering of $(A, 0)$ and α is a covering of (X, A), $\sigma < \alpha$ will mean that for every U of α_2 there is some V of σ_1 with $U \frown A \subset V$. If $\sigma < \alpha$ and if $U \in \alpha_2$, let $q_{\alpha\sigma}U \in \sigma_1$ be chosen so that $U \frown A \subset \theta_{\alpha\sigma}U$. Let $s = U_0 \cdots U_q$ be a simplex of L_α and let $p \in U_0 \frown \cdots \frown U_q \frown A$. Then, for each U_i, $p \in U_i \frown A \subset \theta_{\alpha\sigma}U_i$; therefore $\theta_{\alpha\sigma}U_0, \cdots, \theta_{\alpha\sigma}U_q$ are vertices of a simplex $\theta_{\alpha\sigma}s$ of K_σ. Thus $\theta_{\alpha\sigma}$: $(L_\alpha, 0) \to (K_\sigma, L_\sigma)$ is a simplicial map. If $\theta'_{\alpha\sigma}$ is a second choice of $\theta_{\alpha\sigma}$, then, for the same point p, $p \in \theta'_{\alpha\sigma}U_i$. It follows that $\theta_{\alpha\sigma}s$ and $\theta'_{\alpha\sigma}s$ are faces of a common simplex of K_σ, and hence $\theta^*_{\alpha\sigma} = \theta'^*_{\alpha\sigma}$. Thus $\theta^*_{\alpha\sigma}$ does not depend on the choice of $\theta_{\alpha\sigma}$. Let $\delta_\alpha: H^q(L_\alpha) \to H^{q+1}(K_\alpha, L_\alpha)$ be the coboundary homomorphism for the pair of complexes (K_α, L_α). Let $\delta_{\alpha\sigma} = \delta_\alpha\theta^*_{\alpha\sigma}: H^q(K_\sigma, L_\sigma) \to H^{q+1}(K_\alpha, L_\alpha)$. Let $\delta = \{\delta_{\alpha\sigma}\}$ be the set of these homomorphisms.

To show that $\delta = \{\delta_{\alpha\sigma}\}$ is a simple map we must verify conditions 1) and 2) of B5.

1) Let σ be any covering of $(A, 0)$. Notice that any open set $V \in \sigma_1$ is the intersection with A of some open set U of X. Let γ_2 be the set of all open sets U of X such that $U \frown A \in \sigma_1$. Let γ_1 be the set of all open sets of X. Then $\gamma = (\gamma_1, \gamma_2)$ is a covering of (X, A) such that for every $U \in \gamma_2$, $U \frown A \subset U \frown A \in \sigma_1$. Thus $\sigma < \gamma$.

2) Let $\rho < \sigma < \alpha < \beta$. If $U \in \beta_2$, then $U \subset \pi_{\beta\alpha}U$, and hence $U \frown A \subset (\pi_{\beta\alpha}U) \frown A \subset \theta_{\alpha\sigma}\pi_{\beta\alpha}U \subset \pi_{\sigma\rho}\theta_{\alpha\sigma}\pi_{\beta\alpha}U \in \rho_1$. Thus $\rho < \beta$ and $\pi_{\sigma\rho}\theta_{\alpha\sigma}(\pi_{\beta\alpha} \mid L_\beta)$ is a possible choice for $\theta_{\beta\rho}$. Hence $\theta^*_{\beta\rho} = (\pi_{\beta\alpha} \mid L_\beta)^* \, \theta^*_{\alpha\sigma}\pi^*_{\sigma\rho}$. Therefore, using A3, $\pi^*_{\beta\alpha}\delta_{\alpha\sigma}\pi^*_{\sigma\rho} = \pi^*_{\beta\alpha}\delta_\alpha\theta^*_{\alpha\sigma}\pi^*_{\sigma\rho} = \delta_\beta(\pi_{\beta\alpha}L_\beta)^*\theta^*_{\alpha\sigma}\pi^*_{\sigma\rho} = \delta_\beta\theta^*_{\beta\rho} = \delta_{\beta\rho}$.

Thus $\delta = \{\delta_{\alpha\sigma}\}$ is a simple map of $S^q(A, 0)$ into $S^{q+1}(X, A)$. We also write δ for the induced homomorphism of the limit groups, $\delta: H^q(A, 0) \to H^{q+1}(X, A)$.

We now further analyse the homomorphisms $\delta_{\alpha\sigma}$. Let τ_1 be the set of all open sets $U \frown A$ of A with $U \in \alpha_2$. Then $\tau = (\tau_1, 0) = (\alpha_2 \frown A, 0)$ is a covering of $(A, 0)$. Let $\theta_{\alpha\tau}: (L_\alpha, 0) \to (K_\tau, 0)$ be chosen so that $\theta_{\alpha\tau}U = U \frown A$ for each vertex of L_α; thus $\theta_{\alpha\tau}$ becomes uniquely determined. For each vertex V of K_τ, let $\kappa_{\tau\alpha}V$ be chosen as one of the open sets $U \in \alpha_2$ such that $U \frown A = V$. If $s = V_0 \cdots V_q$ is a simplex of K_τ, let $p \in V_0 \frown \cdots \frown V_q$. Thus for each V_i, $p \in V_i = (\kappa_{\tau\alpha}V_i) \frown A$. Hence $p \in (\kappa_{\tau\alpha}V_0) \frown \cdots \frown (\kappa_{\tau\alpha}V_q) \frown A$. Therefore $\kappa_{\tau\alpha}s$ is a simplex of L_α. Thus $\kappa_{\tau\alpha}: (K_\tau, 0) \to (L_\alpha, 0)$ is a simplicial map. Clearly $\theta_{\alpha\tau}\kappa_{\tau\alpha}$ is the identity map of $(K_\tau, 0)$ on itself and hence $\kappa^*_{\tau\alpha}\theta^*_{\tau\alpha}$ is the identity. If $s = U_0 \cdots U_q$ is a simplex of L_α, let $p \in U_0 \frown \cdots \frown U_q \frown A$. Then, for each U_i, $p \in U_i \frown A = \theta_{\alpha\tau}U_i = (\kappa_{\tau\alpha}\theta_{\alpha\tau}U_i) \frown A$. Hence s and $\kappa_{\tau\alpha}\theta_{\alpha\tau}s$ are faces of a common simplex of L_α; hence, by A4, $\theta^*_{\alpha\tau}\kappa^*_{\tau\alpha}$ is the identity. Hence $\theta^*_{\alpha\tau}$ is an isomorphism and $\kappa^*_{\tau\alpha}$ is its inverse. Thus $\kappa^*_{\tau\alpha}$ is independent of the choice of the map $\kappa_{\tau\alpha}$.

Now if $\sigma < \alpha$, $\sigma < \tau = (\alpha_2 \frown A, 0)$. Then $\delta_{\alpha\sigma} = \delta_{\alpha\tau}\pi^*_{\tau\sigma}$, where $\delta_{\alpha\tau} = \delta_\alpha\theta^*_{\alpha\tau}$. Thus

$$\delta_{\alpha\sigma} = \delta_\alpha\theta^*_{\alpha\tau}\pi^*_{\tau\sigma}$$

where δ_α is a coboundary homomorphism, $\theta^*_{\alpha\tau}$ is an isomorphism and $\pi^*_{\tau\sigma}$ is a projection homomorphism.

E. Verification of the axioms

We have now defined cohomology groups $H^q(X, A)$ for pairs (X, A) where X is any topological space and A is any subset of X. For any continuous map $f: (X, A) \to (Y, B)$ we have defined $f^*: H^q(Y, B) \to H^q(X, A)$. And we have defined homomorphisms $\delta: H^q(A) \to H^{q+1}(X, A)$. We wish to show that the cohomology theory so defined satisfies the seven Eilenberg-Steenrod axioms. The statements of these axioms are the statements of the following seven theorems. Thus in proving these theorems we show that the Čech cohomology theory satisfies the axioms for a cohomology theory.

E1. *If $f: (X, A) \to (X, A)$ is the identity map, then $f^*: H^q(X, A) \to H^q(X, A)$ is the identity homomorphism.*

Proof. Since f is the identity, $f^{-1}\sigma = \sigma$ for every covering σ. An obvious choice for $f_{\sigma\sigma}$ is that defined by $f_{\sigma\sigma}U = U \subset U$ for all $U \in \sigma_1$. Then $f_{\sigma\sigma}:N_\sigma \to N_\sigma$ is the identity map and $f^*_{\sigma\sigma}$ is the identity homomorphism. If $\sigma < \alpha$, $f^*_{\alpha\sigma} = \pi^*_{\alpha\sigma} f^*_{\sigma\sigma} = \pi^*_{\alpha\sigma}$. Thus f^* is the identity map of $S^q(X, A)$ into itself and hence, by B7, $f^*:H^q(X, A) \to H^q(X, A)$ is the identity homomorphism.

E2. *Given* g: $(X, A) \to (Y, B)$ *and* f: $(Y, B) \to (Z, C)$, *then*

$$(fg)^* = g^*f^*:H^q(Z, C) \to H^q(X, A).$$

PROOF. A typical element of the map g^*f^* of $S^q(Z, C)$ into $S^q(X, A)$ is $g^*_{\mu\alpha}f^*_{\alpha\sigma}$ where μ is a covering of (X, A), α is a covering of (Y, B) and σ is a covering of (Z, C). If $U \in \mu_1$, $gU \subset g_{\mu\alpha}U$ and $fgU \subset fg_{\mu\alpha}U \subset f_{\alpha\sigma}g_{\mu\alpha}U$. Moreover, if $U \in \mu_2$, $f_{\alpha\sigma}g_{\mu\alpha}U \in \sigma_2$. Thus $f_{\alpha\sigma}g_{\mu\alpha}$ is a possible choice for $(fg)_{\mu\sigma}$. Hence $(fg)^*_{\mu\sigma} = (f_{\alpha\sigma}g_{\mu\alpha})^* = g^*_{\mu\alpha}f^*_{\alpha\sigma}$. Thus $g^*_{\mu\alpha}f^*_{\alpha\sigma} \in (fg)^*$. Thus $g^*f^* \subset (fg)^*$ and hence, by B4, the two maps induce the same homomorphism of the limit groups, $(fg)^* = g^*f^*:H^q(Z, C) \to H^q(X, A)$.

E3. *If* $f:(X, A) \to (Y, B)$ *then* $f^*\delta = \delta(f \mid A)^*:H^q(B) \to H^{q+1}(X, A)$.

PROOF. A typical element of the map $f^*\delta$ of $S^q(B, 0)$ into $S^{q+1}(X, A)$ is $f^*_{\mu\eta}\delta_{\eta\sigma}$ where μ is a covering of (X, A), η is a covering of (Y, B) and σ is a covering of $(B, 0)$. Using A3 we have

$$\begin{aligned} f^*_{\mu\eta}\delta_{\eta\sigma} &= f^*_{\mu\eta}\delta_\eta\theta^*_{\eta\sigma} \\ &= \delta_\mu(f_{\mu\eta} \mid L_\mu)^*\theta^*_{\eta\sigma}. \end{aligned}$$

Let β be the covering $((f^{-1}\eta_2) \frown A, 0)$ of $(A, 0)$. Let U be any element of β_1. Then, for some $V \in \eta_2$, $U = (f^{-1}V) \frown A$. Since $\sigma < \eta$, $V \frown B \subset W$ for some $W \in \sigma_1$. Then, for the partial map $f \mid A:A \to B$, $(f \mid A)^{-1}W = (f^{-1}W) \frown A \supset (f^{-1}V \frown f^{-1}B) \frown A = f^{-1}V \frown A = U$, and hence $(f \mid A)U \subset W$. Therefore $\sigma < \beta$ and hence $(f \mid A)^*_{\beta\sigma}$ exists in $(f \mid A)^*$. Now let U be any element of μ_2. Then, since $\eta < \mu$, $fU \subset V$ for some $V \in \eta_2$. Then $U \subset f^{-1}V$ and $U \frown A \subset f^{-1}V \frown A \in \beta_1$. Therefore $\beta < \mu$ and $\delta_{\mu\beta}$ exists in the map δ. Then by the definition of δ,

$$\delta_{\mu\beta}(f \mid A)^*_{\beta\sigma} = \delta_\mu\theta^*_{\mu\beta}(f \mid A)^*_{\beta\sigma}.$$

Let $s = U_0 \cdots U_q$ be any simplex of L_μ and, for each vertex U_i of s, let V_i be the vertex $\theta_{\eta\sigma}f_{\mu\eta}U_i$ of K_σ and let W_i be the vertex $(f \mid A)_{\beta\sigma}\theta_{\mu\beta}U_i$ of K_σ. Let $p \in U_0 \frown \cdots \frown U_q \frown A$. Then $p \in U_i \frown A$ and $f(p) \in fU_i \frown fA \subset f_{\mu\eta}U_i \frown B \subset V_i$. Also $p \in U_i \frown A \subset \theta_{\mu\beta}U_i$ and $f(p) \in (f \mid A)\theta_{\mu\beta}U_i \subset W_i$. Thus $f(p) \in V_0 \frown \cdots \frown V_q \frown W_0 \frown \cdots \frown W_q$. Thus $\theta_{\eta\sigma}f_{\mu\eta}s$ and $(f \mid A)_{\beta\sigma}\theta_{\mu\beta}s$ are faces of a common simplex of K_σ. Therefore the simplicial maps $\theta_{\eta\sigma}(f_{\mu\eta} \mid L_\mu)$ and $(f \mid A)_{\beta\sigma}\theta_{\mu\beta}:(L_\mu, 0) \to (K_\sigma, L_\sigma)$ induce the same homomorphisms

$$(f_{\mu\eta} \mid L_\mu)^*\theta^*_{\eta\sigma} = \theta^*_{\mu\beta}(f \mid A)^*_{\beta\sigma}$$

of the cohomology groups. Substituting in the equations above, we see that $f^*_{\mu\eta}\delta_{\eta\sigma} = \delta_{\mu\beta}(f \mid A)^*_{\beta\sigma} \in \delta(f \mid A)^*$ for an arbitrary element $f^*_{\mu\eta}\delta_{\eta\sigma}$ of $f^*\delta$. Thus $f^*\delta \subset \delta(f \mid A)^*$, and hence, by B4, the two maps of $S^q(B, 0)$ into $S^{q+1}(X, A)$ induce the same homomorphism of the limit groups, $f^*\delta = \delta(f \mid A)^*:H^q(B) \to H^{q+1}(X, A)$.

E4. *If* $f:(X, A) \to (Y, B)$ *and* $g:(X, A) \to (Y, B)$ *are homotopic, then* $f^* = g^*:H^q(Y, B) \to H^q(X, A)$.

We use the following lemma.

LEMMA. *Let* I *be the closed interval* $0 \leqq t \leqq 1$ *and let* λ *and* μ *be the maps of* (X, A) *into* $(X \times I, A \times I)$ *defined by* $\lambda(x) = (x, 0)$ *and* $\mu(x) = (x, 1)$ *for* $x \in X$. *Then* $\lambda^* = \mu^*:H^q(X \times I, A \times I) \to H^q(X, A)$.

PROOF. Let ρ be any covering of $(X \times I, A \times I)$. Let x be any point of X. For each point (x, t) of $x \times I$, choose an open set U of X and a proper subinterval W of I open in I so that $(x, t) \in U \times W$, so that $U \times W$ is contained in a set of ρ_1, and also, if $(x, t) \in A \times I$, i.e., if $x \in A$, so that $U \times W$ is contained in a set of ρ_2. The covering of I by these sets W contains a finite irreducible covering. The intervals of this finite covering can be arranged in order $W_0(x)$, $W_1(x), \cdots, W_r(x)$ so that $0 \in W_0(x)$, $1 \in W_r(x)$ and $W_{i-1}(x) \frown W_i(x) \neq 0$ for $i = 1, 2, \cdots, r(x)$. Since W_0 is a proper subinterval of I, $W_0(x) \neq W_{r(x)}(x)$ and hence $r(x) > 0$. Let $U(x)$ be the intersection of the finite number of open sets $U_0, \cdots, U_{r(x)}$ corresponding to the sets $W_0, \cdots, W_{r(x)}$. Let $V(x, i) = U(x) \times W_i(x)$.

The sets $U(x)$ cover X and the sets $U(x)$ with $x \in A$ cover A. Let a subset $[A]$ of A be chosen so that the sets $U(x)$ with $x \in [A]$ are mutually distinct and cover A. Let a subset $[X]$ of X be chosen so that $[X] \supset [A]$ and so that the sets $U(x)$ with $x \in [X]$ cover X. Let α_1 be the set of all the open sets $U(x)$ with $x \in [X]$ and let α_2 be the set of all $U(x)$ with $x \in [A]$. Then $\alpha = (\alpha_1, \alpha_2)$ is a covering of (X, A). Let σ_1 be the set of all the open sets $V(x, i) = U(x) \times W_i(x)$ with $x \in [X]$ and let σ_2 be the set of all $V(x, i)$ with $x \in [A]$. Then $\sigma = (\sigma_1, \sigma_2)$ is a covering of $(X \times I, A \times I)$ and $\rho < \sigma$.

Let $U(x) \in \alpha_1$; then, since $\lambda(x) = (x, 0)$, $\lambda U(x) = U(x) \times 0 \subset U(x) \times W_0(x) = V(x, 0) \in \sigma_1$. If $U(x) \in \alpha_2$, then $x \in [A]$ and $\lambda U(x) = V(x, 0) \in \sigma_2$. Thus $\sigma \prec \alpha$ and we may choose $\lambda_{\alpha\sigma} U(x) = V(x, 0)$. Similarily we may choose $\mu_{\alpha\sigma} U(x) = V(x, r(x))$. Let us use the following simplified notation: Let x stand for the vertex $U(x)$ of K_α and let x^i stand for the vertex $V(x, i)$ of K_σ. With this notation, $\lambda_{\alpha\sigma} x = x^0$ and $\mu_{\alpha\sigma}\, x = x^{r(x)}$.

Let the vertices of K_α be given an arbitrary simple order. With each vertex $x^i = V(x, i)$ of K_σ we associate a number $t(x, i)$, $0 \leqq t(x, i) < 1$, as follows: For every x let $t(x, 0) = 0$. If $i > 0$, $t(x, i)$ is chosen arbitrarily in the non-empty interval $W_{i-1}(x) \frown W_i(x)$. If x^i and y^j are vertices of K_σ we shall say that $x^i < y^j$ if $t(x, i) < t(y, j)$ or, in case $t(x, i) = t(y, j)$, if $x < y$ in K_α. It is easily verified that the vertices of K_σ are thus given a simple order. Notice that $x^i < x^j$ if and only if $i < j$, $x^0 < y^j$ if $j > 0$, $x^0 < y^0$ if $x < y$, and if $x^i < y^j$ then $t(x, i) \leqq t(y, j)$.

Corresponding to each vertex x^j of K_σ we define a homomorphism $c_q(x^j)$: $C_q(K_\alpha) \to C_q(K_\sigma)$ of the integral q-chains of K_α into the integral q-chains of K_σ. For any elementary chain (oriented simplex) $s_q \in C_q(K_\alpha)$ such that $x \notin s_q$ let $c_q(x^j, s_q) = 0$. If $x \in s_q$, suppose that $s_q = +x_0x_1 \cdots x_q$ with the vertices written in ascending order and with $x = x_n$. Then $c_q(x_n^j, s_q)$ is defined to be the elementary chain $x_0^i \cdots x_p^l \cdots x_n^j \cdots x_q^k$ where each vertex x_p of s_q is replaced by the largest x_p^l not exceeding x_n^j; in other words, l is chosen as large as possible with $x_p^l \leqq x_n^j$. In the cases where this choice of vertices is impossible, namely when $j = 0$ and $n < q$, we define $c_q(x_n^0, s_q) = 0$. Note that, if $x_n^j \neq x_m^h$ and if $c_q(x_n^j, s_q)$ and $c_q(x_m^h, s_q)$ are not both zero, then $c_q(x_n^j, s_q) \neq c_q(x_m^h, s_q)$. We postpone verifying that there actually is a simplex of K_σ with the vertices $x_0^i, \cdots, x_q^k$.

For a fixed s_q, the non-zero elementary chains $c_q(x_n^j, s_q)$ can be ordered by x_n^j, i.e., $c_q(x_n^j, s_q) < c_q(x_m^h, s_q)$ if $x_n^j < x_m^h$. Let them be written in ascending order: $c_q^0(s_q), \cdots, c_q^r(s_q), \cdots, c_q^R(s_q)$. The first in this order is $c_q^0(s_q) = c_q(x_q^0, s_q) = x_0^0 x_1^0 \cdots x_q^0 = \lambda_{\alpha\sigma q}(s_q)$, the last is $c_q^R(s_q) = x_0^{r(x_0)} \cdots x_q^{r(x_q)} = \mu_{\alpha\sigma q}(s_q)$. If, for $r > 0$, $c_q^r(s_q) = c_q(x_n^j, s_q) = x_0^i \cdots x_n^j \cdots x_q^k$, then it is clear that $c_q^{r-1}(s_q) = x_0^i \cdots x_n^{j-1} \cdots x_q^k$. The number R is the number of x_n^j with $x_n \in s_q$ and $j > 0$; thus $R = R(s_q) = r(x_0) + r(x_1) + \cdots + r(x_q)$.

Corresponding to each vertex x^j of K_σ we define a homomorphism $d_{q+1}(x^j)$: $C_q(K_\alpha) \to C_{q+1}(K_\sigma)$. If $x \notin s_q$ or if $j = 0$, let $d_{q+1}(x^j, s_q) = 0$. If $x = x_n \in s_q$ and $j > 0$, let $d_{q+1}(x_n^j, s_q) = x_n^{j-1} x_0^i \cdots x_n^j \cdots x_q^k$ where $x_0^i \cdots x_n^j \cdots x_q^k = c_q(x_n^j, s_q)$. Then, if $x_n^j \neq x_m^h$ and if $d_{q+1}(x_n^j, s_q)$ and $d_{q+1}(x_m^h, s_q)$ are not both zero, $d_{q+1}(x_n^j, s_q) \neq d_{q+1}(x_m^h, s_q)$. For fixed s_q the non-zero elementary chains $d_{q+1}(x_n^j, s_q)$ are ordered by x_n^j. Let them be written in ascending order: $d_{q+1}^1(s_q), \cdots, d_{q+1}^R(s_q)$. If $d_{q+1}^r(s_q) = d_{q+1}(x_n^j, s_q) = x_n^{j-1} x_0^i \cdots x_n^j \cdots x_q^k$, then the boundary of $d_{q+1}^r(s_q)$ is $\partial d_{q+1}^r(s_q) = x_0^i \cdots x_n^j \cdots x_q^k - x_0^i \cdots x_n^{j-1} \cdots x_q^k + \sum_{p \neq n} (-1)^{p+1} x_n^{j-1} x_0^i \cdots \hat{x}_p \cdots x_n^j \cdots x_q^k$, where by $\hat{x}_p$ we mean that the vertex with the index p is omitted. Thus

$$\partial d_{q+1}^r(s_q) = \partial d_{q+1}(x_n^j, s_q) = c_q^r(s_q) - c_r^{q-1}(s_q) + \sum_{\substack{p=0 \\ (p \neq n)}}^{q} (-1)^{p+1} d_q(x_n^j, x_0 \cdots \hat{x}_p \cdots x_q).$$

We now verify the existence of $d_{q+1}(x_n^j, s_q) = x_n^{j-1} x_0^i \cdots x_p^l \cdots x_n^j \cdots x_q^k$ with $j > 0$ and $x_n \in s_q$ by showing that the vertices $x_n^{j-1}, x_0^i, \cdots, x_q^k$ of K_σ are vertices of a $(q + 1)$-simplex of K_σ; i.e., that the open sets $V(x_n, j - 1)$, $V(x_0, i)$, $\cdots$, $V(x_q, k)$ have a non-empty intersection in $X \times I$. Since $s_q = x_0 \cdots x_q = U(x_0) \cdots U(x_q)$ is a simplex of K_α, the open sets $U(x_0), \cdots, U(x_q)$ have a non-empty intersection in X; let y be a point of this intersection. We show that the point $(y, t(x_n, j))$ is contained in $V(x_n, j - 1) \frown \cdots \frown V(x_q, k)$. Let $p \neq n$; then l is chosen as large as possible with $x_p^l < x_n^j$. If $l < r(x_p)$, $t(x_p, l) \leqq t(x_n, j) \leqq t(x_p, l + 1)$ with both $t(x_p, l)$ and $t(x_p, l + 1)$ contained in the interval $W_l(x_p)$; if $l = r(x_p)$, $t(x_p, l) \leqq t(x_n, j) < 1$ with both $t(x_p, l)$ and 1 contained in $W_l(x_p)$. Hence, in either case, $t(x_n, j) \in W_l(x_p)$ and $(y, t(x_n, j)) \in U(x_p) \times W_l(x_p) = V(x_p, l)$. Since $j > 0$, $t(x_n, j) \in W_{j-1}(x_n) \frown W_j(x_n)$ and hence $(y, t(x_n, j)) \in U(x_n) \times W_{j-1}(x_n) = V(x_n, j - 1)$ and also $(y, t(x_n, j)) \in U(x_n) \times W_j(x_n) = V(x_n, j)$. Thus the intersection $V(x_n, j - 1) \frown V(x_0, i) \frown \cdots \frown V(x_n, j) \frown \cdots \frown V(x_q, k)$ is not empty and the simplex $x_n^{j-1} x_0^i \cdots x_n^j \cdots x_q^k$ exists in K_σ.

The vertices of $c_q(x_n^j, s_q)$, with $j > 0$, are among the vertices of $d_{q+1}(x_n^j, s_q)$. Hence $c_q(x_n^j, s_q)$ exists when $j > 0$. The point $(y, 0)$ is contained in $V(x_0, 0) \frown \cdots \frown V(x_q, 0)$ and hence $c_q^0(s_q) = c_q(x_q^0, s_q) = x_0^0 \cdots x_q^0$ also exists.

Suppose the simplex $s_q = x_0 \cdots x_q$ is in L_α. Then the open sets $U(x_0), \cdots, U(x_q)$ are in α_2, hence the points $x_0, \cdots, x_q$ are in $[A]$ and the sets $V(x_0, i), \cdots, V(x_q, k)$ are in σ_2. Moreover, since s_q is in L_α, the point y can be chosen in the non-empty intersection $U(x_0) \frown \cdots \frown U(x_q) \frown A$. Then $y \in A$ and $(y, t(x_n, j)) \in A \times I$. Therefore the intersection $V(x_n, j-1) \frown \cdots \frown V(x_q, k) \frown (A \times I)$ is not empty. Hence $x_n^{j-1} \cdots x_q^k$ is a simplex of L_σ. Thus, if s_q is in L_α, $d_{q+1}(x_n^j, s_q)$ is in L_σ.

Finally we define a homomorphism $D_{q+1} : C_q(K_\alpha) \to C_{q+1}(K_\sigma)$ of the integral q-chains of K_α into the integral $(q+1)$-chains of K_σ. For any elementary q-chain s_q of K_α let

$$D_{q+1}(s_q) = \sum_{\substack{j \\ x_n}} d_{q+1}(x_n^j, s_q)$$

or, if we omit the zero terms in the indicated sum,

$$D_{q+1}(s_q) = \sum_{\substack{x_n \in s_q \\ j > 0}} d_{q+1}(x_n^j, s_q) = \sum_{r=1}^{R} d_{q+1}^r(s_q).$$

(In the special case where $s_q = x$ is an elementary 0-chain, $c_0(x^j, x) = x^j$, $d_1(x^j, x) = x^{j-1}x^j$, and $D_1(x) = \sum_{i=1}^{r(x)} x^{i-1}x^i$.) If s_q is in L_α, $d_{q+1}^r(s_q)$ is in L_σ, and hence $D_{q+1}(s_q)$ is in L_σ. Thus D_{q+1} maps $C_q(L_\alpha)$ into $C_{q+1}(L_\sigma)$.

The boundary of $s_q = x_0 \cdots x_q$ is

$$\partial s_q = \sum_{p=0}^{q} (-1)^p x_0 \cdots \hat{x}_p \cdots x_q$$

and therefore

$$D_q \partial s_q = \sum_{p=0}^{q} (-1)^p \sum_{\substack{x_n \in s_q, n \neq p \\ j > 0}} d_q(x_n^j, x_0 \cdots \hat{x}_p \cdots x_q).$$

The boundary of $D_{q+1}(s_q)$ is

$$\partial D_{q+1}(s_q) = \sum_{r=1}^{R} \partial d_{q+1}^r(s_q) = \sum_{\substack{x_n \in s_q \\ j > 0}} \partial d_{q+1}(x_n^j, s_q).$$

Therefore, if we substitute the expression for $\partial d_{q+1}(x_n^j, s_q)$,

$$\partial D_{q+1}(s_q) = \sum_{r=1}^{R} c_q^r(s_q) - \sum_{r=1}^{R} c_q^{r-1}(s_q)$$

$$+ \sum_{\substack{x_n \in s_q \\ j > 0}} \sum_{\substack{p=0 \\ (p \neq n)}}^{q} (-1)^{p+1} d_q(x_n^j, x_0 \cdots \hat{x}_p \cdots x_q)$$

$$= c_q^R(s_q) - c_q^0(s_q) - D_q \partial s_q = \mu_{\alpha\sigma q} s_q - \lambda_{\alpha\sigma q} s_q - D_q \partial s_q.$$

Therefore

$$\partial D_{q+1} = \mu_{\alpha\sigma q} - \lambda_{\alpha\sigma q} - D_q \partial.$$

Applying A8 we have $\overset{*}{\mu}_{\alpha\sigma} = \overset{*}{\lambda}_{\alpha\sigma} : H^q(K_\sigma, L_\sigma) \to H^q(K_\alpha, L_\alpha)$. Therefore $\overset{*}{\lambda}_{\alpha\sigma}\overset{*}{\pi}_{\sigma\rho} = \overset{*}{\mu}_{\alpha\sigma}\overset{*}{\pi}_{\sigma\rho}$, i.e., $\overset{*}{\lambda}_{\alpha\rho} = \overset{*}{\mu}_{\alpha\rho} : H^q(K_\rho, L_\rho) \to H^q(K_\alpha, L_\alpha)$, where ρ is an arbitrary covering of $(X \times I, A \times I)$ and α is a suitably chosen covering of (X, A). Thus the two maps λ^* and μ^* of $S^q(X \times I, A \times I)$ into $S^q(X, A)$ are equivalent. Therefore, by B3, these maps induce the same homomorphism of the limit groups,

$$\lambda^* = \mu^* : H^q(X \times I, A \times I) \to H^q(X, A).$$

This completes the proof of the lemma.

PROOF OF E4. Let f and g be homotopic maps of (X, A) into (Y, B). This means that there exists a map $h: (X \times I, A \times I) \to (Y, B)$ such that $h(x, 0) = f(x)$ and $h(x, 1) = g(x)$ for $x \in X$. Thus, if λ and μ are the maps of (X, A) into $(X \times I, A \times I)$ defined by $\lambda(x) = (x, 0)$ and $\mu(x) = (x, 1)$, then $f = h\lambda$ and $g = h\mu$. Therefore, by E2, $f^* = \lambda^*h^*$ and $g^* = \mu^*h^*$. But, according to the lemma, $\lambda^* = \mu^*$. Therefore

$$f^* = g^* : H^q(Y, B) \to H^q(X, A).$$

E5. *The following cohomology sequence is exact:*

$$0 \to H^0(X, A) \to \cdots \to H^q(X, A) \xrightarrow{j^*} H^q(X) \xrightarrow{i^*} H^q(A) \xrightarrow{\delta} H^{q+1}(X, A) \to \cdots.$$

PROOF. Let σ be any covering of (X, A), α any covering of $(X, 0)$ and μ any covering of $(A, 0)$. Then (α_1, α_1) is a covering of (X, A). Let ρ_2 be the set of all open sets U of X such that $U \frown A \in \lambda_1$ and let ρ_1 be the set of all open sets of X. Then $\rho = (\rho_1, \rho_2)$ is a covering of (X, A). Let $\tau = (\mathfrak{U}, \mathfrak{B})$ be a common refinement of σ, (α_1, α_1) and ρ. Let β be the covering $(\mathfrak{U}, 0)$ of $(X, 0)$ and let μ be the covering $(\mathfrak{B} \frown A, 0)$ of $(A, 0)$. Then $\sigma < \tau$ and it is easily seen that $\alpha < \beta$ and $\lambda < \mu$.

For each complex L we define the (-1)-dimensional cohomology group to be the group consisting only of zero, i.e., $H^{-1}(L) = 0$. Then, by A5, the sequence

$$H^{-1}(L_\tau) \to H^0(K_\tau, L_\tau) \cdots H^q(K_\tau, L_\tau) \xrightarrow{j_\tau^*} H^q(K_\tau) \xrightarrow{i_\tau^*} H^q(L_\tau) \xrightarrow{\delta_\tau} H^{q+1}(K_\tau, L_\tau)$$

is exact. Since μ was defined to be $(\tau_2 \frown A, 0)$, $\mu < \tau$ and $\delta_{\tau\mu} = \delta_\tau\overset{*}{\theta}_{\tau\mu}$ exists and, as we saw in part D, $\overset{*}{\theta}_{\tau\mu}$ is an isomorphism of $H^q(K_\mu, 0)$ onto $H^q(L_\tau, 0)$. Since K_τ and K_β are both nerves of the same covering $\mathfrak{U}$ there is an obvious isomorphic map $\zeta_{\tau\beta} : (K_\tau, 0) \to (K_\beta, 0)$ with $\zeta_{\tau\beta}U = U$ for each $U \in \tau_1$. Then $\overset{*}{\zeta}_{\tau\beta}$ is an isomorphism of $H^q(K_\beta, 0)$ onto $H^q(K_\tau, 0)$. The isomorphisms $(\overset{*}{\zeta}_{\tau\beta})^{-1}$ and $(\overset{*}{\theta}_{\tau\lambda})^{-1}$ transform the above exact sequence into

$$\to H^q(K_\tau, L_\tau) \to H^q(K_\beta, 0) \to H^q(K_\mu, 0) \to H^{q+1}(K_\tau, L_\tau) \to$$

which must also be exact. The homomorphisms of the transformed sequence are $(\overset{*}{\zeta}_{\tau\beta})^{-1}\overset{*}{j}_\tau$, $(\overset{*}{\theta}_{\tau\mu})^{-1}\overset{*}{i}_\tau\overset{*}{\zeta}_{\tau\beta}$ and $\delta_\tau\overset{*}{\theta}_{\tau\mu}$.

Let $\kappa_{\mu\tau} : (K_\mu, 0) \to (L_\tau, 0)$ be defined as in part D so that, for each set $U \in \mu_1$, $\kappa_{\mu\tau}U$ is one of the open sets V such that $V \frown A = U$; then $\overset{*}{\kappa}_{\mu\tau} = (\overset{*}{\theta}_{\tau\mu})^{-1}$. If i: $(A, 0) \to (X, 0)$ is the inclusion map, then $iU = U \subset V = \kappa_{\mu\tau}U = \zeta_{\tau\beta}\, i_\tau\kappa_{\mu\tau}U \in \beta_1$. Thus $\beta < \mu$ and $\zeta_{\tau\beta}i_\tau\kappa_{\mu\tau}U$ is a possible choice for $i_{\mu\beta}U$. Hence $\overset{*}{i}_{\mu\beta} = \overset{*}{\kappa}_{\mu\tau}\overset{*}{i}_\tau\overset{*}{\zeta}_{\tau\beta} = (\overset{*}{\theta}_{\tau\mu})^{-1}\overset{*}{i}_\tau\overset{*}{\zeta}_{\tau\beta}$.

Let $j: (X, 0) \to (X, A)$ be the inclusion map. Then, for each $U \in \beta_1$, $jU = U \subset U \in \tau_1$. Thus $\tau < \beta$ and a possible choice for $j_{\beta\tau}$ is $j_{\beta\tau}U = U$. Then $j_{\beta\tau}\zeta_{\tau\beta}U = U = j_\tau U$ for each $U \in \tau_1$ and hence $\zeta^*_{\tau\beta}\dot{j}^*_{\beta\tau} = \dot{j}^*_\tau$. Therefore $\dot{j}^*_{\beta\tau} = (\zeta^*_{\tau\beta})^{-1}\dot{j}^*_\tau$.

Therefore, using the fact that $\delta_{\tau\mu} = \delta_\tau\theta^*_{\tau\mu}$, we see that the above exact sequence is

$$\to H^q_\tau \xrightarrow{\dot{j}^*_{\beta\tau}} H^q_\beta \xrightarrow{\dot{i}^*_{\mu\beta}} H^q_\mu \xrightarrow{\delta_{\tau\mu}} H^{q+1}_\tau \to.$$

It follows from B8 that the limit sequence

$$H^{-1}(A) \to H^0(X, A) \cdots H^q(X, A) \xrightarrow{j^*} H^q(X) \xrightarrow{i^*} H^q(A) \xrightarrow{\delta} H^{q+1}(X, A) \to \cdots$$

is also exact. Here $H^{-1}(A)$ is the direct limit of a spectrum of groups all of which are zero; thus the only bundle is the zero bundle and hence $H^{-1}(A) = 0$. This completes the proof.

E6. *Let U be an open set whose closure is contained in the interior of A and let f be the inclusion map of $(X - U, A - U)$ into (X, A). Then $f^*: H^q(X, A) \to H^q(X - U, A - U)$ is an isomorphism onto.*

Proof. Let ρ be any covering of (X, A) and let α be any covering of $(X - U, A - U)$. Let λ_1 be the set of all open sets W of X such that $W \frown (X - U) \in \alpha_1$ and let λ_2 be the set of all open sets W of X such that $W \frown (X - U) \in \alpha_2$. Then $\lambda = (\lambda_1, \lambda_2)$ is a covering of (X, A). Let $\mu_1 = \mu_2$ be the pair of open sets: the interior of A and the set $X - \bar{U}$. By hypothesis, $\mu = (\mu_1, \mu_2)$ is a covering of (X, A). Let ν be a common refinement of ρ, λ and μ. Then every open set of ν_1 which meets $\bar{U}$ is contained in the interior of A. Let σ_1 be formed by dropping from ν_1 those sets which meet $\bar{U}$ and are not in ν_2; let $\sigma_2 = \nu_2$. Then $\sigma = (\sigma_1, \sigma_2)$ is still a covering of (X, A) and $\nu < \sigma$. Let $\beta = f^{-1}\sigma$. Clearly $\alpha < \beta$ and $\rho < \sigma$.

Let V be the set of all simplexes $s = U_0 \cdots U_q$ of K_σ such that

$$U_0 \frown \cdots \frown U_q \subset U.$$

Then, if s_1 is any simplex of the star of s, s_1 is also in V. Since each $U_i \frown U \neq 0$, $U_i \in \sigma_2$ and s is in L_σ. Thus V is an open subset of L_σ. The complex K^f_σ (See part D) is the set of all simplexes $s = U_0 \cdots U_q$ of K_σ such that, for some point p of $X - U$, $p = f(p) \in U_0 \frown \cdots \frown U_q$, i.e., such that $U_0 \frown \cdots \frown U_q$ is not contained in U. Thus $K^f_\sigma = K_\sigma - V$ and similarily $L^f_\sigma = L_\sigma - V$. Then, as we have seen, $f^*_{\beta\sigma} = \psi^*_{\beta\sigma}\phi^*_\sigma$ where $\psi^*_{\beta\sigma}$ is an isomorphism onto and ϕ^*_σ is the homomorphism induced by the inclusion map $\phi_\sigma: (K_\sigma - V, L_\sigma - V) \to (K_\sigma, L_\sigma)$. But, by A6, ϕ^*_σ is an isomorphism onto. Hence $f^*_{\beta\sigma}$ is an isomorphism onto. It follows from B6 that f^* is an isomorphism of $H^q(X, A)$ onto $H^q(X - U, A - U)$.

E7. *If X consists of a single point, $H^q(X) = 0$ for $q > 0$.*

Proof. The (finite) directed set of coverings of $(X, 0)$ has a cofinal subset consisting of the single covering σ where σ_1 consists of the one open set X and σ_2 is empty. Thus the spectrum $S^q(X, 0)$ has a cofinal subspectrum consisting only of $H^q(K_\sigma, 0)$. The direct limit of a spectrum consisting of one group is this group itself; hence $H^q(X, 0)$ is isomorphic with $H^q(K_\sigma, 0)$. But K_σ consists of a single vertex; hence, if $q > 0$, $H^q(K_\sigma, 0) = 0$ and therefore $H^q(X, 0) = 0$.

Princeton University

292 C. H. DOWKER

Bibliography

1. C. H. Dowker, *Mapping theorems for non-compact spaces*, Amer. J. Math., 69 (1947), 200–242.
2. S. Eilenberg, *Singular homology theory*, Ann. of Math., 45 (1944), 407–447.
3. S. Eilenberg and N. E. Steenrod, *Axiomatic approach to homology theory*, Proc. Nat. Acad. Sci. U. S. A., 31 (1945), 117–120.
4. S. Eilenberg and N. E. Steenrod, Foundations of Algebraic Topology (unpublished).
5. J. L. Kelly and E. Pitcher, *Exact homomorphism sequences in homology theory*, Ann. of Math., 48 (1947), 682–709.
6. E. H. Spanier, *Cohomology theory for general spaces*, Ann. of Math., 49 (1948), 407–427.

Differential Geometry

Ivan Kolář

1. Introduction. E. Čech was one of the founders of projective differential geometry and almost all of his papers in geometry belong to this field. He worked in differential geometry during two separate periods. The first one lasted from the end of the First World War until approximately the end of the 1930's; the second period was from the end of the Second World War up to his death in 1960. Already in his first papers he followed several new ideas introduced by an excellent Italian mathematician G. Fubini. Afterwards, Čech spent the academic year 1921 – 22 in Torino with Fubini, who greatly appreciated their mutual cooperation. At the end of this stay Fubini offered Čech coauthorship of a monograph on projective differential geometry. This book, written in Italian, appeared in two volumes (about 800 pages) in 1926 and 1927, [17]. To make their results more accessible to the general public, both authors decided to prepare another book on this subject in French, which appeared in 1931, [18]. This was the most famous book on projective differential geometry at the time and it remains a useful textbook today.

At a first glance, projective differential geometry seems to be only one of the generalizations of metric differential geometry in the spirit of the Erlangen Programm by F. Klein. However, the historical role of projective differential geometry is much more significant for at least the following three reasons, all of them being related to the excellent French geometer, E. Cartan. Firstly, unlike metric and affine spaces, which have certain specific properties, projective space seems to be, from several points of view, a typical example of a general Klein space (i.e., of a space endowed with a transitive transformation group). That is why the creation of a general method for investigating submanifolds of Klein spaces by Cartan was directly motivated by the projective geometry, [8]. In the second place, Cartan's definition of a space with projective connection, [7], can be immediately generalized to the case of an arbitrary Klein space, which gave rise to the contemporary idea of a connection on an arbitrary principal fiber bundle. Finally, the problem of projective deformation of surfaces motivated Cartan to a general approach to deformations of submanifolds of Klein spaces, [6], which grew into one of the leading ideas in different branches of differential geometry. Since all of these subjects are virtually related to E. Čech, we shall pay special attention to them.

Several discussions with my colleagues and students convinced me that it is not possible to understand Čech's influence on contemporary differential geometry without having presented at least a few concrete problems from classical projective geometry of surfaces and line congruences. We include them in Sections 3 – 6, but we always simultaneously explain the related original results by Čech. Sections 8 and 9 are devoted mostly to the second period of Čech's geometric activity. In the remaining sections we face the difficult problem of selecting from the further geometrical research motivated and inspired by Čech. We prefer the general problems to the concrete ones, but the references in our list can contribute to making the picture complete.

Unless otherwise specified, all objects and maps are assumed to be infinitely differentiable.

2. Projective spaces. The idea of a projective plane P_2 is related to the central projection in the usual 3-space. That is why the elements of P_2 can be identified with the lines (i.e., one-dimensional linear subspaces) in a 3-dimensional vector space V_3. In general, the projective n-space P_n is defined as the space of all lines in an $(n+1)$-dimensional vector space V_{n+1}. The projective transformations of P_n correspond to the linear isomorphisms of V_{n+1}. The non-zero vectors of V_{n+1} are called the *analytic points* of P_n. Hence an analytic point in V_{n+1} is determined by a point of P_n up to scalar factor. Every $(k+1)$-tuple of linearly independent analytic points $X_0, \ldots, X_k$ defines a linear k-dimensional subspace $[X_0, \ldots, X_k] \subset P_n$. In the case $k = n-1$, we call it a hyperplane. The dual vector space V^*_{n+1} generates the dual projective space P^*_n, the elements of which coincide with the hyperplanes of P_n. If we fix a hyperplane $I_{n-1} \subset P_n$ (called the *improper hyperplane*), then $P_n \setminus I_{n-1}$ is identified with an affine n-space A_n. The affine transformations of A_n correspond to the projective transformations of P_n preserving I_{n-1}. A frame of P_n is an $(n+1)$-tuple $(X_0, \ldots, X_n)$ of linearly independent analytic points. The induced coordinates on $V_{n+1} \setminus \{0\}$ are called the homogeneous coordinates on P_n.

Let X_1, X_2, X_3 be distinct points of a projective line P_1. We have to recall the concept of the *harmonically conjugate point* X_4 of X_3 with respect to the pair X_1, X_2. The most elementary approach is: we take X_3 for the improper point and then X_4 is the center of the affine segment determined by X_1, X_2. This is a symmetric property of both pairs X_1, X_2 and X_3, X_4, which are also said to *harmonically separate each other.*

The lines in P_3 are in bijection with the points of a hyperquadric $Q_4 \subset P_5$ called the *Klein quadric*. In coordinates, this map can be constructed as follows. Given two different points $X = (x_0, x_1, x_2, x_3)$ and $Y = (y_0, y_1, y_2, y_3)$, the determinants

$$(1) \qquad p_{ij} = \begin{vmatrix} x_i & x_j \\ y_i & y_j \end{vmatrix}, \qquad i, j = 0, \dots, 3, \quad i < j,$$

define 6 coordinates $p_{01}, p_{02}, p_{03}, p_{12}, p_{13}, p_{23}$ of a point in P_5. Consider the determinant, the rows of which are the coordinates of X, Y, X, Y. If we develop this zero determinant with respect to the first two rows, we obtain the relation

$$(2) \qquad p_{01}p_{23} - p_{02}p_{13} + p_{03}p_{12} = 0 .$$

This is the equation of a hyperquadric $Q_4 \subset P_5$. Now one verifies easily that rule (1) defines a geometrical bijection between Q_4 and the set of all lines in P_3.

Let us remark that even for arbitrary k and n, $k < n$, there is an analogous identification (established by Plücker) of the k-dimensional linear subspaces in P_n with the points of an algebraic submanifold of a suitable projective space.

3. The projective differential geometry. The classical subjects of projective differential geometry are curves in P_2, curves and surfaces in P_3, and one-, two-, and three-parameter systems of lines in P_3. The latter objects can be defined as one-, two- or three-dimensional submanifolds of the 4-dimensional Klein quadric Q_4. In the classical terminology they are called *ruled surfaces, line congruences* or *line complexes*, respectively. But we have to mention at the very beginning that one of the specific features of Čech's geometric research was that he systematically studied the correspondences between two objects of the same type. Hence we must add here the projective differential geometry of correspondences as a further subject. This will be treated in Section 8 in detail.

The study of the one-dimensional manifolds, i.e., of curves and ruled surfaces, can be based on the theory of ordinary differential equations. These theories were already well developed at the beginning of 20th century, see the monograph by Wilczynski, [70], so that they lie beyond the scope of the present paper. We only have to recall that the ruled surfaces are classified in *developable* and *non-developable* ones. The name developable comes from metric geometry, where the developable surfaces are locally isometric to the plane. In projective geometry the developable surfaces are characterized by the property that the tangent planes (considered as planes in P_3) along each generating line coincide, while for a non-developable surface the tangent plane varies along each generator. One easily finds that a developable surface is formed by the tangents of a space curve in general. The curve itself is a singular subset of the surface and is called the *edge of regression*. Finally we remark that the osculating 2-plane of the Klein image of a non-developable ruled surface intersects the Klein quadric in a conic section, which corresponds to a ruled quadric in P_3 called the *osculating quadric* of the ruled surface in question.

The foundations of the theory of surfaces in P_3 will be discussed in the next sections. Here we recall the concept of the *asymptotic curves* of a surface $S \subset P_3$ only. In general, there are two directions in the tangent plane of a point $X \in S$ such that the osculating plane of every curve on S in such a direction coincides with the tangent plane. If these directions are real, X is called a *hyperbolic point*; if they are imaginary, X is an *elliptic point*. (It was clarified in [36] that one can consider the imaginary elements up to order r on every manifold of class C^r.) Hence, on a surface with all hyperbolic points, there is a net of asymptotic curves which are tangent to the asymptotic directions. If both asymptotic directions coincide, we have a *parabolic point*. A surface with only parabolic points is a (or part of a) developable surface. If every direction is asymptotic, the point is said to be *planar*. A surface with only planar points is a (or part of a) plane. Finally we remark that a net on an arbitrary surface in P_3 is called *conjugate*, if its tangents harmonically separate the asymptotic directions.

We also need the basic facts from the theory of line congruences. Having a line congruence L, we denote by the same symbol $l \in L$, a line in P_3 as well as its image on the Klein quadric Q_4. A simple evaluation shows that, in general, there are two directions in the tangent plane of the Klein image of L at l such that, for each ruled surface in this direction, its tangent planes in P_3 along the line l coincide. If these directions are imaginary for all l, L is said to be *elliptic*. If they are real and different, we have a *hyperbolic congruence* L with two systems of developable surfaces. The points of regression of both systems are said to be the *foci* of L. So we have two foci on each line. In general, each focus generates a surface called the *focal surface* of L. Thus, a generic hyperbolic congruence is locally formed by the common tangents of both focal surfaces. The lines of L determine a correspondence between the focal surfaces. A special class of line congruences, the *W-congruences* (in honor of Weingarten), is characterized by the property that the pairs of asymptotic directions on both focal surfaces correspond to each other. A line congruence is said to be *parabolic*, if the directions of the developable surfaces coincide everywhere. It is easy to see that such a line congruence L is formed by the tangents of one system of the asymptotic curves on a surface, which is the unique focal surface of L.

Another specific feature of Čech's work was that he always systematically studied the dual elements. To keep this approach here, we should include one- and two-parameter systems of planes in P_3 into the classical subjects of projective differential geometry. But it does not give new objects. In general, the envelope of a one-parameter system of planes is a developable surface and the original system is formed by its tangent planes. Quite similarly, the envelope of a two-parameter system of planes is a surface in general.

A third characteristic feature of Čech's geometric activity was that he paid special attention to the contact of submanifolds. The foundations of the theory of contact of algebraic submanifolds, which can approximate the smooth ones in any finite order, are treated systematically in his first book [16] (written in Czech). In two papers from 1928 and 1930, where he studied the contact of curves in projective n-space, he emphasized the problem of increasing the contact by means of the central projection from a suitable center. And he came back to such a subject in his last paper from 1960.

Some further subjects of projective differential geometry arise by generalizing the classical ones. In particular, one studies arbitrary submanifolds in P_n, m-parameter families of linear k-spaces in P_n and certain derived objects. The basic monographs and textbooks in projective differential geometry are [1], [7], [17], [18], [25], [26], [42], [43].

4. Surfaces in projective 3-space. The oldest method for investigating a hyperbolic surface $S \subset P_3$ is a result of the work of Wilczynski. Consider some local parameters $(u,v) \in U \subset \mathbb{R}^2$ on S such that the coordinate curves are asymptotic. Let $X(u,v)$ be an analytic point generating S and $X_u = \dfrac{\partial X}{\partial u}$, $X_v = \dfrac{\partial X}{\partial v}$, $X_{uu} = \dfrac{\partial^2 X}{\partial u^2}$, $X_{vv} = \dfrac{\partial^2 X}{\partial v^2}$. By the definition of the asymptotic directions, it holds that

$$
\begin{aligned}
X_{uu} &= a_1 X + a_2 X_u + a_3 X_v\,, \\
X_{vv} &= a_4 X + a_5 X_u + a_6 X_v\,,
\end{aligned}
\tag{3}
$$

where $a_1, \ldots, a_6 : U \to \mathbb{R}$ are some functions. Since the equations (3) are linear, they determine the map $X : U \to V_4 \setminus \{0\}$ up to a linear isomorphism, so that the corresponding map $U \to P_3$ is defined up to a projective transformation. The geometric situation is preserved, if one changes the parametrization of the asymptotic net

$$\bar{u} = \varphi(u)\,, \qquad \bar{v} = \psi(v) \tag{4}$$

and the scalar factor of X

$$\bar{X}(u,v) = f(u,v)X(u,v)\,. \tag{5}$$

The point $X_{uv} = \dfrac{\partial^2 X}{\partial u \partial v}$ is linearly independent on X, X_u, X_v, so that (X, X_u, X_v, X_{uv}) is a frame associated to $X \in S$, which can be used for evaluating the local geometric constructions related to S. But one can also apply the analytic point of view: the geometric quantities derived from (3), and invariant with respect to the changes (4) and (5), are the geometric objects determined by S.

There is another definition of the asymptotic directions at $X \in S$: the intersection of the tangent plane at X with the surface itself is a curve with double point at X and its tangents are the asymptotic ones. Let us replace the tangent plane with an *osculating quadric*, i.e., a quadric having a second order contact with S at X; such quadrics form a 3-parameter family. The intersection curve then has a triple point at X. There are only three directions in the tangent plane such that there exists an osculating quadric with the property that all three tangents of its intersection curve coincide with one of these directions. These directions were discovered by Darboux and they determine a triple system of *Darboux curves*. The directions conjugate to those of Darboux are called Segre directions and they determine a triple system of *Segre curves*. Furthermore, there is a one-parameter family of osculating quadrics such that the tangents of their intersection curves are the Darboux tangents. This family is called *Darboux pencil*.

We have described all classical geometric objects determined by the *third order element* of S at X. (We shall follow Čech in using this term for an equivalence class of surfaces having third order contact at X. However, the name "3rd order contact element" introduced by Ehresmann, [23], would be more appropriate today.) In particular, it can be proved that the third order element does not determine a line not lying in the tangent plane, which could play the role of the normal of S. We remark that in affine differential geometry the normal is usually defined as the line of centers of the quadrics of the Darboux pencil.

However, already in his PhD Thesis, which was published in 1921, [12], Čech pointed out that the third order element of a surface in P_3 determines some correspondences between associated geometric objects and he proved that these correspondences completely characterize the third order element.

In 1922 Čech wrote an excellent paper [13] (incorporated into this book), in which he explicitely determined the surfaces all of whose Segre curves are planar. The problem required the integration of a rather complicated system of partial differential equations. This paper remains a masterpiece of classical research in differential geometry. In the same year Čech also described all surfaces with planar Darboux curves.

5. The canonical pencil. First we will show how the fourth order element of S distinguishes one quadric in the Darboux pencil, which is called *Lie quadric*. This is the osculating quadric of the ruled surface formed by the asymptotic tangents of one system along the asymptotic curve of the other system. If we interchange both systems of asymptotic curves, we obtain the other family of straight lines on the same quadric. Here we have to add that Čech contributed essentially to the geometrical characterization of all quadrics of the Darboux pencil, [18].

Several geometric facts suggest that the role of the normal of a surface $S \subset P_3$ at a point X should be played by two lines: the first one passing through X and not lying in the tangent plane and the second one lying in the tangent plane and not passing through X. The second line is usually taken for the improper line of the tangent plane. Such two lines are said to be *reciprocal*, if they are polar with respect to the Lie quadric.

So let us start with any two reciprocal line congruences L_1 and L_2 associated with S and let us try to pick up some which are geometrically distinguished. A simple condition is that the developable surfaces of both L_1 and L_2 correspond to each other. This uniquely defines a pair of reciprocal congruences called the *Wilczynski directrices*. (In his original construction, Wilczynski used the osculating 4-spaces of the Klein images of the developable surfaces determined by both asymptotic curves.) On the other hand, the quadrics having fourth order contact with both asymptotics are said to be *principal*; they form a one-parameter family. There is exactly one pair of reciprocal congruences which are polar with respect to all principal quadrics. Those are the *Green edges*.

The first Wilczyński directrix and the first Green edge determine at each point a pencil of lines called *canonical*. The intersection of the canonical pencil with the tangent plane is the *canonical tangent*. The lines reciprocal to the canonical ones form a pencil in the tangent plane, the center of which is called the *canonical point*. If we take the canonical tangent for an improper element, then the Wilczynski directrix and the Green edge determine an affine coordinate system in the canonical pencil. Hence we have defined the concept of *congruence of canonical lines with constant index*. An interesting feature of the projective geometry of surfaces is that several different constructions lead to the congruences of canonical lines with constant index. In principle, each of them can play the role of the first normals of the surface in question.

The *Fubini normal* is the line harmonically conjugate to the Wilczynski directrix with respect to the Green edge and the canonical tangent. Fubini first constructed it by means of the projective linear element (see the next Section) in a way analogous to some of the properties of the metric normal, [27].

Čech deduced in 1921 that the osculating planes of the three Segre curves intersect one other in a single line. This gives another line of constant index in the canonical pencil which is today called *Čech axis*. Further important canonical lines of constant index are e.g. *Bompiani normal*, [2], *Cartan line* or *two Fubini principal normals*, [1].

6. The projective linear element and deformations. Fubini analytically introduced two differential forms F_2 and F_3 on a surface $S \subset P_3$. The form F_2 is quadratic and F_3 cubic. Both F_2 and F_3 are only partially invariant, but the ratio F_3/F_2 is fully invariant and is called the *projective linear element* of S (its geometric definition is rather complicated, see [18]). The projective linear element can be viewed as an analogy of the metric linear element from the Euclidean geometry. For example, the *projective geodesics* can be defined as the extremals of the projective linear element. (But we remark that in projective geometry there are several candidates for the geodesic curves, which is related to the phenomena of the canonical pencil explained in Section 5.)

In the Euclidean 3-space E_3, a deformation of two surfaces S_1, $S_2 \subset E_3$ means an isometry $\varphi : S_1 \to S_2$, or, which is the same, a diffeomorphism preserving the metric linear element. Analogously to the metric case, Fubini defined the *projective deformation* of two surfaces S_1, $S_2 \subset P_3$ as a diffeomorphism $\varphi : S_1 \to S_2$ preserving the projective linear element. However, the projective deformation is a much more complicated subject than the metric one. This motivated E. Cartan to a deeper analysis of the idea of deformation. He clarified that the deformation of submanifolds of an arbitrary Klein space can be defined in a unified way, which depends directly on the transformations of the group in question only.

We recall that a *Klein space* is a manifold M with a transitive left action of a Lie group G. Let S_1, $S_2 \subset M$ be two submanifolds and $\varphi : S_1 \to S_2$ be a diffeomorphism. According to E. Cartan, [6], φ is said to be an *r-th order deformation*, if for every $x \in S_1$ there exists a transformation $\tau_x : M \to M$ of the group G such that for every curve $\gamma : \mathbb{R} \to S_1$, $\gamma(0) = x$, the induced curves $\varphi \circ \gamma$ and $\tau_x \circ \gamma$ have r-th order contact at $0 \in \mathbb{R}$. In E_3 we clearly obtain the classical concept of deformation of surfaces for $r = 1$. It can be proved that the projective deformation of surfaces in P_3 in the sense of Fubini coincides with the second order deformation in the sense of Cartan.

Initially, it was a surprising result that not all surfaces in P_3 are locally deformable. Cartan proved that the projectively deformable surfaces depend on 6 functions of one variable only. Čech, partially in cooperation with Fubini, deduced several geometric results on projective deformations of surfaces, [18]. The general type of the projectively deformable non-ruled surfaces are *R-surfaces*. By definition, such a surface carries an *R-net*, which is a conjugate net with the property that both line congruences of its tangents are W-congruences. The remaining types are R_0*-surfaces*, which can be characterized as a certain limit case of R-surfaces for the R-net tending to one system of the asymptotic curves. Čech and Fubini also deduced that a non-ruled surface admits at most 3-parameter system of projective deformations and they characterized geometrically the surfaces admitting three-, two- or one-parameter system of deformations. Special attention was paid to a surface projectively deformable into itself.

7. Further results on surfaces and related subjects. In 1918, Fubini established an original method for investigating hypersurfaces (i.e., submanifolds of codimension one) in projective n-space, which is based on two differential forms F_2 and F_3 analogous to those from Section 6. In [14] Čech presented a significant contribution to the foundations of that method as well as to its applications. Some of the simplifications by Čech are due to his systematic use of the dual objects. Next he describes the higher dimensional analogies of the Lie quadric. Finally the projective deformations of some hypersurfaces are discussed.

Čech was used to study the special case of ruled surface in connection with general theories. For example he discussed in detail the projective deformation of ruled surfaces, [18]. In 1924 he established a specific method for investigating ruled surfaces in projective n-space. But this method led to several new results even for ruled surfaces in P_3. Such a research was carried on especially by J. Klapka and his pupils. Klapka deduced remarkable results on the flecnodal properties of ruled surfaces, [33]. (In general, there are two *flecnodal points* on each generator of a non-developable ruled surface in which the osculating quadric has the third order contact.)

Čech significantly contributed even to the theory of nets in P_2, which was established by an excellent Roumanian geometer G. Tzitzeica, [66]. Čech realized that several problems from the theory of surfaces and line congruences can be reduced to certain properties of plane nets. His most important results on this subject are collected in Chapter 10 of the book [18]. It is remarkable that quite recently Švec deduced a global result on the elliptic plane nets, [62].

Another general geometric subject is the study of curves on a surface. An interesting testimony of Čech's creative power is his invention of a special concept of *band* for such a situation, [15]. This means that one considers a curve together with the surface elements of an order r. In the metric geometry, one gets the standard situation for $r = 1$, in which everything is very instructive. But the general approach was new even in the affine space A_3, where one has to take $r = 2$. So in [15] Čech was able to solve some problems posed by two famous German geometers G. Pick and W. Blaschke. Later Čech applied such an approach to the curves on a surface in P_3, [18].

The first book by Fubini and Čech, [17], uses mostly the Fubini method and deals with all the basic subjects of projective differential geometry. The original results of both authors represent a significant part of the book. The text contains three appendices, which are written by other top experts in projective differential geometry: G. Tzitzeica discusses some projective deformations, E. Bompiani studies the projective invariants of some surfaces and A. Terracini treats projective differential geometry in higher dimension.

The second book [18] was briefly characterized in the introduction. Here we have to underline that its last three chapters, prepared by Čech, are devoted to the methods of E. Cartan. First, Čech presents a very readable (and precisely formulated, which was not obvious at the time) survey of the theory of differential systems in involution with two independent variables. Then he constructs the canonical frame field of a non-ruled surface in P_3 by using the analytic procedure for specialization of frames invented by Cartan (see Section 11). This is compared with some previous constructions from the book. Next, Cartan's basic results on the projective deformation of surfaces are presented. Finally Čech uses Cartan's methods for deducing an original result: he determines all W-congruences realizing the projective deformation between both focal surfaces.

8. The theory of correspondences. Čech began with the study of arbitrary correspondences between two surfaces in P_3 as early as 1921 in connection with the problem of projective deformation. This research was closely related to some works of Bompiani. Special attention was paid to asymptotic correspondences, i.e., those which transform asymptotic curves into asymtotic curves, [18]. (One verifies easily that every projective deformation is an asymptotic correspondence.)

Next Čech posed to O. Borůvka the problem of investigating arbitrary correspondences between two projective planes, i.e., local diffeomorphisms $\varphi : P_2 \to P'_2$, [3]. It is remarkable that Borůvka already used the Cartan's method of moving frames in 1926. (The graph of φ is a surface in $P_2 \times P'_2$, which is a Klein space with respect to the product of the groups of projective transformations of P_2 and P'_2.) He first clarified that the basic geometrical object of such a correspondence is formed by the *characteristic directions*, which are defined by the following property. In general, there are three directions in the tangent plane of P_2 at a point X such that every curve in such a direction having inflection in X is transformed into a curve with inflection at $\varphi(X)$. Borůvka then classified the fundamental types of correspondences between two projective planes and described some of their geometric properties.

Čech came back to this subject after the second world war. We underline that in all works from his second period Čech systematically used the Cartan method of moving frames. In a series of papers from the late forties and early fifties, [19], he developed a complete theory of correspondences between two projective n-spaces. He introduced an original concept of linearizing transformation, which grew into a powerful tool for investigating suitable approximations of correspondences by means of the tangent projective transformations. This enabled Čech to give a natural classification of special types of correspondences. Some of them were constructed geometrically; for the other ones the existence together with the degree of freedom were deduced. In this connection the projective deformations of a layer of hypersurfaces were studied in detail. Further it was clarified that several concrete results about the correspondences between projective 3-spaces are closely related to the theory of line congruences.

These results were highly appreciated and continued, especially by the Italian school of M. Villa from Bologna, cf. [48], [68].

9. Line congruences. From 1954 on Čech systematically studied the correspondences between line congruences and their projective deformations. In [20] he presents the foundations of such a theory in a projective n-space P_n. The tangent planes of all ruled surfaces of a line congruence (i.e., of a 2-parameter family of 1-dimensional linear subspaces) L in P_n along a line $l \in L$ generate a linear subspace of dimension at most 5; this dimension is called the *character* of L at l. Čech first geometrically describes all individual types in detail. Congruences of character 3 are closest to the line congruences in P_3; they have 2 foci on each line in general. Then he discusses the basic properties of the correspondences between line congruences in P_n.

In the next paper [21] (incorporated into this book) Čech deduced several profound results on developable correspondences (where developable surfaces correspond to each other) between line congruences in P_3. In contradistinction to Terracini, [64], Čech defines the *projective linear element of a line congruence* as a quadruple of differential forms with two equations among them. This enables him to characterize not only the projective deformations, but even several other classes of developable correspondences. In another paper from 1956, Čech geometrically and analytically characterized the projective deformations of W-congruences. The W-congruences with ruled focal surfaces were studied from such a point of view by V. Horák, [30].

A complete theory of correspondences between line congruences in P_n was created by Čech's pupil, A. Švec. In particular, he gave a full solution of the problem of projective deformation of line congruences in P_n; see his monograph [59]. Švec also studied systematically the line congruences with projective connection, which will be treated in the next section. In 1960 Švec published a survey paper on the contribution of Czechoslovak geometers to projective differential geometry of line congruences and of surfaces (in P_n) with conjugate net, [57].

Čech's new ideas in the theory of line congruences strengthened his contacts with several famous geometers in the traditional European centers of projective differential geometry. From Italy we mention B. Segre, [51], and A. Terracini, [65]. In Roumania it was F. Marcus, who studied the linear projective element of a parabolic congruence in P_5, [47]. Several contacts appeared with the Russian geometric school headed by S. P. Finikov. These studies were continued by the pupils of E. Čech. Beside Švec, this work concerns primarily J. Klapka, K. Svoboda, [53], [54], and their schools. Last but not least we have to mention a book on line differential geometry by another excellent Czech geometer V. Hlavatý, [29]. Even though it has no direct relation to Čech's research, its influence on the further development of Czechoslovak geometry should be appreciated.

It is remarkable that recently line congruences have appeared in some other branches of mathematics. In the seventies, V. I. Arnold and his coworkers obtained excellent results on classical line congruences from the viewpoint of the theory of singularities. There are also other interesting relations to the theory of Bäcklund transformations, which represents an important geometric tool for solving several non-linear partial differential equations of mathematical physics. This can be testified by a paper of the last Čech's pupil, B. Cenkl, [11], which is incorporated into this book, or by [63].

10. Manifolds with connection. Let us first remark that over the course of time geometers became interested in "curved" versions of the classical spaces (the latter spaces are said to be flat in such a context). For example, a Riemannian space is a curved version of the Euclidean space of the same dimension, and a manifold with a classical linear connection can be looked at as a curved version of the flat affine space. Directly continuing Čech's research on projective differential geometry of surfaces and line congruences, Švec initiated a systematic study of the related objects with connection. He used the concept of König space, which is much more general, but still too concrete to be explained here from the contemporary point of view. We shall go directly to the general case.

As remarked in the introduction, E. Cartan defined the concept of space with projective connection using only general ideas. That is why Ehresmann gave the name of a space with Cartan connection to the curved version of an arbitrary Klein space. But the primary idea is the contemporary concept of connection on a principal fiber bundle, for which we refer the reader to [34]. Hence the following way of reconstructing the historically older structures is the most suitable for us.

Let $P(B,G)$ be a principal fiber bundle with a structure group G and let Γ be a connection on P. Roughly speaking, the elements of P represent certain frames and the connection Γ defines the tangent vectors of the virtual displacements of the frames. If we have a Klein space M with transformation group G, we can construct the associated fiber bundle $E(B,M,G,P)$, the fibers E_x, $x \in B$, of which are called local spaces in classical situations. Then a section $s : B \to E$ defines a center $s(x)$ in each local space E_x, $x \in B$. So the base B is identified with the manifold $s(B)$ of all centers of the local spaces. Such a quadruple $(P(B,G),\Gamma,M,s)$ is said to be a space with connection, [38]. If we have a curve $\gamma : \mathbb{R} \to B$, then for every $t \in \mathbb{R}$ the connection Γ maps the curve $s \circ \gamma$ into a curve in the local space $E_{\gamma(t)}$. The latter curve is called the development of γ into the local space $E_{\gamma(t)}$. The developments induce a linear map of the tangent spaces $T_xB \to T_{s(x)}E_x$, which coincides with the absolute differential of s with respect to Γ.

A *space with Cartan connection* of type M can now be defined as a space with connection $(P(B,G),\Gamma,M,s)$ satisfying the following two conditions:

a) $\dim B = \dim M$,

b) the absolute differential of s with respect to Γ defines linear isomorphisms $T_xB \to T_{s(x)}E_x$ for all $x \in B$.

The canonical structure of a (flat) space with Cartan connection on the Klein space M itself is the following one. We consider the product principal fiber bundle $M \times G$ over M with the canonical flat connection Γ determined by the product structure. Then $E(M,M,G,M \times G) = M \times M$ and we take for s the diagonal section $M \to M \times M$.

A space with connection $(P(B,G),\Gamma,M,s)$ is said to be an m-dimensional *manifold with connection* of type M, [38], if $m = \dim B < \dim M$ and the absolute differential of s with respect to Γ defines linear monomorphisms $T_xB \to T_{s(x)}E_x$ for all $x \in B$. If we have a submanifold $S \subset B$ of the base of a space with Cartan connection $(P(B,G),\Gamma,M,s)$, then the restrictions of P,Γ and s over S define a manifold with connection. This induced structure represents the starting point for the standard geometric investigation of the submanifold S of the space with Cartan connection. Conversely, one can easily see that every manifold with connection can be locally embedded into a space with Cartan connection of the same type, i.e., it can be locally constructed from a suitable space with Cartan connection in the above way.

In particular, *a surface with projective connection* defines a 2-dimensional manifold with connection of type P_3. Its simplest geometric properties had already been deduced by E. Cartan in [7]. But only Švec in the late fifties and early sixties started with a systematic study of this subject. In [58], which is incorporated into this book, Švec discusses some third and fourth order objects determined by a surface with projective connection. Special attention is paid to different types of generalized Darboux quadrics, a subject treated in the flat case by Čech; see Section 5. Then Švec constructs a canonical frame field of the surface. From the related paper by Cenkl, [9], we can see a complete system of geometric conditions for a surface with projective connection to be holonomic, i.e., locally embedded into the flat projective 3-space. In [35], which is also incorporated into this book, I. Kolář introduces the concept of a surface with projective connection, holonomic up to an order r, and characterizes geometrically the situation up to the sixth order. Then he proves that a generic 6-holonomic surface is holonomic. The projective deformations of surfaces with projective connection were studied by Muracchini, [49]. Another paper generalizing Čech's results to the "curved" case is that of M. Hejný, [28].

Line congruences with projective connection can be defined in the same manner. The foundations of such a theory were established by Švec in [56]. First he presents the most direct definition of a line congruence with projective connection and describes its basic geometric properties. Then he generalizes the differential forms introduced in the flat case by Čech, [21], and studies systematically different kinds of deformations. In particular, the so-called singular projective deformations are completely characterized.

11. The Cartan method of moving frames. A frame in an n-dimensional vector space V_n is an n-tuple $u = (v_1, \ldots, v_n)$ of linearly independent vectors. Let $F(V_n)$ denote the space of all frames of V_n. From a general point of view, an essential property of every frame $u \in F(V_n)$ is that it identifies V_n with $\mathbb{R}^n$ by means of the induced linear coordinates on V_n. Hence u can be interpreted as a map $\tilde{u} : \mathbb{R}^n \to V_n$. The standard left action of the general linear group $GL(n, \mathbb{R})$ on $\mathbb{R}^n$ defines a right action of $GL(n, \mathbb{R})$ on $F(V_n)$ by

$$\widetilde{ug}(x) = \tilde{u}(gx) . \tag{6}$$

This action is simply transitive, i.e. for every u_1, $u_2 \in F(V_n)$ there is exactly one $g \in GL(n, \mathbb{R})$ such that $u_1 g = u_2$. Consider further an m-parameter system of frames

$$u(t_1, \ldots, t_m) = (v_1(t_1, \ldots, t_m), \ldots, v_n(t_1, \ldots, t_m)) \tag{7}$$

with parameters from a domain $D \subset \mathbb{R}_m$, which can be called a *moving frame*. Then the differentials dv_i can be expressed as linear combinations of the vectors of the moving frame itself, i.e., we have

$$dv_i = \sum_{j=1}^{n} \omega_i^j v_j , \qquad i = 1, \ldots, n \tag{8}$$

where ω_i^j are some differential forms on D. These forms are said to be the *relative components of the moving frame* (7). The collection (ω_i^j) is a vector-valued form on D with values in the Lie algebra $\mathfrak{gl}(n, \mathbb{R})$ of $GL(n, \mathbb{R})$. Clearly, one can replace D by an arbitrary manifold Z and (7) by any map $f : Z \to F(V_n)$. In particular, one can consider the identity map of $F(V_n)$. In this case the collection $\omega = (\omega_j^i) : TF(V_n) \to \mathfrak{gl}(n, \mathbb{R})$ is said to be the *relative component of the frame space* $F(V_n)$. Obviously, the collection of the relative components of a moving frame $f : Z \to F(V_n)$ is of the form $\omega \circ Tf : TZ \to \mathfrak{gl}(n, \mathbb{R})$, where ω is the relative component of $F(V_n)$.

If we consider an n-dimensional manifold N and the fiber bundle P^1N of all linear frames in the tangent spaces of N, we obtain a "fibered" version of the previous example, where each fiber P^1_xN is defined by $P^1_xN = F(T_xN)$, $x \in N$. Hence, in modern differential geometry the general idea of frames on geometric spaces of various types is incorporated into the theory of principal and associated fiber bundles; see e.g. [34]. By definition, the structure group G of a principal fiber bundle $P(B, G)$ acts simply transitively on each fiber P_x, $x \in B$. Given a left action of G on a manifold Q, we can construct the associated fiber bundle $E(B, Q, G, P)$. Its elements are the equivalence classes $\{u, z\}$, $u \in P$, $z \in Q$, with respect to the equivalence relation $(u, z) \sim (ug, g^{-1}z)$, $g \in G$. Hence every $u \in P_x$ can be interpreted as a diffeomorphism $\tilde{u} : Q \to E_x$, $\tilde{u}(z) = \{u, z\}$ and (6) holds. Let $\mathfrak{g}$ denote the Lie algebra of G. The role of the relative component of each frame space P_x, $x \in B$, is played by a $\mathfrak{g}$-valued 1-form $\omega_x : TP_x \to \mathfrak{g}$ defined as follows. Every element $A \in TP_x$ is the tangent vector to a curve $u(t)$ for $t = 0$. This curve can be uniquely expressed in the form $u(t) = u(0)g(t)$, where $g(t)$ is a curve on G satisfying $g(0) = e =$ the unit of G. Then we define $\omega_x(A)$ as the tangent vector to $g(t)$ for $t = 0$. If we fix a point $u \in P_x$, the relation $v = ug$, $v \in P_x$, identifies P_x with G and ω_x corresponds to the left Maurer-Cartan form of G. This implies the structure equations

$$(9) \qquad d\omega^\alpha = \sum_{\beta,\gamma} c^\alpha_{\beta\gamma}\omega^\beta \wedge \omega^\gamma\,, \qquad \alpha, \beta, \gamma = 1, \ldots, r = \dim G\,,$$

where $\omega_x = \sum_\alpha \omega^\alpha e_\alpha$ is a decomposition of ω_x with respect to a basis $(e_1, \ldots, e_r)$ of $\mathfrak{g}$ and $c^\alpha_{\beta\gamma}$ are the structure constants of G.

In the rest of this section we restrict ourselves to the case of a Klein space M with transformation group G. The left action of G on M defines a right action of G on the space $\mathit{Diff}\,M$ of all diffeomorphisms of M into itself by

$$(10) \qquad ug(x) = u(gx)\,,$$

$u \in \mathit{Diff}\,M$, $g \in G$, $x \in M$. Taking into account the general ideas from the theory of fiber bundles, we define the *frame space of Klein space* M as a subset $FM \subset \mathit{Diff}\,M$ with the following two properties:

a) G acts simply transitively on FM,

b) the identity map id_M belongs to FM (it plays the auxiliary role of an "absolute" frame from [8]).

For every $u \in FM$, the relation $v = ug$ identifies FM with G. All these identifications define the same structure of a smooth manifold on FM. Then FM can be interpreted as a principal fiber bundle, the base of which consists of a single point. Moreover, let us fix a point $a \in M$. Then the point $u(a) \in M$ will be called the *center of the frame* u. Hence we have a projection "center of frame" $c : FM \to M$, $c(u) = u(a)$.

A well known result from the theory of Lie groups reads that two maps $f_1, f_2 : S \to G$ of a connected manifold S into G are congruent, i.e. there exists a $g \in G$ such that $f_1(x) = gf_2(x)$ for all $x \in S$, if and only if $\omega \circ Tf_1 = \omega \circ Tf_2 : TS \to \mathfrak{g}$, where ω is the left Maurer-Cartan form. This fact, together with the construction of canonical frame fields of submanifolds, creates Cartan's approach to the equivalence problem of submanifolds. Two submanifolds S_1, S_2 of a Klein space M are said to be congruent if there exists a $g \in G$ such that $g(S_1) = S_2$. A simpler problem is to study the equivalence of S_1 and S_2 with respect to a given diffeomorphism $S_1 \to S_2$. Hence we have in fact two embeddings $i_1, i_2 : S \to M$ and we are looking for a $g \in G$ such that $gi_1(x) = i_2(x)$ for all $x \in S$. Assume that for a general class of submanifolds we have geometrically constructed a unique frame field (called *canonical*) along each submanifold S of this class, i.e. a section $\varkappa : S \to FM$ of the projection $c : FM \to M$. Then the two embeddings i_1, $i_2 : S \to M$ of this class are congruent if and only if $\omega \circ T(\varkappa \circ i_1) = \omega \circ T(\varkappa \circ i_2) : TS \to \mathfrak{g}$, where ω is the relative component of the frame space FM. This is usually expressed by saying that the relative components of both canonical frame fields coincide.

E. Cartan developed an analytical procedure for the specialization of frames, which can be applied to submanifolds of any Klein space M. Let us fix a basis $(e_1, \dots, e_n, e_{n+1}, \dots, e_r)$ of $\mathfrak{g}$ in such a way that the vectors $e_{n+1}, \dots, e_r$ lie in the Lie algebra of the stability group of the distinguished point $a \in M$. Then the real valued components

$$\omega^1, \dots, \omega^n \tag{11}$$

of the $\mathfrak{g}$-valued relative component ω of FM depend only on the differentials of the parameters from M, which are said to be *principal*, while the remaining parameters from FM are called *secondary*. Given an m-dimensional submanifold $S \subset M$, we first restrict ourselves to all those frames, the centers of which lie on S; they are called *zero order frames*. This yields some $n - m$ linear relations among the principal forms $\omega^1, \dots, \omega^n$. Next, one usually selects the *first order frames* by the geometric property to be tangent to S, i.e., the equations of the tangent spaces of S are required to be

$$\omega^{m+1} = 0, \ \ldots, \ \omega^n = 0 . \tag{12}$$

Then the basic principal forms are $\omega^1, \ldots, \omega^m$.

Cartan's procedure consists of the exterior differentiation of (12) and the application of structure equations (9). This leads to an explicit expression of some further components of ω as linear combinations of $\omega^1, \ldots, \omega^m$. These equations are said to form the *prolongation* of (12). The next exterior differentiation yields the explicit formulae for variations of the coefficients from the prolongation with respect to the remaining secondary parameters. Hence we can fix some further secondary parameters in an invariant way. Then we can repeat the same procedure step by step. There are some problems with the submanifolds of special types, but if we have a "generic" submanifold, we can obtain its canonical frame field after a finite number of steps. Several concrete examples of such specialization of frames are discussed in a textbook by Favard, [24].

This method for specialization of frames was explained by E. Cartan in [8], but in a manner which is far from being complete and is not considered to be fully satisfactory from a contemporary point of view. Therefore, several authors contributed to the modern version and some were inspired directly by E. Čech. Let us first mention A. Švec, [60], who simultaneously studied a generalization of this method to manifolds with connection. O. Kowalski then essentially contributed to the problem of classifying the orbits involved, which is an important practical question in every specialization, cf. [40], [41]. Further results can be found in a book by G. R. Jensen, [31], and in several papers by R. Sulanke; see e.g. [52]. Y. Bossard was the first who discussed even the case of a non-transitive action, [4]. Other contributions result from the work of J. A. Baddou, D. Bernard, H. Gollek, P. Griffith and last but not least of G. F. Laptěv. But Laptěv's approach will be discussed separately in the next section.

12. Generalizations of the Cartan method. Laptěv pointed out that the prolongation procedure in the Cartan method is independent of the specialization of frames, [44]. (It should be remarked here that this was related with the study of hypersurfaces in projective n-space.) Usually one restricts oneself to the first order frames. Laptěv also observed that in the course of such prolongation procedure one obtains the equations of the infinitesimal action of the group in question, i.e., the fundamental vector fields corresponding to a basis of its Lie algebra. One can thus apply the analytical procedures for finding equivariant maps, which determine the geometric objects of the submanifold. Such a method was used by Laptěv's school for solving several concrete problems on submanifolds of various Klein spaces

In a later paper Laptěv clarified that a similar prolongation procedure can be related with an arbitrary n-manifold N, [45]. If one fixes the principal parameters in the r-th step of his procedure, one obtains the structure equations of a Lie group, in which one can recognize the r-th differential group G_n^r in dimension n, i.e., the group of all invertible r-jets from $\mathbb{R}^n$ into $\mathbb{R}^n$ with source and target 0. Laptěv also applied an analogous procedure to arbitrary principal fiber bundles. Thus Laptěv and his coworkers established a method, called the Laptěv method today, which can be applied to a wide class of geometric problems, cf. [32]. But the original presentation of this method had some formal features of analytical character.

On the other hand, Ehresmann geometrically defined the concept of r-th order frame P^rN of an n-manifold N as the space of all r-jets at 0 of the local diffeomorphisms of $\mathbb{R}^n$ into N, [23]. This is a principal fiber bundle with structure group G_n^r. Then Kobayashi introduced a canonical $\mathbb{R}^n \oplus \mathfrak{g}_n^{r-1}$-valued 1-form on P^rN ($\mathfrak{g}_n^{r-1}$ being the Lie algebra of G_n^{r-1}), which generalizes the classical canonical $\mathbb{R}^n$-valued form on the linear frame bundle P^1N, cf. [34]. In [37] it was clarified that the core of Kobayashi's construction appears if we consider the first prolongation, W^1P, of an arbitrary principal fiber bundle $P(B,G)$. This is the principal fiber bundle $W^1P \to B$ of all 1-jets at $(0,e)$ from the local principal fiber bundle isomorphisms $\mathbb{R}^n \times G \to P$, $n = \dim B$. Such a construction was introduced by Ehresmann in a somewhat different situation, [23], while the principal bundle case was then studied by J. Virsík, [69], P. Libermann, [46], and I. Kolář, [37]. Choosing a basis in $\mathfrak{g}$ identifies a neighborhood of the unit in G with $\mathbb{R}^{\dim G}$ by means of the exponential map. This induces an inclusion $W^1P \subset P^1(P)$ and the restriction of the classical canonical form of $P^1(P)$ to W^1P defines the canonical $\mathbb{R}^n \oplus \mathfrak{g}$-valued form Θ of W^1P. The form of Kobayashi can now be defined by a natural inclusion $P^rM \subset W^1(P^{r-1}M)$.

In [37] the structure equations of Θ are studied and it is proved that the problem leads to the algorithms of Laptěv's type. Some further aspects were clarified by A. Dekrét in [22]. In another paper from 1971, Kolář explained how a Laptěv-like procedure can be applied to the jet prolongations of arbitrary geometric object fields (i.e., of sections of an associated fiber bundle). He proved that this procedure leads to the equations of the infinitesimal actions of the induced groups. Then the absolute differentiation of arbitrary geometric object fields was discussed from the same point of view. In [38] the method is adapted to the study of manifolds with connection in the sense of Section 10. This gives, among others, a conceptual clarification of the gradual holonomization of a surface with projective connection mentioned in Section 10.

It is interesting that the historically oldest subject, which is that of the submanifolds of Klein spaces, could not be treated from this modern point of view earlier than in the final stage. In the first half of [39] the submanifolds of an arbitrary manifold N are discussed, for the case of any transitive Lie pseudogroup on N can be reduced to such a "pure" situation. (Roughly speaking, a Lie pseudogroup on N consists of local diffeomorphisms satisfying a system of partial differential equations.) Even the Klein spaces are reduced to the "pure" case. In the second half of [39] it is clarified that the simplest situation is on those Klein spaces M with the transformation group G, for which there exists an Abelian subgroup in G complementary to the stability group of a point of M (but all spaces from the classical differential geometries are of this type). In such a case it is proved that the prolongation coefficients coincide with the coefficients of the Taylor series expansions of the submanifold in the corresponding local frame. This is quite a practical result, which can simplify various concrete investigations.

13. The theory of submanifolds, differential systems. The general theory of submanifolds in Klein spaces is a very attractive subject, but it is rather complicated, for there are several exceptional cases in every field of differential geometry. Švec worked systematically in this direction in the period 1966-68. In particular, he treated the general problem of deformation, where he obtained some final results for surfaces in certain 3-dimensional spaces. The invariants and tensor invariants of submanifolds of a space with a Lie pseudogroup of transformations were studied by his pupil J. Vanžura, [67]. The deformations of G-structures (i.e., of subbundles of the first order frame bundle) were discussed by another of Švec's pupils J. Bureš, [5].

It is well-known that several problems in differential geometry can be reduced to systems of partial differential equations, cf. [71]. When using the Cartan method of moving frames, one usually meets the so-called exterior differential systems, the theory of which was developed by E. Cartan as well. We remarked in Section 7 that already in 1931 Čech wrote an excellent survey of this theory in the case of two independent variables. Afterwards, he profited from this theory in the greater part of his papers from the second period and he turned the attention of his pupils to it. A further important step forward in the theory of systems of partial differential equations is due to D. C. Spencer and his school. One of the basic advantages is that Spencer's method can be applied to systems of class C^∞, which are rather different from the analytic ones (studied by E. Cartan).

Spencer's methods were used by Švec in the theory of deformations of submanifolds of Klein spaces. In [61] he generalized this problem by replacing the geometric structure by a suitable system of partial differential equations. He deduced that there exists an integer k with the property that every deformation of order greater than k is a formal equivalence (hence a local equivalence in the analytic case). Another pupil of E. Čech, B. Cenkl, studied the general theory of differential operators under Spencer's guidance. His main results concerning the vanishing theorem for an elliptic differential operator were published in [10].

14. Final remarks. We have already explained Čech's influence on world development of projective differential geometry. What remains is to remark that he founded a geometric school in Czechoslovakia, which moved gradually to other branches of differential geometry as well. In the sixties and early seventies, this school was headed by A. Švec (1931 - 1989). In Prague his first coworker was B. Cenkl and then L. Boček, J. Bureš, A. Karger, M. Kočandrle, V. Kohout and J. Vanžura cooperated in the research on related subjects. In 1969 this group was joined by O. Kowalski. In the late seventies A. Švec worked in global differential geometry of surfaces in Euclidean 3-spaces. Later he presented several important contributions to affine differential geometry of surfaces. Some others of Čech's students from the second period are V. Alda, M. Jůza, L. Koubek and Z. Nádeník, [50]. But we also have to appreciate the contacts with other distinguished Prague geometers as K. Havlíček, F. Nožička, A. Urban, Z. Vančura, F. Vyčichlo. In Brno the research in projective differential geometry was continued in the framework of a seminar headed by J. Klapka and later by K. Svoboda, the standard participants of which were J. Bayer, J.Beneš, J. Brejcha, J. Čučka, V. Havel, J. Havelka, V. Horák, J. Kerndl, J. Krejzlík, R. Piska, V. Radochová, L. Seichter, J. Vala, J. Vaněk, J. Veverka, the author and others. There were also several contacts with Slovak geometers as A. Dekrét, P. Grešąk, M. Hejný, F. Husárik, T. Klein, J. Virsík and others.

Today, differential geometry deals frequently with global problems and essentially profits from algebraic topology including Čech homology theories. This fully concerns metric geometry and partially affine geometry, but it is not yet true for projective geometry (a recent paper by Švec, [62], seems to be the only exception). So we have to state, and we feel it is a kind of historical paradox, that there are no contacts between differential geometry and algebraic topology in the research inspired directly by Čech. Of course we have no intention to separate the individual fields of Čech's mathematical activities. The development of mathematics reminds one of a never-finishing symphony. E. Čech played diverse parts in the orchestra and he was excellent in all of them.

REFERENCES

[1] G. BOL, *Projektive Differentialgeometrie, I, II*, Göttingen, 1950, 1954.

[2] E. BOMPIANI, *Nozioni di geometria proiettivo-differenziale relative ad una superficie dello spazio ordinario*, Atti della Reale Accademia dei Lincei (**5**), 33_1 (1924), 85 - 90.

[3] O. BORŮVKA, *Sur les correspondences analytiques entre deux plans projectifs, I, II*, Spisy Přírod. Fak. Masaryk. Univ. Brno **72, 85** (1926, 1927).

[4] Y. BOSSARD, *Pseudogroups de Lie de type fini et méthode du repère mobile*, Ann. Mat. Pura Appl. **105** (1975), 3 - 36.

[5] J. BUREŠ, *Deformation and equivalence G-structures, I*, Czechoslovak Math. J. **22** (1972), 641 - 652.

[6] E. CARTAN, *Sur le problème général de la déformation*, C. R. du Congrès Int. des Math. de Strasbourg en 1920, 397 - 406.

[7] E. CARTAN, *Leçons sur la théorie des espaces à connexion projective*, Paris 1937.

[8] E. CARTAN, *La théorie des groupes finis et continus et la géometrie différentielle*, Paris 1937.

[9] B. CENKL, *L'équation de structure d'un espace à connexion projective*, Czechoslovak Math. J. **14** (1964), 79 - 94.

[10] B. CENKL, *Vanishing theorem for an elliptic differential operator*, J. Differential Geometry **1** (1967), 381 - 418.

[11] B. CENKL, *Geometric deformations of the evolution equations and Bäcklund transformations*, Physica **18D** (1986), 217 - 219.

[12] E. ČECH, *On the third order elements of curves and surfaces in projective space* (Czech), Časopis pěst. mat. **50** (1921), 219 - 249, 305 - 306.

[13] E. ČECH, *Sur les surfaces dont toutes les courbes de Segre sont planes*, Spisy Přírod. Fak. Masaryk. Univ. Brno **11** (1922), 3 - 35.

[14] E. ČECH, *I fondamenti della geometria proiettiva differenzialle secondo il metodo di Fubini*, Annali di Mat. **31** (1922), 251 - 278.

[15] E. ČECH, *Courbes tracées sur une surface dans l'espace affine*, Spisy Přírod. Fak. Masaryk. Univ. Brno **28** (1923).

[16] E. ČECH, *Projective differential geometry* (Czech), Praha 1926, 406pp.

[17] E. ČECH, G. FUBINI, *Geometria proiettiva differenziale I, II*, Bologna 1926,1927.

[18] E. ČECH, G. FUBINI, *Introduction à la géométrie différentielle projective des surfaces*, Paris 1931.

[19] E. ČECH, *Projective differential geometry of correspondences between two spaces, I - VIII* (Russian), Czechoslovak Math. J., **2** (1952), 91 - 107, 109 - 123, 125 - 148, 149 - 166, 167 - 188, 297 - 331, **3** (1953), 123 - 137, **4** (1954), 143 - 174.

[20] E. ČECH, *On point deformations of line congruences* (Russian), Czechoslovak Math. J. **5** (1955), 234 - 273.

[21] E. ČECH, *Transformations développables des congruences des droites*, Czechoslovak Math. J. **6** (1956), 260 - 286.

[22] A. DEKRÉT, *On canonical forms on non-holonomic and semi-holonomic prolongations of principal fibre bundles*, Czechoslovak Math. J. **22** (1972), 653 - 662.

[23] C. EHRESMANN, *Oeuvres complètes et commentées, Partie I-1 et I-2*, Suppléments 1 et 2 au Volume XXIV (1983) des Cahiers de Topologie et Géometrie Différentielle.

[24] J. FAVARD, *Cours de la géometrie différentielle locale*, Paris 1957.

[25] P. S. FINIKOV, *Projective Differential Geometry* (Russian), Moskva-Leningrad 1937.

[26] P. S. FINIKOV, *Theory of congruences* (Russian), Moskva-Leningrad 1950.

[27] G. FUBINI, *Fondamenti della geometria proiettivo-differenziale di una superficie*, Atti della Reale Acad. di Torino **53** (1918), 1032 - 1043.

[28] M. HEJNÝ, *Generalization of the correspondence of relative normal of the surface into space with projective connection*, Acta F. R. N. Universitatis Comenianae Tom. **X**, Fasc V (1966), 1 - 12.

[29] V. HLAVATÝ, *Differentielle Liniengeometrie*, Groningen 1945.

[30] V. HORÁK, *Theorie der Torsen des Kleinschen fünfdimensionalen projektiven Raumes und ihre Applikation auf Segresche W-Kongruenzen des dreidimensionalen projektiven Raumes*, Czechoslovak Math. J. **9** (1959), 590 - 628.

[31] G. R. JENSEN, *Higher order contact of submanifolds of homogeneous spaces*, Lecture Notes in Mathematics 610, Springer-Verlag 1977.

[32] L. JE. JEVTUŠIK, JU. G. LUMISTE, N. M. OSTIANU, A. P. ŠIROKOV, *Differential geometric structures on manifolds* (Russian), Problemy Geometrii 9, Moskva 1979.

[33] J. KLAPKA, *Über Paare von konjugierten Kurven einer Regelfläche*, Publ. Přírod. Fak. Univ. J. E. Purkyně Brno No **393** (1958), 161 - 188.

[34] S. KOBAYASHI, K. NOMIZU, *Foundations of differential geometry I*, New York - London - Sydney 1963.

[35] I. KOLÁŘ, *Order of holonomy of a surface with projective connection*, Časopis Pěst. Mat. **96** (1971), 73 - 80.

[36] I. KOLÁŘ, *Complex velocities on real manifolds*, Czechoslovak Math. J. **21** (1971), 118 - 123.

[37] I. KOLÁŘ, *Canonical forms on the prolongations of principal fibre bundles*, Rev. Roumaine Math. Pures Appl. **16** (1971), 1091 - 1106.

[38] I. KOLÁŘ, *On manifolds with connection*, Czechoslovak Math. J. **23** (1973), 34 - 44.

[39] I. KOLÁŘ, *On the invariant method in differential geometry of submanifolds*, Czechoslovak Math. J. **27** (1977), 96 - 113.

[40] O. KOWALSKI, *A contribution to the Cartan's method of specialization of frames*, Spisy Přírod. Fak. Univ. J. E. Purkyně Brno **4** (1968), 107 - 120.

[41] O. KOWALSKI, *Orbits of transformation groups on certain Grassmann manifolds*, Czechoslovak Math. J. **18** (1968), 144 - 177, 240 - 273.

[42] E. P. LANE, *Projective differential geometry of curves and surfaces*, Chicago 1932.

[43] E. P. LANE, *A treatise on projective differential geometry*, Chicago 1942.

[44] G. F. LAPTĚV, *Differential geometry of embedded submanifolds* (Russian), Trudy MMO **2** (1953), 275 - 382.

[45] G. F. LAPTĚV, *Fundamental infinitesimal structures of higher order on smooth manifolds* (Russian), Trudy Geom. Seminara I (1966), 139 - 289.

[46] P. LIBERMANN, *Sur les prolongements des fibres principaux et des groupoides différentiables banachiques*, in: Analyse globale, Sém. Math. Supérieures No **42** (Montreal 1971), 7 - 108.

[47] F. MARCUS, *L'élement linéaire projectif d'une congruence de droites parabolique dans* S_5, Czechoslovak Math. J. **11** (1961), 57 - 61.

[48] L. MURACCHINI, *Sulle transformazioni puntuali che sono inviluppi di omografie*, Bull UMI **(3)**, **8** (1953), 390 - 398.

[49] L. MURACCHINI, *Sulla applicabilità proiettiva delle superficie negli spazi a connesione proiettiva a tre dimesioni*, Czechoslovak Math. J. **5** (1955), 274 - 288.

[50] Z. NÁDENÍK, *Bertrand curves in fivedimensional space* (Russian), Czechoslovak Math. J. **2** (1952), 57 - 87.

[51] B. SEGRE, *L'élement linéaire projectif d'une congruence quadratique de droites*, Bull. Acad. Roy. Belgique **39** (1953), 481 - 489.

[52] R. SULANKE, *On E. Cartan's method of moving frames*, Colloquia Math. Soc. J. Bolyai, **31**. Differential geometry, North Holland (1982), 681 - 704.

[53] K. SVOBODA, *Cycles de congruences stratifiables dans un espace projectif de dimension impaire*, Ann. Mat. Pura Appl. **57** (1962), 239 - 256.

[54] K. SVOBODA, *Sur la déformation projective des systemes osculateurs d'une congruence de droites*, Czechoslovak Math. J. **20** (1970), 315 - 326.

[55] A. ŠVEC, *Les surfaces R dans les espaces projectifs de dimension impaire*, Czechoslovak Math. J. **9** (1959), 243 - 264.

[56] A. ŠVEC, *Congruences de droites à connection projective*, Ann. Polon. Math. **VIII** (1960), 291 - 322.

[57] A. ŠVEC, *Contribution tchécoslovaque à la géometrie différentielle des congruences de droites et des surfaces à réseau conjugué*, Časopis Pěst. Mat. **85** (1960), 389 - 409.

[58] A. ŠVEC, *Sur la géometrie différentielle d'une surface plongée dans un espace à trois dimension à connexion projective*, Czechoslovak Math. J. **11** (1961), 386 - 397.

[59] A. ŠVEC, *Projective differential geometry of line congruences*, Praha 1964.

[60] A. ŠVEC, *Cartan's method of specialization of frames*, Czechoslovak Math. J. **16** (1966), 552 - 599.

[61] A. ŠVEC, *Submanifolds of Klein spaces*, Czechoslovak Math. J. **19** (1969), 492 - 499.

[62] A. ŠVEC, *On special plane nets*, Czechoslovak Math. J. **40** (1990), 64 - 69.

[63] K. TENENBLAT, L. C. TERNG, *Bäcklund's theorem for n-dimensional submanifolds of R^{2n-1}*, Ann. of Math. **111** (1980), 477 - 490.

[64] A. TERRACINI, *Su alcuni elementi lineari proiettivi*, Ann. R. Scuola Norm. Sup. Pisa (2) **2** (1933), 401 - 428.

[65] A. TERRACINI, *Sull'elemento lineare proiettivo di una congruenza di rette nello spazio a cinque dimensioni*, Atti. Accad. Naz. Lincei. Rend. Cl. Sci. Fis. Mat. Nat. (8) **28** (1960), 1 - 7.

[66] G. TZITZEICA, *Géometrie différentielle projective des réseaux*, Bucarest 1923.

[67] J. VANŽURA, *Invariants of submanifolds*, Czechoslovak Math. J. **19** (1969), 452 - 468.

[68] M. VILLA, *L'applicabilité projective de deux transformations ponctuelles*, Czechoslovak Math. J. **6** (1956), 435 - 443.

[69] J. VIRSÍK, *A generalized point of view to higher order connections on fibre bundles*, Czechoslovak Math. J. **19** (1969), 110 - 142.

[70] E. J. WILCZYNSKI, *Projective differential geometry of curves and ruled surfaces*, Leipzig 1906.

[71] R. L. BRYANT, S. S. CHERN, R. B. GARDNER, H. L. GOLDSCHMIDT, P. A. GRIFFITHS, *Exterior Differential Systems*, Publications, Mathematical Science Research Institute **18**, Springer-Verlag 1991.

SPISY
VYDÁVANÉ
PŘÍRODOVĚDECKOU FAKULTOU
MASARYKOVY UNIVERSITY

PUBLICATIONS
DE LA
FACULTÉ DES SCIENCES
DE L'UNIVERSITÉ MASARYK

vol. 11, 1922, 3 - 35

ON THE SURFACES ALL SEGRE CURVES OF WHICH ARE PLANE CURVES

BY
EDUARD ČECH

Already in 1880*, Darboux defined, at a non-parabolic point of a surface, a remarkable triple of tangents, to which he gave the name *quadratic osculation tangents*. In 1908**, Segre, following a different approach, was led to the same tangents, and also to the tangents conjugate to them. Finally, in 1916***, Fubini defined a differential cubic form, which is invariant with respect to the projective transformations, and which, being equalized to zero, gives the directions considered by Darboux. Following Green, I shall call the quadratic osculation tangents *Darboux tangents*, and the tangents conjugate to them *Segre tangents*. Several definitions of these tangents can be given, e.g. the following one, which I have given in my paper†, which is due to appear soon in *Annali di Matematica.* In the tangent plane of a point under consideration, let us construct the two parabolas each of which has second order contact with one asymptotic curve, and with the other asymptotic tangent being the diameter. The three intersection points of these parabolas, different from the point of the surface, are situated on the Segre tangents.

I shall call *Darboux (Segre) curves* the curves on the surface whose tangents are Darboux (Segre) tangents. The Segre curves have the following important property: *at every point of the surface, the osculating planes of three Segre curves , meeting at this point, pass through a common line*††. Having this recalled, *let L be a surface with all Segre curves* (i.e. Segre curves of all three families) *being plane curves.* One can immediately see that all the lines l, and consequently *all the planes of Segre curves pass through a fixed point s.* The cone W defined as the envelope of planes of Segre curves is *an algebraic cone of third class.* It may be reducible. The surface L itself is, in general, transcendent. Its equations can be obtained in the finite form by means of the quadratures.

**Sur le contact des courbes et des surfaces*, Bull. des Sc. Math. (2) 4, 1880.

**Rend. Acc. Lincei, 17.

****Applicabilità proiettiva di due superficie*, Rend. Circ. Mat. Palermo, 41.

†*L'intorno d'un punto d'una superficie considerato dal punto di vista proiettivo.*

††See my note: „*O trilineárních systémech atd.*" Rozpravy České Akademie, Prague, 30, 1921, no 23, v. 13.

EDUARD ČECH

I start by presenting a simple geometric interpretation of systems of linear partial differential equations, which, I hope, will be useful in various questions of infinitesimal projective geometry. Then I reduce the problem of determination of the surfaces L to the integration of the system

$$\frac{\partial^2\varphi}{\partial u^2} = 6\varphi\frac{\partial\varphi}{\partial v}, \quad \frac{\partial^2\varphi}{\partial v^2} = 6\varphi\frac{\partial\varphi}{\partial u},$$

which can be accomplished by means of the elliptic functions. After that I pass to the determination of the cone W mentioned above. Finally, I give in the finite form, in various cases, the equations of the surfaces studied.*

1. GEOMETRIC INTERPRETATION OF A SYSTEM OF PARTIAL DIFFERENTIAL EQUATIONS**.

Let us consider the system

$$(1)\qquad \begin{aligned} \frac{\partial y_i}{\partial u} &= a_{i1}y_1 + a_{i2}y_2 + a_{i3}y_3 + a_{i4}y_4, \\ \frac{\partial y_i}{\partial v} &= b_{i1}y_1 + b_{i2}y_2 + b_{i3}y_3 + b_{i4}y_4, \end{aligned} \qquad (i = 1,2,3,4)$$

where a_{ik} and b_{ik} are functions of u and v. Assuming that the integrability conditions

$$(2)\qquad \frac{\partial a_{ik}}{\partial v} - \frac{\partial b_{ik}}{\partial u} + \sum_{\nu=1}^{4}(a_{i\nu}b_{\nu k} - a_{\nu k}b_{i\nu}) = 0 \qquad (i,k = 1,2,3,4)$$

of this system are satisfied, the system (1) has four linearly independent solutions

$$(3)\qquad y_i^{(1)}, \quad y_i^{(2)}, \quad y_i^{(3)}, \quad y_i^{(4)} \qquad (i = 1,2,3,4)$$

by means of which the general solution y can be expressed by the formulae

$$(4)\qquad \begin{aligned} y_1 &= y_1^{(1)}c_1 + y_1^{(2)}c_2 + y_1^{(3)}c_3 + y_1^{(4)}c_4, \\ y_2 &= y_2^{(1)}c_1 + y_2^{(2)}c_2 + y_2^{(3)}c_3 + y_2^{(4)}c_4, \\ y_3 &= y_3^{(1)}c_1 + y_3^{(2)}c_2 + y_3^{(3)}c_3 + y_3^{(4)}c_4, \\ y_4 &= y_4^{(1)}c_1 + y_4^{(2)}c_2 + y_4^{(3)}c_3 + y_4^{(4)}c_4, \end{aligned}$$

where c_1, c_2, c_3, c_4 are arbitrary constants. Thus, the fixed solutions (3) being chosen once for ever, let us consider y_1, y_2, y_3, y_4 as the linear forms (4) in the four *indeterminates* c_1, c_2, c_3, c_4. Let us interpret c_1, c_2, c_3, c_4 as homogeneous coordinates of planes in an ordinary space with respect to a fixed tetrahedron of reference.

*I have presented these results, except for the determination of the finite equations, where only the possibility have been mentioned, in a short note published in Rendiconti dell'Accademia dei Lincei.

**It is only for fixing the ideas that I consider two independent and four dependent variables.

ON THE SURFACES ALL SEGRE CURVES OF WHICH ARE PLANE CURVES

Therefore, by (4), y_1, y_2, y_3, y_4 will be homogeneous coordinates of planes with respect to a system varying with u and v, which I shall briefly call *moving system*. *To pass from the moving coordinate system to the fixed one means to integrate the system* (1). Let us assume, for instance, that one particular solution

$$\bar{y}_1, \quad \bar{y}_2, \quad \bar{y}_3, \quad \bar{y}_4 \tag{5}$$

of the system (1) is known. One can immediately see that the quantities (5) are the moving coordinates of a *fixed* plane. Conversely, let

$$Y_1, \quad Y_2, \quad Y_3, \quad Y_4$$

be the moving coordinates of a fixed plane: We get then

$$\bar{y}_i = \varrho(u,v)Y_i,$$

where y_i is a particular solution of the system (1). Substituting into the system (1), we get ϱ by a logarithmic quadrature.

A point is determined by a linear and homogeneous relation among coordinates in the plane. Let us assume that the relation

$$A \equiv \alpha_1(u,v)y_1 + \alpha_2(u,v)y_2 + \alpha_3(u,v)y_3 + \alpha_4(u,v)y_4 = 0 \tag{6}$$

defines a *fixed* point. That is to say that, replacing y_i by the expressions (4), one gets

$$A \equiv \varrho(u,v)(a_1c_1 + a_2c_2 + a_3c_3 + a_4c_4), \tag{7}$$

with the a_i being numbers. Thus, let us differentiate the linear form A, taking into account the system (1). By (7) the result will be

$$\frac{\partial A}{\partial u} = \frac{\partial \log \varrho}{\partial u}A, \quad \frac{\partial A}{\partial v} = \frac{\partial \log \varrho}{\partial v}A.$$

Conversely, if the linear form A satisfies the conditions

$$\frac{\partial A}{\partial u} = \sigma_1(u,v)A, \quad \frac{\partial A}{\partial v} = \sigma_2(u,v)A, \tag{8}$$

it represents a fixed point. More generally, if A is an algebraic form of arbitrary degree and satisfies relations of the form (8), then the envelope of ∞^2 planes $A = 0$ is fixed. The importance of knowing such a form for the integration of the system (1) depends on the linear substitutions transforming it into itself. If they are only in finite number, the integration of the system (1) requires only the logarithmic quadrature

$$\exp\left(\int \sigma_1 du + \sigma_2 dv\right)$$

EDUARD ČECH

and algebraic operations. One can also suppose that the equations are known in moving coordinates of an arbitrary fixed geometric configuration. For instance, let us suppose that the linear substitution

$$y_i = \alpha_{i1}(u,v)z_1 + \alpha_{i2}(u,v)z_2 + \alpha_{i3}(u,v)z_3 + \alpha_{i4}(u,v)z_4 \quad (i = 1,2,3,4) \tag{9}$$

leaves the system unchanged. I have explained elsewhere how the integration problem can be simplified in this case. In other words, our assumption says that the equations (9) represent, in moving coordinates, a fixed homography.

I content myself with these short indications, and I refer the reader to a work by Fano*, where geometric interpretation of a unique equation

$$\frac{d^n y}{dx^n} + a_{n-1}(x)\frac{d^{n-1}y}{dx^{n-1}} + \ldots + a_0(x) = 0.$$

can be found. Anyhow, the consideration of first order systems of equations and the use of tangential coordinates, as I have indicated here, seems to me to be preferable.

2. Analytic conditions for a surface L.

If u, v are asymptotic parameters of a surface S, one can fix, following Wilczynski**, the arbitrary factor of the homogeneous coordinates y_1, y_2, y_3, y_4 in such a way that they satisfy the equations

$$\begin{aligned} &\frac{\partial^2 y}{\partial u^2} + 2b\frac{\partial y}{\partial v} + fy = 0, \\ &\frac{\partial^2 y}{\partial v^2} + 2a'\frac{\partial y}{\partial u} + gy = 0. \end{aligned} \tag{10}$$

Setting

$$y_1 = \frac{\partial y}{\partial u}, \quad y_2 = \frac{\partial y}{\partial}, \quad y_3 = \frac{\partial^2 y}{\partial u \partial v}, \tag{11}$$

one can reduce the system (10) to the form ( ). That is what Wilczynski does. But in some cases, the surface S must be in certain given relations with fixed elements of the space. Therefore, it will be useful to write the system (1) showing clearly these relations. Thus, in what follows, we shall consider a fixed point s in a relation with S: writing the system (1), we shall take as the unknowns $y, y_1 = \frac{\partial y}{\partial u}, y_2 = \frac{\partial y}{\partial v}$, and an arbitrary constant s. Consequently, we shall have the system

$$\begin{aligned} &\frac{\partial y}{\partial u} = y_1, && \frac{\partial y}{\partial v} = y_2, \\ &\frac{\partial y_1}{\partial u} = -2by_2 - fy, && \frac{\partial y_1}{\partial v} = \alpha y + \alpha_1 y_1 + \alpha_2 y_2 + \beta s, \\ &\frac{\partial y_2}{\partial u} = \alpha y + \alpha_1 y_1 + \alpha_2 y_2 + \beta s, && \frac{\partial y_2}{\partial v} = -2a' y_1 - gy, \\ &\frac{\partial s}{\partial u} = 0, && \frac{\partial s}{\partial v} = 0. \end{aligned} \tag{12}$$

*Mathematische Annalen, 53, 1900.

**Five papers in Trans. Amer. Math. Soc. in 1907-09.

ON THE SURFACES ALL SEGRE CURVES OF WHICH ARE PLANE CURVES

The integrability conditions (2) are here

$$\frac{\partial \beta}{\partial v} + \alpha_1 \beta = 0, \qquad \frac{\partial \beta}{\partial u} + \alpha_2 \beta = 0, \tag{13a}$$

$$\frac{\partial \alpha_1}{\partial u} + \alpha + \alpha_1 \alpha_2 - 4a'b = 0, \qquad \frac{\partial \alpha_2}{\partial v} + \alpha + \alpha_1 \alpha_2 - 4a'b = 0, \tag{13b}$$

$$f + 2\frac{\partial b}{\partial v} + \frac{\partial \alpha_2}{\partial u} + {\alpha_2}^2 - 2b\alpha_1 = 0, \qquad g + 2\frac{\partial a'}{\partial u} + \frac{\partial \alpha_1}{\partial v} + {\alpha_1}^2 - 2a'\alpha_2 = 0, \tag{13c}$$

$$2bg - \frac{\partial f}{\partial v} = \frac{\partial \alpha}{\partial u} - \alpha_1 f + \alpha\alpha_2, \qquad 2a'f - \frac{\partial g}{\partial u} = \frac{\partial \alpha}{\partial v} + \alpha\alpha_1 - \alpha_2 g. \tag{13d}$$

Let us set

$$\beta = \frac{1}{\omega}.$$

(13a) gives then

$$\alpha_1 = \frac{\partial \log \omega}{\partial v}, \qquad \alpha_2 = \frac{\partial \log \omega}{\partial u}.$$

The two equations (13b) can be reduced to a single equation

$$\alpha = -\frac{1}{\omega}\frac{\partial^2 \omega}{\partial u \partial v} + 4a'b,$$

and the equations (13c) give

$$\begin{aligned} f &= -2\frac{\partial b}{\partial v} - \frac{1}{\omega}\frac{\partial^2 \omega}{\partial u^2} + 2b\frac{\partial \log \omega}{\partial v}, \\ g &= 2\frac{\partial a'}{\partial u} - \frac{1}{\omega}\frac{\partial^2 \omega}{\partial v^2} + 2a'\frac{\partial \log \omega}{\partial u}. \end{aligned} \tag{14}$$

Thus, one can see that any surface S, referred to its asymptotic lines, can be defined in infinitely many ways by the system of partial differential equations

$$\begin{aligned} &\frac{\partial^2 y}{\partial u^2} + 2b\frac{\partial y}{\partial v} + fy = 0, \\ &\frac{\partial^2 y}{\partial v^2} + 2a'\frac{\partial y}{\partial u} + gy = 0, \\ &\frac{\partial^2 y}{\partial u \partial v} + \left(\frac{1}{\omega}\frac{\partial^2 \omega}{\partial u \partial v} - 4a'b\right) y - \frac{\partial \log \omega}{\partial v}\frac{\partial y}{\partial u} - \frac{\partial \log \omega}{\partial u}\frac{\partial y}{\partial v} - \frac{s}{\omega} = 0, \\ &\frac{\partial s}{\partial u} = \frac{\partial s}{\partial v} = 0, \end{aligned} \tag{15}$$

EDUARD ČECH

with f and g being the expressions (14). To each fixed point of the space, there corresponds one system of the type indicated. Concerning the integrability conditions (13d) of the system (15), they can be written as follows

$$(16)\qquad \begin{aligned} \frac{\partial f}{\partial v}+\frac{\partial^2 b}{\partial v^2}+4\frac{\partial a'}{\partial u}b+2a'\frac{\partial b}{\partial u}&=0,\\ \frac{\partial g}{\partial u}+\frac{\partial^2 a'}{\partial u^2}+2\frac{\partial a'}{\partial v}b+4a'\frac{\partial b}{\partial v}&=0. \end{aligned}$$

Written in the form

$$(17)\qquad \begin{aligned} \frac{\partial^2\omega}{\partial u^2}-2b\frac{\partial\omega}{\partial u}+\left(f+2\frac{\partial b}{\partial v}\right)\omega&=0,\\ \frac{\partial^2\omega}{\partial v^2}+2a'\frac{\partial\omega}{\partial v}+\left(f+2\frac{\partial a'}{\partial u}\right)\omega&=0, \end{aligned}$$

the equations (14) form what Wilczynski calls the *adjoint* system of the system (10). They are satisfied by the coordinates of tangent planes of the surface S. According to what we have said in the previous section, a particular solution of the system (10) gives the moving coordinates of a fixed plane. Similarly, a particular solution of the system (17) gives the moving coordinates of a fixed point. Thus, one can well see the geometric meaning of the equations (14).

The system (15) comes in useful for the investigation of surfaces having some relation to a fixed point of the space. Therefore, Tzitzéica, studying* surfaces with directrices passing through a fixed point, i.e. the surfaces S, as he calls them, used a system which is in fact the system (15).

But, leaving aside these generalities, we shall restrict our attention to the particular question we have in mind here. The expression

$$(18)\qquad \lambda y-\left(a'\frac{\partial b}{\partial v}-b\frac{\partial a'}{\partial v}\right)\frac{\partial y}{\partial u}+\left(a'\frac{\partial b}{\partial u}-b\frac{\partial a'}{\partial u}\right)\frac{\partial y}{\partial v}+6a'b\frac{\partial^2 y}{\partial u\partial v}$$

represents, for different values of λ, different points of the intersection line l of the three osculating planes of Segre curves. For a surface L, the lines l pass through a fixed point s, which we can make use of, for the formation of the system (15). Hence, we can choose λ in such a way that the expression (18) represents precisely the point s. Comparing the expression (18) with the third equation of (15), we find the conditions

$$(19)\qquad \begin{gathered} \frac{\partial\log\omega}{\partial u}=+\frac{1}{6}\frac{\partial}{\partial u}\log\left(\frac{a'}{b}\right),\qquad \frac{\partial\log\omega}{\partial v}=-\frac{1}{6}\frac{\partial}{\partial v}\log\left(\frac{a'}{b}\right),\\ \lambda=6a'b\left(\frac{1}{\omega}\frac{\partial^2\omega}{\partial u\partial v}-4a'b\right) \end{gathered}$$

with the last one determining the factor λ. From the equations (19), we get the condition

$$(20)\qquad \frac{\partial^2}{\partial u\partial v}\log\left(\frac{a'}{b}\right)=0.$$

*Rend. Circ. Mat. Palermo 25, 1908 and 28, 1909

The equation (20) characterizes a particular class of surfaces which were called *isothermoasymptotic* by Fubini. If a surface is isothermoasymptotic, then by a suitable change of parameters of the asymptotic curves we can achieve that

$$a' = b = \varphi(u, v). \tag{21}$$

Thus, let us suppose that the condition (20) is satisfied. Then the equations (19) give simply

$$\omega = \text{constant},$$

and we can, without loss of generality, assume that $\omega = +1$. Taking into account the equations (14), one can write the system (15) for a surface L as follows:

$$\begin{aligned} &\frac{\partial^2 y}{\partial u^2} + 2\varphi\frac{\partial y}{\partial v} - 2\frac{\partial \varphi}{\partial v}y = 0, \\ &\frac{\partial^2 y}{\partial v^2} + 2\varphi\frac{\partial y}{\partial u} - 2\frac{\partial \varphi}{\partial u}y = 0, \\ &\frac{\partial^2 y}{\partial u \partial v} - 4\varphi^2 y = s = \text{constant}. \end{aligned} \tag{22}$$

Conversely, if the integrability conditions (16) are fulfilled, then coordinates of the points of a surface L satisfy the system (22). These conditions, in the actual case, are

$$\frac{\partial^2 \varphi}{\partial u^2} = 6\varphi\frac{\partial \varphi}{\partial v}, \qquad \frac{\partial^2 \varphi}{\partial v^2} = 6\varphi\frac{\partial \varphi}{\partial u}. \tag{23}$$

We are thus led to the study of the system (23). Fortunately, this system can be completely integrated. That is what we are going to do in the next section.

3. Integration of the system $\frac{\partial^2 \varphi}{\partial u^2} = 6\varphi\frac{\partial \varphi}{\partial v}$, $\frac{\partial^2 \varphi}{\partial v^2} = 6\varphi\frac{\partial \varphi}{\partial u}$.

From the equations (23), we get

$$\frac{\partial^2 \varphi}{\partial u^2} - \frac{\partial^2 \varphi}{\partial v^2} + 6\varphi\frac{\partial \varphi}{\partial u} - 6\varphi\frac{\partial \varphi}{\partial v} = 0,$$

which can also be written in the form

$$\left(\frac{\partial}{\partial u} - \frac{\partial}{\partial v}\right)\left(\frac{\partial \varphi}{\partial u} + \frac{\partial \varphi}{\partial v} + 3\varphi^2\right) = 0. \tag{24}$$

In the next, I set for brevity

$$\varepsilon = e^{\frac{2\pi i}{3}}. \tag{25}$$

Beside the equation (24), there are two other analogues

$$\left(\varepsilon\frac{\partial}{\partial u} - \varepsilon^2\frac{\partial}{\partial v}\right)\left(\varepsilon\frac{\partial \varphi}{\partial u} + \varepsilon^2\frac{\partial \varphi}{\partial v} + 3\varphi^2\right) = 0,$$

$$\left(\varepsilon^2\frac{\partial}{\partial u} - \varepsilon\frac{\partial}{\partial v}\right)\left(\varepsilon^2\frac{\partial \varphi}{\partial u} + \varepsilon\frac{\partial \varphi}{\partial v} + 3\varphi^2\right) = 0.$$

EDUARD ČECH

I introduce also the abbreviations

$$x_0 = u + v, \quad x_1 = \varepsilon^2 u + \varepsilon v, \quad x_2 = \varepsilon u + \varepsilon^2 v, \tag{26}$$

so that

$$x_0 + x_1 + x_2 = 0. \tag{27}$$

From the three equations written above, one gets

$$\begin{aligned} \frac{\partial\varphi}{\partial u} + \frac{\partial\varphi}{\partial v} + 3\varphi^2 &= f_0(x_0), \\ \varepsilon\frac{\partial\varphi}{\partial u} + \varepsilon^2\frac{\partial\varphi}{\partial v} + 3\varphi^2 &= f_1(x_1), \\ \varepsilon^2\frac{\partial\varphi}{\partial u} + \varepsilon\frac{\partial\varphi}{\partial v} + 3\varphi^2 &= f_2(x_2), \end{aligned} \tag{28}$$

where f_0, f_1, f_2 are functions of a single variable. Introducing three new functions F_0, F_1, F_2, defined up to an additive constant, by the equations

$$F_0'(x_0) = f_0(x_0), \quad F_1'(x_1) = f_1(x_1), \quad F_2'(x_2) = f_2(x_2), \tag{29}$$

where prime denotes the derivative, and choosing suitably the constants, we have

$$\begin{aligned} 3\varphi &= F_0(x_0) + F_1(x_1) + F_2(x_2), \\ 9\varphi^2 &= F_0'(x_0) + F_1'(x_1) + F_2'(x_2). \end{aligned} \tag{30}$$

Conversely, using the equation (28), we can pass from the equation (30) to the system (23). Hence, we get successively

$$18\varphi\frac{\partial\varphi}{\partial u} = F_0''(x_0) + \varepsilon^2 F_1''(x_1) + \varepsilon F_2''(x_2),$$

$$\begin{aligned} &2\cdot 3\varphi\cdot 3\frac{\partial\varphi}{\partial u} = \\ &\quad = 2[F_0(x_0) + F_1(x_1) + F_2(x_2)]\,[F_0'(x_0) + \varepsilon^2 F_1'(x_1) + \varepsilon F_2'(x_2)], \\ &F_0''(x_0) + \varepsilon^2 F_1''(x_1) + \varepsilon F_2''(x_2) = \\ &\quad = 2[F_0(x_0) + F_1(x_1) + F_2(x_2)]\,[F_0'(x_0) + \varepsilon^2 F_1'(x_1) + \varepsilon F_2'(x_2)], \\ &F_0''(x_0) + \varepsilon F_1''(x_1) + \varepsilon^2 F_2''(x_2) = \\ &\quad = 2[F_0(x_0) + F_1(x_1) + F_2(x_2)]\,[F_0'(x_0) + \varepsilon F_1'(x_1) + \varepsilon^2 F_2'(x_2)], \end{aligned}$$

$$\frac{F_1''(x_1) - F_2''(x_2)}{F_1'(x_1) - F_2'(x_2)} = 2[F_0(x_0) + F_1(x_1) + F_2(x_2)]. \tag{31}$$

Let us return to the equations (28), and consider, for example, the first one

$$\frac{\partial\varphi}{\partial u} + \frac{\partial\varphi}{\partial v} + 3\varphi^2 = f_0(x_0). \tag{32a}$$

Differentiating, we get from this, by virtue of the equations (23),

$$\frac{\partial^2\varphi}{\partial u\partial v} + 6\varphi\left(\frac{\partial\varphi}{\partial u} + \frac{\partial\varphi}{\partial v}\right) = f_0'(x_0). \tag{32b}$$

ON THE SURFACES ALL SEGRE CURVES OF WHICH ARE PLANE CURVES

Further, we get

$$\left(\frac{\partial\varphi}{\partial u}\right)^2+\left(\frac{\partial\varphi}{\partial v}\right)^2+\frac{\partial\varphi}{\partial u}\frac{\partial\varphi}{\partial v}+6\varphi^2\left(\frac{\partial\varphi}{\partial u}+\frac{\partial\varphi}{\partial v}\right)+\varphi\frac{\partial^2\varphi}{\partial u\partial v}=\tfrac{1}{6}f_0''(x_0),$$

$$18\varphi^3\left(\frac{\partial\varphi}{\partial u}+\frac{\partial\varphi}{\partial v}\right)+6\varphi\left(\frac{\partial\varphi}{\partial u}+\frac{\partial\varphi}{\partial v}\right)^2+\frac{\partial^2\varphi}{\partial u\partial v}\left(\frac{\partial\varphi}{\partial u}+\frac{\partial\varphi}{\partial v}+3\varphi^2\right)=\tfrac{1}{12}f_0'''(x_0).$$

The last equation, when compared with (32a) and (32b), gives immediately

$$f_0'''(x_0)=12f_0(x_0)f_0'(x_0). \tag{33}$$

Along the same lines one gets

$$f_1'''(x_1)=12f_1(x_1)f_1'(x_1),\qquad f_2'''(x_2)=12f_2(x_2)f_2'(x_2). \tag{33 bis}$$

It is known that a solution of the equation

$$f'''(x)=12f(x)f'(x)$$

is either the elliptic function $p(x+a)$ with arbitrary periods, or the monoperiodic function

$$\alpha^2\operatorname{cotg}^2(\alpha x+\beta)+\frac{2\alpha^2}{3},$$

with α and β being arbitrary constants, or the rational function

$$\frac{1}{(x+a)^2},$$

with a being an arbitrary constant, or finally a constant function.

Let us exclude for the moment the case where at least one of the three functions f_0, f_1, f_2 is constant. Then we deduce easily from the identity (31) that every period of any one of the three functions f_0, f_1, f_2 is a period of the other two ones. Therefore, there are three possibilities: either

$$f_0(x_0)=p(x_0+a_0),\quad f_1(x_1)=p(x_1+a_1),\quad f_2(x_2)=p(x_2+a_2),$$

or

$$f_i(x_i)=\alpha_i^2\operatorname{cotg}^2\alpha_i(x+a_i)+\frac{2\alpha_i^2}{3}\qquad(i=0,1,2),$$

or finally

$$f_i(x_i)=\frac{1}{x_i+a_i}.$$

The conditions for the constants can be easily found with the aid of the equations (30).

Without performing the complete discusion, which represents no problem, I shall only announce the result. In the first case the solution is

$$\varphi=-\frac{1}{3}[\zeta(x_0+a_0)+\zeta(x_1+a_1)+\zeta(x_2+a_2)], \tag{34}$$

EDUARD ČECH

where

$$a_0 + a_1 + a_2 = 0\,. \tag{35}$$

It depends on four arbitrary constants. The verification can be done by using the addition theorem. In fact, by virtue of this theorem, from (34) and from the identity (27), one gets

$$\varphi^2 = \frac{1}{9}[p(x_0 + a_0) + p(x_1 + a_1) + p(x_2 + a_2)]\,. \tag{36}$$

Differentiating the equation (34), we obtain

$$\frac{\partial^2\varphi}{\partial u^2} = \frac{1}{3}[p'(x_0 + a_0) + \varepsilon p'(x_1 + a_1) + \varepsilon^2 p'(x_2 + a_2)],$$
$$\frac{\partial^2\varphi}{\partial v^2} = \frac{1}{3}[p'(x_0 + a_0) + \varepsilon^2 p'(x_1 + a_1) + \varepsilon p'(x_2 + a_2)],$$

whilst the equation (36) gives

$$2\varphi\frac{\partial\varphi}{\partial u} = \frac{1}{9}[p'(x_0 + a_0) + \varepsilon^2 p'(x_1 + a_1) + \varepsilon p'(x_2 + a_2)],$$
$$2\varphi\frac{\partial\varphi}{\partial v} = \frac{1}{9}[p'(x_0 + a_0) + \varepsilon p'(x_1 + a_1) + \varepsilon^2 p'(x_2 + a_2)].$$

In the second case the solution is

$$\varphi = -\frac{\alpha}{3}[\,\operatorname{cotg}\alpha(x_0 + a_0) + \operatorname{cotg}\alpha(x_1 + a_1) + \operatorname{cotg}\alpha(x_2 + a_2)], \tag{37}$$

where the constant α is arbitrary, and a_0, a_1, a_2 satisfy the identity (35). The verification makes use of the identity

$$\operatorname{cotg} y_0 \operatorname{cotg} y_1 + \operatorname{cotg} y_0 \operatorname{cotg} y_2 + \operatorname{cotg} y_1 \operatorname{cotg} y_2 = 1,$$

which holds for

$$y_0 + y_1 + y_2 = 0.$$

In the third case the solution is

$$\varphi = -\frac{1}{3}\left[\frac{1}{x_0 + a_0} + \frac{1}{x_1 + a_1} + \frac{1}{x_2 + a_2}\right], \tag{38}$$

where the constants a_0, a_1, a_2 satisfy always the condition (35).

It remains to consider the case where at least one of the functions f_i is constant. Using the equations (30), it can be shown that this holds then for at least two of these functions. Therefore three new cases appear:

$$\varphi = -\frac{\alpha}{3}[\,\operatorname{cotg}\alpha(x_i + a_i)], \qquad (i = 0, 1, 2) \tag{39}$$

$$\varphi = -\frac{1}{3}\frac{1}{x_i + a_i}, \qquad (i = 0, 1, 2) \tag{40}$$

$$\varphi = \text{constant}. \tag{41}$$

I denote by $L_1, L_2, \ldots, L_6$ the surfaces L corresponding to the solutions (34),(37), ... ,(41), respectively. Before passing to look for the finite equations of these surfaces, I shall determine the envelope of the planes of Segre curves.

ON THE SURFACES ALL SEGRE CURVES OF WHICH ARE PLANE CURVES

4. THE CONE ENVELOPED BY THE PLANES OF SEGRE CURVES.

In the system (22) we set $s = 0$. The system so specialized

$$\frac{\partial y}{\partial u} = y_1, \qquad \frac{\partial y}{\partial v} = y_2,$$
$$\frac{\partial y_1}{\partial u} = 2\frac{\partial \varphi}{\partial v} y - 2\varphi y_2, \qquad \frac{\partial y_1}{\partial v} = 4\varphi^2 y, \tag{42}$$
$$\frac{\partial y_2}{\partial u} = 4\varphi^2 y, \qquad \frac{\partial y_2}{\partial v} = 2\frac{\partial \varphi}{\partial u} y - 2\varphi y_1,$$

where the function φ satisfies the integrability conditions (23),

$$\frac{\partial^2 \varphi}{\partial u^2} = 6\varphi\frac{\partial \varphi}{\partial v}, \quad \frac{\partial^2 \varphi}{\partial v^2} = 6\varphi\frac{\partial \varphi}{\partial u} \tag{23}$$

admits three linearly independent solutions. I am going to show that this system can be integrated algebraically.

If

$$y = y^{(1)}c_1 + y^{(2)}c_2 + y^{(3)}c_3,$$
$$y_1 = \frac{\partial y^{(1)}}{\partial u}c_1 + \frac{\partial y^{(2)}}{\partial u}c_2 + \frac{\partial y^{(3)}}{\partial u}c_3, \tag{43}$$
$$y_2 = \frac{\partial y^{(1)}}{\partial v}c_1 + \frac{\partial y^{(2)}}{\partial v}c_2 + \frac{\partial y^{(3)}}{\partial v}c_3$$

is the general solution of the system (42), we shall interpret c_1, c_2, c_3 as homogeneous coordinates of planes in the bundle whose center is the fixed point s. Thus, the y, y_1, y_2, which are linear forms in c_1, c_2, c_3, are the moving coordinates in this bundle. A linear and homogeneous expression in y, y_1, y_2 represents a line in the bundle s, and this line varies in general with u, v. In particular, y represents the line projecting the moving point of the surface L from the point s.

With the equations of Segre curves being

$$u - v = \text{constant},$$
$$\varepsilon^2 u - \varepsilon v = \text{constant}, \tag{44}$$
$$\varepsilon u - \varepsilon^2 v = \text{constant},$$

the expression

$$\varepsilon^i y_1 + \varepsilon^{2i} y_2 + \lambda y \qquad (i = 0,\ 1,\ 2) \tag{45}$$

represents the projection (taken from the point s) of an arbitrary point situated on one of the Segre tangents. Let us try to determine λ in such a way that this line is the line of contact of the plane projecting the tangent in question with its envelope, when running along any curve different from the corresponding Segre curve on the surface L. To achieve this it is sufficient to differentiate the expression (45), for

EDUARD ČECH

instance with respect to u, and to express the fact that the expression thus obtained is of the same type as (45). According to (42), one gets

$$\frac{\partial}{\partial u}(\varepsilon^i y_1 + \varepsilon^{2i} y_2 + \lambda y) = \varepsilon^i \lambda y_1 - 2\varepsilon^{2i}\varphi y_2 + \mu y;$$

therefore,

$$\varepsilon^i y_1 + \varepsilon^{2i} y_2 + 2\varphi y \tag{46}$$

is the line of contact of the plane of the Segre curve

$$\varepsilon^{2i} u - \varepsilon^i v = \text{constant} \tag{23}$$

with its envelope.

Having this, one can easily prove the fundamental property of the surfaces L: *the planes of Segre curves envelope an algebraic cone of the third class.* Let us admit, for a moment, this theorem. We know the three tangent planes of this cone passing through the line y, as well as the lines of contact of these planes. From this we deduce that the cone, if it exists, can be represented in the moving coordinates by a cubic form of the type

$$W = {y_1}^3 + {y_2}^3 + 6\varphi y y_1 y_2 + 3(Ay + By_1 + Cy_2)y^2. \tag{47}$$

The only question is, whether we can determine the three functions A, B, C of the variables u, v in such a way that the cone represented by the form W remains fixed in the space. This can be seen by forming, with the aid of the equations (42), the derivatives

$$\frac{1}{3}\frac{\partial W}{\partial u} = \left(\frac{\partial A}{\partial u} + 2\frac{\partial \varphi}{\partial v}B + 4\varphi^2 C\right) y^3 + \left(8\varphi^3 + 3A + \frac{\partial B}{\partial u}\right) y^2 y_1 + \\ + \left(4\varphi\frac{\partial \varphi}{\partial v} - 2\varphi B + \frac{\partial C}{\partial u}\right) y^2 y_2 + 2\left(B + \frac{\partial \varphi}{\partial v}\right) y{y_1}^2 + 2\left(C + \frac{\partial \varphi}{\partial u}\right) y y_1 y_2,$$

$$\frac{1}{3}\frac{\partial W}{\partial v} = \left(\frac{\partial A}{\partial v} + 2\frac{\partial \varphi}{\partial u}C + 4\varphi^2 B\right) y^3 + \left(\frac{\partial B}{\partial v} - 2\varphi C + 4\varphi\frac{\partial \varphi}{\partial u}\right) y^2 y_1 + \\ + \left(\frac{\partial C}{\partial v} + 8\varphi^3 + 3A\right) y^2 y_2 + 2\left(C + \frac{\partial \varphi}{\partial u}\right) y{y_2}^2 + 2\left(B + \frac{\partial \varphi}{\partial v}\right) y y_1 y_2.$$

For the cone W to be fixed it is necessary and sufficient that $\frac{\partial W}{\partial u}$ and $\frac{\partial W}{\partial v}$ be proportional to W. According to the above expressions, this can take place only if it holds identically

$$\frac{\partial W}{\partial u} = \frac{\partial W}{\partial v} = 0. \tag{48}$$

ON THE SURFACES ALL SEGRE CURVES OF WHICH ARE PLANE CURVES

Thus, we have the conditions

$$(49)\qquad B+\frac{\partial\varphi}{\partial v}=0,\quad C+\frac{\partial\varphi}{\partial u}=0,\quad 3A=-\frac{\partial B}{\partial u}-8\varphi^3=-\frac{\partial C}{\partial v}-8\varphi^3,$$

$$(50)\qquad \begin{aligned} &\frac{\partial B}{\partial v}-2\varphi C+4\varphi\frac{\partial\varphi}{\partial u}=0, && \frac{\partial C}{\partial u}-2\varphi B+4\varphi\frac{\partial\varphi}{\partial v}=0,\\ &\frac{\partial A}{\partial u}+2\frac{\partial\varphi}{\partial v}B+4\varphi^2C=0, && \frac{\partial A}{\partial v}+2\frac{\partial\varphi}{\partial u}C+4\varphi^2B=0. \end{aligned}$$

Already from the equations (49), we get without ambiguity

$$(51)\qquad B=-\frac{\partial\varphi}{\partial v},\quad C=-\frac{\partial\varphi}{\partial u},\quad 3A=\frac{\partial^2\varphi}{\partial u\partial v}-8\varphi^3.$$

Substituting these values into the first terms of the equations (50), we find

$$\frac{\partial B}{\partial v}-2\varphi C+4\varphi\frac{\partial\varphi}{\partial u}=-\frac{\partial^2\varphi}{\partial v^2}+6\varphi\frac{\partial\varphi}{\partial u},$$

$$\frac{\partial C}{\partial u}-2\varphi B+4\varphi\frac{\partial\varphi}{\partial v}=-\frac{\partial^2\varphi}{\partial u^2}+6\varphi\frac{\partial\varphi}{\partial v}.$$

$$3\left(\frac{\partial A}{\partial u}+2\frac{\partial\varphi}{\partial v}B+4\varphi^2C\right)=\frac{\partial}{\partial v}\left(\frac{\partial^2\varphi}{\partial u^2}-6\varphi\frac{\partial\varphi}{\partial v}\right)+6\varphi\left(\frac{\partial^2\varphi}{\partial v^2}-6\varphi\frac{\partial\varphi}{\partial u}\right),$$

$$3\left(\frac{\partial A}{\partial v}+2\frac{\partial\varphi}{\partial u}C+4\varphi^2B\right)=\frac{\partial}{\partial u}\left(\frac{\partial^2\varphi}{\partial v^2}-6\varphi\frac{\partial\varphi}{\partial u}\right)+6\varphi\left(\frac{\partial^2\varphi}{\partial u^2}-6\varphi\frac{\partial\varphi}{\partial v}\right).$$

Thus, the equations (50) are also satisfied due to the integrability conditions (23).

Finally, we find that the cone W, the envelope of the planes of Segre curves, is in the moving coordinates represented by the cubic form

$$(52)\qquad W=\left(\frac{\partial^2\varphi}{\partial u\partial v}-8\varphi^3\right)y^3+{y_1}^3+{y_2}^3+6\varphi yy_1y_2-3\frac{\partial\varphi}{\partial v}y^2y_1-3\frac{\partial\varphi}{\partial u}y^2y_2.$$

According to the equations (48), every particular integral of the system (42) makes the expression W constant. From this, it can be easily deduced that in general — more precisely if the expression W admits only a finite number of linear substitutions into itself — the system (42) can be integrated algebraically. In fact, in this case, with the aid of the expression W, one can determine by algebraic operations the moving coordinates of an arbitrary fixed plane of the bundle s not touching the cone W. Let

$$Y\ :\ Y_1\ :\ Y_2$$

EDUARD ČECH

be these coordinates. We know that we can choose λ in such a way that

$$y = \lambda Y, \; y_1 = \lambda Y_1, \; y_2 = \lambda Y_2$$

is a solution of the system (42). In order to determine λ it is sufficient to substitute these expressions, with λ considered as indeterminate, into the form W. Because this form must be constant, we obtain for λ^3 a rational expresion.

At the end of this section, I present the following theorem, which will be verified successively in the separate cases.

For the different types $L_1, L_2, L_3, L_4, L_5, L_6$ of a surface L the cone W

(L_1) *has genus one,*
(L_2) *has a double tangent plane,*
(L_3) *has a stationary tangent plane,*
(L_4 and L_5) *decomposes into a quadratic cone and a pencil the axis of which in the case L_5 is situated on the quadratic cone,*
(L_6) *decomposes into three pencils.*

5. The surface L_6.

Without any loss of generality, we may assume that

$$\varphi = 1. \tag{53}$$

The system (22) takes in this case the form

$$\begin{aligned} &\frac{\partial y}{\partial u} = y_1, && \frac{\partial y}{\partial v} = y_2, \\ &\frac{\partial y_1}{\partial u} = -2y_2, && \frac{\partial y_1}{\partial v} = 4y + s, \\ &\frac{\partial y_2}{\partial u} = 4y + s, && \frac{\partial y_2}{\partial v} = -2y_1. \end{aligned} \tag{54}$$

It can be immediately verified that the three lines (46) are fixed as I have already announced. More precisely, it can be proved that the three points given by the expressions

$$t^i = \varepsilon^i y_1 + \varepsilon^{2i} y_2 - 2y - \frac{s}{2}, \qquad (i = 0, 1, 2) \tag{55}$$

are fixed. In fact, from the equations (54) we get

$$\frac{\partial t_i}{\partial u} = -2\varepsilon^{2i} t_i, \qquad \frac{\partial t_i}{\partial v} = -2\varepsilon^i t_i,$$

as well as

$$t_i = c_i e^{-2x_i}, \tag{56}$$

where the c_i's are constants, and the x_i's are the expressions (26). Let us compute y from the equations (55), and replace the t_i's by the values (56). We find thus the general solution of the system (54)

$$(57) \qquad -6y = c_0 e^{-2x_0} + c_1 e^{-2x_1} + c_2 e^{-2x_2} + \frac{3s}{2}.$$

Taking into account the identity (27), we can see that the surface L_6 is the well known *tetrahedral surface of the* 3rd *order and the* 3rd *class*

$$(58) \qquad y^{(1)} y^{(2)} y^{(3)} = \left(y^{(4)}\right)^3.$$

The planes passing through the three intersection lines

$$y^{(1)} = y^{(2)} = 0, \quad y^{(1)} = y^{(3)} = 0, \quad y^{(2)} = y^{(3)} = 0$$

of the tetrahedron of reference cut the surface along the Segre curves, and those passing through the opposite intersection lines

$$y^{(1)} = y^{(4)} = 0, \quad y^{(2)} = y^{(4)} = 0, \quad y^{(3)} = y^{(4)} = 0$$

cut it along the Darboux curves. Thus, the three families of Darboux curves consist of *conics*. The conics of each family pass through two fixed points.

The surface being correlative with itself, one has also the correlative properties.

6. The surface L_5.

Without any loss of generality, we may assume that

$$(59) \qquad \varphi = -\frac{1}{3(u+v)}.$$

The cubic form W decomposes here as I have already announced:

$$(60) \quad W = \left[y_1 + y_2 + \frac{2y}{3(u+v)}\right] \left[{y_1}^2 - y_1 y_2 + {y_2}^2 - \frac{5y^2}{9(u+v)^2} - \frac{2(y_1+y_2)y}{3(u+v)}\right].$$

The planes of the Segre curves

$$u - v = \text{constant}$$

form a pencil, whilst the planes of the Segre curves of the other two families envelop a quadratic cone represented in the moving coordinates by the equation

$$(61) \qquad {y_1}^2 - y_1 y_2 + {y_2}^2 - \frac{5y^2}{9(u+v)^2} - \frac{2(y_1+y_2)y}{3(u+v)} = 0.$$

EDUARD ČECH

Let us form the system (22):

$$
(62)\qquad
\begin{aligned}
\frac{\partial y}{\partial u} &= y_1, & \frac{\partial y}{\partial v} &= y_1,\\
\frac{\partial y_1}{\partial u} &= \frac{2}{3(u+v)}y_2 + \frac{2}{3(u+v)^2}y, & \frac{\partial y_1}{\partial v} &= \frac{4}{9(u+v)^2}y + s,\\
\frac{\partial y_2}{\partial u} &= \frac{4}{9(u+v)^2}y + s, & \frac{\partial y_2}{\partial v} &= \frac{2}{3(u+v)}y_1 + \frac{2}{3(u+v)^2}y.
\end{aligned}
$$

The expression

$$
(63)\qquad t = y_1 + y_2 + \frac{2y}{3(u+v)} - 3(u+v)s
$$

represents a fixed point. Namely, by virtue of the equations (62), we have

$$
\frac{\partial t}{\partial u} = \frac{\partial t}{\partial v} = \frac{2t}{3(u+v)},
$$

from where

$$
(64)\qquad t = a(u+v)^{2/3},
$$

with a being a constant. It can be easily verified that the line $s\,t$ is situated on the quadratic cone (61), and the corresponding tangent plane is

$$
y : y_1 : y_2 : s = 3(u+v) : -1 : -1 : 0.
$$

The point t belongs to this plane. Another point in this plane is

$$
y_1 - y_2.
$$

Thus, let us try to determine λ in such a way that the point

$$
z = y_1 - y_2 + \lambda t
$$

is fixed. We find

$$
\begin{aligned}
\frac{\partial z}{\partial u} &= -\frac{z}{3(u+v)} + \left[\frac{\partial \lambda}{\partial u} + \frac{3\lambda+1}{3(u+v)}\right] t,\\
\frac{\partial z}{\partial v} &= -\frac{z}{3(u+v)} + \left[\frac{\partial \lambda}{\partial v} + \frac{3\lambda-1}{3(u+v)}\right] t.
\end{aligned}
$$

The simplest solution of the system

$$
\frac{\partial \lambda}{\partial u} + \frac{3\lambda+1}{3(u+v)} = 0, \qquad \frac{\partial \lambda}{\partial v} + \frac{3\lambda-1}{3(u+v)} = 0
$$

is

$$
\lambda = \frac{v-u}{3(v+u)}.
$$

Thus, we can set

$$z = y_1 - y_2 + \frac{v-u}{3(v+u)}t,$$

i.e.

$$z = -\frac{2(u-v)}{9(u+v)^2}y + \frac{2(u+2v)}{3(u+v)}y_1 - \frac{2(2u-v)}{3(u+v)}y_2 + (u-v)s\,. \tag{65}$$

Then we get

$$\frac{\partial z}{\partial u} = \frac{\partial z}{\partial v} = -\frac{z}{3(u+v)},$$

which gives

$$z = \frac{b}{\sqrt[3]{u+v}}, \tag{66}$$

with b being a constant.

The plane

$$y \;:\; y_1 \;:\; y_2 \;:\; s \;=\; 3(u^2-v^2) \;:\; 2(u+2v) \;:\; -2(2u+v) \;:\; 0$$

is the polar plane of the point z with respect to the cone (61), and consequently it is fixed. It contains the points s , t, and, for instance, the point

$$3(uy_1 + vy_2) - 2y\,.$$

Therefore, we can choose λ and μ in such a way that the point

$$r = 3(uy_1 + vy_2) - 2y + \lambda t + \mu s$$

is fixed. We have

$$\begin{aligned}
\frac{\partial r}{\partial u} &= -\frac{r}{3(u+v)} + \left[\frac{\partial\lambda}{\partial u} + \frac{\lambda}{u+v} + \frac{2u+v}{u+v}\right]t+\\
&\quad + \left[\frac{\partial\mu}{\partial u} + \frac{\mu}{3(u+v)} + 6(u+v)\right]s\\
\frac{\partial r}{\partial v} &= -\frac{r}{3(u+v)} + \left[\frac{\partial\lambda}{\partial v} + \frac{\lambda}{u+v} + \frac{u+2v}{u+v}\right]t+\\
&\quad + \left[\frac{\partial\mu}{\partial v} + \frac{\mu}{3(u+v)} + 6(u+v)\right]s.
\end{aligned}$$

The simplest way how to annul the coefficients of t and s in these equations is

$$\lambda = -\frac{u^2+uv+v^2}{u+v}, \qquad \mu = -\frac{18(u+v)^2}{7}\,.$$

EDUARD ČECH

Thus, let us set

$$r = 3(uy_1 + vy_2) - 2y - \frac{u^2 + uv + v^2}{u+v}t - \frac{18(u+v)^2}{7}s,$$

i.e.

$$\begin{aligned} r = &- \frac{2(4u^2 + 7uv + 4v^2)}{3(u+v)^2}y + \frac{2u^2 + 2uv - v^2}{u+v}y_1 - \\ &- \frac{u^2 + 2uv + 2v^2}{u+v}y_2 + \frac{3(u^2 - 5uv + v^2)}{7}s\,. \end{aligned} \tag{67}$$

We have

$$r = \frac{c}{\sqrt[3]{u+v}}, \tag{68}$$

with c being a constant.

Let us sum up the equations (63), (65), and (67), after having them multiplied by the factors

$$\frac{2(u^2 + uv + v^2)}{u+v}, \quad 3(u-v), \quad -2,$$

respectively. Replacing t, z, and r by their values (64), (66), and (68), we obtain the general solution of the system (62)

$$y = \frac{u^2 + uv + v^2}{\sqrt[3]{u+v}}a + \frac{3}{2}\frac{u-v}{\sqrt[3]{u+v}}b - \frac{1}{\sqrt[3]{u+v}}c + \frac{27(u+v)^2}{14}s. \tag{69}$$

Thus the parametric equations of the surface L_5 are

$$x = \frac{u^2 + uv + v^2}{(u+v)^{7/3}}, \quad y = \frac{u-v}{(u+v)^{7/3}}, \quad z = \frac{1}{3(u+v)^{7/3}}.$$

Eliminating u and v, we can write

$$z^8 = (4xz - y^2)^7. \tag{70}$$

The Darboux curves of the family

$$u + v = \text{constant}$$

are conics touching one another at a fixed point.

ON THE SURFACES ALL SEGRE CURVES OF WHICH ARE PLANE CURVES

7. The surface L_4.

Evidently we can suppose

$$\varphi = -\frac{1}{3} \operatorname{cotg}(u+v). \tag{71}$$

I shall follow the same lines as in the preceding case. Even here, the cubic form W decomposes:

$$W = \left[y_1 + y_2 + \frac{2}{3} \operatorname{cotg}(u+v) y\right] \cdot w, \tag{72}$$

where

$$w = {y_1}^2 - y_1 y_2 + {y_2}^2 - \frac{5 + 4\sin^2(u+v)}{9\sin^2(u+v)} y^2 - \frac{2}{3} \operatorname{cotg}(u+v)(y_1+y_2)y. \tag{73}$$

The system (22) has now the form

$$\begin{aligned}
&\frac{\partial y}{\partial u} = y_1, \qquad \frac{\partial y}{\partial v} = y_2,\\
&\frac{\partial y_1}{\partial u} = \frac{2}{3} \operatorname{cotg}(u+v) y_2 + \frac{2}{3} \frac{1}{\sin^2(u+v)} y,\\
&\frac{\partial y_1}{\partial v} = \frac{4}{9} \operatorname{cotg}^2(u+v) y + s,\\
&\frac{\partial y_2}{\partial u} = \frac{4}{9} \operatorname{cotg}^2(u+v) y + s,\\
&\frac{\partial y_2}{\partial v} = \frac{2}{3} \operatorname{cotg}(u+v) y_1 + \frac{2}{3} \frac{1}{\sin^2(u+v)} y.
\end{aligned} \tag{74}$$

The planes of Segre curves of the family

$$u - v = \text{constant}$$

pass through the line joining the points s and t, where

$$t = y_1 + y_2 + \frac{2y}{3} \operatorname{cotg}(u+v) + \lambda s.$$

I want to determine λ in such a way that the point t remains fixed. For this I form

$$\begin{aligned}
\frac{\partial t}{\partial u} &= \frac{2}{3} \operatorname{cotg}(u+v) t + \left[\frac{\partial \lambda}{\partial u} - \frac{2}{3} \lambda \operatorname{cotg}(u+v) + 1\right] s,\\
\frac{\partial t}{\partial v} &= \frac{2}{3} \operatorname{cotg}(u+v) t + \left[\frac{\partial \lambda}{\partial v} - \frac{2}{3} \lambda \operatorname{cotg}(u+v) + 1\right] s.
\end{aligned}$$

For the sake of brevity, let us set

$$j = \int_{\infty}^{\sin^{2/3}(u+v)} \frac{dx}{\sqrt{x(1-x^3)}}. \tag{75}$$

EDUARD ČECH

We get

$$\frac{\partial j}{\partial u} = \frac{\partial j}{\partial v} = \frac{2}{3\sin^{2/3}(u+v)}\,. \tag{76}$$

Having this, we can set

$$\lambda = -\frac{3}{2}j \cdot \sin^{2/3}(u+v).$$

Thus, the point t is fixed, and we have

$$t = y_1 + y_2 + \frac{2}{3}\operatorname{cotg}(u+v)y - \frac{3}{2}j \cdot \sin^{2/3}(u+v)s, \tag{77}$$

$$t = a\sin^{2/3}(u+v), \tag{78}$$

with a being constant.

The plane

$$y : y_1 : y_2 : s = -3\sin(u+v)\cos(u+v) : [1+2\sin^2(u+v)] : [1+2\sin^2(u+v)] : 0$$

is the polar plane of the point t with respect to the quadratic cone (73), and consequently it is fixed. It contains, for instance, the point

$$y_1 - y_2,$$

and being fixed, it contains also the point

$$\left(\frac{\partial}{\partial u} - \frac{\partial}{\partial v}\right)(y_1 - y_2).$$

The line joining this latter point with the point s contains the point

$$2y + \operatorname{cotg}(u+v)t.$$

Thus, let us set

$$r_1 = y_1 - y_2, \qquad r_2 = 2y + \operatorname{cotg}(u+v)t$$

and determine first λ and μ in such a way that the line joining the points s and $\lambda r_1 + \mu r_2$ is fixed. To achieve this, we compute, taking $s = 0$,

$$\begin{aligned}
&\frac{\partial}{\partial u}(\lambda r_1 + \mu r_2) = \\
&\quad = \left[\frac{\partial \lambda}{\partial u} - \frac{1}{3}\operatorname{cotg}(u+v)\cdot\lambda + \mu\right] r_1 + \left[\frac{\partial \mu}{\partial u} + \frac{\lambda}{3} - \frac{1}{3}\operatorname{cotg}(u+v)\cdot\mu\right] r_2, \\
&\frac{\partial}{\partial v}(\lambda r_1 + \mu r_2) = \\
&\quad = \left[\frac{\partial \lambda}{\partial v} - \frac{1}{3}\operatorname{cotg}(u+v)\cdot\lambda - \mu\right] r_1 + \left[\frac{\partial \mu}{\partial v} - \frac{\lambda}{3} - \frac{1}{3}\operatorname{cotg}(u+v)\cdot\mu\right] r_2.
\end{aligned}$$

ON THE SURFACES ALL SEGRE CURVES OF WHICH ARE PLANE CURVES

The second terms of these equations vanish if we set

$$\lambda = \sqrt{3}\sin^{1/3}(u+v)e^{-\frac{u-v}{\sqrt{3}}}, \qquad \mu = \sin^{1/3}(u+v)e^{-\frac{u-v}{\sqrt{3}}},$$

or

$$\lambda = -\sqrt{3}\sin^{1/3}(u+v)e^{\frac{u-v}{\sqrt{3}}}, \qquad \mu = \sin^{1/3}(u+v)e^{\frac{u-v}{\sqrt{3}}}.$$

If we now introduce the two linear forms

$$\begin{aligned} z_1 &= \sin^{1/3}(u+v)e^{-\frac{u-v}{\sqrt{3}}}\,(\sqrt{3}r_1 + r_2) + \lambda_1 s, \\ z_2 &= \sin^{1/3}(u+v)e^{\frac{u-v}{\sqrt{3}}}\,(-\sqrt{3}r_1 + r_2) + \lambda_2 s, \end{aligned} \tag{79}$$

we can determine λ_1 and λ_2 in such a way that the points reperesented by them are fixed. From the very procedure which led us to these expressions, it follows, irrespective of the λ_i's, that in the derivatives $\frac{\partial z_i}{\partial u}, \frac{\partial z_i}{\partial v}$ only s can appear.

One finds the conditions

$$\begin{aligned} \frac{\partial \lambda_1}{\partial u} &= \sqrt{3}\left[\sin^{1/3}(u+v) - \cos\left(u+v-\frac{\pi}{3}\right)j\right] e^{-\frac{u-v}{\sqrt{3}}}, \\ \frac{\partial \lambda_1}{\partial v} &= -\sqrt{3}\left[\sin^{1/3}(u+v) - \cos\left(u+v+\frac{\pi}{3}\right)j\right] e^{-\frac{u-v}{\sqrt{3}}}, \\ \frac{\partial \lambda_2}{\partial u} &= -\sqrt{3}\left[\sin^{1/3}(u+v) - \cos\left(u+v+\frac{\pi}{3}\right)j\right] e^{\frac{u-v}{\sqrt{3}}}, \\ \frac{\partial \lambda_2}{\partial v} &= \sqrt{3}\left[\sin^{1/3}(u+v) - \cos\left(u+v-\frac{\pi}{3}\right)j\right] e^{\frac{u-v}{\sqrt{3}}}. \end{aligned}$$

Thus, we can set

$$\begin{aligned} \lambda_1 &= \left[\frac{3}{2}j\cos(u+v) - 3\sin^{1/3}(u+v)\right] e^{-\frac{u-v}{\sqrt{3}}}, \\ \lambda_2 &= \left[\frac{3}{2}j\cos(u+v) - 3\sin^{1/3}(u+v)\right] e^{\frac{u-v}{\sqrt{3}}}. \end{aligned}$$

Substituting these values into the expressions (79), and replacing r_1 and r_2 by their values, we arrive at the definitive result

$$\begin{aligned} z_1 =& \sin^{1/3}(u+v)e^{-\frac{u-v}{\sqrt{3}}}[(\operatorname{cotg}(u+v)+\sqrt{3})y_1 + \\ &+(\operatorname{cotg}(u+v)-\sqrt{3})y_2 + \frac{2}{3}(\operatorname{cotg}^2(u+v)+3)y - 3s], \end{aligned} \tag{80}$$

$$\begin{aligned} z_2 =& \sin^{1/3}(u+v)e^{\frac{u-v}{\sqrt{3}}}[(\operatorname{cotg}(u+v)-\sqrt{3})y_1 + \\ &+(\operatorname{cotg}(u+v)+\sqrt{3})y_2 + \frac{2}{3}(\operatorname{cotg}^2(u+v)+3)y - 3s], \end{aligned} \tag{81}$$

EDUARD ČECH

$$z_1 = b, \qquad z_2 = c, \tag{82}$$

with b and c being constant. Eliminating y_1 and y_2 from the equations (77), (80) and (81), and replacing t, z_1 and z_2 by the expressions (78) and (82), we obtain finally the general solution of the system (74):

$$\begin{aligned} -4\sin^{1/3}(u+v)y &= 2a\cos(u+v) - be^{\frac{u-v}{\sqrt{3}}} - \\ &- ce^{-\frac{u-v}{\sqrt{3}}} + 3s\left[\cos(u+v)\int\limits_{\infty}^{\sin^{2/3}(u+v)} \frac{dx}{\sqrt{x(1-x^3)}} - 2\sin^{1/3}(u+v)\right]. \end{aligned} \tag{83}$$

Therefore the surface L_4 can be represented by the parametric equations

$$\begin{aligned} x &= \alpha^2, \\ y &= \alpha\sqrt{1-\beta^3}, \\ z &= \alpha\left[\sqrt{1-\beta^3}\int\limits_{\infty}^{\beta} \frac{d\beta}{\sqrt{\beta(1-\beta^3)}} - 2\sqrt{\beta}\right]. \end{aligned} \tag{84}$$

The family $\beta =$ const. or $u + v =$ const. of Darboux curves consists of conics passing through two fixed points. With a convenient metric the surface becomes a surface of revolution*.

8. The surface L_3.

In this case we can achieve that

$$\varphi = -\frac{1}{3}\left[\frac{1}{u+v} + \frac{1}{\varepsilon^2 u + \varepsilon v} + \frac{1}{\varepsilon u + \varepsilon^2 v}\right] = \frac{uv}{u^3+v^3}. \tag{85}$$

Let us start by writing the system (22)

$$\begin{aligned} &\frac{\partial y}{\partial u} = y_1, \qquad \frac{\partial y}{\partial v} = y_2, \\ &\frac{\partial y_1}{\partial u} = \frac{2u(u^3-2v^3)}{(u^3+v^3)^2}y - \frac{2uv}{u^3+v^3}y_2, \qquad \frac{\partial y_1}{\partial v} = \frac{4u^2v^2}{(u^3+v^3)^2}y + s, \\ &\frac{\partial y_2}{\partial u} = \frac{4u^2v^2}{(u^3+v^3)^2}y + s, \qquad \frac{\partial y_2}{\partial v} = -\frac{2v(2u^3-v^3)}{(u^3+v^3)^2}y - \frac{2uv}{u^3+v^3}y_1, \end{aligned} \tag{86}$$

as well as the cubic form W

$$\begin{aligned} W = &-\frac{2(u^6+v^6-3u^3v^3)}{(u^3+v^3)^2}y + {y_1}^3 + {y_2}^3 + \frac{6uv}{u^3+v^3}yy_1y_2 - \\ &- \frac{3u(u^3-2v^3)}{(u^3+v^3)^2}y^2y_1 + \frac{3v(2u^3-v^3)}{(u^3+v^3)^2}y^2y_2. \end{aligned} \tag{87}$$

*On each surface of revolution the parallels form a family of Darboux curves. This can be seen without any calculation using only the Darboux's definition of his tangents.

ON THE SURFACES ALL SEGRE CURVES OF WHICH ARE PLANE CURVES

I have already mentioned that the cone W has a stationary tangent plane.

In order to find the moving coordinates of this plane, the best way is to make use of the equations

$$\frac{\partial W}{\partial y_1} = \frac{\partial W}{\partial y_2} = \frac{\partial^2 W}{\partial {y_1}^2}\frac{\partial^2 W}{\partial {y_2}^2} - \left(\frac{\partial^2 W}{\partial y_1 \partial y_2}\right)^2 = 0,$$

from which we get

(88) $$y \,:\, y_1 \,:\, y_2 \,:\, s \;=\; -(u^3+v^3) \,:\, u^2 \,:\, v^2 \,:\, 0.$$

The line of contact of this tangent plane joins the point s with the point

$$t = (u^3-v^3)y + (u^3+v^3)(uy_1 - vy_2) + \lambda s.$$

Using the equations (86), we produce

$$\frac{\partial t}{\partial u} = \frac{5u^2}{u^3+v^3}t + \left[\frac{\partial \lambda}{\partial u} - \frac{5u^2\lambda}{u^3+v^3} - v(u^3+v^3)\right] s$$
$$\frac{\partial t}{\partial v} = \frac{5u^2}{u^3+v^3}t + \left[\frac{\partial \lambda}{\partial v} - \frac{5v^2\lambda}{u^3+v^3} + u(u^3+v^3)\right] s.$$

For the sake of brevity, we set

(89) $$j = \int_{\infty}^{\frac{u}{v}} \frac{dz}{(z^3+1)^{2/3}},$$

so that

(90) $$\frac{\partial j}{\partial u} = \frac{v}{(u^3+v^3)^{2/3}}, \quad \frac{\partial j}{\partial v} = -\frac{u}{(u^3+v^3)^{2/3}}.$$

Consequently the point

(91) $$t = (u^3-v^3)y + (u^3+v^3)(uy_1 - vy_2) + (u^3+v^3)^{5/3} \cdot js$$

is fixed, and we have

(92) $$t = a(u^3+v^3)^{5/3},$$

with a being constant.

Another point of the fixed plane (88) is evidently the point

$$v^2y_1 - u^2y_2.$$

Therefore, the functions λ and μ can be chosen in such a way that the point

$$z = v^2y_1 - u^2y_2 + \lambda t + \mu s$$

EDUARD ČECH

is fixed. We have

$$\frac{\partial z}{\partial u}=\frac{2u^2(u^3+2v^3)}{u^6-v^6}z+\left[\frac{\partial\lambda}{\partial u}+\frac{3u^2(u^3-3v^3)}{u^6-v^6}\lambda-\frac{2uv^2(u^3+2v^3)}{(u^3+v^3)^2(u^3-v^3)}\right]t+$$
$$+\left[\frac{\partial\mu}{\partial u}-\frac{2u^2(u^3+2v^3)}{u^6-v^6}\mu+\frac{2uv^2(u^3+2v^3)}{(u^3-v^3)(u^3+v^3)^{1/3}}\cdot j-u^2\right]s,$$
$$\frac{\partial z}{\partial v}=-\frac{2v^2(2u^3+v^3)}{u^6-v^6}z+\left[\frac{\partial\lambda}{\partial v}+\frac{3v^2(3u^3-v^3)}{u^6-v^6}\lambda+\frac{2u^2v(2u^3+v^3)}{(u^3+v^3)^2(u^3-v^3)}\right]t+$$
$$+\left[\frac{\partial\mu}{\partial v}+\frac{2v^2(2u^3+v^3)}{u^6-v^6}\mu-\frac{2u^2v(2u^3+v^3)}{(u^3-v^3)(u^3+v^3)^{1/3}}\cdot j+v^2\right]s.$$

I leave to the reader to verify that the choice

$$\lambda=-\frac{2u^2v^2}{(u^3+v^3)^2},\qquad \mu=\frac{2u^2v^2}{(u^3+v^3)^{1/3}}j+u^3-v^3$$

annuls the coefficients of s and t in the expressions for $\frac{\partial z}{\partial u}$ and $\frac{\partial z}{\partial v}$. I shall use these values of λ and μ, but, for the sake of simplicity, I introduce the expression

$$z_1=-\frac{z}{u^3-v^3}\,.$$

We have

$$z_1=\frac{v^2y_1+u^2y_2}{u^3+v^3}+\frac{2u^2v^2}{(u^3+v^3)^2}y-s, \tag{93}$$

and

$$\frac{\partial z_1}{\partial u}=-\frac{u^2}{u^3+v^3}z_1,\qquad \frac{\partial z_1}{\partial v}=-\frac{v^2}{u^3+v^3}z_1,$$

from where

$$z_1=\frac{b}{\sqrt[3]{u^3+v^3}}, \tag{94}$$

with b being constant.

Besides the plane (88), there is only one other tangent plane of the cone W, passing through the point z_1. We can find that the moving coordinates of this new fixed plane are

$$y\,:\,y_1\,:\,y_2\,:\,s\,=\,(u^6-v^6)\,:\,2u^2(u^3+2v^3)\,:\,-2v^2(2u^3+v^3)\,:\,0.$$

This plane contains evidently the point

$$2y-uy_1-vy_2.$$

Thus, let us set

$$r=2y-uy_1-vy_2+\lambda z_1+\mu s.$$

ON THE SURFACES ALL SEGRE CURVES OF WHICH ARE PLANE CURVES

We find

$$\frac{\partial r}{\partial u} = -\frac{u^2}{u^3+v^3}r + \left(\frac{\partial \lambda}{\partial u} + v\right) z_1 + \left(\frac{\partial \mu}{\partial u} + \frac{u^2}{u^3+v^3}\mu\right) s,$$

$$\frac{\partial r}{\partial v} = -\frac{v^2}{u^3+v^3}r + \left(\frac{\partial \lambda}{\partial v} + u\right) z_1 + \left(\frac{\partial \mu}{\partial v} + \frac{v^2}{u^3+v^3}\mu\right) s.$$

Consequently, the choice

$$\lambda = -uv, \qquad \mu = 0$$

makes the point r fixed. Therefore

$$r = -\frac{u(u^3+2v^3)}{u^3+v^3}y_1 - \frac{v(2u^3+v^3)}{u^3+v^3}y_2 + 2\frac{u^6+u^3v^3+v^6}{(u^3+v^3)^2}y + uvs, \tag{95}$$

$$t = \frac{c}{\sqrt[3]{u^3+v^3}}, \tag{96}$$

with c being constant.

The elimination of y_1 and y_2 from the equations (91), (92) and (95) gives us finally, if we replace t, z_1, and r by their values (92), (94) and (96), the general solution of the system (86)

$$y = \frac{1}{2\sqrt[3]{u^3+v^3}}\left\{a(u^3-v^3) + 3buv + c - s\left[(u^3-v^3)\int_{\infty}^{\frac{u}{v}} \frac{dx}{(x^3+1)^{2/3}} - 2uv\sqrt[3]{u^3+v^3}\right]\right\}. \tag{97}$$

Thus, the surface L_3 can be represented by means of the parameters α and β by the equations

$$\begin{aligned} x &= \alpha^3, \\ y &= \frac{\alpha^2\beta}{\sqrt[3]{(\beta^3-1)^2}}, \\ z &= \frac{2\alpha^3\beta}{\beta^3-1}\sqrt[3]{\beta^3+1} - \alpha^3\int_{\infty}^{\beta}\frac{dx}{(x^3+1)^{2/3}}. \end{aligned} \tag{98}$$

One can see that the planes of the pencil

$$\frac{z}{x} = \text{constant}$$

cut the surface along cubic curves for which the point

$$x = y = z = 0$$

is a common point of regression.

EDUARD ČECH

9. The surface L_2.

In this section, I shall write for brevity

$$(99) \quad \xi_0 = \operatorname{cotg}(u+v), \qquad \xi_1 = \operatorname{cotg}(\varepsilon^2 u + \varepsilon v), \qquad \xi_2 = \operatorname{cotg}(\varepsilon u + \varepsilon^2 v),$$

so that we have

$$(100) \qquad \xi_1\xi_2 + \xi_2\xi_0 + \xi_0\xi_1 = 0$$

and

$$(101) \qquad \frac{\partial \xi_i}{\partial u} = -\varepsilon^{2i}(1+\xi_i^2), \qquad \frac{\partial \xi_i}{\partial v} = -\varepsilon^{i}(1+\xi_i^2) \qquad (i = 0,1,2).$$

We may suppose that

$$(102) \qquad \varphi = -\frac{1}{3}(\xi_o + \xi_1 + \xi_2).$$

The equations (22), in the case we are dealing with, are

$$(103) \qquad \begin{aligned}
\frac{\partial y}{\partial u} &= y_1, \\
\frac{\partial y_1}{\partial u} &= \frac{2}{3}(\xi_0^2 + \varepsilon\xi_1^2 + \varepsilon^2\xi_2^2)y + \frac{2}{3}(\xi_0 + \xi_1 + \xi_2)y_2, \\
\frac{\partial y_2}{\partial u} &= \frac{4}{9}(\xi_0^2 + \xi_1^2 + \xi_2^2 + 2)y + s, \\
\frac{\partial y}{\partial v} &= y_2, \\
\frac{\partial y_1}{\partial v} &= \frac{4}{9}(\xi_0^2 + \xi_1^2 + \xi_2^2 + 2)y + s, \\
\frac{\partial y_2}{\partial v} &= \frac{2}{3}(\xi_0^2 + \varepsilon^2\xi_1^2 + \varepsilon\xi_2^2)y + \frac{2}{3}(\xi_0 + \xi_1 + \xi_2)y_1.
\end{aligned}$$

Let us write also the cubic form W

$$(104) \qquad \begin{aligned}
W = &-\frac{1}{27}[10(\xi_0^3 + \xi_1^3 + \xi_2^3) - 6(\xi_0 + \xi_1 + \xi_2) + 24\xi_0\xi_1\xi_2]y^3 + \\
&+ y_1^3 + y_2^3 - 2(\xi_0 + \xi_1 + \xi_2)yy_1y_2 - (\xi_0^2 + \varepsilon\xi_1^2 + \varepsilon^2\xi_2^2)^2 y^2 y_1 - \\
&- (\xi_0^2 + \varepsilon^2\xi_1^2 + \varepsilon\xi_2^2)y^2 y_2.
\end{aligned}$$

The cubic cone W has a double tangent plane whose moving coordinates, as the reader can easily verify, are

$$(105) \qquad y : y_1 : y_2 : s = -3 : (\xi_0 + \varepsilon^2\xi_1 + \varepsilon\xi_2) : (\xi_0 + \varepsilon\xi_1 + \varepsilon^2\xi_2) : 0.$$

Let us set

$$(106) \qquad \delta_1 = +1, \qquad \delta_2 = -1.$$

ON THE SURFACES ALL SEGRE CURVES OF WHICH ARE PLANE CURVES

Thus, the lines of contact of the double tangent (105) contain the points

$$\text{(107)}\quad \begin{aligned} t_i =&(\xi_0+\xi_1+\xi_2+\delta_i\sqrt{3})[(\xi_0+\varepsilon^2\xi_1+\varepsilon\xi_2)y+3y_1\\ &+(\xi_0+\varepsilon\xi_1+\varepsilon^2\xi_2)[(\xi_0+\varepsilon\xi_1+\varepsilon^2\xi_2)y+3y_2], \end{aligned} \qquad (i=1,2)$$

respectively. We have

$$\begin{aligned} \frac{\partial t_i}{\partial u} &= -\frac{(\xi_0+\xi_1+\xi_2-\delta_i\sqrt{3})^2}{3(\xi_0+\varepsilon\xi_1+\varepsilon^2\xi_2)}t_i+3(\xi_0+\varepsilon\xi_1+\varepsilon^2\xi_2)s\\ \frac{\partial t_i}{\partial v} &= -\frac{(\xi_0+\varepsilon^2\xi_1+\varepsilon\xi_2)(\xi_0+\xi_1+\xi_2-\delta_i\sqrt{3})}{3(\xi_0+\varepsilon\xi_1+\varepsilon^2\xi_2)}t_i+\\ &\quad +3(\xi_0+\xi_1+\xi_2+\delta_i\sqrt{3})s. \end{aligned} \qquad (\text{i} = 1,2)$$

From this, it follows

$$\text{(108)}\quad t_i = 3e^{-\tau_i}\int se^{\tau_i}[(\xi_0+\varepsilon\xi_1+\varepsilon^2\xi_2)du+(\xi_0+\xi_1+\xi_2+\delta_i\sqrt{3})dv], \qquad (i=1,2)$$

where I have set

$$\text{(109)}\quad \begin{aligned} \tau_i = \int &\frac{\xi_0+\xi_1+\xi_2-\delta_i\sqrt{3}}{3(\xi_0+\varepsilon\xi_1+\varepsilon^2\xi_2)}[(\xi_0+\xi_1+\xi_2-\delta_i\sqrt{3})du+\\ &+(\xi_0+\varepsilon^2\xi_1+\varepsilon\xi_2)dv]\,. \end{aligned} \qquad (i=1,2)$$

For the evaluation of these integrals we shall use the identity (100), which can be written in the form

$$\begin{aligned} (\xi_0+\xi_1+\xi_2+\sqrt{3})\ (\xi_0+\xi_1+\xi_2-\sqrt{3}) &=\\ = (\xi_0+\varepsilon\xi_1+\varepsilon^2\xi_2)\ (\xi_0+\varepsilon^2\xi_1+\varepsilon\xi_2)&. \end{aligned}$$

This enables to introduce two new variables α and β by setting

$$\text{(110)}\quad \begin{aligned} \alpha &= \frac{\xi_0+\xi_1+\xi_2+\sqrt{3}}{\xi_0+\varepsilon^2\xi_1+\varepsilon\xi_2} = \frac{\xi_0+\varepsilon\xi_1+\varepsilon^2\xi_2}{\xi_0+\xi_1+\xi_2-\sqrt{3}},\\ \beta &= \frac{\xi_0+\xi_1+\xi_2+\sqrt{3}}{\xi_0+\varepsilon\xi_1+\varepsilon^2\xi_2} = \frac{\xi_0+\varepsilon^2\xi_1+\varepsilon\xi_2}{\xi_0+\xi_1+\xi_2-\sqrt{3}}. \end{aligned}$$

EDUARD ČECH

From this, we can deduce the following formulae

$$\xi_0 = \frac{1+2(\alpha+\beta)+\alpha\beta}{\sqrt{3}(\alpha\beta-1)}, \qquad \xi_1 = \frac{1+2(\varepsilon^2\alpha+\varepsilon\beta)+\alpha\beta}{\sqrt{3}(\alpha\beta-1)},$$

$$\xi_2 = \frac{1+2(\varepsilon\alpha+\varepsilon^2\beta)+\alpha\beta}{\sqrt{3}(\alpha\beta-1)}, \qquad \xi_0+\xi_1+\xi_2+\sqrt{3} = \frac{2\sqrt{3}\alpha\beta}{\alpha\beta-1},$$

$$\xi_0+\xi_1+\xi_2-\sqrt{3} = \frac{2\sqrt{3}}{\alpha\beta-1}, \qquad \xi_0+\varepsilon\xi_1+\varepsilon^2\xi_2 = \frac{2\sqrt{3}\alpha}{\alpha\beta-1},$$

$$\xi_0+\varepsilon^2\xi_1+\varepsilon\xi_2 = \frac{2\sqrt{3}\beta}{\alpha\beta-1},$$

$$d\xi_i = -\frac{2}{\sqrt{3}}\frac{(\varepsilon^{2i}+\beta+\varepsilon^i\beta^2)\,d\alpha+(\varepsilon^i+\alpha+\varepsilon^{2i}\alpha^2)\,d\beta}{(\alpha\beta-1)^2} \qquad (i=0,1,2),$$

$$du = \frac{\sqrt{3}}{2}\left(-\frac{d\alpha}{\alpha^3-1}+\frac{\beta d\beta}{\beta^3-1}\right), \qquad dv = \frac{\sqrt{3}}{2}\left(\frac{\alpha d\alpha}{\alpha^3-1}-\frac{d\beta}{\beta^3-1}\right).$$

Using these formulae, we can write the equation (109) in the form

$$d\tau_1 = \frac{d\alpha}{\alpha(\alpha^3-1)}, \qquad d\tau_2 = \frac{\beta^2 d\beta}{\beta^3-1},$$

from where

$$\tau_1 = \log\frac{(\alpha^3-1)^{1/3}}{\alpha}, \qquad \tau_2 = \log(\beta^3-1).$$

Then the equation (108) gives

$$t_1 = \frac{\alpha}{(\alpha^3-1)^{1/3}}\cdot a_1 + s\cdot\frac{9\alpha}{(\alpha^3-1)^{1/3}}\int_\infty^\alpha \frac{dx}{(x^3-1)^{2/3}}, \tag{111}$$

$$t_2 = \frac{1}{(\beta^3-1)^{1/3}}\cdot a_2 + s\cdot\frac{9}{(\beta^3-1)^{1/3}}\int_\infty^\beta \frac{dx}{(x^3-1)^{2/3}}, \tag{112}$$

with a_1 and a_2 being constant. It is also convenient to introduce the variables α and β into the equation (107)

$$t_1 = \frac{6\alpha}{\alpha\beta-1}\left[\frac{2(\alpha+\beta^2)}{\alpha\beta-1}y+\sqrt{3}\cdot\beta y_1+\sqrt{3}\cdot y_2\right], \tag{113}$$

$$t_2 = \frac{6}{\alpha\beta-1}\left[\frac{2(\alpha^2+\beta)}{\alpha\beta-1}y+\sqrt{3}\cdot y_1+\sqrt{3}\cdot\alpha y_2\right]. \tag{114}$$

ON THE SURFACES ALL SEGRE CURVES OF WHICH ARE PLANE CURVES

Let us come back to the cubic form W, from which we shall form the Hessian H

$$H = \begin{vmatrix} \frac{\partial^2 W}{\partial y^2}, & \frac{\partial^2 W}{\partial y \partial y_1}, & \frac{\partial^2 W}{\partial y \partial y_2} \\ . & . & . \\ . & . & . \end{vmatrix}.$$

We have

$$\begin{aligned}\frac{1}{8}H =& Ay^3 + Byy_1y_2 + Cy^2y_1 + Dy^2y_2 + Eyy_2{}^2 - \\ &- 6({\xi_0}^2 + \varepsilon^2{\xi_1}^2 + \varepsilon{\xi_2}^2)(\xi_0 + \xi_1 + \xi_2)yy_1{}^2 - \\ &- 9({\xi_0}^2 + \varepsilon{\xi_1}^2 + \varepsilon^2{\xi_2}^2)y_1{}^2y_2 - 9(\xi_0 + \varepsilon\xi_1 + \varepsilon^2\xi_2)y_1y_2{}^2.\end{aligned}$$

I shall not need the coefficients A, B, C, D, E.

According to a well known theorem, we can choose ψ in such a way that it holds

$$\frac{1}{8}H + \psi W = \left[-3\frac{\partial W}{\partial y} + (\xi_0 + \varepsilon^2\xi_1 + \varepsilon\xi_2)\frac{\partial W}{\partial y_1} + (\xi_0 + \varepsilon\xi_1 + \varepsilon^2\xi_2)\frac{\partial W}{\partial y_2}\right] \cdot \frac{z}{3},$$

with z being a linear form in y, y_1, y_2. Setting $y = 0$ in the previous identity, we find $\psi = 9$, from which it follows

$$\begin{aligned} z =& -2({\xi_0}^2 + {\xi_1}^2 + {\xi_2}^2 + 5)y - 3(\xi_0 + \varepsilon\xi_1 + \varepsilon^2\xi_2)y_1 - \\ &- 3(\xi_0 + \varepsilon^2\xi_1 + \varepsilon\xi_2)y_2. \end{aligned} \tag{115}$$

Thus, we get

$$\begin{aligned}\frac{\partial z}{\partial u} &= -(\xi_0 + \varepsilon^2\xi_1 + \varepsilon\xi_2)\left(\frac{z}{3} + 3s\right), \\ \frac{\partial z}{\partial v} &= -(\xi_0 + \varepsilon\xi_1 + \varepsilon^2\xi_2)\left(\frac{z}{3} + 3s\right).\end{aligned}$$

Hence

$$z = -9s + be^{-tsize\frac{\omega}{3}}, \tag{116}$$

with b being constant. I have set

$$\omega = \int (\xi_0 + \varepsilon^2\xi_1 + \varepsilon\xi_2)du + (\xi_0 + \varepsilon\xi_1 + \varepsilon^2\xi_2)dv.$$

If we again introduce here the variables α and β, we get immediately, according to the formulae written above,

$$d\omega = 3\left[\frac{(\alpha^2 - \beta)d\alpha}{(\alpha\beta - 1)(\alpha^3 - 1)} + \frac{(\beta^2 - \alpha)d\beta}{(\alpha\beta - 1)(\beta^3 - 1)}\right].$$

The integration gives

$$\omega = \log\frac{(\alpha\beta - 1)^3}{(\alpha^3 - 1)(\beta^3 - 1)},$$

which substituted into (116) gives

$$z = b\frac{\sqrt[3]{(\alpha^3-1)(\beta^3-1)}}{\alpha\beta-1} - 9s. \tag{117}$$

If we introduce α and β into the equation (115), we have

$$z = \frac{6}{\alpha\beta-1}\left[-\frac{2(\alpha^2\beta^2+1)}{\alpha\beta-1}y - \sqrt{3}\cdot\alpha y_1 - \sqrt{3}\cdot\beta y_2\right]. \tag{118}$$

The elimination of y_1 and y_2 from (113), (114) and (118) gives, with respect to the equations (111), (112) and (117), the general solution of the system (103):

$$\begin{aligned} y = -\frac{1}{12(\alpha\beta-1)}\Bigg[& a_1\frac{\alpha^2-\beta}{\sqrt[3]{\alpha^3-1}} + a_2\frac{\beta^2-\alpha}{\sqrt[3]{\beta^3-1}} + b\sqrt[3]{(\alpha^3-1)(\beta^3-1)} + \\ & + 9s\left(\frac{\alpha^2-\beta}{\sqrt[3]{\alpha^3-1}}\int_{\infty}^{\alpha}\frac{dx}{(x^3-1)^{2/3}} + \frac{\beta^2-\alpha}{\sqrt[3]{\beta^3-1}}\int_{\infty}^{\beta}\frac{dx}{(x^3-1)^{2/3}} - \alpha\beta + 1\right)\Bigg]. \end{aligned} \tag{119}$$

Therefore, the parametric equations of the surface L_2 are

$$\begin{aligned} x &= \frac{\alpha^2-\beta}{\sqrt[3]{(\alpha^3-1)^2(\beta^3-1)}}, \\ y &= \frac{\beta^2-\alpha}{\sqrt[3]{(\alpha^3-1)(\beta^3-1)^2}}, \\ z &= x\int_{\infty}^{\alpha}\frac{d\lambda}{(\lambda^3-1)^{2/3}} + y\int_{\infty}^{\beta}\frac{d\lambda}{(\lambda^3-1)^{2/3}} - \frac{\alpha\beta-1}{\sqrt[3]{(\alpha^3-1)(\beta^3-1)}}. \end{aligned} \tag{120}$$

10. The surface L_1.

In the case which remains, the computations to be done are no more so simple as in the preceding cases. Therefore, I am using an indirect method, which is completely successful in the algebraic part of the problem. However, the results presented before allow to expect that the quadratures to be accomplished can be written in a simpler way.

As in the section 3, I shall use the abbreviations

$$x_0 = u + v, \quad x_1 = \varepsilon^2 u + \varepsilon v, \quad x_2 = \varepsilon u + \varepsilon^2 v, \quad x_0 + x_1 + x_2 = 0. \tag{121}$$

Without any loss of generality, we can assume that

$$\varphi = -\frac{1}{3}(\zeta x_0 + \zeta x_1 + \zeta x_2). \tag{122}$$

The cubic form W is

$$W = \left(\frac{\partial^2\varphi}{\partial u\partial v} - 8\varphi^3\right)y^3 + {y_1}^3 + {y_2}^3 + 6\varphi y y_1 y_2 - 3\frac{\partial\varphi}{\partial v}y^2 y_1 - 3\frac{\partial\varphi}{\partial u}y^2 y_2. \tag{123}$$

ON THE SURFACES ALL SEGRE CURVES OF WHICH ARE PLANE CURVES

Let us compute the two invariants S and T of this form. According to the Salmon's formulae, we get

(124) $$S = \varphi\frac{\partial^2\varphi}{\partial u\partial v} - 9\varphi^4 - \frac{\partial\varphi}{\partial u}\frac{\partial\varphi}{\partial v},$$

(125) $$T = \left(\frac{\partial^2\varphi}{\partial u\partial v}\right)^2 - 36\varphi^3\frac{\partial^2\varphi}{\partial u\partial v} + 216\varphi^6 - 4\left(\frac{\partial\varphi}{\partial u}\right)^3 - 4\left(\frac{\partial\varphi}{\partial u}\right)^3 + 36\varphi^2\frac{\partial\varphi}{\partial u}\frac{\partial\varphi}{\partial v}.$$

If we substitute into the second term of (124) the value (122), we have

$$\begin{aligned}-9S = (\zeta x_0 + \zeta x_1 + \zeta x_2)(p'x_0 + p'x_1 + p'x_2) + (px_0 + px_1 + px_2)^2 + \\ + (px_0 + \varepsilon px_1 + \varepsilon^2 px_2)(px_0 + \varepsilon^2 px_1 + \varepsilon px_2).\end{aligned}$$

Let us assume that x_2 is an arbitrary fixed value, and consider x_1 as a function of x_0

$$x_1 = -x_0 - x_2\,.$$

Thus, the second term is a doubly periodic function of x_0, the poles of which can be only

$$x_0 = 0, \qquad x_0 = -x_2.$$

In a deleted neighborhood of the point $x_0 = 0$, we have

$$\begin{aligned}\zeta x_0 &= \frac{1}{x_0} - \frac{g_2}{60}{x_0}^3 - \frac{g_3}{140}{x_0}^5 + \dots,\\ px_0 &= \frac{1}{{x_0}^2} + \frac{g_2}{20}{x_0}^2 + \frac{g_3}{28}{x_0}^4 + \dots,\end{aligned}$$

from where we obtain

$$\zeta x_0 + \zeta x_1 + \zeta x_2 = \frac{1}{x_0} + px_2\cdot x_0 + \frac{1}{2}p'x_2\cdot {x_0}^2 + \left(\frac{1}{6}p''x_2 - \frac{g_2}{60}\right){x_0}^3 + \dots,$$

$$px_0 + px_1 + px_2 = \frac{1}{{x_0}^2} + 2px_2 + p'x_2\cdot x_0 + \left(\frac{1}{2}p''x_2 + \frac{g_2}{20}\right){x_0}^2 + \dots,$$

$$\begin{aligned}&px_0 + \varepsilon^i px_1 + \varepsilon^{2i} px_2 =\\ &= \frac{1}{{x_0}^2} - px_2 + \varepsilon^i p'x_2\cdot x_0 + \left(\frac{\varepsilon^i}{2}p''x_2 + \frac{g_2}{20}\right){x_0}^2 + \dots, \qquad (i = 1, 2)\end{aligned}$$

$$p'x_0 + p'x_1 + p'x_2 = -\frac{2}{{x_0}^3} + \left(-p''x_2 + \frac{g_2}{10}\right)x_0 + \dots;$$

$$\begin{aligned}&(\zeta x_0 + \zeta x_1 + \zeta x_2)(p'x_0 + p'x_1 + p'x_2) =\\ &= -\frac{2}{{x_0}^4} - \frac{2px_2}{{x_0}^2} - \frac{p'x_2}{x_0} + \left(-\frac{4}{3}p''x_2 + \frac{4g_2}{30}\right) + \dots,\end{aligned}$$

$$(px_0 + px_1 + px_2)^2 = \frac{1}{{x_0}^4} + \frac{4px_2}{{x_0}^2} + \frac{2p'x_2}{x_0} + \left(4p^2x_2 + p''x_2 + \frac{g_2}{10}\right) + \dots,$$

$$\begin{aligned}&(px_0 + \varepsilon px_1 + \varepsilon^2 px_2)(px_0 + \varepsilon^2 px_1 + \varepsilon px_2) =\\ &= \frac{1}{{x_0}^4} - \frac{2px_2}{{x_0}^2} - \frac{p'x_2}{x_0} + \left(4p^2x_2 - \frac{1}{2}p''x_2 + \frac{g_2}{10}\right) + \dots,\end{aligned}$$

EDUARD ČECH

and finally

$$-9S=\left(5p^2x_2-\frac{5}{6}p''x_2+\frac{g_2}{3}\right)+\cdots\equiv\frac{9}{12}g_2+\dots.$$

One can see that the function remains finite. For the reason of symmetry, the function is also finite for $x_0=-x_2$. S is constant, and we can see that

$$S=-\frac{g_2}{12}. \tag{126}$$

As far as T is concerned, it can be written first in the form

$$\begin{aligned}T=&\left(\frac{\partial^2\varphi}{\partial u\partial v}\right)^2-4\left[\left(\frac{\partial\varphi}{\partial u}\right)^3+\left(\frac{\partial\varphi}{\partial v}\right)^3+27\varphi^6\right]-\\&-36\varphi^2\left(\varphi\frac{\partial^2\varphi}{\partial u\partial v}-9\varphi^4-\frac{\partial\varphi}{\partial u}\frac{\partial\varphi}{\partial v}\right)=\left(\frac{\partial^2\varphi}{\partial u\partial v}\right)^2-\\&-4\left[\left(\frac{\partial\varphi}{\partial u}\right)^3+\left(\frac{\partial\varphi}{\partial v}\right)^3+27\varphi^6\right]+3g_2\varphi^2,\end{aligned}$$

which is, according to (122), equal to

$$\begin{aligned}T=&\frac{1}{9}[(p'x_0+p'x_1+p'x_2)^2-4(p^3x_0+p^3x_1+p^3x_2+6px_0px_1px_2)+\\&+3g_2(px_0+px_1+px_2)].\end{aligned}$$

In a deleted neighborhood of $x_0=0$, we have

$$\begin{aligned}&(p'x_0+p'x_1+p'x_2)^2=\\&=\frac{4}{{x_0}^6}+\left(4p''x_2-\frac{2g_2}{5}\right)\cdot\frac{1}{{x_0}^2}+\frac{2p'''x_2}{x_0}+\left(\frac{2}{3}p^{IV}x_2-\frac{4g_3}{7}\right)+\dots,\\&-4(p^3x_0+p^3x_1+p^3x_2+6px_0px_1px_2)=\\&=-\frac{4}{{x_0}^6}-\left(24p^2x_2+\frac{3g_2}{5}\right)\frac{1}{{x_0}^2}-\frac{24px_2\cdot p'x_2}{x_0}+\\&+\left(-8p^3x_2-12px_2\cdot p''x_2-\frac{3g_3}{7}\right)+\dots,\\&3g_2(px_0+px_1+px_2)=\frac{3g_2}{{x_0}^2}+6g_2xp^2+\dots,\end{aligned}$$

from where

$$\begin{aligned}9T=&\frac{4(p''x_2-6p^2x_2+\frac{1}{2}g_2)}{{x_0}^2}+\frac{2(p''x_2-12px_2\cdot p'x_2)}{x_0}+\\&+\left(\frac{2}{3}p^{IV}x_2-8p^3x_2-12px_2\cdot p''x_2+6g_2px_2-g_3\right)+\dots\\&\equiv-9g_3+\dots.\end{aligned}$$

ON THE SURFACES ALL SEGRE CURVES OF WHICH ARE PLANE CURVES

As before, this implies

$$T = -g_3 . \tag{127}$$

It was a priori clear that the form W could be reduced by a linear substitution to a form with constant coefficients. Now, we can say that it is possible to reduce it by the substitution

$$\begin{aligned} y &= y^{(1)}c_1 + y^{(2)}c_2 + y^{(3)}c_3, \\ y_1 &= {y_1}^{(1)}c_1 + {y_1}^{(2)}c_2 + {y_1}^{(3)}c_3, \\ y_2 &= {y_2}^{(1)}c_1 + {y_2}^{(2)}c_2 + {y_2}^{(3)}c_3, \end{aligned} \tag{128}$$

to the form

$$4{c_1}^3 - g_2c_1{c_3}^2 - g_3{c_3}^3 - {c_2}^2c_3, \tag{129}$$

and that, moreover, the determinant

$$\Delta = \begin{vmatrix} y^{(1)} & y^{(2)} & y^{(3)} \\ {y_1}^{(1)} & {y_1}^{(2)} & {y_1}^{(3)} \\ {y_2}^{(1)} & {y_2}^{(2)} & {y_2}^{(3)} \end{vmatrix} \tag{130}$$

is *constant*. The equations (128) represent the general solution of the system (22), in which we have set $s = 0$. Especially, we have

$$ {y_1}^{(i)} = \frac{\partial y^{(i)}}{\partial u}, \quad {y_2}^{(i)} = \frac{\partial y^{(i)}}{\partial v}. \qquad (i = 1,2,3) \tag{131}$$

Let us consider an auxiliary plane π, and, for the lines lying in it, let us define homogeneous coordinates. For arbitrary values of u and v, let us consider the three lines of the plane π whose homogeneous coordinates are

$$p(x_i) \ : \ p'(x_i) \ : \ 1. \qquad (i = 0,1,2)$$

The identity

$$x_0 + x_1 + x_2 = 0$$

shows that the three lines have a point in common. Taking this point as the image of the point (u, v) of the surface L_1, one obtains a plane representation in which to the *Darboux curves* there correspond the tangents of a curve of the third class which can be supposed to be identical with the intersection of the plane π with the cone W. One can also immediately see that the properties of this representation are preserved if we replace x_i by $mx_i + a_i$, where m, a_0, a_1, a_2 are constants satisfying

$$a_0 + a_1 + a_2 = 0,$$

and in this case only.

EDUARD ČECH

Thus, the projection of L_1 from the point s possesses precisely the indicated properties, only that this time they are the *Segre curves* to which there correspond the tangents of the curve of the third class. This shows that the *fixed* coordinates of the planes of Segre curves have the form

$$c_1 : c_2 : c_3 = p(z_i) : p'(z_i) : 1, \qquad (i = 0, 1, 2)$$

where

$$\begin{aligned} z_0 &= m(x_1 + a_1 - x_2 - a_2) \\ z_1 &= m(x_2 + a_2 - x_0 - a_0) \qquad a_0 + a_1 + a_2 = 0. \\ z_2 &= m(x_0 + a_0 - x_1 - a_1) \end{aligned} \tag{132}$$

m, a_0, a_1, a_2 are constants which need not be computed. It is only important to notice the identity

$$z_0 + z_1 + z_2 = 0\,. \tag{133}$$

Using the fact that the *moving* coordinates of the three planes satisfy the relation $y = 0$, from the first of the equations (128) we obtain the three relations

$$p(z_i)y^{(1)} + p'(z_i)y^{(2)} + y^{(3)} = 0\,, \qquad (i = 0, 1, 2)$$

which are reduced to two relations determining the ratios

$$y^{(1)} : y^{(2)} : y^{(3)}.$$

One obtains

$$y^{(1)} = -2Ay^{(2)}, \quad y^{(3)} = -\frac{1}{3}By^{(2)}, \tag{134}$$

where I have set

$$A = \zeta z_0 + \zeta z_1 + \zeta z_2, \tag{135}$$

$$B = p' z_0 + p' z_1 + p' z_2 - 8A^3. \tag{136}$$

Introducing the values (134) into the determinant (130), we find

$$\Delta = \frac{2}{3} \begin{vmatrix} \frac{\partial A}{\partial u} & \frac{\partial A}{\partial v} \\ \frac{\partial B}{\partial u} & \frac{\partial B}{\partial v} \end{vmatrix} \cdot \left(y^{(2)}\right)^3.$$

Taking into account the equations (132), then, according to (135) and (136), we have*

$$\begin{vmatrix} \frac{\partial A}{\partial u} & \frac{\partial A}{\partial v} \\ \frac{\partial B}{\partial u} & \frac{\partial B}{\partial v} \end{vmatrix} = k \begin{vmatrix} pz_0 + \varepsilon^2 pz_1 + \varepsilon pz_2, & pz_0 + \varepsilon pz_1 + \varepsilon^2 pz_2 \\ p'' z_0 + \varepsilon^2 p'' z_1 + \varepsilon p'' z_2, & p'' z_0 + \varepsilon p'' z_1 + \varepsilon^2 p'' z_2 \end{vmatrix} =$$

$$= 6k \left\| \begin{matrix} 1 & \varepsilon^2 & \varepsilon \\ 1 & \varepsilon & \varepsilon^2 \end{matrix} \right\| \cdot \left\| \begin{matrix} pz_0 & pz_1 & pz_2 \\ p^2 z_0 & p^2 z_1 & p^2 z_2 \end{matrix} \right\| = k_1 D,$$

*In what follows I shall denote by $k, k_1, k_2 \ldots$ the constants which we need not know.

ON THE SURFACES ALL SEGRE CURVES OF WHICH ARE PLANE CURVES

where I have set

(137) $$D = (pz_1 - pz_2)(pz_2 - pz_0)(pz_0 - pz_1).$$

Because Δ is constant, we have

(138) $$y^{(2)} = k_2 D^{-\frac{1}{3}}.$$

Thus, we know the general integral (128) of the equations (22) when setting in them $s = 0$, provided that we neglect the values of the numbers m, a_0, a_1, a_2. But these values are of no importance if we want only to obtain the finite equations of a surface L_1. But it is necessary to integrate the equations (22) for s different from zero. For this I shall use the variation of constants as I did in the preceding cases.

Using the fact that the determinant of the equations (128) is constant, we find

$$c_3 = 6k_3 D^{-\frac{2}{3}} \left(Q_3 y + \frac{\partial A}{\partial v} y_1 - \frac{\partial A}{\partial u} y_2 \right),$$
$$c_1 = -k_3 D^{-\frac{2}{3}} \left(Q_1 y + \frac{\partial B}{\partial v} y_1 - \frac{\partial B}{\partial u} y_2 \right).$$

The first equation of (128) gives then

$$c_2 = 2k_3 D^{-\frac{2}{3}} \left[Q_2 y + \left(\frac{\partial A}{\partial v} B - \frac{\partial B}{\partial v} A \right) y_1 - \left(\frac{\partial A}{\partial u} B - \frac{\partial B}{\partial u} A \right) y_2 \right].$$

We do not need the values of Q_1, Q_2, Q_3. If $y, y_1 = \dfrac{\partial y}{\partial u}, y_2 = \dfrac{\partial y}{\partial v}$ is the general solution of the equations (22), then c_1, c_2, c_3 are no more constants. Nevertheless, the expressions $\dfrac{\partial c_i}{\partial u}, \dfrac{\partial c_i}{\partial v}$, linear and homogeneous in y_1, y_2, y_3, s, vanish necessarily if we set $s = 0$. Thus, we find without any computation

$$\frac{\partial c_1}{\partial u} = k_3 D^{-\frac{2}{3}} \frac{\partial B}{\partial u} \cdot s, \qquad \frac{\partial c_1}{\partial v} = -k_3 D^{-\frac{2}{3}} \frac{\partial B}{\partial v} \cdot s,$$
$$\frac{\partial c_2}{\partial u} = -2k_3 D^{-\frac{2}{3}} \left(\frac{\partial A}{\partial u} B - \frac{\partial B}{\partial u} A \right) s, \qquad \frac{\partial c_2}{\partial v} = 2k_3 D^{-\frac{2}{3}} \left(\frac{\partial A}{\partial v} B - \frac{\partial B}{\partial v} A \right) s,$$
$$\frac{\partial c_3}{\partial u} = -6k_3 D^{-\frac{2}{3}} \frac{\partial A}{\partial u} \cdot s, \qquad \frac{\partial c_3}{\partial v} = 6k_3 D^{-\frac{2}{3}} \frac{\partial A}{\partial v} \cdot s.$$

Therefore, let us introduce three functions of the three variables z_0, z_1, z_2, only two of which are independent because of the identity $z_0 + z_1 + z_2 = 0$. Let

(139)
$$\omega_1 = \int D^{-\frac{2}{3}} \left[\left(\frac{\partial B}{\partial z_1} - \frac{\partial B}{\partial z_2} \right) dz_0 + \left(\frac{\partial B}{\partial z_2} - \frac{\partial B}{\partial z_0} \right) dz_1 + \left(\frac{\partial B}{\partial z_0} - \frac{\partial B}{\partial z_1} \right) dz_2 \right],$$

(140)
$$\omega_2 = \int D^{-\frac{2}{3}} \left\{ B \left[\left(\frac{\partial A}{\partial z_1} - \frac{\partial A}{\partial z_2} \right) dz_0 + \left(\frac{\partial A}{\partial z_2} - \frac{\partial A}{\partial z_0} \right) dz_1 + \left(\frac{\partial A}{\partial z_0} - \frac{\partial A}{\partial z_1} \right) dz_2 \right] \right.$$
$$\left. - A \left[\left(\frac{\partial B}{\partial z_1} - \frac{\partial B}{\partial z_2} \right) dz_0 + \left(\frac{\partial B}{\partial z_2} - \frac{\partial B}{\partial z_0} \right) dz_1 + \left(\frac{\partial B}{\partial z_0} - \frac{\partial B}{\partial z_1} \right) dz_2 \right] \right\},$$

EDUARD ČECH

(141)
$$\omega_3 = \int D^{-\frac{2}{3}} \left[\left(\frac{\partial A}{\partial z_1} - \frac{\partial A}{\partial z_2} \right) dz_0 + \left(\frac{\partial A}{\partial z_2} - \frac{\partial A}{\partial z_0} \right) dz_1 + \left(\frac{\partial A}{\partial z_0} - \frac{\partial A}{\partial z_1} \right) dz_2 \right],$$

where the lower limits of the integrals are fixed arbitrarily. Thus, we have

$$c_1 = k_4 \omega_1 s + b_1,$$
$$c_2 = -2k_4 \omega_2 s + b_2,$$
$$c_3 = -6k_4 \omega_3 s + b_3,$$

with b_1, b_2, b_3 being arbitrary constants.

Consequently, the general solution of the system (22) is

$$y = D^{-\frac{1}{3}} [(a_1 A + a_2 B + a_3) + k_5 s(A\omega_1 + \omega_2 - B\omega_3)], \tag{142}$$

where a_1, a_2, a_3 are constants. The surface L_1 can be represented by means of two parameters by the formulae

$$\begin{aligned} x &= A, \\ y &= B, \\ z &= A\omega_1 - B\omega_3 + \omega_2 . \end{aligned} \tag{143}$$

Czechoslovak Mathematical Journal v. 6 (81), 1956, 260 – 286.

DEVELOPABLE TRANSFORMATIONS OF LINE CONGRUENCES

EDUARD ČECH, PRAGUE

(Received November 19, 1955)

This paper contains a detailed account of the results announced in a lecture delivered by the author on February 1, 1955 in Torino (see Università e Politecnico di Torino, Rendiconti del Seminario Matematico, vol. 14, 1954–55, pp. 55–66).

1. Let L be a congruence in the three dimensional projective space S_3 generated by the line

$$g = [A_1 A_2] \tag{1.1}$$

which depends (analytically) on two parameters u, v. Then

$$[A_1 \quad A_2 \quad \mathrm{d}A_1 \quad \mathrm{d}A_2] \tag{1.2}$$

is a quadratic form in $\mathrm{d}u$, $\mathrm{d}v$, and we shall suppose that it has discriminant different from zero, so that the congruence L is *non-parabolic*. We neglect questions of reality, and consequently we can suppose that the form (2) is proportional to $\mathrm{d}u\mathrm{d}v$. Thus, the developables contained in L are given by $u =$ const., and by $v =$ const., and we shall call u, v *developable parameters* of the congruence L. These parameters are not determined without ambiguity, and they can be replaced either by

$$u_1 = f(u)\ , \qquad v_1 = g(v) \tag{1.3}$$

or by

$$u_2 = g(v)\ , \qquad v_2 = f(u)\ ; \tag{1.4}$$

but we shall avoid the transformations (4) by supposing that the congruence L is *oriented*. This means that we shall distinguish the *first* and the *second family of developables*, the first family being given by the equation $v =$ const., and the second family by the equation $u =$ const.

The point of regression of the developable of the congruence L described by the line (1.1) for $v =$ const. will be called the *first focus* of L, and the plane touching this developable along the line (1.1) will be called the *first focal plane* of L. For $u =$ const. one obtains the *second focus* and the *second focal plane*. One can suppose that A_1 is the first focus, and A_2 is the second one. Then one has

$$\left[A_1 A_2 \frac{\partial A_1}{\partial u}\right] = 0\ , \qquad \left[A_1 A_2 \frac{\partial A_2}{\partial v}\right] = 0\ .$$

EDUARD ČECH

The first (second) focal plane joins the line g with the line $g_2(g_1)$ which is the tangent to the curve described by the second (first) focus for $v =$ const. (for $u =$ const.). It holds

$$g_1 = \left[A_1 \frac{\partial A_1}{\partial v}\right] , \qquad g_2 = \left[A_2 \frac{\partial A_2}{\partial u}\right] .$$

The lines g_1 and g_2 are the *Laplace transforms* of the line g.

In what follows, we shall make use of the moving frame

$$A_1, A_2, A_3, A_4 , \tag{1.5}$$

where $A_1(A_2)$ is the first (second) focus, and $A_3(A_4)$ lies on the line $g_1(g_2)$. The scalar factors of the points (1.5) are subject to the condition

$$[A_1 A_2 A_3 A_4] = 1 . \tag{1.6}$$

The first focal plane is $[A_1A_2A_4]$, the second is $[A_1A_2A_3]$. The frame (1.5) depends on the two *principal parameters* u, v, and on 5 *secondary parameters*, which we leave at this moment completely arbitrary. We make use of the usual notation

$$\mathrm{d}A_i = \omega_{i1}A_1 + \omega_{i2}A_2 + \omega_{i3}A_3 + \omega_{i4}A_4 \qquad (i = 1,2,3,4) . \tag{1.7}$$

Let us set, in addition,

$$\omega_{13} = \omega_1 , \qquad \omega_{24} = \omega_2 , \tag{1.8}$$

and let us notice that by virtue of (1.6) it holds

$$\omega_{11} + \omega_{22} + \omega_{33} + \omega_{44} = 0 . \tag{1.9}$$

It can be proved without difficulty that the conditions we have imposed upon the choice of the frame (1.5) are expressed analytically by the equations

$$\begin{aligned} \omega_{14} = 0 , \qquad & \omega_{23} = 0 , \\ [\omega_{12}\omega_2] = 0 , \qquad & [\omega_{21}\omega_1] = 0 . \end{aligned} \tag{1.10}$$

From (1.10) one can deduce by the exterior differentiation

$$[\omega_{34}\omega_1] = 0 , \qquad [\omega_{43}\omega_2] = 0 .$$

Thus, one can set

$$\omega_{12} = \alpha_1\omega_2, \quad \omega_{21} = \alpha_2\omega_1, \quad \omega_{34} = \beta_2\omega_1, \quad \omega_{43} = \beta_1\omega_2. \tag{1.11}$$

Then the equations (1.7) get the form

$$\begin{aligned} \mathrm{d}A_1 &= \omega_{11}A_1 + \alpha_1\omega_2A_2 + \omega_1A_3 , \\ \mathrm{d}A_2 &= \alpha_2\omega_1A_1 + \omega_{22}A_2 + \omega_2A_4 , \\ \mathrm{d}A_3 &= \omega_{31}A_1 + \omega_{32}A_2 + \omega_{33}A_3 + \beta_2\omega_1A_4 , \\ \mathrm{d}A_4 &= \omega_{41}A_1 + \omega_{42}A_2 + \beta_1\omega_2A_3 + \omega_{44}A_4 . \end{aligned} \tag{1.12}$$

One can see that it holds

$$[\omega_1 \quad dv] = 0\ , \qquad [\omega_2 \quad du] = 0\ ; \tag{1.13}$$

and besides

$$[\omega_1\omega_2] \neq 0\ , \tag{1.14}$$

because the line (1.1) depends essentially on the two parameters u, v.

It is also useful to notice the relations

$$[d\omega_1] = [\omega_{11} - \omega_{33} \quad \omega_1], \qquad [d\omega_2] = [\omega_{22} - \omega_{44} \quad \omega_2]\ . \tag{1.15}$$

Simultaneously with the point frame (1.5), it is convenient to consider the plane frame

$$E_1,\ E_2,\ E_3,\ E_4\ , \tag{1.16}$$

where

$$E_1 = [A_2A_3A_4],\ E_2 = -[A_1A_3A_4],\ E_3 = [A_1A_2A_4],\ E_4 = -[A_1A_2A_3]. \tag{1.17}$$

It can be seen without difficulty that corresponding to the equations (1.12) are the equations

$$\begin{aligned} &dE_1 + \omega_{11}E_1 + \alpha_2\omega_1E_2 + \omega_{31}E_3 + \omega_{41}E_4 = 0\ , \\ &dE_2 + \alpha_1\omega_2E_1 + \omega_{22}E_2 + \omega_{32}E_3 + \omega_{42}E_4 = 0\ , \\ &dE_3 + \omega_1E_1 + \omega_{33}E_3 + \beta_1\omega_2E_4 = 0\ , \\ &dE_4 + \omega_2E_2 + \beta_2\omega_1E_3 + \omega_{44}E_4 = 0\ . \end{aligned} \tag{1.18}$$

It is also useful to write the equations related to the line frame associated to the point frame, which obviously are

$$\begin{aligned} d[A_1A_2] &= (\omega_{11} + \omega_{22})[A_1A_2] + \omega_2[A_1A_4] - \omega_1[A_2A_3]\ , \\ d[A_1A_3] &= \omega_{32}[A_1A_2] + (\omega_{11} + \omega_{33})[A_1A_3] + \beta_2\omega_1[A_1A_4] + \alpha_1\omega_2[A_2A_3]\ , \\ d[A_2A_4] &= -\omega_{41}[A_1A_2] + \alpha_2\omega_1[A_1A_4] + \beta_1\omega_2[A_2A_3] + (\omega_{22} + \omega_{44})[A_2A_4]\ , \\ d[A_1A_4] &= \omega_{42}[A_1A_2] + \beta_1\omega_2[A_1A_3] + (\omega_{11} + \omega_{44})[A_1A_4] + \alpha_1\omega_2[A_2A_4] + \\ &\quad + \omega_1[A_3A_4]\ , \\ d[A_2A_3] &= -\omega_{31}[A_1A_2] + \alpha_2\omega_1[A_1A_3] + (\omega_{22} + \omega_{33})[A_2A_3] + \beta_2\omega_1[A_2A_4] - \\ &\quad - \omega_2[A_3A_4]\ , \\ d[A_3A_4] &= -\omega_{41}[A_1A_3] + \omega_{31}[A_1A_4] - \omega_{42}[A_2A_3] + \omega_{32}[A_2A_4] + \\ &\quad + (\omega_{33} + \omega_{44})[A_3A_4]\ . \end{aligned} \tag{1.19}$$

In order to study the existence questions of the theory of congruences, which form the main subject of this paper, and other questions which will represent their continuation, it is necessary to write the integrability conditions of the system (1.12), which can be obtained by the exterior differentiation of the equations (1.11), and which have the form

$$\begin{aligned} [\omega_{32}\omega_1] + [d\alpha_1 + \alpha_1(2\omega_{22} - \omega_{11} - \omega_{44}) \quad \omega_2] &= 0\ , \\ [d\alpha_2 + \alpha_2(2\omega_{11} - \omega_{22} - \omega_{33}) \quad \omega_1] + [\omega_{41}\omega_2] &= 0\ , \\ -[\omega_{41}\omega_1] + [d\beta_1 + \beta_1(\omega_{22} + \omega_{33} - 2\omega_{44}) \quad \omega_2] &= 0\ , \\ [d\beta_2 + \beta_2(\omega_{11} + \omega_{44} - 2\omega_{33}) \quad \omega_1] - [\omega_{32}\omega_2] &= 0\ . \end{aligned} \tag{1.20}$$

EDUARD ČECH

We have introduced the hypothesis that the congruence L under consideration is oriented, but we have also made such a choice of notations that the change of orientation can be analytically expressed by a simple substitution, namely

$$\begin{pmatrix} A_1 & A_2 & A_3 & A_4 & E_1 & E_2 & E_3 & E_4 & \omega_1 & \omega_2 & \alpha_1 & \alpha_2 & \beta_1 & \beta_2 \\ A_2 & A_1 & A_4 & A_3 & E_2 & E_1 & E_4 & E_3 & \omega_2 & \omega_1 & \alpha_2 & \alpha_1 & \beta_2 & \beta_1 \end{pmatrix} . \tag{1.21}$$

The duality is expressed by the substitution

$$\begin{pmatrix} A_1 & A_2 & A_3 & A_4 & E_1 & E_2 & E_3 & E_4 & \omega_1 & \omega_2 & \alpha_1 & \alpha_2 & \beta_1 & \beta_2 \\ E_3 & E_4 & E_1 & E_2 & A_3 & A_4 & A_1 & A_2 & -\omega_1 & -\omega_2 & \beta_1 & \beta_2 & \alpha_1 & \alpha_2 \end{pmatrix} . \tag{1.22}$$

2. Following E. Cartan we use the symbol δ in order to indicate the differentiation related to the change of the secondary parameters only, so that $\delta u = \delta v = 0$, and we set $\omega_{ik}(\delta) = e_{ik}$. Because the number of the secondary parameters is equal to 5, there are 5 independent linear combinations of their differentials, namely $e_{11} - e_{22}$, $e_{11} - e_{33}$, $e_{11} - e_{44}$, e_{31}, e_{42}. The relations (1.15) and (1.20) show that

$$\begin{aligned} \delta\omega_1 &= (e_{11} - e_{33})\omega_1 , & \delta\omega_2 &= (e_{22} - e_{44})\omega_2 , \\ \delta\alpha_1 &= (e_{11} - 2e_{22} + e_{44})\alpha_1 , & \delta\alpha_2 &= (e_{22} - 2e_{11} + e_{33})\alpha_2 , \\ \delta\beta_1 &= -(e_{22} + e_{33} - 2e_{44})\beta_1 , & \delta\beta_2 &= -(e_{11} + e_{44} - 2e_{33})\beta_2 . \end{aligned} \tag{2.1}$$

Hence, it follows that the differential forms

$$\varphi = \alpha_1\alpha_2\omega_1\omega_2, \quad \varphi^* = \beta_1\beta_2\omega_1\omega_2, \quad F_1 = \alpha_1\beta_1\frac{\omega_2^3}{\omega_1}, \quad F_2 = \alpha_2\beta_2\frac{\omega_1^3}{\omega_2}, \tag{2.2}$$

related by

$$\varphi \cdot \varphi^* = F_1 \cdot F_2 , \tag{2.3}$$

are invariant ($\delta\varphi = \delta\varphi^* = \delta F_1 = \delta F_2 = 0$) as well as the equations

$$\beta_2\omega_1^2 + \alpha_1\omega_2^2 = 0 , \qquad \alpha_2\omega_1^2 + \beta_1\omega_2^2 = 0 . \tag{2.4}$$

We shall call

φ the *point form*
φ^* the *plane form*
F_1 the *first focal form*
F_2 the *second focal form*

of the congruence L under consideration. The reasons for this terminology will be clear in n° 5. We shall also call the collection of the four forms (2.2) and the two equations (2.4) *linear projective element* of the congruence L. Besides, it is clear that in general the equations (2.4) are uniquely determined if the forms (2.2) are known. But this is no more true if $\alpha_1 = \beta_2 = 0$ or if $\alpha_2 = \beta_1 = 0$. It follows from (1.21) that the change of orientation of L leads to the substitution

$$\begin{pmatrix} \varphi & \varphi^* & F_1 & F_2 \\ \varphi & \varphi^* & F_2 & F_1 \end{pmatrix}, \tag{2.5}$$

and from (1.22) that the duality leads to the substitution

$$\begin{pmatrix} \varphi & \varphi^* & F_1 & F_2 \\ \varphi^* & \varphi & F_1 & F_2 \end{pmatrix}. \tag{2.6}$$

In 1933 A. TERRACINI introduced, under the name of linear projective element of a congruence L, a fractional differential form, which in our notations is equal to

$$\frac{(\alpha_1\omega_2^2 + \beta_2\omega_1^2)(\alpha_2\omega_1^2 + \beta_1\omega_2^2)}{4\omega_1\omega_2} = \tfrac{1}{4}(\varphi + \varphi^* + F_1 + F_2). \tag{2.7}$$

See A. Terracini *Su alcuni elementi lineari proiettivi*, Ann. della R. Scuola Norm. Sup. di Pisa, ser. II, vol. 2, 1933, pp. 401–428; see also A. Terracini, *Osservazioni sulla geometria proiettiva differenziale delle congruenze di rette*, Atti del R. Istituto Veneto vol. 94, 1934, pp. 75–86. The differential form (2.7) remains unchanged under the substitutions (2.5) and (2.6), and it determines *in general* the forms (2.2) up to these substitutions.

We shall not have much opportunity to consider other invariants of the congruence L than the forms (2.2) and the equations (2.4). Let us remark only that the classical Wälsch invariant (*Sur le premier invariant différentiel projectif des congruences rectignes*, Comptes Rendus Paris, vol. 118, 1894, pp. 736–738) is equal to

$$W = \frac{\alpha_1\alpha_2}{\beta_1\beta_2} = \frac{\varphi}{\varphi^*}. \tag{2.8}$$

Each of the equations $\alpha_1 = 0$, $\alpha_2 = 0$, $\beta_1 = 0$, $\beta_2 = 0$ is obviously invariant. It can be easily seen[1]) that, regardless of the orientation, there are the following 10 *types of non-parabolic congruences*:

Type I: $\alpha_1\alpha_2\beta_1\beta_2 \neq 0$ congruences possessing two non-developable focal surfaces.

Type II: $\alpha_1\beta_1\beta_2 \neq 0 = \alpha_2$ or $\alpha_2\beta_1\beta_2 \neq 0 = \alpha_1$ congruences possessing one non-developable focal surface and one non-rectilinear directrix.

Type II*: $\alpha_1\alpha_2\beta_1 \neq 0 = \beta_2$ or $\alpha_1\alpha_2\beta_2 \neq 0 = \beta_1$ correlative to the type II.

Type III: $\alpha_1\beta_2 \neq 0, \alpha_2 = \beta_1 = 0$ or $\alpha_2\beta_1 \neq 0, \alpha_1 = \beta_2 = 0$ congruences possessing one non-developable focal surface and one rectilinear directrix.

Type IV: $\alpha_2\beta_2 \neq 0, \alpha_1 = \beta_1 = 0$ or $\alpha_1\beta_1 \neq 0, \alpha_2 = \beta_2 = 0$ congruences possessing one developable focal surface and one non-rectilinear directrix.

Type V: $\beta_1\beta_2 \neq 0, \alpha_1 = \alpha_2 = 0$ congruences possessing two non-rectilinear directrices.

Type V*: $\alpha_1\alpha_2 \neq 0, \beta_1 = \beta_2 = 0$ correlative to the type V.

Type VI: $\beta_2 \neq 0, \alpha_1 = \alpha_2 = \beta_1 = 0$ or $\beta_1 \neq 0, \alpha_1 = \alpha_2 = \beta_2 = 0$ congruences possessing one rectilinear directrix and one non-rectilinear directrix.

---

[1])It is necessary to remark that the equations (1.20) show that the relations $\alpha_1 = \beta_2 = 0$ imply $\omega_{32} = 0$, and the relations $\alpha_2 = \beta_1 = 0$ imply $\omega_{41} = 0$.

EDUARD ČECH

Type VI*: $\alpha_2 \neq 0, \alpha_1 = \beta_1 = \beta_2 = 0$ or $\alpha_1 \neq 0, \alpha_2 = \beta_1 = \beta_2 = 0$ correlative to the type VI.

Type VII: $\alpha_1 = \alpha_2 = \beta_1 = \beta_2 = 0$ non-parabolic linear congruences, i.e. congruences possessing two rectilinear directrices.

The congruences of the types IV, VI, VI* and VII can be decomposed into ∞^1 pencils of lines whose centers describe a curve d. The plane of the pencil whose center is situated at the point A of d is not tangent to the curve d at the point A (otherwise the congruence would be parabolic). In the case of the type VI the planes of all the pencils pass through a fixed line. In the case of the type VI* the curve d is a straight line.

If $\alpha_1\beta_2 \neq 0$, the first focus A_1 describes a non-developable surface (A_1) called the *first focal surface*. The asymptotic curves of (A_1) are given by the first equation in (2.4), because from (1.12) one can deduce that

$$[A_1A_2A_3 \quad \mathrm{d}^2A_1] = \alpha_1\omega_2^2 + \beta_2\omega_1^2 ,$$

and $[A_1A_2A_3]$ is the tangent plane of the surface (A_1) at the point A_1. Similarly, for $\alpha_2\beta_1 \neq 0$ the second focus A_2 describes the non-developable *second focal surface* (A_2) whose asymptotic curves are given by the second equation in (2.4). We must not forget that the tangent plane to the *first* focal surface is the *second* focal plane.

The differential forms φ, φ^* are well known. Neglecting infinitely small quantities of the second order, we have according to (1.12)

$$(A_1 + \mathrm{d}A_1)_{\omega_1=0} = A_1 + \alpha_1\omega_2A_2 ,$$
$$(A_2 + \mathrm{d}A_2)_{\omega_2=0} = A_2 + \alpha_2\omega_1A_1 ,$$

so that the anharmonic ratio of the four points

$$A_1,\ A_2,\ (A_1 + \mathrm{d}A_1)_{\omega_1=0},\ (A_2 + \mathrm{d}A_2)_{\omega_2=0}$$

of the line (1.1) is equal to the form φ. Similarly we can deduce from (1.18) that the anharmonic ratio of the four planes

$$E_3,\ E_4,\ (E_3 + \mathrm{d}E_3)_{\omega_1=0},\ (E_4 + \mathrm{d}E_4)_{\omega_2=0}$$

passing through the line (1.1) is equal to the form φ^*. From this, it follows, applying a result obtained by A. Terracini in 1927 (*Un'osservazione sugli invarianti di un'equazione di Laplace*, Boll. Un. Mat. Ital., vol. 6, 1927, pp. 57–60), that the form φ is equal to $h\ du\ dv$, where h is the first Laplace-Darboux invariant of the equation which expresses $\dfrac{\partial^2 A_2}{\partial u \partial v}$ as a linear combination of $\dfrac{\partial A_2}{\partial u}$, $\dfrac{\partial A_2}{\partial v}$, A_2, and simultaneously the second invariant of the equation which expresses $\dfrac{\partial^2 A_1}{\partial u \partial v}$ as a linear combination of $\dfrac{\partial A_1}{\partial u}$, $\dfrac{\partial A_1}{\partial v}$, A_1. Similarly, the form φ^* is equal to h^*dudv, where h^* is the first invariant of the equation which expresses $\dfrac{\partial^2 E_4}{\partial u \partial v}$ as a linear combination of $\frac{\partial E_4}{\partial u}$, $\frac{\partial E_4}{\partial v}$, E_4, and the second invariant of the equation which expresses $\frac{\partial^2 E_3}{\partial u \partial v}$ as a linear combination of $\frac{\partial E_3}{\partial u}$, $\frac{\partial E_3}{\partial v}$, E_3.

Let us set

$$G_1 = -\frac{\alpha_1\, \omega_2^2}{\beta_2\, \omega_1^2}\,, \qquad G_2 = -\frac{\alpha_2\, \omega_1^2}{\beta_1\, \omega_2^2}\,, \tag{2.9}$$

so that

$$\begin{aligned} F_1 &= -\varphi^* \cdot G_1 = -\frac{\varphi}{G_2}\,, \\ F_2 &= -\frac{\varphi}{G_1} = -\varphi^* \cdot G_2\,. \end{aligned} \tag{2.10}$$

Let us suppose that $\alpha_1\beta_2 \neq 0$, having thus the first focal surface (A_1) non-developable with the asymptotic curves given by the first of the equations (2.4) which can be written in the form $G_1 = 1$. In the pencil of tangents to the surface (A_1) (center of the pencil A_1, plane of the pencil E_4) let us consider the involution J_1 whose double lines are the line $[A_1 A_2]$ and its Laplace transform $[A_1 A_3]$. It can be easily seen that G_1 is the anharmonic ratio of the following four elements of the involution J_1: the double line $[A_1 A_2]$, the double line $[A_1 A_3]$, the pair consisting of the asymptotic tangents of the surface (A_1), and the pair of the involution J_1 containing the tangent $[A_1\ dA_1] = [A_1\ \alpha_1\omega_2 A_2 + \omega_1 A_3]$. If one has $\alpha_2\beta_1 \neq 0$, then G_2 can be interpreted in an analogical way with the aid of the involution J_2 in the pencil of tangents to the surface (A_2), the double lines of J_2 being $[A_1 A_2]$ and $[A_2 A_4]$.

3. Beside the congruence L, we shall consider another non-parabolic congruence L' in a projective space S_3'. The relative position of the two spaces S_3 and S_3' is irrelevant for the problems we are here concerned with. We introduce for L' the analogical notation to that employed for L, indicating with the primes all the expressions concerning L'. It will also be useful to set

$$\tau_{ik} = \omega'_{ik} - \omega_{ik}\,, \tag{3.1}$$

so that we have e.g. (see (1.9) and (1.10))

$$\tau_{11} + \tau_{22} + \tau_{33} + \tau_{44} = 0\,, \tag{3.2}$$

$$\tau_{14} = 0\,, \qquad \tau_{23} = 0\,. \tag{3.3}$$

Thus, let T be a transformation (line $\rightarrow$ line) between L and L'. We shall restrict our considerations to the *developable transformations*, i.e. we suppose that T maps each developable surface in L into a developable surface in L'. Therefore, we can suppose that, for an arbitrary line g of L, the two lines g and $g' = Tg$ correspond to the same values of u and v with the parameters u and v being developable simultaneously for L and L'. It can be easily seen that the frame

$$A_1',\ A_2',\ A_3',\ A_4' \tag{3.4}$$

EDUARD ČECH

can be subject to the condition $\omega_1' = \omega_1$, $\omega_2' = \omega_2$, which can also be written in the form

$$\tau_{13} = 0\,, \qquad \tau_{24} = 0\,. \tag{3.5}$$

Let us explicitly notice the equations

$$\omega_{12}' = \alpha_1'\omega_2,\ \omega_{21}' = \alpha_2'\omega_1,\ \omega_{34}' = \beta_2'\omega_1,\ \omega_{43}' = \beta_1'\omega_2, \tag{3.6}$$

which correspond to the equations (1.11). From (3.5) we deduce by the exterior differentiation

$$[\tau_{11} - \tau_{33} \quad \omega_1] = 0\,, \qquad [\tau_{22} - \tau_{44} \quad \omega_2] = 0\,. \tag{3.7}$$

Having arbitrarily chosen the values of u and v, we shall call a homography H $(S_3 \to S_3')$ a *tangent homography* resp. an *osculating homography* to the transformation T (at the line g of L corresponding to the chosen values of u and v) if H realizes an analytic contact of the first resp. second order between the two congruences L and L'. We mean here the contact in the sense of line geometry which turns to be a point contact if we use the classical representation of lines by the points of a hyperquadric in the 5-dimensional space. The analytic conditions for a tangent homography H, when choosing conveniently the scalar factor of H, are

$$H[A_1A_2] = [A_1'A_2']\,, \qquad H\mathrm{d}[A_1A_2] = \mathrm{d}[A_1'A_2'] + \vartheta[A_1'A_2']\,. \tag{3.8}$$

For an osculating homography it is necessary to add a further condition

$$H\mathrm{d}^2[A_1A_2] = \mathrm{d}^2[A_1'A_2'] + 2\vartheta\mathrm{d}[A_1'A_2'] + (\cdot)[A_1'A_2']\,. \tag{3.9}$$

If we take into account the first equation of (1.19) as well as the first one of (3.5), we can reformulate the condition (3.8) as follows

$$H[A_1A_2] = [A_1'A_2'], H[A_2A_3] = [A_2' \quad A_3' + \lambda_1A_1'], H[A_1A_4] = [A_1' \quad A_4' + \lambda_2A_2'], \tag{3.10}$$

where

$$\vartheta = \lambda_1\omega_1 + \lambda_2\omega_2 - (\tau_{11} + \tau_{22})\,. \tag{3.11}$$

We can see that the tangent homographies H exist for any developable T. (By the way, it is evident that they cannot exist if T is not developable.) They have the form

$$\begin{aligned} HA_1 &= \varrho A_1'\,,\\ HA_2 &= \varrho^{-1}A_2'\,,\\ HA_3 &= \varrho(A_3' + \lambda_1A_1') + \mu_1A_2'\,,\\ HA_4 &= \varrho^{-1}(A_4' + \lambda_2A_2') + \mu_2A_1'\,, \end{aligned} \tag{3.12}$$

where the quantities $\varrho \neq 0$, λ_1, λ_2, μ_1, μ_2 are completely arbitrary, so that for given u and v there are ∞^5 tangent homographies. It is useful to remark that a change of the sign of ϱ is of merely formal significance. More precisely, one can see that the substitution

$$\begin{pmatrix} \varrho & \lambda_1 & \lambda_2 & \mu_1 & \mu_2 \\ -\varrho & \lambda_1 & \lambda_2 & -\mu_1 & -\mu_2 \end{pmatrix} \tag{3.13}$$

changes only the inessential scalar factor of H. The expression of H in the plane coordinates is

$$\begin{aligned}
HE_1 &= \varrho^{-1}(E_1' - \lambda_1 E_3') - \mu_2 E_4' \,, \\
HE_2 &= \varrho(E_2' - \lambda_2 E_4') - \mu_1 E_3' \,, \\
HE_3 &= \varrho^{-1} E_3' \,, \\
HE_4 &= \varrho E_4' \,,
\end{aligned} \tag{3.14}$$

and in the line coordinates we have

$$\begin{aligned}
H[A_1A_2] &= [A_1'A_2'] \,, \\
H[A_1A_3] &= \varrho^2[A_1'A_3'] + \varrho\mu_1[A_1'A_2'] \,, \\
H[A_2A_4] &= \varrho^{-2}[A_2'A_4'] - \varrho^{-1}\mu_2[A_1'A_2'] \,, \\
H[A_1A_4] &= [A_1'A_4'] + \lambda_2[A_1'A_2'] \,, \\
H[A_2A_3] &= [A_2'A_3'] - \lambda_1[A_1'A_2'] \,, \\
H[A_3A_4] &= [A_3'A_4'] + [\lambda_1\lambda_2 - \mu_1\mu_2][A_1'A_2'] + \lambda_1[A_1'A_4'] - \\
&\quad - \lambda_2[A_2'A_3'] - \varrho\mu_2[A_1'A_3'] + \varrho^{-1}\mu_1[A_2'A_4'] \,.
\end{aligned} \tag{3.15}$$

It is clear that, for T developable, a fixed orientation of L induces that of L'. When changing the orientation of L, and consequently also that of L', we have the substitution (1.21) to which it is now necessary to add the substitution

$$\begin{pmatrix} \varrho & \lambda_1 & \lambda_2 & \mu_1 & \mu_2 \\ \varrho^{-1} & \lambda_2 & \lambda_1 & \mu_2 & \mu_1 \end{pmatrix} . \tag{3.16}$$

To the substitution (1.22), expressing the duality, it is now necessary to add the substitution

$$\begin{pmatrix} \varrho & \lambda_1 & \lambda_2 & \mu_1 & \mu_2 \\ \varrho^{-1} & -\lambda_1 & -\lambda_2 & -\mu_2 & -\mu_1 \end{pmatrix} . \tag{3.17}$$

Further, let us remark that the notion of tangent (or osculating) homography is self dual and does not depend on the orientation.

From the equations (1.19), it follows

$$\begin{aligned}
&\mathrm{d}^2[A_1A_2] = (\mathrm{d}\overline{\omega_{11} + \omega_{22}} + \overline{\omega_{11} + \omega_{22}}^2 + \omega_{31}\omega_1 + \omega_{42}\omega_2)[A_1A_2] + \\
&+(\mathrm{d}\omega_2 + \overline{2\omega_{11} + \omega_{22} + \omega_{44}} \cdot \omega_2)[A_1A_4] - (\mathrm{d}\omega_1 + \overline{\omega_{11} + 2\omega_{22} + \omega_{33}} \cdot \omega_1) \,. \\
&[A_2A_3] + (\beta_1\omega_2^2 - \alpha_2\omega_1^2)[A_1A_3] + (\alpha_1\omega_2^2 - \beta_2\omega_1^2)[A_2A_4] + 2\omega_1\omega_2[A_3A_4] \,.
\end{aligned} \tag{3.18}$$

For the congruence L' we have an analogical equation. Taking into account (3.11) and (3.15), we obtain

$$\begin{aligned}
&H\mathrm{d}^2[A_1A_2] = \mathrm{d}^2[A_1'A_2'] + 2\vartheta\mathrm{d}[A_1'A_2'] + (\cdot)[A_1'A_2'] - \\
&-(\tau_{11} - \tau_{33} - 2\lambda_1\omega_1)\omega_1[A_2'A_3'] + (\tau_{22} - \tau_{44} - 2\lambda_2\omega_2)\omega_2[A_1'A_4'] + \\
&+\{(\alpha_2' - \varrho^2\alpha_2)\omega_1^2 - (\beta_1' - \varrho^2\beta_1)\omega_2^2 - 2\varrho\mu_2\omega_1\omega_2\}[A_1'A_3'] + \\
&+\{(\beta_2' - \varrho^{-2}\beta_2)\omega_1^2 - (\alpha_1' - \varrho^{-2}\alpha_1)\omega_2^2 + 2\varrho^{-1}\mu_1\omega_1\omega_2\}[A_2'A_4'] \,.
\end{aligned} \tag{3.19}$$

EDUARD ČECH

Finally, let us recall the equations (3.7), which enable us to set

$$\tau_{11} - \tau_{33} = f_1\omega_1 , \qquad \tau_{22} - \tau_{44} = f_2\omega_2 . \tag{3.20}$$

Comparing (3.19) and (3.9), we arrive at the fundamental conclusion that a tangent homography H is osculating if and only if

$$\alpha_1' = \varrho^{-2}\alpha_1,\ \alpha_2' = \varrho^2\alpha_2,\ \beta_1' = \varrho^2\beta_1,\ \beta_2' = \varrho^{-2}\beta_2, \tag{3.21}$$

$$2\lambda_1 = f_1,\ 2\lambda_2 = f_2,\ \mu_1 = 0,\ \mu_2 = 0. \tag{3.22}$$

4. The transformation T (necessarily developable) is called, following G. Fubini and E. Cartan, the *projective deformation* if, for any choice of the parameters u and v, there exists at least one osculating homography H. We have proved that *a necessary and sufficient condition for a developable transformation T of a non-parabolic congruence L to be a projective deformation is the existence of a quantity ϱ^2 satisfying the equations* (3.21). It is obvious that the type (see n° 2) of the congruence L is invariant with respect to the projective deformations. If L, L' are two linear congruences (type VII), the equations (3.21) are identically satisfied, so that any developable transformation $L \to L'$ is in this case a projective deformation, and for any choice of u and v the osculating homography depends in addition on an arbitrary parameter $\varrho^2 \neq 0$. On the other hand, for all the other types, provided that the equations (3.21) are solvable, they determine ϱ^2 without any ambiguity in such a way that, given u and v, the osculating homography is uniquely determined. In the case of the type VI or VI*, the equations (3.21) can be reduced to a single equation, and we find again that any devélopable transformation is a projective deformation. This is no more true for the other types, and for the solution of the relevant existence problems we refer the reader to a paper which will represent a continuation of the present one.

Let us further notice that an osculating homography, which we shall denote by H_0, is given by

$$\begin{aligned} H_0A_1 &= \varrho A_1' , \\ H_0A_2 &= \varrho^{-1}A_2' , \\ H_0A_3 &= \varrho(A_3' + \tfrac{1}{2}f_1A_1') , \\ H_0A_4 &= \varrho^{-1}(A_4' + \tfrac{1}{2}f_2A_2') , \end{aligned} \tag{4.1}$$

where ϱ is to be determined according to (3.21). The quantities f_1, f_2 are given by (3.20).

Eliminating ϱ^2 from the equations (3.21), we can immediately see that *a necessary and sufficient condition characterizing a projective deformation of a non-parabolic congruence is the invariance of the projective linear element.*

There are four equations in (3.21), and they contain an auxiliary quantity ϱ^2. Thus, we can see that the projective deformation imposes a *triple condition* on the congruence L. Because a congruence depends only on *two* arbitrary functions of two variables, we are led to expect that a congruence is in general projectively indeformable, which is after all well known. Later on we shall see (see n° 5) that the study of tangent homographies enables to introduce more general classes of developable transformations than projective deformations, which are such that each congruence L admits an infinite number of them depending on six arbitrary functions of one variable.

Let us start with a definition. We consider two moving points $A(t)$ and $A'(t)$ depending on a common parameter such that there is given a correspondence between the curve C described by $A(t)$ and the curve C' described by $A'(t)$. Let us suppose also that the two points $A(0)$ and $A'(0)$ coincide, and that the two curves C and C' have geometric contact (at least) of the first order at the point $A(0)$. Then we have

$$A'(0) = cA(0), \; \mathrm{d}A' = cj\mathrm{d}A + \vartheta A \quad \text{for } t = 0 ,$$

where ϑ is of no interest to us. It is the quantity j we are interested in, and which will be called the *coefficient of dilatation* of the contact between C and C' under consideration. It is a projective invariant (more generally, j is invariant with respect to any transformation of the space which is regular at the point $A(0)$), and $j = 1$ if and only if the contact under consideration is analytic.

Now, let us consider a developable transformation T $(L \to L')$, and, for two chosen values of u and v, let (3.12) be a homography tangent to T. If $\alpha_1 \neq 0 \neq \alpha'_1$, then for $v =$ const. (or $\omega_1 = 0$) the point A_1 describes a curve C_1 with g being its tangent, and the point A'_1 describes a curve C'_1 with g' being its tangent. Because the homography H maps A_1 to A'_1, and g to g', the two curves HC_1 and C'_1 have at A'_1 a geometric contact. Let j_1 be the coefficient of dilatation of this contact. (We shall say that H realizes a geometric contact between C_1 and C'_1 with the coefficient of dilatation j_1.) We can find without difficulty that

$$j_1 = \frac{\varrho^2 \alpha'_1}{\alpha_1} . \tag{4.2}$$

Similarly if $\alpha_2 \neq 0 \neq \alpha'_2$, H realizes a geometric contact of the first order between the curves described for $u =$ const. (or $\omega_2 = 0$) by the points A_2 and A'_2 whose coefficient of dilatation is

$$j_2 = \frac{\alpha'_2}{\varrho^2 \alpha_2} . \tag{4.3}$$

Passing to the spaces S_3^*, $S_3'^*$ correlative to the spaces S_3, S'_3 (e. g. the points of S_3^* are the planes in S_3), we have two further coefficients of dilatation

$$j_1^* = \frac{\beta'_1}{\varrho^2 \beta_1} , \tag{4.4}$$

$$j_2^* = \frac{\varrho^2 \beta'_2}{\beta_2} ; \tag{4.5}$$

j_1^* can be obtained (if $\beta_1 \neq 0 \neq \beta'_1$) by considering the motion of the planes E_3 and E'_3 for $\omega_1 = 0$, and j_2^* (if $\beta_2 \neq 0 \neq \beta'_2$) by considering the motion of the planes E_4 and E'_4 for $\omega_2 = 0$.

It is important to remark that the quantities j_1, j_2, j_1^*, j_2^* depend on ϱ^2 only, while the tangent homography H under consideration depends, in addition, on λ_1, λ_2, μ_1, μ_2. Thus, H maps the line $[A_1A_2]$ to the line $[A'_1A'_2]$ by means of a projectivity π:

$$\pi A_1 = \varrho A'_1 , \qquad \pi A_2 = \varrho^{-1} A'_2 , \tag{4.6}$$

and the pencil of planes with the axis $[A_1A_2]$ to the pencil of planes with the axis $[A'_1A'_2]$ by means of a projectivity π^*:

$$\pi^*E_3 = \varrho^{-1}E'_3 \,, \qquad \pi^*E_4 = \varrho E'_4 \,. \tag{4.7}$$

The projectivity π maps A_1 to A'_1, and A_2 to A'_2. π^* maps E_3 to E'_3, and E_4 to E'_4. For the given values of u and v there are still ∞^1 projectivities π, and ∞^1 projectivities π^*. When choosing ϱ^2, the two projectivities are well determined. Conversely, the choice of one of them determines ϱ^2, and, consequently, also the second one.

The mutual relation of the two projectivities π and π^* can be described geometrically. The transformation T maps each line g of L to a line g' of L'. This is neither a point transformation nor a plane transformation. But if we choose ϱ^2 as a function of u and v, we can transform the points of each line g of L by means of the corresponding projectivity (4.6), and the planes passing through g by means of the projectivity (4.7). Thus, we obtain, on the one hand, a *point extension* of T, which I denote by $T(\varrho^2)$, and which maps (for any choice of u and v) the point $x_1A_1 + x_2A_2$ of S_3 to the point $\varrho x_1 A'_1 + \varrho^{-1}x_2A'_2$ of S'_3, and, on the other hand, a *plane extension* of T, which I denote by $T^*(\varrho^2)$, and which maps (for any choice of u and v) the plane $x_1E_3 + x_2E_4$ of S_3 to the plane $\varrho^{-1}x_1E'_3 + \varrho x_2E'_4$ of S'_3. The question is to describe geometrically the relation between the two transformations $T(\varrho^2)$ and $T^*(\varrho^2)$. To this end, let us consider a non-developable ruled surface R of the congruence L, and the corresponding non-developable ruled surface R' of the congruence L' obtained by substituting for u and v arbitrary functions of a parameter t such that $[\omega_1 \,\mathrm{d}t] \neq 0 \neq [\omega_2 \,\mathrm{d}t]$. Thus if $X = x_1A_1 + x_2A_2$ is an arbitrary point of R, and πX is the corresponding point of R', we can easily see that if X^* is the tangent plane of R at the point X, then π^*X^* is the tangent plane of R' at the point πX. In fact, the two tangent planes are obviously

$$[A_1 \quad A_2 \quad \mathrm{d}(x_1A_1 + x_2A_2)] = [A_1 \quad A_2 \quad x_1\omega_1A_3 + x_2\omega_2A_4] = x_2\omega_2E_3 - x_1\omega_1E_4 \,,$$
$$[A'_1 \quad A'_2 \quad \mathrm{d}(\varrho x_1A'_1 + \varrho^{-1}x_2A'_2)] = [A'_1 \quad A'_2 \quad \varrho x_1\omega_1A'_3 + \varrho^{-1}x_2\omega_2A'_4] =$$
$$= \varrho^{-1}x_2\omega_2E'_3 - \varrho x_1\omega_1E'_1 \,.$$

Let us add that it holds

$$[A_1 \quad A_2 \quad \mathrm{d}^2(x_1A_1 + x_2A_2)] = (2\omega_1\mathrm{d}x_1 + x_1\mathrm{d}\omega_1 + x_1\,\overline{\omega_{11} + \omega_{33}}\,\omega_1 +$$
$$+ x_2\overline{\alpha_2\omega_1^2 + \beta_1\omega_2^2})[A_1A_2A_3] + (2\omega_2\mathrm{d}x_2 + x_2\mathrm{d}\omega_2 + x_2\,\overline{\omega_{22} + \omega_{44}}\,\omega_2 +$$
$$+ x_1\overline{\alpha_1\omega_2^2 + \beta_2\omega_1^2})[A_1A_2A_4] \,,$$

so that the differential equation of asymptotic curves of the surface R is

$$\begin{gathered} 2\omega_1\omega_2(x_2\mathrm{d}x_1 - x_1\mathrm{d}x_2) - x_1^2(\alpha_1\omega_2^2 + \beta_2\omega_1^2)\omega_1 + x_2^2(\alpha_2\omega_1^2 + \beta_1\omega_2^2)\omega_2 + \\ + x_1x_2(\omega_2\mathrm{d}\omega_1 - \omega_1\mathrm{d}\omega_2 + \overline{\omega_{11} - \omega_{22} + \omega_{33} - \omega_{44}} \cdot \omega_1\omega_2) = 0 \,. \end{gathered} \tag{4.8}$$

DEVELOPABLE TRANSFORMATIONS OF LINE CONGRUENCES

For the asymptotic curves of R' we obtain an analogical equation

$$2\omega_1\omega_2(x_2\mathrm{d}x_1 - x_1\mathrm{d}x_2) - \varrho^2 x_1^2(\alpha_1'\omega_2^2 + \beta_2'\omega_1^2)\omega_1 + \varrho^{-2}x_2^2(\alpha_2'\omega_1^2 + \beta_1'\omega_2^2)\omega_2 + \\ + x_1x_2\left(\omega_2\mathrm{d}\omega_1 - \omega_1\mathrm{d}\omega_2 + \overline{4\frac{\mathrm{d}\varrho}{\varrho} + \omega_{11} - \omega_{22} + \omega_{33} - \omega_{44} + \tau_{11} - \tau_{22} +} \right. \\ \left. \overline{+\tau_{33} - \tau_{44}} \cdot \omega_1\omega_2\right) = 0 \, .$$

The subtraction gives us the equation

$$x_1^2(\overline{\alpha_1 - \varrho^2\alpha_1'} \cdot \omega_2^2 + \overline{\beta_2 - \varrho^2\beta_2'} \cdot \omega_1^2)\omega_1 - x_2^2(\overline{\alpha_2 - \varrho^{-2}\alpha_2'} \cdot \omega_1^2 + \\ + \overline{\beta_1 - \varrho^{-2}\beta_1'} \cdot \omega_2^2)\omega_2 + x_1x_2\left(4\frac{\mathrm{d}\varrho}{\varrho} + \tau_{11} - \tau_{22} + \tau_{33} - \tau_{44}\right)\omega_1\omega_2 = 0 \, , \tag{4.9}$$

the study of which yields consequences which I consider in this paper only incompletely, restricting myself to the case where T is a projective deformation (see n° 7).

5. For the sake of brevity, I consider in this n° only the case of the type I

$$\alpha_1\alpha_2\beta_1\beta_2 \neq 0 \neq \alpha_1'\alpha_2'\beta_1'\beta_2' \, , \tag{5.1}$$

where the congruence L possesses two non-developable focal surfaces (A_1), (A_2), and the same holds for L'.

We have seen that the four equations (3.21) give a system of necessary and sufficient conditions for a developable transformation T to be a projective deformation. We are going to consider those T which satisfy only a part of the conditions (3.21). First of all, let us remark that by virtue of our assumption (5.1) all the four coefficients of dilatation j_1, j_2, j_1^*, j_2^* [see (4.2) - (4.5)] are well defined, and $\neq 0$. One of them can be prescribed arbitrarily, the quantity $\rho^2 \neq 0$ being at our disposal. We know that $j_1 = 1$ is the condition for the tangent homography H given by (3.12) to realize an analytic contact of the 1st order $A_1 \to A_1'$ for $\omega_1 = 0$, $j_2 = 1$ is the condition for H to realize an analytic contact of the 1st order $A_2 \to A_2'$, for $\omega_2 = 0$, $j_1^* = 1$ is the condition for H to realize an analytic contact of the 1st order $E_3 \to E_3'$ for $\omega_1 = 0$, and $j_2^* = 1$ is the condition for H to realize an analytic contact of the 1st order $E_4 \to E_4'$ for $\omega_2 = 0$. Now, we can deduce from (1.12), (1.18), and (3.12), (3.14)

$$\begin{aligned} HA_1 &= \varrho A_1', & H\mathrm{d}A_1 &= \mathrm{d}(\varrho A_1') + (\cdot)A_1' + (\mu_1\omega_1 + \overline{\varrho^{-1}\alpha_1 - \varrho\alpha_1'} \cdot \omega_2)A_2', \\ HA_2 &= \varrho^{-1}A_2', & H\mathrm{d}A_2 &= \mathrm{d}(\varrho^{-1}A_2') + (\cdot)A_2' + (\mu_2\omega_2 + \overline{\varrho\alpha_2 - \varrho^{-1}\alpha_2'} \cdot \omega_1)A_1', \\ HE_3 &= \varrho^{-1}E_3', & H\mathrm{d}E_3 &= \mathrm{d}(\varrho^{-1}E_3') + (\cdot)E_3' + (\mu_2\omega_1 - \overline{\varrho\beta_1 - \varrho^{-1}\beta_1'} \cdot \omega_1)E_4', \\ HE_4 &= \varrho E_4', & H\mathrm{d}E_4 &= \mathrm{d}(\varrho E_4') + (\cdot)E_4' + (\mu_1\omega_2 - \overline{\varrho^{-1}\beta_2 - \varrho\beta_2'} \cdot \omega_2)E_3', \end{aligned}$$

from which it follows (*without the limitations* $\omega_1 = 0$ *or* $\omega_2 = 0$)

$$\left.\begin{aligned} j_1 &= 1, & \mu_1 &= 1 \\ j_2 &= 1, & \mu_2 &= 0 \\ j_1^* &= 1, & \mu_2 &= 0 \\ j_2^* &= 1, & \mu_1 &= 0 \end{aligned}\right\} \begin{matrix} \text{is the condition} \\ \text{for } H \text{ to realize} \\ \text{an analytic contact} \\ \text{of the first order} \end{matrix} \left\{\begin{aligned} A_1 &\to A_1', \\ A_2 &\to A_2', \\ E_3 &\to E_3', \\ E_4 &\to E_4'. \end{aligned}\right.$$

EDUARD ČECH

Therefore: $\varphi = \varphi'$ *if and only if, for arbitrarily given* u *and* v, *there exists a homography realizing an analytic contact of the first order* $A_1 \to A_1'$, *and simultaneously an analytic contact of the first order* $A_2 \to A_2'$ (one can see without difficulty that such a homography is necessarily tangent for T, i. e. it also realizes an analytic contact of the first order $[A_1A_2] \to [A_1'A_2']$); $\varphi^* = \varphi^{*\prime}$ *if and only if there exists a homography realizing an analytic contact of the first order* $E_3 \to E_3'$, *and simultaneously an analytic contact of the first order* $E_4 \to E_4'$. Besides: $F_1 = F_1'$ *if and only if there exists a homography simultaneously realizing the three analytic contacts of the first order* $A_1 \to A_1'$, $E_3 \to E_3'$, $[A_1A_2] \to [A_1'A_2']$; $F_2 = F_2'$ *if and only if there exists a homography simultaneously realizing the three analytic contacts of the first order* $A_2 \to A_2'$, $E_4 \to E_4'$, $[A_1A_2] \to [A_1'A_2']$. One can see that the condition $\varphi = \varphi'$ deals only with the relation of the foci A_1 and A_2, the condition $\varphi^* = \varphi^{*\prime}$ deals with the focal planes E_3 and E_4, the condition $F_1 = F_1'$ deals with the first focus A_1 and the first focal plane E_3, and the condition $F_2 = F_2'$ deals with the second focus A_2 and the second focal plane E_4. This explains the reason which has led us to call φ a point form, φ^* a plane form, F_1 a first, and F_2 a second focal form respectively. Accordingly, we are going to introduce terminology for particular classes of developable transformations of a non-parabolic congruence: *point deformations* are characterized by $\varphi = \varphi'$, *plane deformations* are characterized by $\varphi^* = \varphi^{*\prime}$, *focal deformations of the first* resp. *second kind* are characterized by $F_1 = F_1'$ resp. $F_2 = F_2'$. Besides, we shall introduce the name of *asymptotic deformation of the first* resp. *second kind* for the developable transformations T inducing an asymptotic transformation $(A_1) \to (A_1')$ resp. $(A_2) \to (A_2')$ of the first resp. second focal surface. Analytically the asymptotic deformations of the first (second) kind are characterized by the condition $\alpha_1\beta_2' = \alpha_1'\beta_2$ $(\alpha_2\beta_1' = \alpha_2'\beta_1)$, or else, in the notation (2.8), $G_1 = G_1'$ $(G_2 = G_2')$. We have thus introduced six particular categories of developable transformations T, which can also be characterized as those satisfying two equations chosen from the four equations (3.21). The transformations T belonging simultaneously to all these six categories coincide with the projective deformations.

Because the congruence L depends only on *two* arbitrary functions of two variables, one cannot prescribe arbitrarily, as functions of u and v, all the *three* differential forms φ, φ^*, F_1, which according to (2.3) determine without ambiguity also the fourth form F_2. But we are going to show that we can prescribe completely arbitrarily two relations among these forms. One can easily see that it is possible without loss of generality to choose the frame (1.5) in such a way that there is

$$\omega_1 = \mathrm{d}v\,, \qquad \omega_2 = \mathrm{d}u\,, \tag{5.2}$$

and consequently

$$\varphi = \alpha_1\alpha_2 \mathrm{d}u\mathrm{d}v, \quad \varphi^* = \beta_1\beta_2 \mathrm{d}u\mathrm{d}v, \quad F_1 = \alpha_1\beta_1 \frac{\mathrm{d}u^3}{\mathrm{d}v}. \tag{5.3}$$

We are looking for the congruences L satisfying two independent relations of the form

$$\Phi(\alpha_1\alpha_2, \beta_1\beta_2, \alpha_1\beta_1, u, v) = 0, \qquad \Psi(\alpha_1\alpha_2, \beta_1\beta_2, \alpha_1\beta_1, u, v) = 0. \tag{5.4}$$

DEVELOPABLE TRANSFORMATIONS OF LINE CONGRUENCES

The question is to integrate the Pfaff system

$$(1.10) + (1.11) + (5.2) \tag{5.5}$$

under the condition (5.4), and with the assumption that the coefficients of (5.3) are $\neq 0$. Differentiating (5.4), we obtain two relations of the form

$$\begin{aligned} \lambda_1 \mathrm{d}(\alpha_1\alpha_2) + \lambda_2 \mathrm{d}(\beta_1\beta_2) + \lambda_3 \mathrm{d}(\alpha_1\beta_1) + \lambda_4 \mathrm{d}u + \lambda_5 \mathrm{d}v = 0 \ , \\ \mu_1 \mathrm{d}(\alpha_1\alpha_2) + \mu_2 \mathrm{d}(\beta_1\beta_2) + \mu_3 \mathrm{d}(\alpha_1\beta_1) + \mu_4 \mathrm{d}u + \mu_5 \mathrm{d}v = 0 \ , \end{aligned} \tag{5.6}$$

where λ_i, μ_i are known functions of the 5 variables $\alpha_1\alpha_2$, $\beta_1\beta_2$, $\alpha_1\beta_1$, u, v such that the rank of the matrix

$$\begin{pmatrix} \lambda_1 & \lambda_2 & \lambda_3 \\ \mu_1 & \mu_2 & \mu_3 \end{pmatrix}$$

is equal to 2 even if the relations (5.4) hold. The exterior differentiation of the Pfaff system under consideration gives the equations

$$[\omega_{11} - \omega_{33}\mathrm{d}v] = 0 \ , \qquad [\omega_{22} - \omega_{44}\mathrm{d}u] = 0 \tag{5.7}$$

and the equations (1.20), where the quantities α_1, α_2, β_1, β_2, and their differentials satisfy the relations (5.4) and (5.6). Setting

$$\begin{aligned} D\alpha_1 &= \mathrm{d}\alpha_1 + \alpha_1(\omega_{22} - \omega_{11}), & D\alpha_2 &= \mathrm{d}\alpha_2 - \alpha_2(\omega_{22} - \omega_{11}), \\ D\beta_1 &= \mathrm{d}\beta_1 - \beta_1(\omega_{22} - \omega_{11}), & D\beta_2 &= \mathrm{d}\beta_2 + \beta_2(\omega_{22} - \omega_{11}), \end{aligned}$$

the equations (5.6) take the form

$$\begin{gathered} \lambda_1(\alpha_1 D\alpha_2 + \alpha_2 D\alpha_1) + \lambda_2(\beta_1 D\beta_2 + \beta_2 D\beta_1) + \lambda_3(\alpha_1 D\beta_1 + \beta_1 D\alpha_1) + \lambda_4 \mathrm{d}u + \\ + \lambda_5 \mathrm{d}v = 0 \ , \\ \mu_1(\alpha_1 D\alpha_2 + \alpha_2 D\alpha_1) + \mu_2(\beta_1 D\beta_2 + \beta_2 D\beta_1) + \mu_3(\alpha_1 D\beta_1 + \beta_1 D\alpha_1) + \mu_4 \mathrm{d}u + \\ + \mu_5 \mathrm{d}v = 0 \ , \end{gathered}$$

and the equations (1.20) the form

$$\begin{gathered} [\omega_{32}\omega_1] + [D\alpha_1 \quad \omega_2] = 0 \ , \qquad [D\alpha_2 \quad \omega_1] + [\omega_{41}\omega_2] = 0 \ , \\ -[\omega_{41}\omega_1] + [D\beta_1 - \beta_1(\omega_{11} - \omega_{33}) \quad \omega_2] = 0 \ , \\ [D\beta_2 - \beta_2(\omega_{22} - \omega_{44}) \quad \omega_1] - [\omega_{32}\omega_2] = 0 \ . \end{gathered}$$

One can easily verify that the determinant

$$\begin{vmatrix} \mathrm{d}v & 0 & 0 & -\mathrm{d}u & 0 & 0 \\ 0 & \mathrm{d}u & -\mathrm{d}v & 0 & 0 & 0 \\ \mathrm{d}u & 0 & 0 & 0 & \lambda_1\alpha_2 + \lambda_3\beta_1 & \mu_1\alpha_2 + \mu_3\beta_1 \\ 0 & \mathrm{d}v & 0 & 0 & \lambda_1\alpha_1 & \mu_1\alpha_1 \\ 0 & 0 & \mathrm{d}u & 0 & \lambda_2\beta_2 + \lambda_3\alpha_1 & \mu_2\beta_2 + \mu_3\alpha_1 \\ 0 & 0 & 0 & \mathrm{d}v & \lambda_2\beta_1 & \mu_2\beta_1 \end{vmatrix} =$$

$$\begin{gathered} = (\lambda_1\mu_2 - \lambda_2\mu_1)(\alpha_1\beta_1\mathrm{d}u^4 - \alpha_2\beta_2\mathrm{d}v^4) + (\lambda_2\mu_3 - \lambda_3\mu_2)\beta_1\mathrm{d}v^2(\alpha_1\mathrm{d}u^2 + \beta_2\mathrm{d}v^2) + \\ + (\lambda_3\mu_1 - \lambda_1\mu_3)\alpha_1\mathrm{d}v^2(\beta_1\mathrm{d}u^2 + \alpha_2\mathrm{d}v^2) \end{gathered}$$

EDUARD ČECH

can not vanish identically. From this, it follows that the Pfaff system (5.5) is in involution, and we arrive at the result that *for the congruences L of type I we can prescribe arbitrarily, besides the evident relation* (1.3), *two other independent relations, which can also depend on u and v, among the differential forms* (2.2), *which are supposed to be expressed as functions of u, v, $\mathrm{d}u$, and $\mathrm{d}v$* (u and v are developable parameters). *Such congruences L always exist and depend on six arbitrary functions of one variable.*

This theorem is rather general, and it is convenient to mention remarkable particular cases of its. E. g., *every congruence of the type I admits developable transformations which* [1] *are simultaneously point and plane deformations,* [2] *are simultaneously focal deformations of the first and the second kind,* [3] *induce asymptotic transformations of both the focal surfaces. In each of these three cases, the developable transformations of L under consideration depend on six arbitrary functions of one variable.* The last example can be generalized. In fact, *if we consider two differential equations of the form*

$$\left(\frac{\mathrm{d}v}{\mathrm{d}u}\right)^2 = f_1(u,v)\,, \tag{5.8}$$

$$\left(\frac{\mathrm{d}v}{\mathrm{d}u}\right)^2 = f_2(u,v)\,, \tag{5.9}$$

then there exist congruences L of type I for which u and v are developable parameters, and for which (5.8) *gives the asymptotic curves of the first focal surface, and* (5.9) *those of the second one. These congruences depend on six arbitrary functions of one variable.* It is interesting that this includes as a very special case a new proof of a classical result by E. Cartan that the R-congruences depend on six arbitrary functions of one variable. In fact, it is known that the R-congruences can be defined as congruences of the type I possessing developable parameters u, v such that on the both focal surfaces the differential equation of asymptotic curves is $\mathrm{d}u^2 - \mathrm{d}v^2 = 0$.

6. Let L and L' be two congruences of type I, and let us consider a developable transformation $T(L \to L')$. We have determined the tangent homographies (3.12) of T depending on u, v, and, in addition, on five other parameters $\varrho^2 \neq 0$, λ_1, λ_2, μ_1, μ_2. If we now fix ϱ^2, we can consider the point extension $T(\varrho^2)$ of T (see n° 4), and the question is to characterize geometrically the tangent homographies H belonging to the chosen value of ϱ^2. To this end, it suffices to make use of what follows from (1.12) and (3.12).

$$\begin{aligned}
&\varrho x_1 A_1' + \varrho^{-1} x_2 A_2' = H(x_1 A_1 + x_2 A_2)\,,\\
&\mathrm{d}(\varrho x_1 A_1' + \varrho^{-1} x_2 A_2') = H\mathrm{d}(x_1 A_1 + x_2 A_2) +\\
&+\left\{\varrho x_1\left(\frac{\mathrm{d}\varrho}{\varrho} + \tau_{11} - \lambda_1\omega_1\right) + x_2(\overline{\varrho^{-1}\alpha_2' - \varrho\alpha_2}\cdot\omega_1 - \mu_2\omega_2)\right\} A_1' +\\
&+\left\{x_1(-\mu_1\omega_1 + \overline{\varrho\alpha_1' - \varrho^{-1}\alpha_1}\cdot\omega_2) + \varrho^{-1}x_2\left(-\frac{\mathrm{d}\varrho}{\varrho} + \tau_{22} - \lambda_2\omega_2\right)\right\} A_2'\,.
\end{aligned}$$

Hence, we can easily deduce that if the point $x_1 A_1 + x_2 A_2$ describes in the space S_3 a curve C, then the transform C' of C by means of $T(\varrho^2)$ has the property that at the point $\varrho x_1 A_1' + \varrho^{-1} x_2 A_2'$, which is the transform of $x_1 A_1 + x_2 A_2$, the tangents of the two curves C' and HC lie in a plane containing the line $[A_1' A_2']$. We can easily find that this is a characteristic property of the tangent homographies corresponding to the value ϱ^2 under consideration.

We can look for a condition under which the curves C' and HC have at the point $\varrho x_1 A_1' + \varrho^{-1} x_2 A_2'$ an analytic contact of the first order. We find

$$\begin{vmatrix} \varrho x_1 & \varrho x_1 \left(\frac{\mathrm{d}\varrho}{\varrho} + \tau_{11} - \lambda_1 \omega_1 \right) + x_2 (\overline{\varrho^{-1}\alpha_2' - \varrho \alpha_2} \cdot \omega_1 - \mu_2 \omega_2) \\ \varrho x_2 & x_1 (-\mu_1 \omega_1 + \overline{\varrho \alpha_1' - \varrho^{-1} \alpha_1} \cdot \omega_2) + \varrho^{-1} x_2 \left(-\frac{\mathrm{d}\varrho}{\varrho} + \tau_{22} - \lambda_2 \omega_2 \right) \end{vmatrix} = 0 ,$$

or

$$\begin{aligned} \varrho x_1^2 (-\mu_1 \omega_1 + \overline{\varrho \alpha_1' - \varrho^{-1} \alpha_1} \cdot \omega_2) - \varrho^{-1} x_2^2 (\overline{\varrho^{-1} \alpha_2' - \varrho \alpha_2} \cdot \omega_1 - \mu_2 \omega_2) - \\ - x_1 x_2 \left(2 \frac{\mathrm{d}\varrho}{\varrho} + \tau_{11} - \tau_{22} - \lambda_1 \omega_1 + \lambda_2 \omega_2 \right) = 0 . \end{aligned} \tag{6.1}$$

If we choose $\mu_1 = \mu_2 = 0$, and determine λ_1, λ_2 in such a way that there is

$$2 \frac{\mathrm{d}\varrho}{\varrho} + \tau_{11} - \tau_{22} - \lambda_1 \omega_1 + \lambda_2 \omega_2 = 0 , \tag{6.2}$$

then, using (4.2) and (4.3), the equation (6.1) can be written in the form

$$(j_1 - 1) \alpha_1 \omega_2 x_1^2 - (j_2 - 1) \alpha_2 \omega_1 x_2^2 = 0 . \tag{6.3}$$

We have thus arrived at a tangent homography, which we denote by $K(\varrho^2)$, and which for the given u, v, and ϱ^2 is determined without ambiguity. Thus we have

$$\begin{aligned} K(\varrho^2) A_1 &= \varrho A_1' , & K(\varrho^2) A_2 &= \varrho^{-1} A_2' , \\ K(\varrho^2) A_3 &= \varrho (A_3' + \lambda_1 A_1') , & K(\varrho^2) A_4 &= \varrho^{-1} (A_4' + \lambda_2 A_2') , \end{aligned} \tag{6.4}$$

where it is necessary to determine λ_1 and λ_2 in such a way that the equation (6.2) is satisfied. From the preceding considerations, we obtain the following geometric characterization of the tangent homography $K(\varrho^2)$: For the curves C satisfying $\omega_2 = 0$, there is at the point $\varrho x_1 A_1' + \varrho^{-1} x_2 A_2'$ an analytic contact of the 1st order between C' and $K(\varrho^2)C$ either (if $j_2 = 1$) *always* or (if $j_2 \neq 1$) *if and only if* the curve C passes through the point A_1. For the curves C satisfying $\omega_1 = 0$, there is at the point $\varrho x_1 A_1' + \varrho^{-1} x_2 A_2'$ an analytic contact of the 1st order between C' and $K(\varrho^2)C$ either (if $j_1 = 1$) always or (if $j_1 \neq 1$) if and only if C passes through the point A_2. In the particular case where T is a *point deformation*, we can fix ϱ^2 uniquely by requiring

$$j_1 = j_2 = 1 . \tag{6.5}$$

The corresponding homography $K(\varrho^2)$ is called the *homography pointwise associated* with T, and is denoted by K_0. The equation (6.3) turns into an identity, so that K_0 realizes an analytic contact of the 1st order $S_3 \to S_3'$ (with respect to

EDUARD ČECH

the point transformation $T(\varrho^2)$). See E. Čech, *Projective differential geometry of correspondences between two spaces V* (Russian), Czech. Math. J., vol. 2(77)1952, pp. 167–188; L. Muracchini, *Sulle transformazioni punctuali che sono inviluppi di omografie*, Bol. Un. Mat. Ital. (3)8, 1953, pp. 390–398; E. Čech, *On point deformations of line congruences* (Russian with a detailed French summary), Czech. Math. J., vol. 5(80), 1955, pp. 234–273.)

It is advisable to announce also the results correlative to the preceding ones, based on the consideration of the plane extension $T^*(\varrho^2)$ of T. Instead of $K(\varrho^2)$, we have now the tangent homography $K^*(\varrho^2)$:

$$\begin{aligned} K^*(\varrho^2)A_1 &= \varrho A_1' , & K^*(\varrho^2)A_2 &= \varrho^{-1}A_2' , \\ K^*(\varrho^2)A_3 &= \varrho(A_3' + \lambda_1^* A_1') , & K^*(\varrho^2)A_4 &= \varrho^{-1}(A_4' + \lambda_2^* A_2') , \end{aligned} \tag{6.6}$$

where λ_1^* and λ_2^* satisfy the equation

$$2\frac{\mathrm{d}\varrho}{\varrho} + \tau_{33} - \tau_{44} + \lambda_1^*\omega_1 - \lambda_2^*\omega_2 = 0 . \tag{6.7}$$

The equation correlative to (6.3) is

$$(j_1^* - 1)\beta_1\omega_2 x_1^2 - (j_2^* - 1)\beta_2\omega_1 x_2^2 = 0 . \tag{6.8}$$

In the particular case where T is a *plane deformation* we can fix ϱ^2 uniquely by requiring

$$j_1^* = j_2^* = 1 . \tag{6.9}$$

The corresponding homography $K^*(\varrho^2)$ is called the *homography planewise associated* with T, and is denoted by K_0^*.

We have supposed in this n° that the two congruences L and L' are of the type I. But it can be easily seen that the results remain valid when we suppose only that the type of L is the same as that of L'. Only if we have e. g. $\alpha_1 = \alpha_1' = 0$, we must set $j_1 = 1$ irrespective of the chosen value of ϱ^2. Similarly, we must set $j_2 = 1$ for $\alpha_2 = \alpha_2' = 0$, $j_1^* = 1$ for $\beta_1 = \beta_1' = 0$, and $j_2^* = 1$ for $\beta_2 = \beta_2' = 0$.

Let us ask furthermore, under which conditions, for a given value of ϱ^2, the two homographies $K(\varrho^2)$ and $K^*(\varrho^2)$ coincide. We can easily find the condition

$$4\frac{\mathrm{d}\varrho}{\varrho} + \tau_{11} - \tau_{22} + \tau_{33} - \tau_{44} = 0 . \tag{6.10}$$

The equation (6.10) is completely integrable if and only if

$$\alpha_1'\alpha_2' - \alpha_1\alpha_2 = \beta_1'\beta_2' - \beta_1\beta_2 ,$$

which can be written in the form

$$\varphi^{*\prime} - \varphi' = \varphi^* - \varphi . \tag{6.11}$$

If, for example, the congruence L is a W-congruence, we have $\varphi^* - \varphi = 0$, and the condition (6.11) is satisfied if and only if L' is also a W-congruence. If the condition (6.11) is satisfied, then (6.10) determines ϱ^2 up to a constant. For an initial value of the line (1.1) we can choose the projectivity π arbitrarily. This determines uniquely all the ∞^2 π's.

DEVELOPABLE TRANSFORMATIONS OF LINE CONGRUENCES

7. Let us pass to the study of the *projective deformation* $T(L \to L')$. We can easily see that the frame (3.4) can be specialized in such a way that the equations (4.1) of the osculating homography H_0 take the simple form

$$H_0A_1 = A_1', \quad H_0A_2 = A_2', \quad H_0A_3 = A_3', \quad H_0A_4 = A_4' , \tag{7.1}$$

so that, besides (3.5), we have on the one hand

$$\alpha_1' = \alpha_1, \quad \alpha_2' = \alpha_2, \quad \beta_1' = \beta_1, \quad \beta_2' = \beta_2 ,$$

or put differently

$$\tau_{12} = \tau_{21} = \tau_{34} = \tau_{43} = 0 , \tag{7.2}$$

and on the other hand

$$\tau_{11} - \tau_{33} = 0 , \qquad \tau_{22} - \tau_{44} = 0 . \tag{7.3}$$

By the exterior differentiation we deduce from (7.2)

$$\begin{aligned} &[\tau_{32}\omega_1] - \alpha_1[\tau_{11} - \tau_{22} \quad \omega_2] = 0 , \\ &\beta_2[\tau_{11} - \tau_{22} \quad \omega_1] + [\tau_{32}\omega_2] = 0 , \\ &[\tau_{41}\omega_1] - \beta_1[\tau_{11} - \tau_{22} \quad \omega_2] = 0 , \\ &\alpha_2[\tau_{11} - \tau_{22} \quad \omega_1] + [\tau_{41}\omega_2] = 0 , \end{aligned} \tag{7.4}$$

and from (7.3)

$$[\tau_{31}\omega_1] = 0 , \qquad [\tau_{42}\omega_2] = 0 . \tag{7.5}$$

The equations (7.4) enable us to set

$$\tau_{22} - \tau_{11} = c_1\omega_1 - c_2\omega_2 , \tag{7.6}$$

$$\tau_{32} = \beta_2c_2\omega_1 + \alpha_1c_1\omega_2 , \qquad \tau_{41} = \alpha_2c_2\omega_1 + \beta_1c_1\omega_2 . \tag{7.7}$$

A projective deformation T is simultaneously a point deformation and a plane deformation,[2]) so that, besides the osculating homography H_0, it is convenient to consider in addition the pointwise associated homography K_0 and the planewise associated homography K_0^*. For these homographies we have the equations (6.2), (6.4), (6.6), and (6.7), where it is necessary to set $\varrho^2 = 1$, so that according to (7.3) and (7.6) we have

$$\begin{aligned} &K_0A_1 = A_1' , \ K_0A_2 = A_2' , \ K_0A_3 = A_3' - c_1A_1' , \ K_0A_4 = A_4' - c_2A_2' , \\ &K_0^*A_1 = A_1' , \ K_0^*A_2 = A_2' , \ K_0^*A_3 = A_3' + c_1A_1' , \ K_0^*A_4 = A_4' + c_2A_2' . \end{aligned} \tag{7.8}$$

[2])The conditions for T to be simultaneously a point deformation and a plane deformation consist in the existence of such values of $\varrho_1^2 \neq 0$ and $\varrho_2^2 \neq 0$ that

$$\alpha_1' = \varrho_1^{-2}\alpha_1 , \quad \alpha_2' = \varrho_1^2\alpha_2 , \quad \beta_1' = \varrho_2^2\beta_1 , \quad \beta_2' = \varrho_2^{-2}\beta_2 .$$

T is a projective deformation if and only if $\varrho_1^2 = \varrho_2^2$.

EDUARD ČECH

The equations of our three homographies in the plane coordinates are

$$\begin{aligned} H_0E_1 &= E_1', \ H_0E_2 = E_2', \ H_0E_3 = E_3', \ H_0E_4 = E_4', \\ K_0E_1 &= E_1' + c_1E_3', \ K_0E_2 = E_2' + c_2E_4', \ K_0E_3 = E_3', \ K_0E_4 = E_4', \\ K_0^*E_1 &= E_1' - c_1E_3', \ K_0^*E_2 = E_2' - c_2E_4', \ K_0^*E_3 = E_3', \ K_0^*E_4 = E_4'. \end{aligned} \tag{7.9}$$

In general, the three homographies H_0, K_0, and K_0^* coincide only for the points of the line $[A_1A_2]$. If they are identical, we say that T is a *singular projective deformation*. The analytic condition for this particular situation is $c_1 = c_2 = 0$ or $\tau_{11} - \tau_{22} = 0$. The notion of a singular projective deformation was introduced already in 1920 (E. Cartan, *Sur le problème général de la déformation*, Comptes Rendus du Congrès Intern. des Math. de Strasbourg en 1920, pp. 397–406). Nevertheless, it can still happen that for T non-singular there exist points outside the line $[A_1A_2]$ for which H_0, K_0, and K_0^* coincide. Then I say that T is a *semi-singular projective deformation*. The points A such that H_0A, K_0A, and K_0^*A coincide form then necessarily one of the two focal planes. If it is the first (second) focal plane $E_3(E_4)$, I speak about semi-singular projective deformation of the first (second) kind. The analytic condition is $c_2 = 0$ or $[\tau_{22} - \tau_{11} \quad \omega_1] = 0$ for the first kind, and $c_1 = 0$ or $[\tau_{22} - \tau_{11} \quad \omega_2] = 0$ for the second kind. Comparing (7.1) and (7.8) with (7.9), we can immediately see that the notion of singular or semi-singular projective deformation is correlative with itself.

According to (1.12) and (7.1), it follows from (7.3), (7.6), and (7.7)

$$\begin{aligned} A_1' &= H_0A_1 , \qquad \mathrm{d}A_1' = H_0\mathrm{d}A_1 + \tau_{11}A_1' , \\ \mathrm{d}^2A_1' &= H_0\mathrm{d}^2A_1 + 2\tau_{11}dA_1' + (\beta_2c_2\omega_1^2 + 2\alpha_1c_1\omega_1\omega_2 - \alpha_1c_2\omega_2^2)A_2' + (\cdot)A_1' , \end{aligned} \tag{7.10}$$

$$\begin{aligned} A_2' &= H_0A_2 , \qquad \mathrm{d}A_2' = H_0\mathrm{d}A_2 + \tau_{22}A_2' , \\ \mathrm{d}^2A_2' &= H_0\mathrm{d}^2A_2 + 2\tau_{22}dA_2' + (-\alpha_2c_1\omega_1^2 + 2\alpha_2c_2\omega_1\omega_2 + \beta_1c_1\omega_2^2)A_1' + (\cdot)A_2' . \end{aligned} \tag{7.11}$$

Let us suppose that $\alpha_1\beta_2 \neq 0$, so that the first focus A_1 describes a non-developable surface (A_1) (the first focal surface). The equation (7.10) demonstrates the well known fact that if the projective deformation T is singular, then H_0 realizes an analytic contact of the second order between the surfaces (A_1) and (A_1'). If T is not singular, this contact is only of the first order, nevertheless, it is of the second order for the curves lying on (A_1) whose tangent at the point A_1 under consideration satisfies the equation

$$\beta_2c_2\omega_1^2 + 2\alpha_1c_1\omega_1\omega_2 - \alpha_1c_2\omega_2^2 = 0 . \tag{7.12}$$

Let us call the *characteristic tangents* of the surface (A_1) with respect to the projective deformation T the two tangents satisfying (7.12). Knowing that (2.4) gives the asymptotic curves of (A_1), we can see that the two tangents (7.12) are conjugate or coincide with one asymptotic tangent. Moreover, we can see that, for a semi-singular projective deformation of the first kind, the pair of characteristic tangents is given by the equation $\omega_1\omega_2 = 0$, and consequently it consists of the line $[A_1A_2]$ and its Laplace transform $[A_1A_3]$, while for a semi-singular projective deformation

DEVELOPABLE TRANSFORMATIONS OF LINE CONGRUENCES

of the second kind the tangents (7.12) separate harmonically $[A_1A_2]$ and $[A_1A_3]$. If $\alpha_2\beta_1 \neq 0$, we have completely analogous results concerning the characteristic tangents

$$-\alpha_2 c_1 \omega_1^2 + 2\alpha_2 c_2 \omega_1 \omega_2 + \beta_1 c_1 \omega_2^2 = 0 \tag{7.13}$$

of the second focal surface (A_2).

The ∞^2 projectivities which are obtained by transforming each line of points $[A_1A_2]$ by means of the corresponding osculating homography H_0, when taken together, form the pointwise extension $T(1)$ of T, which can be called the *principal pointwise extension* of the projective deformation T. (Correlatively, we have *principal planewise extension* $T^*(1)$ of T.) Now, let R be a non-developable ruled surface contained in the congruence L, and R' its transform by $T(1)$. From (4.9), when setting $\varrho^2 = 1$ and making use of (7.2) and (7.3), it follows that the asymptotic curves of R' correspond to those of R if and only if there is on R

$$\tau_{11} - \tau_{22} = 0\,. \tag{7.14}$$

Consequently, the singular projective deformations can be characterized by the property that they are *totally asymptotic*, i. e., they transform asymptotically *every* non-developable ruled surface contained in L. This result is already known (E. Čech, *Projective differential geometry of correspondences between two spaces* IV (Russian), Czech. Math. J., vol. 2(77), 1952, pp. 149–166). If the projective deformation T is not singular, then the equation (7.14) defines a decomposition of the congruence L into ∞^1 ruled surfaces, which we shall call *canonical decomposition* of L with respect to T. If T is not semi-singular, then the ruled surfaces of L satisfying (7.14) are non-developable, and their characteristic property is that they correspond asymptotically to their images by means of the principal pointwise extension $T(1)$ of T.

The geometric characterization of the canonical decomposition, which we are just going to present, is meaningless in the case when the projective deformation T is semi-singular, for then the ruled surfaces of L satisfying (7.14) are developable. In order to fill this gap, let us consider, say, the first family of developables given analytically by $\omega_1 = 0$, and let us denote by $\equiv$ the equalities valid for $\omega_1 = 0$ only. The projective deformation T is not subject to any condition in this moment. Let D be a developable of the first family of L such that $\omega_1 = 0$ on D, and let t be a point transformation $D \to D'$ contained in $T(1)$. Let M be a homography $S_3 \to S_3'$ tangent to t along a generator $[A_1A_2]$. One can evidently choose the scalar factor of M in such a way that

$$MA_1 = A_1', \qquad MA_2 = A_2'. \tag{7.15}$$

Moreover, let

$$MA_3 = p_1A_1' + p_2A_2' + p_3A_3' + p_4A_4'\,,$$
$$MA_4 = q_1A_1' + q_2A_2' + q_3A_3' + q_4A_4'\,.$$

But we have

$$M(x_1A_1 + x_2A_2) = x_1A_1' + x_2A_2'\,,$$
$$M\,\mathrm{d}(x_1A_1 + x_2A_2) \equiv \mathrm{d}(x_1A_1' + x_2A_2') - x_1\tau_{11}A_1' - x_2\tau_{22}A_2' + x_2\omega_2(q_1A_1' +$$
$$+ q_2A_2' + q_3A_3' + \overline{q_4 - 1}\cdot A_4')\,,$$

from which it follows easily (see (7.6)) that a necessary and sufficient condition for M to be tangent to t along $[A_1A_2]$ is $q_1 = 0$, $q_2 = -c_2$, $q_3 = 0$, $q_4 = 1$, or

$$MA_4 = A_4' - c_2A_2' \,, \tag{7.16}$$

from where

$$M\mathrm{d}(x_1A_1 + x_2A_2) \equiv \mathrm{d}(x_1A_1' + x_2A_2') - \tau_{11}(x_1A_1' + x_2A_2') \,.$$

Then we find

$$\begin{aligned} &M\mathrm{d}^2(x_1A_1 + x_2A_2) \equiv \mathrm{d}^2(x_1A_1' + x_2A_2') - 2\tau_{11}\mathrm{d}(x_1A_1' + x_2A_2') - \\ &- \{(\mathrm{d}\tau_{11} - \tau_{11}^2)x_1 + (\tau_{41} - p_1\beta_1\omega_2)\omega_2x_2\}\, A_1' - \\ &- \{\mathrm{d}\tau_{11} - \tau_{11}^2 - (\mathrm{d}c_2 - \tau_{42} + c_2 \cdot \overline{\omega_{22} - \omega_{44}} + p_2 \cdot \overline{\beta_1 - c_2^2} \cdot \omega_2)\omega_2\}x_2A_2' + \\ &+ (p_3 - 1)\beta_1\omega_2^2x_2A_3' - (p_4\beta_1 - 2c_2)\omega_2^2x_2A_4' \,. \end{aligned}$$

Thus, the homography M is (see (7.7)) osculating to the transformation t along the generator $[A_1A_2]$ of D if and only if there is, firstly, $p_1 = c_1$, $p_3 = 1$, $\beta_1p_4 = 2c_2$, where

$$\beta_1MA_3 = \beta_1(A_3 + c_1A_1 + p_2A_2) + 2c_2A_4 \,, \tag{7.17}$$

and secondly,

$$\mathrm{d}c_2 + c_2(\omega_{22} - \omega_{44}) - \tau_{42} + (\beta_1p_2 - c_2^2)\omega_2 \equiv 0 \,. \tag{7.18}$$

The equations (7.15), (7.16), and (7.17) give the point expression of the homography M osculating to T along $[A_1A_2]$.

We leave aside the case $\beta_1 = 0$, where the developable coincides with the plane E_3 [according to (1.18), for $\beta_1 = 0$ we have $\mathrm{d}E_3 \equiv (.)E_3$, which says that if $\beta_1 = 0$, then for $\omega_1 = 0$ the plane E_3 is fixed]. The equation (7.18) determines p_2 uniquely, and the homography M osculating to t along $[A_1A_2]$ is unique.

In the plane coordinates we obtain

$$\begin{gathered} ME_1 = E_1' - c_1E_3' \,, \qquad ME_2 = E_2' - (p_2 + c_2p_4)E_3' + c_2E_4' \,, \\ ME_3 = E_3' \,, \qquad ME_4 = E_4' - p_4E_3' \,, \qquad \beta_1p_4 = 2c_2 \,, \end{gathered}$$

from where it follows

$$\begin{aligned} &ME_3 = E_3' \,, \\ &M\mathrm{d}E_3 \equiv \mathrm{d}E_3' + (\tau_{11} + 2c_2\omega_2)\, E_3' \,, \\ &M\mathrm{d}^2E_3 \equiv \mathrm{d}^2E_3' + 2(\tau_{11} + 2c_2\omega_2)\mathrm{d}E_3' + (\cdot)E_3' + 4c_2\omega_2^2E_4' \,. \end{aligned} \tag{7.19}$$

Now, let t^* be the transformation of the ∞^1 tangent planes to the developable D induced by the point transformation t of D. It follows from (7.19) that the homography M, which is osculating to t, is in general only *tangent* to t^*. In order to be *osculating* to t^*, it is necessary and sufficient that $c_2 = 0$. But D being a developable of the first family of L and $c_2 = 0$ are the conditions for the projective deformation T to be semi-singular of the first kind or, which is the same thing, the conditions for the canonical decomposition of L with respect to T to be the decomposition into developables of the first family.

Excluding always the singular projective deformations, and restricting our considerations to the case $\alpha_1\beta_2 \neq 0$, where there exists a non-developable first focal surface (A_1) of L, we can consider the decomposition of (A_1) into ∞^1 curves C induced by the canonical decomposition of L. For the tangent t_0 of C at the point A_1 of (A_1) we have $\tau_{11} - \tau_{22} = 0$ or $c_1\omega_1 - c_2\omega_2 = 0$. Comparing this with (7.2), we can immediately see that t_0 is the harmonic conjugate of $[A_1A_2]$ with respect to the pair of characteristic tangents of (A_1). If $\alpha_2\beta_1 \neq 0$, we have an analogical result for the second focal surface (A_2).

It is important to determine the generality of the semi-singular projective deformations. We shall do it here only for the congruences of the type I. The other types will be treated in a subsequent paper. We can obviously restrict ourselves to the case $c_1 \neq 0 = c_2$ of semi-singular projective deformations of the first kind. It can be easily seen that it is possible to specialize the frames in such a way that $\alpha_1 = \beta_1 = c_1 = 1$. The question is to discuss the Pfaff system

$$\begin{gathered}\omega_{14} = \omega_{23} = 0\,, \quad \omega_{12} = \omega_2\,, \quad \omega_{43} = \omega_2\,, \quad \omega_{21} = \alpha_2\omega_1\,, \quad \omega_{34} = \beta_2\omega_1\,, \\ \tau_{14} = \tau_{23} = \tau_{13} = \tau_{24} = \tau_{12} = \tau_{21} = \tau_{34} = \tau_{43} = 0\,, \\ \tau_{11} - \tau_{33} = \tau_{22} - \tau_{44} = 0\,, \qquad \tau_{22} - \tau_{11} = \omega_1\,,\end{gathered} \tag{7.20}$$

with the assumption $\alpha_2\beta_2 \neq 0$. The exterior differentiation gives

$$\begin{aligned}[\omega_{32}\omega_1] - [\omega_{11} - 2\omega_{22} + \omega_{44} \quad \omega_2] &= 0\,, \\ [\omega_{41}\omega_1] - [\omega_{22} + \omega_{33} - 2\omega_{44} \quad \omega_2] &= 0\,, \\ [d\alpha_2 + \alpha_2(2\omega_{11} - \omega_{22} - \omega_{33}) \quad \omega_1] + [\omega_{41}\omega_2] &= 0\,, \\ [d\beta_2 + \beta_2(\omega_{11} - 2\omega_{33} + \omega_{44}) \quad \omega_1] - [\omega_{32}\omega_2] &= 0\,,\end{aligned} \tag{7.21}$$

$$[\tau_{32} - \omega_2 \quad \omega_1] = 0\,, \quad [\tau_{41} - \omega_2 \quad \omega_1] = 0\,, \quad [\tau_{32}\omega_2] = 0\,, \quad [\tau_{41}\omega_2] = 0\,, \tag{7.22}$$

$$[\omega_{11} - \omega_{33} \quad \omega_1] = 0\,, \tag{7.23}$$

$$[\tau_{31}\omega_1] = [\tau_{42}\omega_2] = 0\,. \tag{7.24}$$

Taking into account the equations (7.22), we set

$$\tau_{32} = \tau_{41} = \omega_2\,. \tag{7.25}$$

From this we deduce by the exterior differentiation

$$\begin{aligned}[\beta_2\tau_{42} - \omega_{32} \quad \omega_1] - [\tau_{31} - 2\omega_{22} + \omega_{33} + \omega_{44} - \omega_1 \quad \omega_2] &= 0\,, \\ [\alpha_2\tau_{42} - \omega_{41} \quad \omega_1] - [\tau_{31} + \omega_{11} + \omega_{22} - 2\omega_{44} - \omega_1 \quad \omega_2] &= 0\,.\end{aligned} \tag{7.26}$$

The integrability conditions for the Pfaff system (7.20) + (7.25) are (7.21) + (7.23) + (7.24) + (7.26). Therefore, the system is in involution, and we arrive at the result that the *semi-singular projective deformations* (of the congruences of type I) *depend on nine arbitrary functions of one variable.*

Czechoslovak Mathematical Journal v. 11 (86), 1961, 386 – 397.

ON THE DIFFERENTIAL GEOMETRY OF A SURFACE EMBEDDED IN A THREE DIMENSIONAL SPACE WITH PROJECTIVE CONNECTION

ALOIS ŠVEC

(Received April 1, 1960)

The notions of Darboux quadrics and Wilczynski directrices are generalized for the case of surfaces in a space with projective connection. The canonical local frame is determined.

I. THE DARBOUX QUADRICS OF A SURFACE

1. Let be given a surface π in a three dimensional space with projective connection. In my paper [1] I have shown that the study of its properties is equivalent to the study of a manifold $P^2_{0,3}$, which I shall briefly call *surface with projective connection* (or even more briefly surface). The local frames of a surface with projective connection can be chosen in such a way that the connection is given by the equations

$$
\begin{aligned}
(1)\qquad & \mathrm{d}A_0 = \omega_0^0 A_0 + \omega^1 A_1 + \omega^2 A_2\,,\\
& \mathrm{d}A_1 = \omega_1^0 A_0 + \omega_1^1 A_1 + \omega_1^2 A_2 + (1-h)\,\omega^2 A_3\,,\\
& \mathrm{d}A_2 = \omega_2^0 A_0 + \omega_2^1 A_1 + \omega_2^2 A_2 + (1+h)\,\omega^1 A_3\,,\\
& \mathrm{d}A_3 = \omega_3^0 A_0 + \omega_3^1 A_1 + \omega_3^2 A_2 + \omega_3^3 A_3\,;\\
& \omega^a = f_1^a(u,v)\mathrm{d}u + f_2^a(u,v)\mathrm{d}v\,, \qquad \omega_i^j = a_i^j\omega^1 + b_i^j\omega^2\,,\\
& \omega_0^0 + \omega_1^1 + \omega_2^2 + \omega_3^3 = 0\,, \qquad [\omega^1\omega^2] \neq 0\,.
\end{aligned}
$$

The admissible changes of the forms ω^a and of the local frames are

$$(2)\qquad \omega^1 = r\bar{\omega}^1\,, \qquad \omega^2 = s\bar{\omega}^2\,,$$

$$
\begin{aligned}
(3)\qquad & A_0 = \alpha_0^0\bar{A}_0\,, \quad A_1 = \alpha_1^0\bar{A}_0 + r^{-1}\alpha_0^0\bar{A}_1\,, \quad A_2 = \alpha_2^0\bar{A}_0 + s^{-1}\alpha_0^0\bar{A}_2\,,\\
& A_3 = \alpha_3^0\bar{A}_0 + \alpha_3^1\bar{A}_1 + \alpha_3^2\bar{A}_2 + r^{-1}s^{-1}\alpha_0^0\bar{A}_3\,; \qquad (\alpha_0^0)^4 = r^2s^2\,.
\end{aligned}
$$

It is evident that one can even suppose that the asymptotic parameters are chosen on the surface, and

$$(4)\qquad \omega^1 = \mathrm{d}u\,, \qquad \omega^2 = \mathrm{d}v\,.$$

Changing the asymptotic parameters

$$u = u(\bar{u}) , \qquad v = v(\bar{v}) . \tag{5}$$

one obtains

$$du = u'd\bar{u} , \quad dv = v'd\bar{v} \quad \left(\text{i. e. } r = u' = \frac{du}{d\bar{u}} , \quad s = v' = \frac{dv}{d\bar{v}}\right) . \tag{6}$$

The direct computation shows that the transformation laws of the functions a_1^2, a_2^1, b_1^2, b_2^1 with the simultaneous application of (3) and (6) are

$$\bar{a}_1^2 = \frac{u'^2}{v'} a_1^2 , \quad \bar{b}_2^1 = \frac{v'^2}{u'} b_2^1 , \tag{7}$$

$$\begin{aligned} \bar{b}_1^2 &= u'b_1^2 - u'(\alpha_0^0)^{-1}\alpha_1^0 + u'v'(1-h)(\alpha_0^0)^{-1}\alpha_3^2 , \\ \bar{a}_2^1 &= v'a_2^1 - v'(\alpha_0^0)^{-1}\alpha_2^0 + u'v'(1+h)(\alpha_0^0)^{-1}\alpha_3^1 . \end{aligned} \tag{8}$$

From the equations (8) it follows that the local frames can be specialized in such a way that one has

$$b_1^2 = a_2^1 = 0 . \tag{9}$$

Because the admissible changes of local frames satisfy always the conditions (9), one obtains

$$\alpha_1^0 = (1-h)v'\alpha_3^2 , \quad \alpha_2^0 = (1+h)u'\alpha_3^1 . \tag{10}$$

Finally let us set

$$a_1^2 = \beta , \qquad b_2^1 = \gamma \tag{11}$$

so that I shall have local frames for which

$$\begin{aligned} dA_0 &= \omega_0^0 A_0 + du A_1 + dv A_2 , \\ dA_1 &= \omega_1^0 A_0 + \omega_1^1 A_1 + \beta du A_2 + (1-h) dv A_3 , \\ dA_2 &= \omega_2^0 A_0 + \gamma dv A_1 + \omega_2^2 A_2 + (1+h) du A_3 , \\ dA_3 &= \omega_3^0 A_0 + \omega_3^1 A_1 + \omega_3^2 A_2 + \omega_3^3 A_3 . \end{aligned} \tag{12}$$

The equations (7) take the form

$$\bar{\beta} = \frac{u'^2}{v'} \beta , \qquad \bar{\gamma} = \frac{v'^2}{u'} \gamma . \tag{13}$$

In what follows I shall use the notation

$$a = a_0^0 - a_1^1 - a_2^2 + a_3^3 , \qquad b = b_0^0 - b_1^1 - b_2^2 + b_3^3 . \tag{14}$$

ON THE DIFFERENTIAL GEOMETRY OF A SURFACE ...

2. Let us choose a fixed point A_0 of the studied surface π, and consider its local space $P_3(A_0)$. We can introduce the local coordinates of the analytic points in $P_3(A_0)$ with respect to the corresponding local basis by the relation

$$X = x^0 A_0 + x^1 A_1 + x^2 A_2 + x^3 A_3 \equiv x^i A_i \ . \tag{15}$$

I call every quadric of the local space $P_3(A_0)$ containing the element of the second order of the development of an arbitrary curve γ of the surface π passing through the point A_0 (the development in $P_3(A_0)$ is meant) *osculating quadric* Q of the surface π at the point A_0. The curve γ being given by the equation

$$v = v(u) \ , \tag{16}$$

the notation $v' = \dfrac{dv}{du}$ etc. will surely not be confused with (6). For the development γ^* of the curve γ one shall have

$$A = (A)_0 + u(A')_0 + \tfrac{1}{2}u^2(A'')_0 + \tfrac{1}{6}u^3(A''')_0 + \dots \tag{17}$$

where

$$\begin{aligned} (A)_0 &= A_0 \ , \\ (A')_0 &= (a_0^0 + b_0^b v')A_0 + A_1 + v'A_2 \ , \\ (A'')_0 &= (a_{0u}^0 + a_{0v}^0 v' + b_{0u}^0 v' + b_{0v}^0 v' + b_0^0 v'' + \overline{a_0^0 + b_0^0 v'}^2 + \\ &\quad + a_1^0 + b_1^0 v' + a_2^0 v' + b_2^0 v'^2)A_0 + \\ &\quad + (a_0^0 + b_0^0 v' + a_1^1 + b_1^1 v' + \gamma v'^2)A_1 + \\ &\quad + (v'' + a_0^0 v' + b_0^0 v'^2 + \beta + a_2^2 v' + b_2^2 v'^2)A_2 + 2v'A_3 \ . \end{aligned} \tag{18}$$

Let us consider in $P_3(A_0)$ the quadric

$$(X, X) \equiv c_{ij} x^i x^j = 0 \ , \quad c_{ij} = c_{ji} \ ; \quad i, j = 0, \dots, 3 \ . \tag{19}$$

If (19) is an osculating quadric, there must be

$$(A, A) = 0 \tag{20}$$

identically in u^0, u^1, u^2 for every v', v''. Applying here (17), one obtains

$$\begin{aligned} ((A)_0, (A)_0) &= 0 \ , \\ ((A)_0, (A')_0) &= 0 \ , \\ ((A)_0, (A'')_0) + ((A')_0, (A')_0) &= 0 \ . \end{aligned} \tag{21}$$

From (19) it follows that $(A_i, A_j) = c_{ij}$. Substituting (18) into ($21_{1,2}$) it results

$$c_{00} = c_{01} = c_{02} = 0 \ . \tag{22}$$

The substitution into (21) gives

$$2v'(A_0, A_3) + (A_1, A_1) + v'^2(A_2, A_2) + 2v'(A_1, A_2) = 0$$

so that

$$c_{11} = c_{22} = c_{03} + c_{12} = 0 \;. \tag{23}$$

Thus the equation of a general osculating quadric is (I set $c_{12} = 1$. In the case where $c_{12} = 0$ the quadric in question is singular.)

$$x^0x^3 - x^1x^2 - c_{13}x^1x^3 - c_{23}x^2x^3 = \tfrac{1}{2}c_{33}(x^3)^2 \;. \tag{24}$$

On the surface π let us look for curves (16) passing through A_0 and such that their development in $P_3(A_0)$ has with the quadric (23) a contact of the third order. For such a curve the equation (20) must be satisfied identically in u^0, u^1, u^2, u^3, so that one must have

$$((A)_0, (A''')_0) + 3((A')_0, (A'')_0) = 0 \;. \tag{25}$$

The direct computation gives

$$\begin{aligned}(A''')_0 &= (\cdot)A_0 + (\cdot)A_1 + (\cdot)A_2 + \\ &+[(3+h)v'' + (1+h)\beta + (2a_0^0 + a_1^1 + a_2^2 + 2a_3^3 - ha_1^1 + ha_2^2)v' + \\ &+(2b_0^0 + b_1^1 + b_2^2 + 2b_3^3 - hb_1^1 + hb_2^2)v'^2 + \gamma v'^3]\,A_3 \;.\end{aligned} \tag{26}$$

Hence it results by substituting into (25)

$$\begin{aligned}&hv'' + (h-2)\beta + (2a + \overline{a_2^2 - a_1^1} \cdot h - 6c_{13})v' + \\ &+(2b + \overline{b_2^2 - b_1^1} \cdot h - 6c_{23})v'^2 - (h+2)\gamma v'^3 = 0 \;.\end{aligned} \tag{27}$$

If the torsion h of the surface π is $\neq 0$, then there exist on the surface π curves passing through A_0 having an arbitrarily given tangent, and a contact of the third order with an arbitrarily chosen osculating quadric (24). All these curves admit the parametric representation

$$v = v_0 + uv' + \tfrac{1}{2}u^2v'' + \tfrac{1}{6}u^3v''' + \dots \;,$$

where v' and v'' are related by the equation (27). For $h = 0$ (27) is reduced to

$$\beta + (3c_{13} - a)v' + (3c_{23} - b)v'^2 + \gamma v'^2 = 0 \;. \tag{28}$$

For every osculating quadric Q (24) there exist at the point A_0 of the surface π three tangents with the following property: the development of every curve touching one of them has a contact of the third order with Q. The quadrics the three corresponding tangents of which are apolar with respect to the asymptotic tangents form the pencil

$$x^0x^3 - x^1x^2 - \tfrac{1}{3}ax^1x^3 - \tfrac{1}{3}bx^2x^3 = \tfrac{1}{2}c_{33}(x^3)^2 \;. \tag{29}$$

The three tangents mentioned above generate therefore on π a 3-layer of *Darboux curves*

$$\beta du^3 + \gamma dv^3 = 0 \;. \tag{30}$$

3. On the surface π given by the equations (12) let us consider a curve γ (16) touching at the point A_0 the asymptotic curve $v = \text{const.}$ Consequently one has $v' = 0$ at A_0. At each point of the curve γ let us consider the tangent to the asymptotic curve $u = \text{const.}$ passing through this point. Let us develop the ruled surface thus obtained in the local space $P_3(A_0)$. I am going to find the equation of the quadric Q (19) one regulus of which has a contact of the second order in the line space with the development of the mentioned surface. The quadric Q must contain all the points of the line $\{A_0, A_2\}$, i. e. there must be $(A_0 + tA_2, A_0 + tA_2) = 0$ identically in t, which gives

$$(A_0, A_0) = (A_0, A_2) = (A_2, A_2) = 0 . \tag{31}$$

The differentiation (with respect to u) of these equations gives

$$(A_0, A_1) = 0 , \qquad (A_1, A_2) + \gamma v'(A_0, A_1) + (1+h)(A_0, A_3) = 0 , \\ \gamma v'(A_1, A_2) + (1+h)(A_2, A_3) = 0 . \tag{32}$$

Setting there $v' = 0$ one obtains

$$(A_0, A_1) = 0, \quad (A_1, A_2) + (1+h)(A_0, A_3) = 0 , \quad (A_2, A_3) = 0 . \tag{33}$$

The new differentiation of the equations (32), an application of (31) + (33), and $v' = 0$ give

$$(A_1, A_1) = 0 , \quad \left(a + \frac{h_u}{1+h}\right)(A_1, A_2) - 2(1+h)(A_1, A_3) = 0 , \\ (\gamma v'' + \overline{1+h} \cdot a_3^1 - a_2^0)(A_0, A_3) - (1+h)(A_3, A_3) = 0 . \tag{34}$$

For the studied quadric one has (31), (33), and (34), so that its equation is

$$(1+h)x^1x^2 - x^0x^3 + \tfrac{1}{2}\left(a + \frac{h_u}{1+h}\right)x^1x^3 = \\ = \frac{1}{2(1+h)}(\overline{1+h}\, a_3^1 - a_2^0 + \gamma v'')(x^3)^2 . \tag{35}$$

Let us consider the asymptotic curve $v = \text{const.}$ passing through the point A_0 and the curve (16) touching it (this means that $v' = 0$). For the development of these two curves in the local space $P_3(A_0)$ one has — see (17) and (18) —

$$A = A_0 + u(a_0^0 A_0 + A_1) + \\ + \tfrac{1}{2}u^2(\overline{a_{0u}^0 + (a_0^0)^2 + a_1^0} \cdot A_0 + \overline{a_0^0 + a_1^1} \cdot A_1 + \beta A_2) + \dots \tag{36}$$

or

$$A = A_0 + u(a_0^0 A_0 + A_1) + \tfrac{1}{2}u^2(\overline{a_{0u}^0 + (a_0^0)^2 + a_1^0} \cdot A_0 + \overline{a_0^0 + a_1^1} \cdot A_1 + \\ + \overline{v'' + \beta} \cdot A_2) + \tfrac{1}{6}u^3\{(\cdot)A_0 + (\cdot)A_1 + (\cdot)A_2 + \\ + \beta(1 + h + \overline{3+h} \cdot \nu)A_3\} + \dots$$

so that their Smith-Mehmke invariant is

$$1+\nu=\frac{\beta+v''}{\beta} \tag{37}$$

which gives

$$v''=\beta\nu\ . \tag{38}$$

The quadric (35) related with the curve (16) having with the asymptotic curve $v=$ const. the contact invariant $1+\nu$ (I denote this quadric by $Q_v(\nu)$) has the equation

$$\begin{aligned}(1+h)x^1x^2-x^0x^3+\frac{1}{2}\left(a+\frac{h_u}{1+h}\right)x^1x^3=\\ =\frac{1}{2(1+h)}(\overline{1+h}\cdot a_3^1-a_2^0+\nu\beta\gamma)(x^3)^2\ .\end{aligned} \tag{39}$$

If the curve (16) has a contact of the second order with the asymptotic $c=$ const., one has $v''=0$, i. e. $\nu=0$, and I obtain *the Lie quadric* $Q_v(0)$. If the curve (16) has an inflection at the point A_0, one has $v''=-\beta$, i. e. $\nu=-1$, and I obtain *the Wilczynski-Bompiani quadric* $Q_v(-1)$. If the tangent plane $\{A_0,A_1,A_2\}$ has at the point A_0 a contact of the third order with the curve (16), one has by virtue of (36_2) $\nu=-\dfrac{1+h}{3+h}$, and one obtains *the Fubini quadric* $Q_v\left(-\dfrac{1+h}{3+h}\right)$.

4. The interchange of the asymptotic curves can be expressed by the substitution

$$\left\updownarrow\begin{matrix}1 & u & a: & \beta & h & a\\ 2 & v & b: & \gamma & -h & b\end{matrix}\right\updownarrow\ . \tag{40}$$

Thus e. g. the expression $a_1^0+h-\beta a$ will be replaced by $b_2^0-h-\gamma b$.

Consequently the quadric $Q_u(\nu)$ has the equation

$$\begin{aligned}(1-h)x^1x^2-x^0x^3+\frac{1}{2}\left(b-\frac{h_v}{1-h}\right)x^2x^3=\\ =\frac{1}{2(1-h)}(\overline{1-h}\cdot b_3^2-b_1^0+\nu\beta\gamma)(x^3)^2\ .\end{aligned} \tag{41}$$

I give the name of *Darboux quadric* $Q(\nu,\lambda)$ to the quadric $Q_v(\nu)+\lambda Q_u(\nu)=0$, whose equation is

$$\begin{aligned}(1+h+\lambda-h\lambda)x^1x^2-(1+\lambda)x^0x^3+\\ +\frac{1}{2}\left(a+\frac{h_u}{1+h}\right)x^1x^3+\frac{1}{2}\lambda\left(b-\frac{h_v}{1-h}\right)x^2x^3=\\ =\frac{1}{2}\left\{a_3^1+\lambda b_3^2-\frac{a_2^0}{1+h}-\frac{\lambda b_1^0}{1-h}+\nu\beta\gamma\left(\frac{1}{1+h}+\frac{\lambda}{1-h}\right)\right\}(x^3)^2\ .\end{aligned} \tag{42}$$

The pencils of quadrics $Q(\nu,\infty)$ and $Q(\nu,0)$ coincide with the pencils (41) and (39) respectively. The quadric $Q(\nu,\lambda)$ is singular and contains the tangent plane $\{A_0, A_1, A_2\}$ (whose equation is $x^3 = 0$) if and only if $\lambda = \dfrac{h+1}{h-1}$. This demonstrates sufficiently the geometric meaning of the numbers ν and λ for every quadric $Q(\nu,\lambda)$, for in each of the pencils $Q(\nu,\infty)$ and $Q(\nu,0)$ there are three geometrically determined distinguished quadrics, and the same holds for the pencil $Q(\nu,\lambda)$ (with ν being fixed).

I shall call the quadrics of the pencil $Q(0,\lambda)$ *Lie quadrics* of the surface π. I shall call the pencil of quadrics $Q(\nu,1)$, whose quadrics are

$$\begin{aligned}(43)\qquad & x^1x^2 - x^0x^3 + \frac{1}{4}\left(a + \frac{h_u}{1+h}\right)x^1x^3 + \frac{1}{4}\left(b - \frac{h_v}{1-h}\right)x^2x^3 = \\ & = \frac{1}{4}\left(a_3^1 + b_3^2 - \frac{a_2^0}{1+h} - \frac{b_1^0}{1-h} + 2\frac{\nu\beta\gamma}{1-h^2}\right)(x^3)^2 \,;\end{aligned}$$

principal Darboux pencil. The quadric $Q(0,1)$, which is both the Lie quadric and the principal Darboux quadric, and whose equation is

$$\begin{aligned}(44)\qquad & x^1x^2 - x^0x^3 + \frac{1}{4}\left(a + \frac{h_u}{1+h}\right)x^1x^3 + \frac{1}{4}\left(b - \frac{h_v}{1-h}\right)x^2x^3 = \\ & = \frac{1}{4}\left(a_3^1 + b_3^2 - \frac{a_2^0}{1+h} - \frac{b_1^0}{1+h}\right)(x^3)^2\end{aligned}$$

will be called *principal Lie quadric.*

5. Let be given a surface π (12), and let us consider the dual local frames

$$(45)\qquad E^0 = [A_1A_2A_3]\,,\qquad E^1 = -[A_0A_2A_3]\,,\qquad E^2 = [A_0A_1A_3]\,,\qquad E^3 = -[A_0A_1A_2]$$

and the frames related to them

$$(46)\qquad F^3 = E^3\,,\quad F^2 = -(1+h)E^2\,,\quad F^1 = -(1-h)E^1\,,\quad F^0 = E^0\,.$$

The frames (46), of course, do not satisfy any more the condition $[F^3F^2F^1F^0] = 1$.

The dualization π^* of the surface π is given by the equations

$$
\begin{aligned}
(47)\ \mathrm{d}F^3 &= (-a_3^3\mathrm{d}u - b_3^3\mathrm{d}v)F^3 + \mathrm{d}uF^2 + \mathrm{d}vF^1\,,\\
\mathrm{d}F^2 &= (1+h)(a_3^2\mathrm{d}u + b_3^2\mathrm{d}v)F^3 +\\
&+\left\{\left(\frac{h_u}{1+h} - a_2^2\right)\mathrm{d}u + \left(\frac{h_v}{1+h} - b_2^2\right)\mathrm{d}v\right\}F^2 -\\
&-\frac{1+h}{1-h}\beta\mathrm{d}uF^1 + (1+h)\mathrm{d}vF^0\,,\\
\mathrm{d}F^1 &= (1-h)(a_3^1\mathrm{d}u + b_3^1\mathrm{d}v)F^3 - \frac{1-h}{1+h}\gamma\mathrm{d}vF^2 +\\
&+\left\{\left(-\frac{h_u}{1-h} - a_1^1\right)\mathrm{d}u + \left(-\frac{h_v}{1-h} - b_1^1\right)\mathrm{d}v\right\}F^1 + (1-h)\mathrm{d}uF^0,\\
\mathrm{d}F^0 &= (-a_3^0\mathrm{d}u - b_3^0\mathrm{d}v)F^3 + \frac{1}{1+h}(a_2^0\mathrm{d}u + b_2^0\mathrm{d}v)F^2 +\\
&+\frac{1}{1-h}(a_1^0\mathrm{d}u + b_1^0\mathrm{d}v)F^1 + (-a_0^0\mathrm{d}u - b_0^0\mathrm{d}v)F^0\,.
\end{aligned}
$$

Comparing them with (12), one obtains the substitution of particular expressions appearing in (12), which takes place when passing from the surface π to its dualization. If I introduce the dual local coordinates by the relation

$$(48)\qquad \eta = \eta_3F^3 + \eta_2F^2 + \eta_1F^1 + \eta_0F^0 \equiv \eta_iF^i\,,$$

the quadric $Q^*(\bar{\nu}, \lambda)$ will have the equation

$$
\begin{aligned}
(49)\qquad &(1-h+\lambda+h\lambda)\eta_1\eta_2 - (1+\lambda)\eta_0\eta_3 -\\
&-\frac{1}{2}\left(a + \frac{h_u}{1+h}\right)\eta_0\eta_2 - \frac{1}{2}\left(b - \frac{h_v}{1-h}\right)\eta_0\eta_1 =\\
&= -\frac{1}{2}\left\{a_3^1 + \lambda b_3^2 - \frac{a_2^0}{1+h} - \frac{\lambda b_1^0}{1-h} - \bar{\nu}\beta\gamma\left(\frac{1}{1-h} + \frac{\lambda}{1+h}\right)\right\}\eta_0^2\,.
\end{aligned}
$$

The local coordinates

$$(50)\qquad \xi = \xi_3E^3 + \xi_2E^2 + \xi_1E^1 + \xi_0E^0 \equiv \xi_iE^i$$

are related with η_i by the equations

$$(51)\qquad \xi_3 = \eta_3\,,\quad \xi_2 = -(1+h)\eta_2\,,\quad \xi_1 = -(1-h)\eta_1\,,\quad \xi_0 = \eta_0\,,$$

from where one can easily get the equation of the quadric $Q^*(\bar{\nu}, \lambda)$ in the local coordinates ξ_i. The point equation of that quadric is

$$
\begin{aligned}
(52)\qquad &(1+\lambda)x^1x^2 - \frac{1-h+\lambda+h\lambda}{1-h^2}x^0x^3 +\\
&+\frac{1}{2(1+h)}\left(a + \frac{h_u}{1+h}\right)x^1x^3 + \frac{\lambda}{2(1-h)}\left(b - \frac{h_v}{1-h}\right)x^2x^3 =\\
&= \frac{1}{2(1+\lambda)}(a_{00}a_{12} - 2a_{01}a_{02})(x^3)^2\,,
\end{aligned}
$$

where

$$a_{00} = a_3^1 + \lambda b_3^2 - \frac{a_2^0}{1+h} - \frac{\lambda b_1^0}{1-h} - \bar{\nu}\beta\gamma\left(\frac{1}{1-h} + \frac{\lambda}{1+h}\right),$$
$$a_{01} = \frac{\lambda}{2(1-h)}\left(b - \frac{h_v}{1-h}\right), \quad a_{02} = \frac{1}{2(1+h)}\left(a + \frac{h_u}{1+h}\right),$$
$$a_{12} = \frac{1-h+\lambda+h\lambda}{1-h^2}.$$

If I restrict myself to the surfaces without torsion (then I have $h = 0$), the quadrics $Q(\nu,\lambda)$ and $Q^*(\bar{\nu},\lambda)$ will have the equations

$$(53) \qquad (1+\lambda)(x^1x^2 - x^0x^3) + \tfrac{1}{2}ax^1x^3 + \tfrac{1}{2}\lambda bx^2x^3 = \\ = \tfrac{1}{2}(a_3^1 + \lambda b_3^2 - a_2^0 - \lambda b_1^0 + \nu \cdot \overline{1+\lambda} \cdot \beta\gamma)(x^3)^2,$$

$$(54) \qquad (1+\lambda)(x^1x^2 - x^0x^3) + \tfrac{1}{2}ax^1x^3 + \tfrac{1}{2}\lambda bx^2x^3 = \\ = \frac{1}{2}\left(a_3^1 + \lambda b_3^2 - a_2^0 - \lambda b_1^0 - \bar{\nu} \cdot \overline{1+\lambda} \cdot \beta\gamma - \frac{1}{2}\frac{ab\lambda}{1+\lambda}\right)(x^3)^2,$$

respectively.

The quadrics $Q(\nu,\lambda)$ and $Q^*(\bar{\nu},\lambda)$ coincide if and only if

$$(55) \qquad \nu + \bar{\nu} = \frac{ab\lambda}{(1+\lambda)^2\beta\gamma}.$$

6. In my paper [2] I have determined, with the aid of a procedure due to E. Čech for the case of surfaces in a line space, a net of quadrics associated with each point of the surface under consideration. Among them there is a quadric generalizing in an evident way the Lie quadric. It has the equation (cf. [2], (38))

$$(56) \qquad x^1x^2 - x^0x^3 + \frac{1}{4}\left(a - \frac{h_u}{1+h}\right)x^1x^3 + \frac{1}{4}\left(b + \frac{h_v}{1-h}\right)x^2x^3 = \\ = \frac{1}{4}\left(\overline{1-h} \cdot a_3^1 + \overline{1+h} \cdot b_3^2 - a_2^0 - b_1^0 + \varrho_{1v} + \varrho_{2u} - \tfrac{1}{2}ab - \frac{h_uh_v}{2(1-h^2)}\right)(x^3)^2,$$

where

$$(57) \qquad \varrho_1 = -\frac{h_u}{2(1+h)} - \tfrac{1}{2}a, \qquad \varrho_2 = \frac{h_v}{2(1-h)} - \tfrac{1}{2}b.$$

This quadric differs in general from the principal Lie quadric (44).

ALOIS ŠVEC

II. Canonical frame of a surface

7. The equations (10) have the following geometric meaning: The line $\{A_0, A_3\}$ being chosen, the line $\{A_1, A_2\}$ is determined without ambiguity. One can easily find its geometric construction. The apolar line to $\{A_0, A_3\}$ with respect to the quadric $Q(\nu, \lambda)$ does not depend on ν, and its equations are

$$x^3 = (1+\lambda)x^0 - \frac{1}{2}\left(a + \frac{h_u}{1+h}\right)x^1 - \frac{1}{2}\lambda\left(b - \frac{h_v}{1-h}\right)x^2 = 0\,. \tag{58}$$

The line apolar to $\{A_0, A_3\}$ with respect to the quadric $Q(\nu, \infty)$ (resp. $Q(\nu, 0)$) cuts the asymptotic tangent $\{A_0, A_1\}$ (resp. $\{A_0, A_2\}$) at the point A_1 (resp. A_2).

In order to determine the canonical local frame it is sufficient to determine a geometrically remarkable line passing through A_0, and to situate the point A_3 on it and on the principal Lie quadric. In what follows I restrict myself to the surfaces without torsion, and I identify the line $\{A_0, A_3\}$ with the generalized Wilczynski directrix.

8. On the surface π given by the equations (12) (where one sets $h = 0$) let us consider the asymptotic curve $v =$ const. passing through the point A_0, and at each of its points let us consider its tangent. Let us then develop the ruled surface thus obtained in the local space $P_3(A_0)$. I am going to find a linear line complex in $P_3(A_0)$ having a contact of the fourth order with the ruled surface in question. If the linear complex has the form

$$[X, Y] \equiv a_{ij}[x^i, y^j] = 0\,, \tag{59}$$

then the condition for possessing the desired property is evidently

$$\frac{\partial^i}{\partial u^i}[A_0, A_1] = 0 \quad \text{for } i = 0, \ldots, 4\,.$$

A direct computation gives

$$\begin{gathered}[A_0, A_1] = 0\,, \quad [A_0, A_2] = 0\,, \quad [A_1, A_2] + [A_0, A_3] = 0\,, \\ 2[A_1, A_3] - a[A_1, A_2] = 0\,, \quad \alpha_1[A_1, A_2] + \beta[A_2, A_3] = 0\,,\end{gathered} \tag{60}$$

where

$$\alpha_1 = a_3^2 - a_1^0 - \tfrac{1}{2}a_u + \tfrac{1}{2}a(a_3^3 - a_2^2) - \tfrac{1}{4}a^2\,. \tag{61}$$

Because $[A_i, A_j] = a_{ij}$, one obtains the equation of the studied linear complex in the form

$$q^{03} - q^{12} - \tfrac{1}{2}aq^{13} + \frac{\alpha_1}{\beta}q^{23} = 0\,. \tag{62}$$

Interchanging the asymptotic curves, one obtains another linear complex

$$q^{03} + q^{12} - \tfrac{1}{2}bq^{23} + \frac{\alpha_2}{\gamma}q^{13} = 0\,, \tag{63}$$

where

$$\alpha_2 = b_3^1 - b_2^0 - \tfrac{1}{2}b_v + \tfrac{1}{2}b(b_3^3 - b_1^1) - \tfrac{1}{4}b^2 . \tag{64}$$

In the pencil determined by the complexes (62) and (63) there are special linear complexes

$$\begin{aligned} &q^{12} + \frac{1}{2}\left(\frac{\alpha_2}{\gamma} + \tfrac{1}{2}a\right) q^{13} - \frac{1}{2}\left(\frac{\alpha_1}{\beta} + \tfrac{1}{2}b\right) q^{23} = 0 , \\ &q^{03} + \frac{1}{2}\left(\frac{\alpha_2}{\gamma} - \tfrac{1}{2}a\right) q^{13} - \frac{1}{2}\left(\frac{\alpha_1}{\beta} - \tfrac{1}{2}b\right) q^{13} = 0 , \end{aligned} \tag{65}$$

whose axes are

$$\begin{aligned} &p_{03} = 1 , \quad p_{02} = -\frac{1}{2}\left(\frac{\alpha_2}{\gamma} + \tfrac{1}{2}a\right) , \quad p_{01} = -\frac{1}{2}\left(\frac{\alpha_1}{\beta} + \tfrac{1}{2}b\right) , \\ &\qquad p_{12} = p_{13} = p_{23} = 0 , \\ &p_{12} = 1 , \quad p_{02} = -\frac{1}{2}\left(\frac{\alpha_2}{\gamma} - \tfrac{1}{2}a\right) , \quad p_{01} = \frac{1}{2}\left(\frac{\alpha_1}{\beta} - \tfrac{1}{2}b\right) , \\ &\qquad p_{03} = p_{13} = p_{23} = 0 , \end{aligned} \tag{66}$$

which are the lines

$$\begin{aligned} k_1 &= \left\{A_0 , -\frac{1}{2}\left(\frac{\alpha_1}{\beta} + \tfrac{1}{2}b\right) A_1 - \frac{1}{2}\left(\frac{\alpha_2}{\gamma} + \tfrac{1}{2}a\right) A_2 + A_3\right\} , \\ k_2 &= \left\{A_1 - \frac{1}{2}\left(\frac{\alpha_2}{\gamma} - \tfrac{1}{2}a\right) A_0 , A_2 - \frac{1}{2}\left(\frac{\alpha_1}{\beta} - \tfrac{1}{2}b\right) A_0\right\} . \end{aligned} \tag{67}$$

I call them *Wilczynski directrices.*

Starting from the equations mentioned above one can easily find by a direct computation that the Wilczynski directrices of the dualization π^* of the surface π are the lines

$$\begin{aligned} k_1^* &= \left\{E^3 , \frac{1}{2}\left(\frac{\alpha_1}{\beta} - \tfrac{1}{2}b\right) E^2 + \frac{1}{2}\left(\frac{\alpha_2}{\gamma} - \tfrac{1}{2}a\right) E^1 + E^0\right\} , \\ k_2^* &= \left\{E^2 + \frac{1}{2}\left(\frac{\alpha_2}{\gamma} + \tfrac{1}{2}a\right) E^3 , E^1 + \frac{1}{2}\left(\frac{\alpha_1}{\beta} + \tfrac{1}{2}b\right) E^3\right\} , \end{aligned} \tag{68}$$

so that they coincide with the lines (67):

$$k_1 \equiv k_2^* , \qquad k_2 \equiv k_1^* .$$

9. Let us assume now that the point A_3 is situated in the intersection of the Wilczynski directrix k_1 (67_1) with the principal Lie quadric (44) whose equation is

$$x^1x^2 - x^0x^3 + \tfrac{1}{4}ax^1x^3 + \tfrac{1}{4}bx^2x^3 = \tfrac{1}{4}(a_3^1 + b_3^2 - a_2^0 - b_1^0)(x^3)^2 . \tag{69}$$

ALOIS ŠVEC

Thus one has

$$a_u = 2(a_3^2 - a_1^0) + a(a_3^3 - a_2^2) - \tfrac{1}{2}a^2 + b\beta \,, \qquad (70)$$
$$b_v = 2(b_3^1 - b_2^0) + b(b_3^3 - b_1^1) - \tfrac{1}{2}b^2 + a\gamma \,,$$
$$a_3^1 - a_2^0 + b_3^2 - b_1^0 = 0 \,.$$

The geometric points A_0, A_1, A_2, A_3 being completely determined by the above mentioned conditions, it is even possible to change the asymptotic parameters (5) or (6) as well as the local bases

$$A_0 = \alpha_0^0 \bar{A}_0 , \quad A_1 = r^{-1}\alpha_0^0 \bar{A}_1 \,, \quad A_2 = s^{-1}\alpha_0^0 \bar{A}_2 \,, \quad A_3 = r^{-1}s^{-1}\alpha_0^0 \bar{A}_3 \,; \qquad (71)$$
$$(\alpha_0^0)^4 = r^2 s^2 \,.$$

Writing

$$\alpha_0^0 = \alpha \,, \quad \frac{\alpha_u}{\alpha} = \alpha_U \,, \quad \frac{\alpha_v}{\alpha} = \alpha_V \,, \qquad (72)$$

one obtains the transformation equations for the functions a_j^i and b_j^i corresponding to the substitutions (5) + (71):

$$\begin{aligned} \bar{a}_0^0 &= r(a_0^0 - \alpha_U) \,, & \bar{b}_0^0 &= s(b_0^0 - \alpha_V) \,; \qquad (73)\\ \bar{a}_1^0 &= r^2 a_1^0 \,, & \bar{b}_1^0 &= rs b_1^0 \,; \\ \bar{a}_1^1 &= r a_1^1 + r^{-1}r' - \alpha_U r \,, & \bar{b}_1^1 &= s(b_1^1 - \alpha_V) \,; \\ \bar{a}_2^0 &= rs a_2^0 \,, & \bar{b}_2^0 &= s^2 b_2^0 \,; \\ \bar{a}_2^2 &= r(a_2^2 - \alpha_V) \,, & \bar{b}_2^2 &= s b_2^2 + s^{-1}s' - \alpha_V s \,; \\ \bar{\beta} &= r^2 s^{-1}\beta \,, & \bar{\gamma} &= r^{-1}s^2\gamma \,; \\ \bar{a}_3^0 &= r^2 s a_3^0 \,, & \bar{b}_3^0 &= rs^2 b_3^0 \,; \\ \bar{a}_3^1 &= rs a_3^1 \,, & \bar{b}_3^1 &= s^2 b_3^1 \,; \\ \bar{a}_3^2 &= r^2 a_3^2 \,, & \bar{b}_3^2 &= rs b_3^2 \,; \\ \bar{a}_3^3 &= r a_3^3 + r^{-1}r' - \alpha_U r \,, & \bar{b}_3^3 &= s b_3^3 + s^{-1}s' - \alpha_V s \,. \end{aligned}$$

Starting from these relations it is already possible to find mechanically the complete system of projective differential invariants of the surface considered. Of course, it is necessary to take into account the relations (70).

References

[1] A. Švec, *L'application des variétés à connexion à certains problèmes de la géométrie différentielle*, Czech. Math. J., **10(85)**, (1960), 523–550.

[2] A. Švec, *Les quadriques de Lie d'une surface plongée dans un espace tridimensionnel à connexion projective*, Czech. Math. J., **11(86)**, (1961), 134–142.

Časopis pro pěstování matematiky, roč. 96 (1971), Praha

ORDER OF HOLONOMY OF A SURFACE WITH PROJECTIVE CONNECTION

IVAN KOLÁŘ, Brno

(Received October 6, 1969)

A submanifold in a space with Cartan connection, see [3], represents a natural generalization of a submanifold in the corresponding homogeneous space. É. CARTAN himself showed in the case of a surface in a 3-dimensional space with projective connection, [1], that his method of specialization of frames can also be applied to the investigation of these submanifolds. A. ŠVEC pointed out, cf. [5], that such a submanifold can be considered as a separate structure. From this point of view, a surface in a 3-space with projective connection is called a manifold of type $P^2_{0,3}$, or, shortly, a surface with projective connection. Naturally, differential geometry of a surface $\mathcal{P}$ with projective connection differs from differential geometry of a surface in projective 3-space P_3. In this paper, we want to show that the difference between $\mathcal{P}$ and a surface in P_3 can be also measured in individual orders. If we use the computational procedures by É. Cartan, then the difference in order k between $\mathcal{P}$ and a surface in P_3 is characterized by the difference between the formulae of the $(k-1)$-st prolongation for $\mathcal{P}$ and the formulae of the $(k-1)$-st prolongation for a surface in P_3. Conversely, if these formulae coincide, then we say that $\mathcal{P}$ is holonomic of order k, or, shortly, k-holonomic. Dealing with the first prolongation, we show the invariance of the condition for 2-holonomy also in a formal computional way, but we do not repeat it for higher orders, since we present a direct invariant definition of k-holonomy for an arbitrary manifold with connection in [4].

At every order, we geometrize the corresponding conditions for holonomy by means of some properties of some geometric objects of $\mathcal{P}$. In general, the geometric objects of $\mathcal{P}$ differ from the geometric objects of a surface in P_3. But if $\mathcal{P}$ is k-holonomic and if we take into account how one evaluates the geometric objects of order k of $\mathcal{P}$, then we are led to the following proposition: $\mathcal{P}$ is k-holonomic if and only if all its geometric objects of order k are analogous to geometric objects of order k of a surface in P_3. We present an exact formulation of this assertion for an arbitrary manifold with connection as well as its proof in [4]. Our considerations end at the sixth order, since a 6-holonomic non-special surface with projective connection has integrable connection, so that it is locally isomorphic to a surface in P_3 and is holonomic of any order. The totality of our geometric conditions gives a necessary and sufficient geometric condition that a surface with projective connection be locally equivalent to a surface in P_3, which is a problem solved by B. Cenkl, [2]. In contradistinction to this paper, our conditions are organized according to individual orders.

1. Consider a surface $\mathscr{P}$ with projective connection together with the manifold $\mathscr{F}_{12}$ of all frames associated with $\mathscr{P}$, which depend on 12 secondary parameters. Let the connection be given by

$$\mathrm{d}A_i = \omega_i^j A_j \tag{1}$$

where ω_j^i are differential forms on $\mathscr{F}_{12}$ satisfying

$$\omega_i^i = 0\,. \tag{2}$$

The structure equations are

$$\mathrm{d}\omega_i^j = \omega_i^k \wedge \omega_k^j + 2R_i^j\omega^1 \wedge \omega^2 \tag{3}$$

and it holds

$$R_i^i = 0\,. \tag{4}$$

(We write $\omega_0^1 = \omega^1$, $\omega_0^2 = \omega^2$ as usual.)

2. The frame field $\mathscr{F}_{10}$ of the first order is determined by the usual relation

$$\omega_0^3 = 0\,. \tag{5}$$

The exterior differentiation of (5) yields

$$\omega^1 \wedge \omega_1^3 + \omega^2 \wedge \omega_2^3 + 2R_0^3\omega^1 \wedge \omega^2 = 0\,, \tag{6}$$

which is equivalent to

$$\omega_1^3 = a_1\omega^1 + (a_2 - R_0^3)\,\omega^2\,, \quad \omega_2^3 = (a_2 + R_0^3)\,\omega^1 + a_3\omega^2\,. \tag{7}$$

Prolonging (7) and fixing the principal parameters, we obtain

$$\begin{aligned} &\delta a_1 + a_1(e_0^0 - 2e_1^1 + e_3^3) - 2a_2e_1^2 = 0\,,\\ &\delta a_2 + a_2(e_0^0 - e_1^1 - e_2^2 + e_3^3) - a_1e_2^1 - a_3e_1^2 = 0\,,\\ &\delta a_3 + a_3(e_0^0 - 2e_2^2 + e_3^3) - 2a_2e_2^1 = 0\,,\\ &\delta R_0^3 + R_0^3(e_0^0 - e_1^1 - e_2^2 + e_3^3) = 0\,, \end{aligned} \tag{8}$$

so that R_0^3 is a relative invariant. If it holds

$$R_0^3 = 0\,, \tag{9}$$

then $\mathscr{P}$ will be called 2-*holonomic*. Furthermore, we can deduce from (8) that the quantity

$$h = \frac{R_0^3}{\sqrt{[(a_2)^2 - a_1 a_3]}} \tag{10}$$

is an absolute invariant.

Restricting ourselves to the investigation of hyperbolic surfaces, we can specialize the frames by $a_2 = 1$, $a_1 = a_3 = 0$ and we get the frame field $\mathscr{F}_7$ of the second order. When comparing with [6], we find that h is the torsion of $\mathscr{P}$ and we have deduced the following geometric assertion: *$\mathscr{P}$ is 2-holonomic at a point if and only if the conjugate tangents at this point form an involution.*

3. From now on, we shall suppose $\mathscr{P}$ is 2-holonomic, so that we have

$$\omega_1^3 = \omega^2, \quad \omega_2^3 = \omega^1 \tag{11}$$

and the prolongation of (11) yields

$$\begin{aligned} 2\omega_1^2 &= b_1\omega^1 + (b_2 - R_0^2 + R_1^3)\,\omega^2, \\ -\omega_0^0 + \omega_1^1 + \omega_2^2 - \omega_3^3 &= (b_2 + R_0^2 - R_1^3)\,\omega^1 + (b_3 - R_0^1 + R_2^3)\,\omega^2, \\ 2\omega_2^1 &= (b_3 + R_0^1 - R_2^3)\,\omega^1 + b_4\omega^2. \end{aligned} \tag{12}$$

If it holds

$$R_1^3 = R_0^2, \quad R_2^3 = R_0^1, \tag{13}$$

then $\mathscr{P}$ will be said to be 3-*holonomic*.

Now we give a geometric interpretation of (13). Consider the ruled surface $\mathscr{L}_1$ generated by the tangent lines to the asymptotic curves $\omega^2 = 0$ along an asymptotic curve $\omega^1 = 0$ as well as the ruled surface $\mathscr{L}_2$ generated symmetrically. It is easy to see that the quadrics having the first order (line) contact with both $\mathscr{L}_1$ and $\mathscr{L}_2$ form the pencil

$$2x^0x^3 - 2x^1x^2 + (b_2 - R_0^2 + R_1^3)\,x^1x^3 + (b_3 + R_0^1 - R_2^3)\,x^2x^3 = a_{33}(x^3)^2 \tag{14}$$

where x^0, x^1, x^2, x^3 are the local coordinates. On the other hand, consider a quadric Q having the second order contact with $\mathscr{P}$, then there are exactly three directions in which $\mathscr{P}$ has the third order contact with Q. These directions are apolar with respect to the asymptotic directions if and only if Q belongs to the following pencil

$$\begin{aligned} 2x^0x^3 - 2x^1x^2 + [b_2 + \tfrac{1}{3}(R_0^2 - R_1^3)]\,x^1x^3 \,+ \\ + [b_3 - \tfrac{1}{3}(R_0^1 - R_2^3)]\,x^2x^3 = \bar{a}_{33}(x^3)^2, \end{aligned} \tag{15}$$

cf. [6], p. 389. Comparing (13), (14), (15), we can conclude: *A 2-holonomic surface $\mathscr{P}$ is 3-holonomic if and only if both preceding constructions give the same pencil of quadrics (of Darboux).*

4. In what follows, $\mathscr{P}$ will be supposed to be 3-holonomic and non-ruled. Standard procedure shows that we can further specialize the frames by $b_1 = b_4 = 2$, $b_2 = b_3 = 0$ and we get the frame field $\mathscr{F}_3$ of the third order. Prolonging the equations

$$(16) \qquad \omega_1^2 = \omega^1 , \quad \omega_2^1 = \omega^2 , \quad \omega_1^1 + \omega_2^2 = 0 , \quad \omega_0^0 + \omega_3^3 = 0 ,$$

we obtain

$$(17) \qquad \begin{aligned} \omega_0^0 - 3\omega_1^1 &= c_1\omega^1 + (c_2 + R_0^1 - R_1^2)\,\omega^2 , \\ \omega_3^2 - \omega_1^0 &= (c_2 - R_0^1 + R_1^2)\,\omega^1 + (c_3 - R_1^1 - R_2^2)\,\omega^2 , \\ \omega_3^1 - \omega_2^0 &= (c_3 + R_1^1 + R_2^2)\,\omega^1 + (c_4 + R_0^2 - R_2^1)\,\omega^2 , \\ \omega_0^0 + 3\omega_1^1 &= (c_4 - R_0^2 + R_2^1)\,\omega^1 + c_5\omega^2 . \end{aligned}$$

If it holds

$$(18) \qquad R_1^2 = R_0^1 , \quad R_2^1 = R_0^2 , \quad R_1^1 + R_2^2 = 0 ,$$

then $\mathscr{P}$ will be called *4-holonomic*.

The osculating quadric of the ruled surface $\mathscr{L}_1$ or $\mathscr{L}_2$ considered in item 3 has the equation

$$(19) \qquad 2x^0x^3 - 2x^1x^2 + (c_3 - R_1^1 - R_2^2)\,(x^3)^2 = 0$$

or

$$(20) \qquad 2x^0x^3 - 2x^1x^2 + (c_3 + R_1^1 + R_2^2)\,(x^3)^2 = 0$$

respectively, so that both quadrics coincide if and only if $R_1^1 + R_2^2 = 0$. In the sequel we suppose that this condition holds and (19) = (20) will be called the quadric of Lie.

Let $\mathscr{K}_1$ and $\mathscr{K}_2$ be two line congruences associated with $\mathscr{P}$ in such a way that the lines of $\mathscr{K}_1$ pass through the corresponding point of $\mathscr{P}$ but do not lie in the tangent plane and the lines of $\mathscr{K}_2$ lie in the corresponding tangent plane of $\mathscr{P}$ but do not pass through the point of contact. Then $\mathscr{K}_1$ and $\mathscr{K}_2$ are said to be reciprocal, if their lines are conjugate with respect to the quadric of Lie. If $\mathscr{K}_1$ is generated by the straight line $[A_0, pA_1 + qA_2 + A_3]$, then the reciprocal $\mathscr{K}_2$ is generated by $[qA_0 + A_1, pA_0 + A_2]$ and the focal nets of both congruences coincide if and only if

$$(21) \qquad p = -\tfrac{1}{2}(c_2 - R_0^1 + R_2^2) , \quad q = -\tfrac{1}{2}(c_4 + R_0^2 - R_2^1) ;$$

these lines will be called the first or the second directrix of Wilczynski respectively.

By the principal quadrics of $\mathscr{P}$ we mean those quadrics which have contact of the fourth order with the asymptotic curves of $\mathscr{P}$; they form the following pencil

$$(22)\qquad 2x^1x^2 - 6x^0x^3 + c_1x^1x^3 + c_5x^2x^3 = a_{33}(x^3)^2 .$$

There exists exactly one pair of reciprocal congruences whose lines are also conjugate with respect to (22); these lines will be called the edges of Green. The first edge of Green is

$$(23)\qquad [A_0, \tfrac{1}{4}c_5A_1 + \tfrac{1}{4}c_1A_2 + A_3] .$$

The curves of Segre are given by $(\omega^1)^3 - (\omega^2)^3 = 0$. The first axis of Čech is the common line of intersection of the osculating planes of the curves of Segre, which is

$$(24)\qquad [A_0, \tfrac{1}{6}(c_5 - c_2 - R^1_0 + R^2_1)\, A_1 + \tfrac{1}{6}(c_1 - c_4 + R^2_0 - R^1_2)\, A_2 + A_3] .$$

The lines (21), (23), (24) belong to the same (canonical) pencil if and only if $R^2_1 = R^1_0$, $R^1_2 = R^2_0$. Thus, *a 3-holonomic surface $\mathscr{P}$ is 4-holonomic if and only if the osculating quadrics of ruled surfaces $\mathscr{L}_1$ and $\mathscr{L}_2$ coincide and if the directrix of Wilczynski, the edge of Green and the axis of Čech belong to the same pencil.*

5. Suppose $\mathscr{P}$ is 4-holonomic. The remaining secondary parameters can be fixed by $c_2 = c_3 = c_4 = 0$ and we get the canonical frame field $\mathscr{F}$. Then we have

$$(25)\qquad \omega^0_0 - 3\omega^1_1 = c_1\omega^1 , \quad \omega^2_3 = \omega^0_1 , \quad \omega^1_3 = \omega^0_2 , \quad \omega^0_0 + 3\omega^1_1 = c_5\omega^2 .$$

Prolonging (25), we obtain

$$(26)\qquad \begin{aligned} -\mathrm{d}c_1 - c_1(\omega^0_0 - \omega^1_1) - 4\omega^0_1 + 3\omega^2 &= e_1\omega^1 + (e_2 + R^0_0 - 3R^1_1 - c_1R^1_0)\,\omega^2 , \\ 2\omega^0_2 &= (e_2 - R^0_0 + 3R^1_1 + c_1R^1_0)\,\omega^1 + (e_3 + R^2_3 - R^0_1)\,\omega^2 , \\ 2\omega^0_3 &= (e_3 - R^2_3 + R^0_1)\,\omega^1 + (e_4 + R^1_3 - R^0_2)\,\omega^2 , \\ 2\omega^0_1 &= (e_4 - R^1_3 + R^0_2)\,\omega^1 + (e_5 + R^0_0 + 3R^1_1 - c_5R^2_0)\,\omega^2 , \\ -\mathrm{d}c_5 - c_5(\omega^0_0 + \omega^1_1) - 4\omega^0_2 + 3\omega^1 &= (e_5 - R^0_0 - 3R^1_1 + c_5R^2_0)\,\omega^1 + e_6\omega^2 . \end{aligned}$$

If it holds

$$(27)\qquad R^0_0 - 3R^1_1 = c_1R^1_0 , \quad R^2_3 = R^0_1 , \quad R^1_3 = R^0_2 , \quad R^0_0 + 3R^1_1 = c_5R^2_0 ,$$

then $\mathscr{P}$ will be called 5-*holonomic.*

The first normal of Fubini is the line harmonically conjugate to the canonical tangent with respect to the directrix of Wilczynski and the edge of Green, which is

$$(28)\qquad [A_0, \tfrac{1}{2}c_5A_1 + \tfrac{1}{2}c_1A_2 + A_3] .$$

The developable surfaces of this congruence intersect a conjugate net on $\mathscr{P}$ if and only if

$$c_1 R_0^1 + c_5 R_0^2 - 2R_0^0 = 0 . \tag{29}$$

The second focal surface of the congruence of the tangents to a family of curves of Segre is without torsion if and only if

$$c_1 R_0^1 - c_5 R_0^2 + 6R_1^1 = 0 . \tag{30}$$

Furthermore, consider the envelope of the quadrics of Lie

$$x^1 x^2 - x^0 x^3 = 0 . \tag{31}$$

It is easy to see that the characteristic points, i.e. the vertices of the tetrahedron of Demoulin, are determined by (31) and by

$$a_3^0 (x^3)^2 - (x^1)^2 = 0 , \quad b_3^0 (x^3)^2 - (x^2)^2 = 0 . \tag{32}$$

The transversals of the tetrahedron of Demoulin intersect the lines $[A_0 A_3]$ and $[A_1 A_2]$ at

$$\begin{aligned} D_1 &= (\sqrt{a_3^0 b_3^0}, 0, 0, 1) , \quad D_3 = (0, \sqrt{a_3^0}, \quad \sqrt{b_3^0}, 0) \\ D_2 &= (-\sqrt{a_3^0 b_3^0}, 0, 0, 1) , \quad D_4 = (0, \sqrt{a_3^0}, -\sqrt{b_3^0}, 0) , \end{aligned} \tag{33}$$

where the pairs D_1, D_3 and D_2, D_4 lie on the same transversal. If $\xi A_0 + \eta A_3$ or $\lambda A_1 + \mu A_2$ are the coordinates on $[A_0 A_3]$ or $[A_1 A_2]$, then the pair D_1, D_2 or D_3, D_4 has the equation

$$\xi^2 - a_3^0 b_3^0 \eta^2 = 0 \tag{34}$$

or

$$b_3^0 \lambda^2 - a_3^0 \mu^2 = 0 \tag{35}$$

respectively. On the other hand, the focal planes of the congruence of the first directrices of Wilczynski intersect $[A_1 A_2]$ at the points

$$a_1^0 \lambda^2 + (b_1^0 - a_2^0) \lambda\mu - b_2^0 \mu^2 = 0 . \tag{36}$$

Thus, *the pairs* (35), (36) *and* A_1, A_2 *belong to the same involution* if and only if

$$a_1^0 a_3^0 = b_2^0 b_3^0 . \tag{37}$$

The foci of the congruence of the first directrices of Wilczynski are determined by

$$\xi^2 + \xi\eta(a_2^0 + b_1^0) + \eta^2(a_2^0 b_1^0 - a_1^0 b_2^0) = 0 . \tag{38}$$

The tangent plane of the surface (A_1) or (A_2) intersects $[A_0A_3]$ at $T_1 = b_1^0A_0 + A_3$ or $T_2 = a_2^0A_0 + A_3$ respectively. Let T_3 be the harmonically conjugate of A_0 with respect to T_1, T_2, let T_4 be the harmonically conjugate of A_3 with respect to A_0, T_3 and let T be the harmonically conjugate of T_4 with respect to A_0, A_3, then the pair A_0, T is given by

$$(a_2^0 + b_1^0)\,\xi\eta + a_2^0b_1^0\eta^2 = 0\,. \tag{39}$$

The pairs (34), (38), (39) *belong to the same involution* if and only if

$$a_1^0b_2^0 = a_3^0b_3^0\,. \tag{40}$$

(37) and (40) imply

$$a_3^0 = \varepsilon b_2^0\,, \quad b_3^0 = \varepsilon a_1^0\,, \quad \varepsilon = \pm 1\,. \tag{41}$$

On the other hand, the relations $R_3^2 = R_1^0$, $R_3^1 = R_2^0$ are equivalent to $b_2^0 = a_3^0$, $b_3^0 = a_1^0$, cf. (26). The additional condition $\varepsilon = 1$ for (41) is equivalent to *the following condition concerning orientation.* Let F_1, F_2 be the foci of the congruence of the first directrices of Wilczynski taken in such order that the orientation on $[A_0A_3]$ determined by the ordered triple (A_0, F_1, F_2) coincides with the orientation (A_0, D_1, D_2). Let F_{i+2}, $i = 1, 2$, be the point of intersection of the focal plane passing through F_i with $[A_1A_2]$. Then the orientation (A_1, F_3, F_4) coincides with the orientation (A_1, D_3, D_4) if and only if $\operatorname{sgn} a_3^0 = \operatorname{sgn} b_2^0$, i.e. $\varepsilon = 1$. – Thus *we have deduced necessary and sufficient geometric conditions that a 4-holonomic surface $\mathcal{P}$ be 5-holonomic.*

6. Suppose $\mathcal{P}$ is 5-holonomic. Analogous considerations as above suggest the following definition. If it holds

$$\begin{gathered} -c_1R_0^0 + c_1R_1^1 - 4R_1^0 + 3R_0^2 = e_1R_0^1 + e_2R_0^2\,, \\ 2R_2^0 = e_2R_0^1 + e_3R_0^2\,, \quad 2R_3^0 = e_3R_0^1 + e_4R_0^2\,, \quad 2R_1^0 = e_4R_0^1 + e_5R_0^2\,, \\ -c_5R_0^0 - c_5R_1^1 - 4R_2^0 + 3R_0^1 = e_5R_0^1 + e_6R_0^2\,, \end{gathered} \tag{42}$$

then $\mathcal{P}$ will be said to be *6-holonomic.*

It is easy to see that

$$2R_3^0 = e_3R_0^1 + e_4R_0^2 \tag{43}$$

holds if and only if *the surface (A_3) is without torsion* and that

$$2R_0^2 = e_2R_0^1 + e_3R_0^2\,, \quad 2R_1^0 = e_4R_0^1 + e_5R_0^2 \tag{44}$$

are satisfied if and only if *both focal surfaces of the congruence of the first directrices of Wilczynski are without torsion.* Now, taking $(42_{1,5})$ modulo $(27_{1,4})$, (43), (44), we obtain two linear homogeneous equations in R_0^1, R_0^2. Since R_0^1, R_0^2 are the last independent components of the curvature tensor, there are many possibilities how to geometrize these equations; we choose the simpliest way: If the determinant D of this system does not vanish, then $(43_{1,5})$ holds if and only if $R_0^1 = R_0^2 = 0$, i.e. if *the torsion tensor of $\mathcal{P}$ vanishes.*

7. Summarizing the preceding considerations, we get the following result, which concludes our investigation in general case. *If a non-ruled surface $\mathcal{P}$ with $D \neq 0$ is 6-holonomic, then its curvature tensor vanishes, so that its connection is integrable.*

References

[1] *É. Cartan:* Leçons sur la théorie des espaces à connexion projective, Paris 1937.

[2] *B. Cenkl:* L'équation de structure d'un espace à connexion projective, Czechoslovak Math. J., *14* (*89*), 1964, 79—94.

[3] *C. Ehresmann:* Les connexions infinitésimales dans un espace fibré différentiable, Colloque de Topologie, Bruxelles 1950, 29—55.

[4] *I. Kolář:* Order of Holonomy and Geometric Objects of Manifolds with Connection, Comm. Math. Univ. Carolinae *10*, 1969, 559—565.

[5] *A. Švec:* L'application des variétés à connexion à certains problèmes de la géométrie différentielle, Czechoslovak Math. J., *10* (*85*), 1960, 523—550.

[6] *A. Švec:* Sur la géométrie différentielle d'une surface plongée dans un espace à trois dimensions à connexion projective, ibid, *11* (*86*), 1961, 386—397.

Author's address: Brno, Janáčkovo náměstí 2a, (Matematický ústav ČSAV v Brně).

Physica 18D (1986) 217–219
North-Holland, Amsterdam

GEOMETRIC DEFORMATIONS OF THE EVOLUTION EQUATIONS AND BÄCKLUND TRANSFORMATIONS

Bohumil CENKL
Department of Mathematics, Northeastern University, 360 Huntington Ave., Boston, Ma 02115, USA

Extended abstract

The sine-Gordon equation $(2\omega)_{xt} + \sin(2\omega) = 0$ describes a surface with constant negative curvature. Sasaki [3], demonstrated that pseudospherical surfaces are associated with many evolution equations. The surfaces of negative curvature (not necessarily constant), which can be thought of as deformations of the pseudospherical surfaces, are defined by solutions of a system of equations. We call such a system a deformed sine-Gordon system. The geometry of such systems has been extensively studied in the theory of line congruences [2]. It turns out that the more appropriate setting for the study of the geometry in full generality is not the group of euclidean motions but the projective group. In that case we talk about W-systems. It is clear that such a theory is an important special case of the study of the σ-models. The significant feature of the deformed sine-Gordon and W-systems is the existence of Bäcklund transformations and the validity of the permutability theorem. Hence, some of the systems can be integrated by the inverse scattering transform and possess N-soliton solutions. Motivated by the relationship between the evolution equations and pseudospherical surfaces [3], it is natural to consider the deformation of the evolution equations together with the deformation of pseudospherical surfaces. The deformed equations and systems inherit some of the properties of the deformed sine-Gordon system.

The deformation of an evolution equation can be considered in more general setting. Let X, T be two nonsingular $n \times n$ matrix valued functions on $\mathbf{R}^2$ and let $v_x = Xv, v_t = Tv$ be the system of linear equations, $(x, t) \in \mathbf{R}^2$. Then $T_x - X_t - [X, T] = 0$ is an evolution equation in the formalism of [1]. Considering the one form $\phi = X\,dx + T\,dt$ with values in the Lie algebra $SL(n, \mathbf{R})$ the evolution equation $d\phi - \phi \wedge \phi = 0$ simply says that the curvature of the connection ϕ is zero. Now we deform the connection ϕ and get an $SL(n, \mathbf{R})$-valued one form σ with the curvature $d\sigma - \sigma \wedge \sigma = \Sigma$ no longer equal to zero. But we can enlarge the group $SL(n, \mathbf{R})$ and find a flat connection Ω which induces σ. Then the system $d\Omega - \Omega \wedge \Omega = 0$ is again suitable for a study by the inverse scattering transform. It is clear that this is possible when Ω depends on a parameter which plays the role of an eigenvalue. Such a parameter is provided, in many cases, by the parameter in the family of Bäcklund transformations on the space of solutions of the deformed system $d\Omega - \Omega \wedge \Omega = 0$. The equation $d\sigma - \sigma \wedge \sigma = \Sigma$ is called a deformed equation.

The main purpose of this note is to bring out some of the classical results from the theory of line congruences and to relate them to the theory of integrable systems. We restrict ourselves to the simpler euclidean case.

Deformed sine-Gordon systems and B-systems

Let $K = -1/\rho^2$, $\rho > 0$, be a smooth function on $\mathbf{R}^2$. Let $\rho_1 = (\log\sqrt{\rho})_x$, $\rho_2 = (\log\sqrt{\rho})_t$. Suppose that c_1, c_2, ω are unknown functions on $\mathbf{R}^2$. De-

0167-2789/86/$03.50 © Elsevier Science Publishers B.V.
(North-Holland Physics Publishing Division)

fine an SL(2, **R**)-valued one form on $\mathbf{R}^2$,

$$\sigma=\begin{pmatrix}\sigma_1 & \sigma_2\\ \sigma_3 & -\sigma_1\end{pmatrix},\quad 2\sigma_1=-\rho c_1\,dx+\rho c_2\cos 2\omega\,dt,$$

$$2\sigma_2=-\left(2\omega_x+\frac{c_1}{c_2}\rho_2\sin 2\omega\right)dx+c_2\left(\rho-\frac{\rho_1}{c_1}\right)\sin 2\omega\,dt,$$

$$2\sigma_3=\left(2\omega_x+\frac{c_1}{c_2}\rho_2\sin 2\omega\right)dx+c_2\left(\rho+\frac{\rho_1}{c_1}\right)\sin 2\omega\,dt.$$

Denote

$$\omega^1=\sigma_2+\sigma_3,$$
$$\omega^2=-2\sigma_1,\quad \omega_1^2=\sigma_2-\sigma_3,$$
$$\omega_1^3=c_1\,dx-c_2\cos 2\omega\,dt,\quad \omega_2^3=c_2\sin 2\omega\,dt.$$

The deformed sine-Gordon equation has the form

$$d\sigma-\sigma\wedge\sigma=\frac{1}{2}(K+1)\begin{pmatrix}0 & -1\\ 1 & 0\end{pmatrix}\omega^1\wedge\omega^2. \qquad (\sigma)$$

Let $\Omega=(\omega_i^j)$, $\omega_i^j+\omega_j^i=0$, $i,j=1,2,3$. The deformed sine-Gordon system has the form

$$d\Omega-\Omega\wedge\Omega=0. \qquad (\Omega_K)$$

The system (Ω_K) depends on a particular choice of the function K.

Theorem. The deformed sine-Gordon system (Ω_K) is equivalent to the system

$$\frac{\partial(2\omega)}{\partial x\,\partial t}+c_1c_2\sin 2\omega-\left(\frac{c_1}{c_2}\rho_2\sin 2\omega\right)_t-\left(\frac{c_2}{c_1}\rho_1\sin 2\omega\right)_x=0,$$

$$d\omega_1^3-\omega_1^2\wedge\omega_2^3=0,\quad d\omega_2^3+\omega_1^2\wedge\omega_1^3=0.$$

The Bäcklund transformation $\beta_{K,K'}$ which maps the system (Ω_K) into a system $(\Omega_{K'})$ and the solutions of (Ω_K) into the solutions of $(\Omega_{K'})$ has the form

$$\omega'_x-\omega_x=c_1\left(\cot\zeta+\frac{\rho}{\tau}\right)\sin(\omega'+\omega)-\frac{c_1}{c_2}\rho_2\sin\ 2\omega,$$

$$\omega'_t+\omega_t=c_2\left(\cot\zeta-\frac{\rho}{\tau}\right)\sin(\omega'-\omega)+\frac{c_1}{c_2}\rho_1\sin 2\omega,$$

$$(\log\tau)_x=-(\log\psi)_x-c_1\left(\cot\zeta+\frac{\rho}{\tau}\right)\cos(\omega'+\omega),$$

$$(\log\tau)_t=-(\log\psi)_t-c_2\left(\cot\zeta-\frac{\rho}{\tau}\right)\cos(\omega'-\omega),$$

$$(\cot\zeta)_x=\frac{\rho}{\tau}(\log\psi)_x+c_1\left(\frac{\rho}{\tau}\cot\zeta+\frac{1}{\sin^2\zeta}\right)\cos(\omega'+\omega),$$

$$(\cot\zeta)_t=-\frac{\rho}{\tau}(\log\psi)_t-c_2\left(\frac{\rho}{\tau}\cot\zeta-\frac{1}{\sin^2\zeta}\right)\cos(\omega'-\omega). \qquad (\beta)$$

Such transformations are parametrized by the solutions of the Laplace equation $\psi_{xt}+\rho_2\psi_x+\rho_1\psi_t+c_1c_2\psi\cos 2\omega=0$. The Bianchi permutability theorem holds for such Bäcklund transformations.

This result can be found, in a different form, in [2].

Let $\rho=f(x)+g(t)$, where f and g are functions of x and t respectively. The deformed sine-Gordon system specializes. We call such a special system a B-system. The above Bäcklund transformations (β) do not map such a B-system into itself in general.

Theorem. There is a subsystem of the Bäcklund transformations (β) which map the solutions of a B-system into the solutions of the same B-system (i.e. ρ is invariant under such transformations). The Bianchi permutability theorem holds. The B-systems are completely integrable.

The classical part of this theorem is due to Bianchi, [2]. The complete integrability is proved by the inverse scattering transform, using the existence of the Bäcklund transformations.

Deformed modified Korteweg–de Vries equation

Analogously as in [1] and [3], we define an $SL(2,\mathbf{R})$-valued one-form σ on $\mathbf{R}^2$:

$$\sigma=\begin{pmatrix}\sigma_1 & \sigma_2\\ \sigma_3 & -\sigma_1\end{pmatrix},$$

$$\sigma_1=\lambda\,\mathrm{d}x+A\,\mathrm{d}t,$$

$$\sigma_2=q\,\mathrm{d}x+B\,\mathrm{d}t,$$

$$\sigma_3=r\,\mathrm{d}x+C\,\mathrm{d}t,$$

where λ is a real parameter, r, q are functions of (x,t) and A, B, C are functions of (x,t,λ). Let $\omega^1=\sigma_2+\sigma_3$, $\omega^2=-2\sigma_1$. Let $K<0$ be a smooth function on $\mathbf{R}^2$. We assume that σ satisfies the equation (σ). Assume that $r=q$. Then equation (σ) is equivalent to the system

$$R_x=vP,\quad P_x=-4vKR+2KQ,$$

$$Q_x=2v_t-2\lambda^2P,$$

where $v=r=q$, $\lambda P=C-B$, $Q=C+B$, $\lambda R=A$. When P and R are eliminated the last equation takes the form

$$v_t=\frac{\lambda^2}{v}R_x+(vR)_x+\left[\frac{1}{K}\left(\frac{R_x}{4v}\right)_x\right]_x. \qquad (v)$$

Considering the expansion of R in terms of λ^2, $R=R_0\lambda^{2n}+R_1\lambda^{2n-2}+\cdots+R_n$, substituting into (v) and requiring that the resulting equation is independent of λ we get $R_{0,x}=0$ and the recursion formulas for $M_m=v^{-1}\partial R_m/\partial x$, $m=1,2,\ldots,n$, in the following form: Let

$$S=-\tfrac{1}{4}\mathrm{D}K^{-1}\mathrm{D}-v^2-v_x\mathrm{D}^{-1}v,$$

$$\mathrm{D}=\frac{\partial}{\partial x},\quad \mathrm{D}^{-1}f=\int f\,\mathrm{d}x.$$

Then $M_m=SM_{m-1}$. The deformed modified Korteweg–de Vries nth order equation has the form

$$v_t+M_n=0.$$

For $n=1$ we get $v_t-6v^2v_x-((1/K)v_{xx})_x=0$. The deformed system is constructed by following the general scheme mentioned earlier.

References

[1] M.J. Ablowitz, D.J. Kaup, A.C. Newell and H. Segur, Nonlinear Evolution Equations of Physical Significance, Phys. Rev. Lett. 31 (1973) 125–217.

[2] S.P. Finikow, Theorie der Kongruenzen (Akademie-Verlag, Berlin, 1959).

[3] R. Sasaki, Soliton equations and pseudospherical surfaces, Nuclear Physics B154 (1979) 343–357.

Professor Čech and Didactics of Mathematics

Emil Kraemer

Professor E. Čech was primarily an outstanding mathematician and university teacher. He recognized, however, the importance of teaching mathematics at elementary and high schools. During his university studies, he paid attention to papers dealing with problems of teaching elementary mathematics. He made mathematical formulations in these papers more precise and removed logical gaps in their proofs. From 1937 on his interest in teaching mathematics markedly increased. He was not satisfied with mathematical textbooks, which were usually used in Czechoslovakia at that time.

In 1938 he published his first didactic paper in Volume 68 (1938-39) of the journal *Časopis pro pěstování matematiky a fyziky*. It concerned teaching combinatorics and calculus of probabilities at Czech and Slovak high schools. During the following years E. Čech paid attention to problems of teaching mathematics in the first four years of Czech and Slovak grammar schools. The study at these grammar schools lasted eight years; the students were from 11 to 19 years old. He audited classes of mathematics at grammar schools and discussed the possibilities of improving the teaching process with the teachers. The results obtained were presented in his lectures and seminars delivered at Masaryk University in Brno. These lectures took place once a week. Usually, 40 to 50 grammar school teachers were present. The lectures were forcedly stopped in the autumn 1939 when all Czech universities were closed down. Czech universities remained closed till 1945, i.e., during the existence of so-called Protektorat Böhmen und Mähren, which was commanded by Hitler's Deutsches Reich.

From 1939 till 1944, E. Čech fully concentrated his attention on the problems of teaching mathematics. During this period he wrote arithmetic textbooks for the first three years and geometry textbooks for the first four years of Czech eight years' grammar schools. He was in contact with many grammar school teachers, who taught from the manuscripts of his textbooks; Čech subsequently discussed with them their experience. He incorporated the obtained information into the final versions of the textbooks. His aim was to motivate pupils towards a deeper understanding and mastery of the topic. The textbooks contained non-traditional methods, particularly in geometry, as well as many interesting and unusual examples and exercises. They were accompanied by two short papers entitled "Notes on arithmetic (geometry) textbooks for the first three years of high schools", which were published in 1944.

In the period 1945 – 1947, E. Čech's textbooks were revised and published. They formed the basic source for writing new mathematical textbooks for the sixth, seventh, eighth and the ninth class of the compulsory nine years' school, which was introduced in Czechoslovakia after 1948. These textbooks published in the period 1949 – 1951 were written by E. Čech together with his colleagues — former high school teachers. After 1948 E. Čech played a significant role among the authors of mathematics textbooks for four years' grammar schools.

When writing his textbooks E. Čech took into account the existing literature on the art of teaching mathematics. Before 1939, this literature was mostly of the North American origin, after 1945 it came mostly from the Soviet Union. Reacting to the shortage of such literature in Czechoslovakia, E. Čech translated into the Czech language the book "Methodology of arithmetic" written in 1949 by the Soviet author J. S. Berezanskaja. He also initiated the translation from the Russian of the book "Solving arithmetic problems" by N. N. Nikitin. Results of his didactic studies, own views and discussions with teachers of mathematics were summarized and explained in several papers. The most important of them are two essays on teaching geometry and arithmetic in the first years of eight years' grammar schools. The first one was published in the above mentioned mathematical-physical journal (Volume 70, 1940-41); the second one in the journal Střední škola (Volume 23, 1942-43). E. Čech also presented and explained his views on teaching mathematics in many lectures and seminars, which he organized after the war.

Most remarkable were his seminars which took place from 1947 till 1954 at Charles University. In these seminars, many problems of teaching mathematics in the sixth to the ninth class of the unified (non-differentiated) basic nine years' school were analyzed. Among the discussed topics there were, for example, a solution of reasoning problems without using algebra, solving problems containing parameters, teaching geometry, etc. Professor Čech paid great attention to the level of expressing ideas in mathematical textbooks. He also stressed the importance of correct formulations by teachers and pupils in the classroom. His explanations in seminars were always scientifically well founded, comprehensive and well thought-out. His seminars contributed to deepening both professional and didactic education of mathematics teachers at basic schools. It was important because at that time most of the basic school teachers did not have university education and they sometimes had very gifted pupils in their classes.

Four of E. Čech's books are oriented toward teaching. They were published in Czech. Three of them were destined to the broader public: Co je a nač je vyšší matematika, 1942 (What is higher mathematics and what is it good for), Elementární funkce, 1944 (Elementary functions) and Čísla a početní výkony, 1954 (Numbers and numerical operations). The fourth one was a university textbook Základy analytické geometrie I a II, 1951 - 1952 (Foundations of analytic geometry I and II).

Finally, one should mention the fact that it was professor Čech who initiated the establishment of the Mathematical Olympiad in 1951. The Mathematical Olympiad is a nation-wide competition for pupils of basic and high schools. It has existed for over forty years. During this period of time it helped to increase pupils' interest in mathematics.

Subject Index

Acknowledgement

The paper *E. Čech, On Bicompact Spaces* is reprinted from Annals of Mathematics 38, 1937 with kind permission of E. Čech's estates.

The paper *B. Pospíšil, Remark on Bicompact Spaces* is reprinted from Annals of Mathematics 38, 1937.

The paper *I. Gelfand and A. Kolmogoroff, On Rings of Continuous Functions on Topological Spaces* is reprinted from Comptes Rendus (Doklady) de l'Académie des Sciences de l'URSS 22, 1939.

The paper *I. Glicksberg, Stone–Čech Compactifications of Products* is reprinted from Transactions of Amer. Math. Soc. 90, 1959 with kind permission of American Mathematical Society.

The paper *W. Rudin, Homogeneity Problems in the Theory of Čech Compactifications* is reprinted from Duke Math. J. 23, 1956 with kind permission of W. Rudin.

The paper *I. I. Parovičenko, On a Universal Bicompactum of Weight* $\aleph$ is reprinted from Doklady Akad. Nauk. SSSR 150, 1963.

The paper *Z. Frolík, Non-Homogeneity of* $\beta P - P$ is reprinted from Comment. Math. Univ. Carolinae 8, 1967 with kind permission of Z. Frolík's estates.

The paper *K. Kunen, Weak P-points in* N^* is reprinted from Colloquia Math. Soc. J. Bolyai 23, 1978 with kind permission of K. Kunen.

The paper *E. Čech, On the Dimension of Perfectly Normal Spaces* is reprinted from Bull. Intern. Acad. Tcheque Sci. 33, 1932 with kind permission of E. Čech's estates.

The paper *E. Čech, A Contribution to the Theory of Dimension* is reprinted from Časopis Pěst. Mat. Fys. 62, 1933. with kind permission of E. Čech's estates.

The paper *O. V. Lokucievskij, On the Dimension of Bicompacta* is reprinted from Doklady Akad. Nauk SSSR 67,1949.

The paper *C. H. Dowker, Inductive Dimension of Completely Normal Spaces* is reprinted from Quart. J. Math. Oxford Ser. (2) 4, 1953 with kind permission of Oxford University Press.

The paper *C. H. Dowker and W. Hurewicz, Dimension of Metric Spaces* is reprinted from Fundamenta Mathematicae 43, 1956 with kind permission of the journal Fundamenta Mathematicae.

The paper *P. Vopěnka, On the Dimension of Compact Spaces* is reprinted from Czechoslovak Math. J. 8, 1958 with kind permission of P. Vopěnka.

The paper *V. V. Filippov, On Compacta with Unequal Dimension ind and dim* is reprinted from Doklady Akad. Nauk. SSSR 192, 1970 with kind permission of V. V. Filippov.

The paper *E. Pol and R. Pol, A Hereditarily Normal Strongly Zero-Dimensional Space with a Subspace of Positive Dimension and an N-Compact Space of Positive Dimension* is reprinted from Fundamenta Mathematicae 97, 1977 with kind permission of E. and R. Pol.

The paper *E. Čech, General Homology Theory in an Arbitrary Space* is reprinted from Fundamenta Mathematicae 10, 1932 with kind permission of E. Čech's estates.

The paper *E. Čech, Betti Groups of an Infinite Complex* is reprinted from Fundamenta Mathematicae 25, 1935 with kind permission of E. Čech's estates.

The paper *E. Čech, Multiplications On a Complex* is reprinted from Annals of Math. 37, 1936 with kind permission of E. Čech's estates.

The paper *S. Lefschetz, On Generalized Manifolds* is reprinted from American J. of Math. 55, 1933 with kind permission of John Hopkins University Press.

The paper *C. H. Dowker, Čech Cohomology Theory and the Axioms* is reprinted from Annals of Math. 51, 1950 with kind permission of the journal Annals of Mathematics.

The paper *E. Čech, On the Surfaces All Segre Curves of Which Are Plane Curves* is reprinted from Publ. Fac. Sci. Univ. Masaryk 11, 1922 with kind permission of E. Čech's estates.

The paper *E. Čech, Developable Transformations of Line Congruences* is reprinted from Czechoslovak Math. J. 6, 1956 with kind permission of E. Čech's estates.

The paper *A. Švec, On the Differential Geometry of a Surface Embedded in a Three Dimensional Space With Projective Connection* is reprinted from Czechoslovak Math. J. 11, 1961 with kind permission of E. Švec's estates.

The paper *B. Cenkl, Geometric Deformations of the Evolution Equations and Bäcklund Transformations* is reprinted from Physica 18D, 1986 with kind permission of B. Cenkl.